MECHANICS

MECHANICS

Wallace Arthur
Saul K. Fenster

FAIRLEIGH DICKINSON UNIVERSITY

HOLT, RINEHART AND WINSTON, INC.
New York · Chicago · San Francisco · Atlanta
Dallas · Montreal · Toronto · London · Sydney

FOR LOIS AND ROBERTA

PREFACE

This undertaking is an outgrowth of the teaching of mechanics to physics, mathematics, and engineering undergraduates. The authors have found certain recurring conceptual difficulties, quite disparate with the mathematical sophistication of the student. To overcome these difficulties, largely associated with the premature introduction of rigid-body mechanics and the failure to build a secure foundation in particle mechanics, we have set out to follow a rather deliberate organizational scheme. In so doing, we have attempted to underscore the significance of four major idealizations introduced in the following order: the single particle, the multi-particle system, the rigid body, and the deformable continuum of matter. In developing Newton's laws, linear and angular momentum, energy, and oscillations, the authors have sought to minimize the intrusion of concepts related to extended bodies, feeling that the point mass is an excellent vehicle for the presentation of fundamentals.

The authors have taken every opportunity to illustrate the universality of mechanics and the unity of science by applying the laws of mechanics to a variety of systems (atomic, electromagnetic, nuclear, and so on) and in giving emphasis to the recurrence of certain formulations such as the wave equation, the differential equation of the linear oscillator, and the differential equation describing first-order processes.

While the instructor is the proper judge of the subject matter coverage which best suits a particular course, it is suggested that the kinematics and dynamics of a particle referred to moving coordinate systems, and the more advanced parts of the chapters on oscillations of a particle, general motion, and central force motion, can be omitted in a single semester course, without loss of continuity.

The final chapter is given principally to the development of Lagrange's equations. Its relative placement in the text should not be construed as a relegation to lesser importance but rather as an expression of what the authors believe to be a suitable sequence of subject matter, pointing finally to quantum mechanics. Some instructors may find it desirable to introduce selected sections of Chapter 14 earlier, as for example, after Chapter 7.

It is likely that mechanics at the level of this text introduces the first real application of the mathematics to which the student has been exposed. Consequently, the authors have deliberately not skimmed through the calculus, differential equations, and vector and tensor analysis, retaining a considerable number of intermediate steps. Furthermore, the material is so arranged and developed that tensor analysis may be employed at the option of the instructor, but is not a prerequisite to an understanding of the physics.

Finally, an effort has been made to increase the level of difficulty in proceeding from Chapter 1 to Chapter 14 so that the student whose background from a first exposure to mechanics is not very secure can give special emphasis to certain preliminary concepts while others with a stronger foundation may omit the more elementary material. The authors have not, however, sought to sacrifice what they feel is a logical sequence of subject matter for the sake of this gradation.

Many individuals have assisted the authors over the past several years as this book evolved from a rough set of notes to its present form. To the many students who were exposed to the notes upon which this text is based, the authors express appreciation for the many improvements resulting from classroom usage. To Dr. A. Cahit Ugural and Dr. Peter Walsh of Fairleigh Dickinson University, thanks are due for their valuable suggestions in connection with Chapters 13 and 10, respectively. The valuable review of the entire manuscript by Dr. T. Vernon Frazier of The University of Nevada led to a number of clarifications and other improvements. To Mrs. Helen Stanek the authors express their special gratitude for her cheerful and patient efforts in typing and editorial assistance throughout the several drafts of this work. For assistance in the typing, thanks are also due Mrs. Frances Skelton, Mrs. Anne Zumbrunnen, Mrs. Ruth Cluggish, Miss Cynthia Buzzetta, Mrs. Jean Violante, and Miss Christina Rappaport. Many fine texts have undoubtedly influenced the authors as students and teachers of mechanics. Among these we especially acknowledge the works of Goldstein, Shames, Housner and Hudson, Lindsay, and Fowles, cited in the Bibliography. Acknowledgment is also given Electro Technology for permission to reprint several of the figures in Chapters 12 and 13 from the August 1964 issue.

Teaneck, New Jersey
November 1968

WALLACE ARTHUR
SAUL K. FENSTER

CONTENTS

CHAPTER 11 DYNAMICS OF A SYSTEM OF PARTICLES **343**

CHAPTER 12 DYNAMICS OF A RIGID BODY **404**

CHAPTER 13 MECHANICS OF DEFORMABLE CONTINUA   **499**

MECHANICS

CHAPTER 1

BASIC CONCEPTS

1.1 INTRODUCTION

Mechanics is the study of the motion of objects caused by mutual interactions as well as by external influences. Traditionally, the field of mechanics relates to single particles, systems of particles, rigid bodies, and deformable continua. Its foundations are Newton's three laws of motion and various force laws, such as Hooke's law, Newton's law of gravitation, and Coulomb's law of electrostatics.

Predictions based upon the laws of motion have been tested experimentally and their accuracy demonstrated for wide ranges of parameters. Serious deviations have been found, however, for particle speeds approaching the speed of light and for distances on the order of atomic dimensions or less.

Because Newton's laws are well understood and their formalisms so fully developed, modifications in them are made to broaden their areas of applicability. By understanding classical mechanics, a student prepares himself to pursue the fields of space physics, relativistic mechanics, statistical mechanics, as well as quantum mechanics, acoustics, elasticity, and fluid mechanics, all of which can be logically traced to the laws of Newton.

Classical mechanics may be divided into a number of subdisciplines. The study of motion without regard to the forces that cause it is called *kinematics*, often referred to as the geometry of motion. It is the objective of *kinetics* or *dynamics* to relate motion to the influences causing it. *Statics*, the study of equilibrium, deals with systems subjected to no net force or torque, and is a special case of dynamics.

The complete solution of a problem in dynamics requires the determination of the position and velocity of all the objects constituting a system and of all the unknown forces that act.

Consider the description of motion as it relates to a familiar object. Our senses inform us of how a baseball travels. An exact definition, however, of the arc of the ball, its rotation in flight, and its deformation under impact, is not so simple. A "grounder to third" means many different things; even the phrase "a two-hopper wide of third" is not very precise. We prefer the latter, however, because it is more descriptive (once we have seen a number of "two-hoppers wide of third"). Nevertheless, there are many such "two-hoppers"—fast, slow, wide, and not so wide. To say that a ball is hit with an initial velocity of 68 mph parallel to the ground, bouncing 28 ft and 79 ft from home plate, directed 12 degrees toward second base from third, and attaining a maximum height of 4 ft 6 in. is still more descriptive, although most baseball enthusiasts would not understand this picture nearly as well as that of a "two-hopper wide of third." Just as one must learn the jargon of baseball, so must the jargon of mechanics be learned. In baseball we can afford the luxury of more colorful, but less precise description, because we have stored in our memory bank additional information, and because exactness is of little consequence. In the problems we usually encounter in mechanics, greater care must be exercised and less left to the imagination.

It is of great importance to select a suitable model of the dynamical system; this serves to simplify analysis while leading to meaningful results. With regard to selection, we first admit that an exact description may be impossible. Nevertheless, as long as we cannot or do not care to measure better than we can predict, the model used is a good one. The most important models in mechanics embrace several idealizations. These are the particle or multiparticle system, the continuum of matter, and the rigid body.

1.2 THE PARTICLE

When the motion of an object of finite dimension can be represented by the motion of a geometric point, it is valid to employ the concept of the particle or point mass. The earth and sun may be so regarded in studying planetary motion. Obviously, in a treatment of terrestrial tides, only the sun and moon may be regarded as point masses in the earth–sun–moon system.

1.3 RIGID BODY

A rigid body is a system of particles each of which is constrained to maintain a fixed distance with respect to every other particle, although the entire system may itself experience motion. Only in the absence of applied forces can such a situation actually occur, because all real objects suffer deformation under load. Nevertheless, this model is a satisfactory representation where forces are moderate or deformation of little consequence.

1.4 CONTINUUM

Consider a large collection of particles, such as 1 gram-molecular weight of helium. In order to describe the system in the finest of detail, the position and velocity of each of its 6×10^{23} molecules are required. Given the three components of position and velocity of each molecule at a specific time, the state of the system

at any subsequent time can be predicted by the simultaneous application of Newton's laws of motion for each particle. Since this is an obviously impossible task,
the aggregate of particles must either be treated on a statistical basis, as is done
in *statistical mechanics*, or as a *continuous distribution* or *continuum* of matter.
Associated with each point in the continuum is a mass density averaged over a
volume large enough to assure a meaningful result.

1.5 LENGTH

The description of a physical system usually requires the introduction of the concept of size, distance, or length. Although a formal definition of length is not
possible, we can, nevertheless, intuit what it is. By international agreement,[1] the
standard of length, the meter, is taken to be 1,650,763.73 wavelengths in vacuo of
the unperturbed transition $2p_{10}$-$5d_5$ in the atom krypton-86. This is a readily
reproducible color, characteristic of the element. Formerly, the standard was the
distance between parallel lines scribed on a platinum–iridium bar kept in Sèvres,
France, at standard temperature and pressure. All other lengths are considered
in terms of the standard meter.

We understand space to be three-dimensional, that is, requiring three independent lengths measured from a reference point or origin to describe a position.
A *coordinate system* provides a means by which points in space may be conveniently
located. The *Cartesian coordinate system* [2] is the one most frequently employed in
mechanics, Fig. 1.1. The positive sense of X, Y, and Z measured from the origin
is such that if the $+X$-axis is rotated into the $+Y$-axis, similarly to a right-handed
screw, the $+Z$-axis points in the direction of advance of the screw. The position
of any point in the universe may thus be described in this *right-handed* coordinate
system.

When the coordinates of a particle change with time, one speaks of motion
having occurred. One-dimensional or rectilinear motion takes place along a
straight line. In this case it is convenient to choose one of the Cartesian axes as
the direction of motion.

1.6 TIME

The concept of time is another which we cannot properly define, and yet we sense
it and depend upon it. Two particles, each starting together from one position
and moving along an identical path, may not arrive at another point simultaneously. Time is vital for our giving precise meaning to an ordering of events, and
while we rely upon our intuition in "defining" it, this, in no way, precludes the
establishment of standards for measuring it.

---

[1] 11th International Conference on Weights and Measures, Paris, 1960.
[2] *The Geometry of René Descartes*, Dover Publications, Inc., New York, 1954.

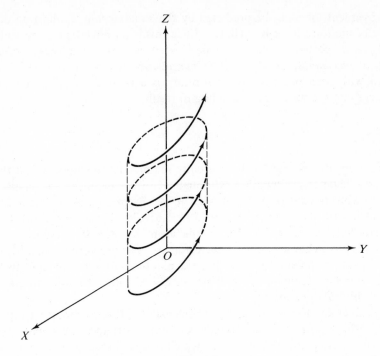

Fig. 1.1

The standard of time (adopted by the 11th International Conference) is the second, 1/31,556,925.9747, of the tropical year, 1900. Formerly, the mean solar year was the standard, defined as the mean time required for the earth to orbit the sun, 365.17 days. The old definition lacks precision because the solar year is not quite constant, whereas the year 1900, having elapsed, is a fixed standard, although, it might be argued, not readily reproducible. In the near future, an atomic clock, using as a time base a readily reproducible molecular frequency, such as that of ammonia, may be used as the standard, as was advocated by the 11th International Conference.[3] Such a clock does in fact exist.

1.7 FORCE AND MASS

One of the great contributions in the evolution of physics was Newton's recognition that force, rather than sustaining a state of motion, is responsible for altering such states. In terms of everyday experience, force perhaps relates more commonly to the deformation which it produces, such as that of a spring.

In engineering, force is often treated in much the same way as we have treated

---

[3] The 13th General Conference of Weights and Measures in 1967 adopted a specific transition of the isotope Cesium-133 as the standard of time.

length and time, that is, as a fundamental quantity. On the other hand, the physicist usually treats mass or inertia, that which we associate with the resistance to changes of motion, as fundamental and undefinable in an absolute sense. With the advent of relativity and quantum physics, momentum is also so regarded, because, in so doing, many aspects of advanced work are simplified.

The standard of mass (adopted by the 11th International Conference) is the international kilogram, a bar of platinum alloy stored in Sèvres, France. The mass of any other object is determined by comparison with the standard.

1.8 UNITS AND DIMENSIONS

We begin by giving some emphasis to the distinction between the terms *dimension* and *unit*. The former refers to the qualitative description of a physical quantity, for example, mass, length, or time. Reference to such quantities imparts no knowledge of magnitude, as one would gain, for example, from the words "one meter." Units such as the meter, kilogram, and second are based upon reference to appropriate standards and offer quantitative specification for the description of physical phenomena.

Table 1.1 Systems of Units

SYSTEM	LENGTH	TIME	MASS	FORCE
Absolute Metric (CGS)	Centimeter	Second	Gram	Dyne *
Technical English	Foot	Second	Slug *	Pound
Absolute Metric (MKS)	Meter	Second	Kilogram	Newton *

* Derived unit

The previously detailed standards of mass, length, and time form what is known as a *primary set* of units. The units of all physical quantities may be expressed in terms of any of the primary sets, these serving as building blocks for quantitative expression. (The primary sets employed in this text are given in Table 1.1.) Thus length divided by time, representing speed or velocity, possesses units of meters per second in the MKS system, and is written

$$v = \frac{L}{t} \tag{1.1}$$

Such equations, basic in the study of mechanics, are *dimensionally homogeneous;* the *dimensions* on either side of the equality sign are equal. Furthermore, mathematical manipulation cannot disturb the homogeneity of dimensions and, as a consequence, we have available a simple check. Nonagreement of the dimensions on either side of an equation at any stage in a development indicates that an error has been committed. On the other hand, agreement does not guarantee correctness.

The frequent appearance of certain arrangements of the units of a primary set leads quite naturally to the establishment of secondary or derived units. These are always reducible to a primary set. Thus the quantity *work*, usually expressed in terms of joules in the MKS system, possesses units of kilogram-meter^2 per second^2:

$$1 \frac{\text{kg-m}^2}{\text{sec}^2} = 1 \text{ J}$$

There is a great temptation to invent secondary units, and they exist in great profusion. Because of the wide variety of units available and currently in use, it is important that different sets not become "mixed up" in the same equation. At the start of a given problem, a set of units should be selected and all given quantities converted to this set. Many errors may be avoided by adhering to this practice.

Newton's second law of motion offers an excellent vehicle for discussion of primary and derived units. If one begins with the statement that force is proportional to the product of mass and acceleration, an *equation* results if a constant of proportionality k is introduced:

$$F = kma \tag{1.2}$$

If a primary set of units possessing dimensions of mass, length, and time is selected, as, for example, the MKS system, then clearly force is a derived quantity possessing the units kg-m/sec^2 and $k = 1$. The combination 1 kg-m/sec^2 is termed the newton. On the other hand, engineers often select as a primary set of dimensions, force, length, and time, possessing units of pounds, feet, and seconds, for example. In this system, the derived unit is the slug, equal to that mass which is accelerated at 1 ft/sec^2 when acted upon by a force of 1 lb. In this system, k is likewise unity.

In the "real world," as distinct from the classroom, numerical answers are invariably required. Some of the exercises at the end of each chapter require answers in literal form, but many others call for numerical results. These have been included because it has not been demonstrated that all students of physics and engineering can perform numerical work without more than an occasional error, and also because it is often of great value in itself to get an idea of the magnitude of things.

It is usually advantageous to retain the derived results in literal form until the last step. By keeping the symbols, one can often distinguish similarities in different physical systems and also determine the manner in which various quantities contribute to the experimental error. To illustrate the first point, we note that all particles of the same ratio of charge to mass precess in a uniform magnetic field with the same angular frequency. This can be readily seen from the equations in literal form, as shown in Chapter 7.

To amplify the second point, concerning error, consider the displacement s of a particle experiencing constant acceleration, a. If the particle starts from rest,

$$s = \tfrac{1}{2}at^2 \qquad\qquad (1.3)$$

where t represents time.

The error in displacement is given by

$$\Delta s = \tfrac{1}{2}a2t\,\Delta t + \tfrac{1}{2}\Delta a t^2 \qquad\qquad (1.4)$$

and the relative error in s is

$$\frac{\Delta s}{s} = \frac{2\Delta t}{t} + \frac{\Delta a}{a} \qquad\qquad (1.5)$$

Thus a 1 percent error in the measurement of acceleration results in a 1 percent error in displacement, whereas the same relative error in time leads to a 2 percent error in displacement. This analysis depends, for its execution, upon retaining the literal form of equations.

1.9 VECTORS

The techniques of vector algebra and vector calculus lend themselves well to the solution of complex problems in mechanics. This is because vectors offer ease of spatial representation as well as simple computational operations.

Vector quantities such as force, velocity, displacement, and acceleration possess both magnitude and direction and are characterized by their commutative property of addition[4]; that is, the order of addition may be changed without affecting the final result.

Not all quantities possessing magnitude and direction are vectors. Angular displacements, for example, may be characterized by magnitude and direction but are not vectors, because the sum of two or more displacements is not, in general, commutative. A *scalar* quantity is completely specified by magnitude only. Examples are mass, energy, and time.

A vector may be represented by specifying its magnitude and direction. Alternatively, its projections or components along the axes of a coordinate system, such as the Cartesian system, may be given. Thus the vector **A** may be written (A_x, A_y, A_z), where A_x is the component of **A** parallel to the X-axis. In this book, vectors are denoted by boldface type, whereas scalars (including vector magnitudes) are represented by ordinary type.

A vector may be depicted graphically by a directed line segment proportional to its magnitude with an arrow at one end to denote its direction, as shown in Fig.

---

[4] A more comprehensive definition involves the properties of a vector under an orthogonal transformation. Many so-called vectors satisfy most of the transformation properties and are treated as vectors, although in one or more operations they do not transform as vectors.

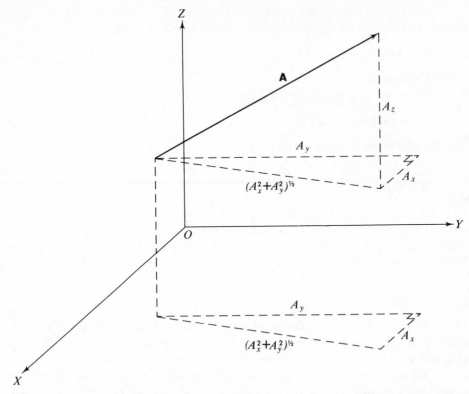

Fig. 1.2

1.2. *Free vectors*, while having *fixed direction*, may be regarded as acting at any convenient point, whereas *fixed vectors* are restricted to act at only one point in space. A *sliding vector* may act anywhere along its line of action, an infinite colinear line.

1.10 FIELDS

A *field* is used to represent a physical quantity that is a function of position in a given region. Temperature, a scalar, is a field, because its value depends upon location. The standard of length, the meter, is a scalar, but is not a field because it is the same everywhere and exists independently of any position in the universe. Wind velocity, a vector, is a field, depending as it does upon location.

1.11 VECTOR ALGEBRA

We now discuss the manner in which vectors relate to scalars and other vectors in various algebraic operations. Two vectors are equal when the components of each are identical. Since vectors have three independent components, equality implies three scalar equations. Thus, if $\mathbf{A} = \mathbf{B}$,

$$A_x = B_x \qquad A_y = B_y \qquad A_z = B_z \tag{1.6}$$

As a corollary, a vector is zero only when each of its components is zero. The zero vector or *null vector* is written **0**. It has zero magnitude and undetermined or arbitrary direction.

Vectors may be added, subtracted, and multiplied, provided, of course, that these operations are suitably defined.

1.12 ADDITION AND SUBTRACTION OF VECTORS

The resultant or sum of a number of vectors has components, each of which is the sum of the components of the individual vectors. Thus if

$$\mathbf{R} = \mathbf{A}_1 + \mathbf{A}_2 + \mathbf{A}_3 + \cdots + \mathbf{A}_n = \sum_{i=1}^{n} \mathbf{A}_i \tag{1.7}$$

then

$$R_x = A_{1x} + A_{2x} + A_{3x} + \cdots + A_{nx} = \sum_{i=1}^{n} A_{ix} \tag{1.8a}$$

$$R_y = A_{1y} + A_{2y} + A_{3y} + \cdots + A_{ny} = \sum_{i=1}^{n} A_{iy} \tag{1.8b}$$

$$R_z = A_{1z} + A_{2z} + A_{3z} + \cdots + A_{nz} = \sum_{i=1}^{n} A_{iz} \tag{1.8c}$$

To subtract vectors we need only reverse the direction of the vector components to be subtracted and add. That is,

$$\mathbf{R} = \mathbf{A}_1 - \mathbf{A}_2 = \mathbf{A}_1 + (-\mathbf{A}_2) \tag{1.9}$$

$$R_x = A_{1x} + (-A_{2x}) \qquad \text{etc.} \tag{1.10}$$

Since adding components of vectors is a scalar operation, the order of addition is immaterial (the techniques of scalar operations being the same as those for numbers). Thus

$$\mathbf{A}_1 + \mathbf{A}_2 = \mathbf{A}_2 + \mathbf{A}_1 \tag{1.11}$$

Vector addition is therefore commutative, a feature which distinguishes vectors from quantities such as finite angular displacement. These have magnitude and direction but do not obey the vector laws of addition and hence are not vectors.

Similarly, the order in which vectors are associated before addition does not alter the final result. Thus

$$\mathbf{A}_1 + (\mathbf{A}_2 + \mathbf{A}_3) = (\mathbf{A}_1 + \mathbf{A}_2) + \mathbf{A}_3 = \mathbf{A}_2 + (\mathbf{A}_3 + \mathbf{A}_1) \tag{1.12}$$

Vectors may be added or subtracted graphically by using a *polygon* of vectors. For example, given the vectors $\mathbf{A}_1$ and $\mathbf{A}_2$, Fig. 1.3(a), $\mathbf{A}_1$ added to $\mathbf{A}_2$ produces the vector $\mathbf{A}_1 + \mathbf{A}_2$, Fig. 1.3(b), or the vector $\mathbf{A}_2 + \mathbf{A}_1$, Fig. 1.3(c). Thus a parallelo-

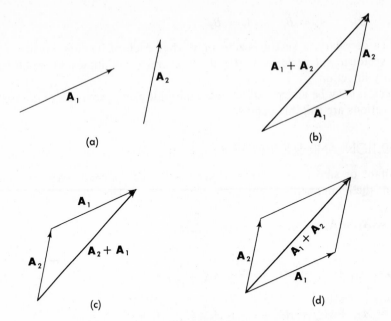

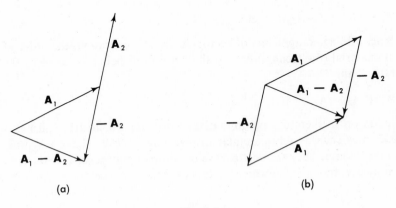

Fig. 1.3

gram of sides A_1 and A_2 may be formed, Fig. 1.3(d). The diagonal shown is the sum $A_1 + A_2$; A_1, A_2, and $A_1 + A_2$ form a simple polygon, a triangle.

Consider $A_1 - A_2$ by the construction of Fig. 1.4(a) and (b); $A_1 - A_2$ is the other diagonal of the parallelogram employed to obtain $A_1 + A_2$.

Fig. 1.4

In adding more than two vectors, say $A_1 + A_2 + A_3 + A_4$, we may add A_3 to $(A_1 + A_2)$, then A_4 to the vector $A_1 + A_2 + A_3$. The constructions are shown

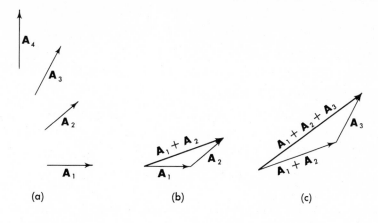

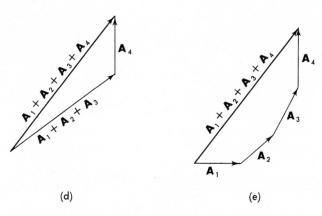

(d)

(e)

Fig. 1.5

in Fig. 1.5. The intermediate steps may be eliminated by placing the vectors tip to tail. The resultant vector runs from the start of the first to the arrowhead of the last.

1.13 MULTIPLICATION OF VECTORS BY SCALARS

Multiplying a vector by a scalar is equivalent to multiplying each of its components by the same scalar:

$$a\mathbf{A} = (aA_x, aA_y, aA_z) \tag{1.13}$$

Only the vector magnitude is thus affected. Multiplication of a vector by a scalar is commutative, distributive, and associative:

$$a\mathbf{A} = \mathbf{A}a \tag{1.14}$$

$$(a + b)\mathbf{A} = a\mathbf{A} + b\mathbf{A} \qquad (1.15)$$

$$a(\mathbf{A}_1 + \mathbf{A}_2) = a\mathbf{A}_1 + a\mathbf{A}_2 \qquad (1.16)$$

and

$$ab\mathbf{A} = a(b\mathbf{A}) = (ab)\mathbf{A} = b(a\mathbf{A}) \qquad (1.17)$$

A vector and a scalar may not, however, be added together, because they are not of the same rank. A scalar requires only one number for complete description, whereas three are required to describe a vector.

1.14 VECTOR MULTIPLICATION

There are several ways of multiplying two vectors, each of which has a special meaning in mechanics. Vector multiplications used frequently in this text are the dot product and the cross product.

The dot product, a scalar quantity, is defined as the product of the magnitude of the two vectors, multiplied by the cosine of the smaller angle between them. Thus

$$\mathbf{A} \cdot \mathbf{B} = AB \cos \psi \qquad (0 \leq \psi \leq \pi) \qquad (1.18)$$

where

$$A = |\mathbf{A}| = (A_x^2 + A_y^2 + A_z^2)^{1/2}$$

and

$$B = |\mathbf{B}| = (B_x^2 + B_y^2 + B_z^2)^{1/2}$$

The cross product, a vector quantity, is defined as a vector whose magnitude equals the product of the vector magnitudes and the sine of the smaller included angle:

$$|\mathbf{A} \times \mathbf{B}| = |\mathbf{A}||\mathbf{B}| \sin \psi \qquad (0 \leq \psi \leq \pi) \qquad (1.19)$$

The direction of the cross product is normal to the plane of the two vectors such that if the first is rotated into the second through the angle ψ, the positive direction is the same as the advance of a right-handed screw so rotated.

For a geometric interpretation of the dot and cross products refer to Fig. 1.6. Applying the law of cosines,

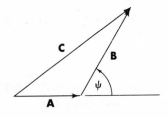

Fig. 1.6

$$C^2 = A^2 + B^2 - 2AB \cos (\pi - \psi)$$

From vector algebra,

$$\mathbf{C} = \mathbf{A} + \mathbf{B}$$

and

$$\mathbf{C} \cdot \mathbf{C} = C^2 = (\mathbf{A} + \mathbf{B}) \cdot (\mathbf{A} + \mathbf{B}) = \mathbf{A} \cdot \mathbf{A} + \mathbf{B} \cdot \mathbf{B} + 2\mathbf{A} \cdot \mathbf{B}$$

Thus

$$2\mathbf{A} \cdot \mathbf{B} = -2AB \cos (\pi - \psi) = 2AB \cos \psi$$

and

$$\cos \psi = \frac{\mathbf{A} \cdot \mathbf{B}}{AB} \qquad (1.20)$$

where ψ is the angle between $\mathbf{A}$ and $\mathbf{B}$.

Thus the law of cosines is consistent with the definition of the scalar product. In addition, Eq. (1.20) is useful in determining the angle between two vectors.

The area of the triangle of Fig. 1.6 equals one half the product of the magnitude of $\mathbf{A}$ and the component of $\mathbf{B}$ perpendicular to $\mathbf{A}$. The vector $\mathbf{B}$ may be resolved into two components, one along $\mathbf{A}$ and one normal to $\mathbf{A}$, or

$$\mathbf{B} = (B \cos \psi, B \sin \psi)$$

Therefore,

$$|\mathbf{A} \times \mathbf{B}| = |\mathbf{A}| \, |\mathbf{B}| \sin \psi = 2 \text{ (area of triangle } ABC) \qquad (1.21)$$

From Eqs. (1.18) and (1.19) it is clear that the dot product of two perpendicular vectors is zero, and the cross product of two vectors in the same direction is zero. Note that the vector and scalar products are independent of any coordinate system, requiring only a knowledge of the vector magnitudes and the angular relationship between the vectors.

The concept of unit vectors, that is, vectors of magnitude 1, is very useful in mechanics. Consider the unit vectors $\hat{\mathbf{i}}$, $\hat{\mathbf{j}}$, and $\hat{\mathbf{k}}$, directed along the positive X-, Y-, and Z-axes, respectively. Since the included angle between identical vectors is zero, it follows that

$$\hat{\mathbf{i}} \cdot \hat{\mathbf{i}} = \hat{\mathbf{j}} \cdot \hat{\mathbf{j}} = \hat{\mathbf{k}} \cdot \hat{\mathbf{k}} = 1$$

and

$$\hat{\mathbf{i}} \times \hat{\mathbf{i}} = \hat{\mathbf{j}} \times \hat{\mathbf{j}} = \hat{\mathbf{k}} \times \hat{\mathbf{k}} = 0$$

Since the angle between the Cartesian axes is $\pi/2$ radians, the definition of the scalar product now gives

$$\hat{\mathbf{i}} \cdot \hat{\mathbf{j}} = \hat{\mathbf{i}} \cdot \hat{\mathbf{k}} = \hat{\mathbf{j}} \cdot \hat{\mathbf{k}} = \hat{\mathbf{j}} \cdot \hat{\mathbf{i}} = \hat{\mathbf{k}} \cdot \hat{\mathbf{i}} = \hat{\mathbf{k}} \cdot \hat{\mathbf{j}} = 0$$

From Eq. (1.19) and the right-hand rule,

$$\hat{i} \times \hat{j} = \hat{k} \qquad \hat{j} \times \hat{k} = \hat{i} \qquad \hat{k} \times \hat{i} = \hat{j}$$
$$\hat{j} \times \hat{i} = -\hat{k} \qquad \hat{k} \times \hat{j} = -\hat{i} \qquad \hat{i} \times \hat{k} = -\hat{j}$$

Note that the dot product is commutative but the cross product is not, so that

$$\mathbf{A} \cdot \mathbf{B} = \mathbf{B} \cdot \mathbf{A}$$

but

$$\mathbf{A} \times \mathbf{B} \neq \mathbf{B} \times \mathbf{A}$$

$$\mathbf{A} \times \mathbf{B} = -(\mathbf{B} \times \mathbf{A})$$

The component of a vector along one of the coordinate axes may be determined from Fig. 1.7:

$$A_x = A \cos \phi$$

but

$$\mathbf{A} \cdot \hat{i} = |\mathbf{A}| \cos \phi = A_x$$

Therefore, $\mathbf{A} \cdot \hat{i}$ gives the component of $\mathbf{A}$ along the X-direction.

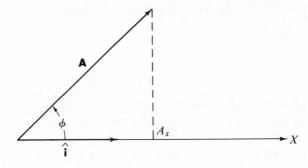

Fig. 1.7

This procedure may be extended to any direction. Thus the component of a vector along any direction is determined by taking the dot product of the vector with a unit vector along the given direction.

Writing vectors $\mathbf{A}$ and $\mathbf{B}$ in terms of their components and the unit vectors,

$$\mathbf{A} = A_x \hat{i} + A_y \hat{j} + A_z \hat{k}$$

and

$$\mathbf{B} = B_x \hat{i} + B_y \hat{j} + B_z \hat{k}$$

Clearly,

$$\mathbf{A} \cdot \mathbf{B} = A_x B_x + A_y B_y + A_z B_z \qquad (1.22)$$

The cross product may be expressed

$$\mathbf{A} \times \mathbf{B} = (A_y B_z - A_z B_y)\hat{\mathbf{i}} + (A_z B_x - A_x B_z)\hat{\mathbf{j}} + (A_x B_y - A_y B_x)\hat{\mathbf{k}} \qquad (1.23)$$

By expansion of the determinant the reader should show that

$$\mathbf{A} \times \mathbf{B} = \begin{vmatrix} \hat{\mathbf{i}} & \hat{\mathbf{j}} & \hat{\mathbf{k}} \\ A_x & A_y & A_z \\ B_x & B_y & B_z \end{vmatrix} \qquad (1.24)$$

The vector $\mathbf{A}$ may be expressed in another form. Consider

$$\mathbf{A} = A_x\hat{\mathbf{i}} + A_y\hat{\mathbf{j}} + A_z\hat{\mathbf{k}} = A\left(\frac{A_x}{A}\hat{\mathbf{i}} + \frac{A_y}{A}\hat{\mathbf{j}} + \frac{A_z}{A}\hat{\mathbf{k}}\right) = A\hat{\mathbf{e}}_A \qquad (1.25)$$

where $\hat{\mathbf{e}}_A$ is a unit vector in the direction of $\mathbf{A}$. The term A_x/A represents the *direction cosine* of $\mathbf{A}$ relative to the X-axis, that is, the cosine of the angle between $\mathbf{A}$ and X (or $\hat{\mathbf{i}}$). Thus a vector may be specified by its magnitude and direction cosines. By taking the dot product of $\mathbf{A}$ with itself, the following familiar result of analytic geometry is obtained:

$$\mathbf{A}\cdot\mathbf{A} = A^2\left[\left(\frac{A_x}{A}\right)^2 + \left(\frac{A_y}{A}\right)^2 + \left(\frac{A_z}{A}\right)^2\right]$$

or

$$\left(\frac{A_x}{A}\right)^2 + \left(\frac{A_y}{A}\right)^2 + \left(\frac{A_z}{A}\right)^2 = l_x^2 + l_y^2 + l_z^2 = 1$$

where $l_x = \cos(\mathbf{A}, X) = A_x/A$, and so on. Thus the sum of the squares of the direction cosines of a line equals 1.

The unit vector $\hat{\mathbf{e}}_A$ may be expressed in terms of the direction cosines of any line parallel to $\mathbf{A}$ and the unit vectors $\hat{\mathbf{i}}$, $\hat{\mathbf{j}}$, and $\hat{\mathbf{k}}$:

$$\hat{\mathbf{e}}_A = l_x\hat{\mathbf{i}} + l_y\hat{\mathbf{j}} + l_z\hat{\mathbf{k}} \qquad (1.26)$$

Illustrative Example 1.1

Given the vectors

$$\mathbf{A} = (5, 4, -3) \qquad \text{and} \qquad \mathbf{B} = (3, -4, 5)$$

determine (a) $6\mathbf{A} - 3\mathbf{B}$, (b) $A^2 + B^2$, (c) $\mathbf{A}\cdot\mathbf{B}$, (d) the angle between $\mathbf{A}$ and $\mathbf{B}$, (e) $4\mathbf{A} \times \mathbf{B} - 20\mathbf{B}$, (f) the direction cosines of $\mathbf{A}$, and (g) the component of $\mathbf{B}$ in the direction of $\mathbf{A}$.

SOLUTION

$$6\mathbf{A} = 6(5, 4, -3) = (30, 24, -15)$$

$$3\mathbf{B} = 3(3, -4, 5) = (9, -12, 15)$$

and

$$6\mathbf{A} - 3\mathbf{B} = (30 - 9, 24 + 12, -18 - 15) = (21, 36, -33)$$

$$= 21\hat{\mathbf{i}} + 36\hat{\mathbf{j}} - 33\hat{\mathbf{k}}$$

$$A^2 = (\mathbf{A} \cdot \mathbf{A}) = 5^2 + 4^2 + (-3)^2 = 50$$

and

$$B^2 = (\mathbf{B} \cdot \mathbf{B}) = (3)^2 + (-4)^2 + (5)^2 = 50$$

Therefore,

$$A^2 + B^2 = 100$$

and

$$\mathbf{A} \cdot \mathbf{B} = (5)(3) + (4)(-4) + (-3)(5) = -16$$

$$\cos \alpha = \frac{\mathbf{A} \cdot \mathbf{B}}{|\mathbf{A}||\mathbf{B}|} = \frac{-16}{\sqrt{50}\,\sqrt{50}} = -0.320$$

or

$$\alpha = 108.7 \text{ degrees}$$

is the angle between $\mathbf{A}$ and $\mathbf{B}$.

$$\mathbf{A} \times \mathbf{B} = \begin{vmatrix} \hat{\mathbf{i}} & \hat{\mathbf{j}} & \hat{\mathbf{k}} \\ 5 & 4 & -3 \\ 3 & -4 & 5 \end{vmatrix} = (20 - 12)\hat{\mathbf{i}} + (-9 - 25)\hat{\mathbf{j}} + (-20 - 13)\hat{\mathbf{k}}$$

$$= 8\hat{\mathbf{i}} - 34\hat{\mathbf{j}} - 32\hat{\mathbf{k}}$$

and

$$4\mathbf{A} \times \mathbf{B} - 20\mathbf{B} = 32\hat{\mathbf{i}} - 136\hat{\mathbf{j}} - 128\hat{\mathbf{k}} - 60\hat{\mathbf{i}} + 80\hat{\mathbf{j}} - 100\hat{\mathbf{k}}$$

$$= -28\hat{\mathbf{i}} - 56\hat{\mathbf{j}} - 228\hat{\mathbf{k}}$$

$$l_x = \frac{A_x}{A} = \frac{5}{\sqrt{50}} \qquad l_y = \frac{A_y}{A} = \frac{4}{\sqrt{50}} \qquad l_z = \frac{A_z}{A} = \frac{-3}{\sqrt{50}}$$

To check this result,

$$l_x^2 + l_y^2 + l_z^2 = \frac{25}{50} + \frac{16}{50} + \frac{9}{50} = 1$$

The component of $\mathbf{B}$ in the direction of $\mathbf{A}$ is

$$B_A = \mathbf{B} \cdot \hat{\mathbf{e}}_A = \mathbf{B} \cdot \frac{\mathbf{A}}{A}$$

where $\hat{\mathbf{e}}_A$ is a unit vector in the direction of $\mathbf{A}$. Thus

$$B_A = (3\hat{\mathbf{i}} - 4\hat{\mathbf{j}} + 5\hat{\mathbf{k}}) \cdot \left(\frac{5}{\sqrt{50}}\hat{\mathbf{i}} + \frac{4}{\sqrt{50}}\hat{\mathbf{j}} - \frac{3}{\sqrt{50}}\hat{\mathbf{k}} \right) = \frac{-16}{\sqrt{50}}$$

1.15 DERIVATIVE OF A VECTOR

A vector may be a function of one or more scalars and vectors. As might be expected, the vectors of greatest importance in mechanics are functions of time and position variables, and we shall therefore concern ourselves primarily with the derivative of a vector function with respect to a scalar. Consider a vector $\mathbf{A}$ represented in two alternative ways:

$$\mathbf{A} = A\hat{\mathbf{e}}_A$$

and

$$\mathbf{A} = A_x\hat{\mathbf{i}} + A_y\hat{\mathbf{j}} + A_z\hat{\mathbf{k}}$$

Suppose that $\mathbf{A}$ depends upon a scalar parameter u. The increment of $\mathbf{A}$ as u proceeds from u to $u + \Delta u$ is

$$\Delta\mathbf{A} = \mathbf{A}(u + \Delta u) - \mathbf{A}(u)$$

$$= A(u + \Delta u)\hat{\mathbf{e}}_A(u + \Delta u) - A(u)\hat{\mathbf{e}}_A(u) \tag{1.27a}$$

or

$$\Delta\mathbf{A} = A_x(u + \Delta u)\hat{\mathbf{i}} + A_y(u + \Delta u)\hat{\mathbf{j}} + A_z(u + \Delta u)\hat{\mathbf{k}}$$

$$- A_x(u)\hat{\mathbf{i}} - A_y(u)\hat{\mathbf{j}} - A_z(u)\hat{\mathbf{k}} \tag{1.27b}$$

where $\hat{\mathbf{i}}$, $\hat{\mathbf{j}}$, and $\hat{\mathbf{k}}$ are constant.

It is simpler to determine the change of $\mathbf{A}$ from Eq. (1.27b). Thus

$$\Delta\mathbf{A} = [A_x(u + \Delta u) - A_x(u)]\hat{\mathbf{i}} + [A_y(u + \Delta u) - A_y(u)]\hat{\mathbf{j}}$$

$$+ [A_z(u + \Delta u) - A_z(u)]\hat{\mathbf{k}} \tag{1.28}$$

The derivative of $\mathbf{A}$ with respect to u, $d\mathbf{A}/du$, is found by dividing Eq. (1.28) by Δu and taking the limit as Δu approaches zero, following a procedure quite similar to that of scalar calculus:

$$\frac{d\mathbf{A}}{du} = \lim_{\Delta u \to 0} \frac{\Delta\mathbf{A}}{\Delta u} = \lim_{\Delta u \to 0} \left\{ \left[\frac{A_x(u + \Delta u) - A_x(u)}{\Delta u} \right]\hat{\mathbf{i}} \right.$$

$$\left. + \left[\frac{A_y(u + \Delta u) - A_y(u)}{\Delta u} \right]\hat{\mathbf{j}} + \left[\frac{A_z(u + \Delta u) - A_z(u)}{\Delta u} \right]\hat{\mathbf{k}} \right\} \tag{1.29a}$$

or

$$\frac{d\mathbf{A}}{du} = \lim_{\Delta u \to 0} \frac{\Delta A_x}{\Delta u}\hat{\mathbf{i}} + \lim_{\Delta u \to 0} \frac{\Delta A_y}{\Delta u}\hat{\mathbf{j}} + \lim_{\Delta u \to 0} \frac{\Delta A_z}{\Delta u}\hat{\mathbf{k}} \tag{1.29b}$$

Therefore,

$$\frac{d\mathbf{A}}{du} = \frac{dA_x}{du}\hat{\mathbf{i}} + \frac{dA_y}{du}\hat{\mathbf{j}} + \frac{dA_z}{du}\hat{\mathbf{k}} \tag{1.30}$$

If the coordinate axes do not maintain a fixed orientation in space, the unit vectors along the coordinate directions may no longer be regarded as constants in taking the derivative.

Consider now the derivative with respect to the scalar u of the vector $\mathbf{A} = A\hat{\mathbf{e}}_A$, in which the magnitude of $\mathbf{A}$ and the direction of $\hat{\mathbf{e}}_A$ are assumed to vary with u. Applying the definition of a derivative, we have

$$\frac{d\mathbf{A}}{du} = \frac{d(A\hat{\mathbf{e}}_A)}{du} = \lim_{\Delta u \to 0} \left[\frac{A(u + \Delta u)\hat{\mathbf{e}}_A(u + \Delta u) - A(u)\hat{\mathbf{e}}_A(u)}{\Delta u} \right] \qquad (1.31)$$

The above may be rewritten

$$\frac{d\mathbf{A}}{du} = \lim_{\Delta u \to 0} \left[\frac{(A + \Delta A)(\hat{\mathbf{e}}_A + \Delta \hat{\mathbf{e}}_A) - A\hat{\mathbf{e}}_A}{\Delta u} \right] \qquad (1.32)$$

Expanding, taking the limit, and discarding higher-order terms, Eq. (1.32) becomes

$$\frac{d\mathbf{A}}{du} = \frac{dA}{du}\hat{\mathbf{e}}_A + A\frac{d\mathbf{e}_A}{du} \qquad (1.33)$$

The differential of a unit vector may be viewed as in Fig. 1.8, in which the direction of $\hat{\mathbf{e}}_A$ is observed to change as u changes. As Δu approaches zero, $\hat{\mathbf{e}}_A(u)$, $\hat{\mathbf{e}}_A(u + \Delta u)$, and $\Delta \mathbf{e}_A$ form a right triangle as shown, in which $\Delta \mathbf{e}_A$ is at right angles to $\hat{\mathbf{e}}_A$. The derivatives of the unit vectors $\hat{\mathbf{i}}$, $\hat{\mathbf{j}}$, and $\hat{\mathbf{k}}$ with respect to time, a scalar of special importance, are explored in Chapter 2.

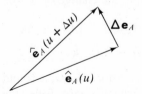

Fig. 1.8

The results derived above are applicable to differentiation of any product of a scalar and a vector:

$$\frac{d}{du}(a\mathbf{A}) = \frac{d}{du}[a(A_x\hat{\mathbf{i}})] + \frac{d}{du}[a(A_y\hat{\mathbf{j}})] + \frac{d}{du}[a(A_z\hat{\mathbf{k}})] \qquad (1.34a)$$

$$\frac{d}{du}(a\mathbf{A}) = \frac{da}{du}(A_x\hat{\mathbf{i}}) + \frac{da}{du}(A_y\hat{\mathbf{j}}) + \frac{da}{du}(A_z\hat{\mathbf{k}})$$

$$+ a\frac{d}{du}(A_x\hat{\mathbf{i}}) + a\frac{d}{du}(A_y\hat{\mathbf{j}}) + a\frac{d}{du}(A_z\hat{\mathbf{k}}) \qquad (1.34b)$$

or

$$\frac{d}{du}(a\mathbf{A}) = \left(\frac{da}{du}\right)\mathbf{A} + a\left(\frac{d\mathbf{A}}{du}\right) \qquad (1.34c)$$

Consider now the derivative of a dot product:

$$\frac{d}{du}(\mathbf{A}\cdot\mathbf{B}) = \frac{d}{du}(A_xB_x + A_yB_y + A_zB_z) \tag{1.35a}$$

$$\frac{d}{du}(\mathbf{A}\cdot\mathbf{B}) = \left(\frac{dA_x}{du}\right)B_x + \left(\frac{dA_y}{du}\right)B_y + \left(\frac{dA_z}{du}\right)B_z$$

$$+ A_x\left(\frac{dB_x}{du}\right) + A_y\left(\frac{dB_y}{du}\right) + A_z\left(\frac{dB_z}{du}\right) \tag{1.35b}$$

or

$$\frac{d}{du}(\mathbf{A}\cdot\mathbf{B}) = \frac{d\mathbf{A}}{du}\cdot\mathbf{B} + \mathbf{A}\cdot\frac{d\mathbf{B}}{du} \tag{1.35c}$$

For the cross product,

$$\frac{d}{du}(\mathbf{A}\times\mathbf{B}) = \frac{d}{du}(A_yB_z - A_zB_y)\hat{\mathbf{i}} + \frac{d}{du}(A_zB_x - A_xB_z)\hat{\mathbf{j}} + \frac{d}{du}(A_xB_y - A_yB_x)\hat{\mathbf{k}} \tag{1.36}$$

Considering only the X-component,

$$\left[\frac{d}{du}(\mathbf{A}\times\mathbf{B})\right]_x = \left(\frac{dA_y}{du}\right)B_z - \left(\frac{dA_z}{du}\right)B_y + A_y\left(\frac{dB_z}{du}\right) - A_z\left(\frac{dB_y}{du}\right) \tag{1.37a}$$

or

$$\left[\frac{d}{du}(\mathbf{A}\times\mathbf{B})\right]_x = \left(\frac{d\mathbf{A}}{du}\times\mathbf{B} + \mathbf{A}\times\frac{d\mathbf{B}}{du}\right)_x \tag{1.37b}$$

where $\hat{\mathbf{i}}$ has been regarded as constant in direction as well as magnitude. Since the derivatives for the Y- and Z-components proceed in a similar fashion, we may generalize the result as

$$\frac{d}{du}(\mathbf{A}\times\mathbf{B}) = \frac{d\mathbf{A}}{du}\times\mathbf{B} + \mathbf{A}\times\frac{d\mathbf{B}}{du} = -\frac{d}{du}(\mathbf{B}\times\mathbf{A}) \tag{1.38}$$

Note that the derivative requires that the terms involving $\mathbf{A}$ and $\mathbf{B}$ retain their original order.

Illustrative Example 1.2
Given

$$\mathbf{A} = 12t^2\hat{\mathbf{i}} - 4t\hat{\mathbf{j}} + \frac{1}{t^2}\hat{\mathbf{k}}$$

$$\mathbf{B} = 3t\hat{\mathbf{i}} - 11t^3\hat{\mathbf{j}} + 2t^2\hat{\mathbf{k}}$$

and

$$C = 5t^{-1}$$

determine

$$\frac{d}{dt}[C(\mathbf{A}\cdot\mathbf{B})] \quad \text{and} \quad \frac{d}{dt}[C(\mathbf{A}\times\mathbf{B})]$$

SOLUTION

$$\mathbf{A}\cdot\mathbf{B} = 36t^3 + 44t^4 + 2$$

$$C(\mathbf{A}\cdot\mathbf{B}) = 180t^2 + 220t^3 + 10t^{-1}$$

$$\frac{d}{dt}[C(\mathbf{A}\cdot\mathbf{B})] = 360t + 660t^2 - 10t^{-2}$$

$$\mathbf{A}\times\mathbf{B} = \begin{vmatrix} \hat{\mathbf{i}} & \hat{\mathbf{j}} & \hat{\mathbf{k}} \\ 12t^2 & -4t & t^{-2} \\ 3t & -11t^3 & 2t^2 \end{vmatrix}$$

$$= (-8t^3 + 11t)\hat{\mathbf{i}} + (3t^{-1} - 24t^4)\hat{\mathbf{j}} + (-132t^5 + 12t^2)\hat{\mathbf{k}}$$

$$C(\mathbf{A}\times\mathbf{B}) = (-40t^2 + 55)\hat{\mathbf{i}} + (15t^{-2} - 120t^3)\hat{\mathbf{j}} + (-660t^4 + 60t)\hat{\mathbf{k}}$$

$$\frac{d}{dt}[C(\mathbf{A}\times\mathbf{B})] = -80t\hat{\mathbf{i}} + (-30t^{-3} - 360t^2)\hat{\mathbf{j}} + (-2640t^3 + 60)\hat{\mathbf{k}}$$

1.16 INTEGRATION OF A VECTOR

The integration of a vector, which is a function of a single scalar, proceeds as ordinary scalar integration; it is the inverse of differentiation. Given a vector $\mathbf{B}(u) = B_x(u)\hat{\mathbf{i}} + B_y(u)\hat{\mathbf{j}} + B_z(u)\hat{\mathbf{k}}$,

$$\int \mathbf{B}(u)\, du = \hat{\mathbf{i}}\int B_x(u)\, du + \hat{\mathbf{j}}\int B_y(u)\, du + \hat{\mathbf{k}}\int B_z(u)\, du + \mathbf{C} \tag{1.39}$$

where $\mathbf{C}$ is a vector constant of integration. Each integral is thus evaluated on the basis of the usual rules of integration.

Similarly, the integral of the function $\mathbf{G}f(u)$ ($\mathbf{G}$ being a constant vector) is simply

$$\int \mathbf{G}f(u)\, du = \mathbf{G}\int f(u)\, du \tag{1.40}$$

1.17 SPATIAL DERIVATIVE OF A SCALAR — THE DEL OPERATOR

Consider a scalar field, such as temperature, represented by the function $\phi = \phi(x, y, z)$. Employing the usual definition, the total differential of ϕ is expressed

$$d\phi = \frac{\partial\phi}{\partial x}\, dx + \frac{\partial\phi}{\partial y}\, dy + \frac{\partial\phi}{\partial z}\, dz \tag{1.41}$$

If now we define the gradient of ϕ,

$$\text{grad } \phi = \frac{\partial \phi}{\partial x}\hat{\mathbf{i}} + \frac{\partial \phi}{\partial y}\hat{\mathbf{j}} + \frac{\partial \phi}{\partial z}\hat{\mathbf{k}} \tag{1.42}$$

it is clear that $d\phi$ may be represented as the following scalar product:

$$d\phi = (\text{grad } \phi)\cdot\mathbf{dr} \tag{1.43}$$

In the above expression, $\mathbf{dr} = dx\,\hat{\mathbf{i}} + dy\,\hat{\mathbf{j}} + dz\,\hat{\mathbf{k}}$, the differential change of position vector $\mathbf{r} = x\hat{\mathbf{i}} + y\hat{\mathbf{j}} + z\hat{\mathbf{k}}$, which locates any point $P(x, y, z)$ in the scalar field with respect to the origin of a convenient set of coordinate axes.

To give geometric interpretation to the gradient, imagine a set of spatial surfaces, each of which includes all points of constant ϕ. Since ϕ represents a physical field, it is continuous and single-valued at a particular instant of time. The surfaces are thus unique and nonintersecting. Any displacement $\mathbf{dr}$ on such a surface occurs without an associated change of ϕ, and consequently

$$d\phi = (\text{grad } \phi)\cdot\mathbf{dr} = 0 \tag{1.44}$$

Since neither $\text{grad } \phi$ nor $\mathbf{dr}$ is zero, we conclude that these vectors are orthogonal. The direction of $\text{grad } \phi$ is thus normal to a constant ϕ surface. Referring to Eq. (1.43), it is clear that the maximum absolute values of $d\phi$ occur where $\text{grad } \phi$ and $\mathbf{dr}$ are parallel. The gradient thus determines the magnitude and direction of the maximum spatial change in the function $\phi(x, y, z)$.

It is customary to employ the following shorthand notation for the gradient operation:

$$\text{gradient} = \hat{\mathbf{i}}\frac{\partial}{\partial x} + \hat{\mathbf{j}}\frac{\partial}{\partial y} + \hat{\mathbf{k}}\frac{\partial}{\partial z} = \nabla \tag{1.45}$$

The inverted Greek delta is referred to as "del" or "nabla." Del is a vector *operator*, producing as it does a vector when it is applied to a scalar quantity. It is not a vector in and of itself.

1.18 THE LINE INTEGRAL AND THE CURL OF A VECTOR

Consider now the integral of the scalar product of a vector $\mathbf{B}(x, y, z)$ and $\mathbf{dr}$ between the limits $P_1(x_1, y_1, z_1)$ and $P_2(x_2, y_2, z_3)$:

$$\int_{P_1}^{P_2} \mathbf{B}\cdot\mathbf{dr} = \int_{P_1}^{P_2} (B_x\hat{\mathbf{i}} + B_y\hat{\mathbf{j}} + B_z\hat{\mathbf{k}})\cdot(dx\hat{\mathbf{i}} + dy\hat{\mathbf{j}} + d_z\hat{\mathbf{k}})$$

$$= \int_{P_1}^{P_2} B_x(x, y, z)\,dx + \int_{P_1}^{P_2} B_y(x, y, z)\,dy + \int_{P_1}^{P_2} B_z(x, y, z)\,dz \tag{1.46}$$

Since the components of $\mathbf{B}$ are, in general, each functions of x, y, and z, the integration requires for its execution more than a knowledge of the limits. The path or curve of integration from P_1 to P_2 must also be specified. Integrals such as that given above are thus termed *line integrals*. To illustrate, consider the vector function $\mathbf{B} = y^2\hat{\mathbf{i}}$, which is to be integrated from the origin to $(1, 1, 1)$:

$$\int_{(0,0,0)}^{(1,1,1)} \mathbf{B}\cdot d\mathbf{r} = \int_{(0,0,0)}^{(1,1,1)} y^2\, dx$$

Unless the functional relationship between y and x is known, the above integral cannot be evaluated, because an infinite number of curves join P_1 and P_2.

A special case of particular importance in mechanics is one in which the scalar product $\mathbf{B}\cdot d\mathbf{r}$ is equal to an exact differential,

$$\mathbf{B}\cdot d\mathbf{r} = d\xi = \mathbf{grad}\ \xi\cdot d\mathbf{r} \tag{1.47}$$

and therefore $\mathbf{B} = \mathbf{grad}\ \xi = \nabla\xi$. The integration thus depends only upon the limits; the function ξ is path-independent:

$$\int_{P_1}^{P_2} \mathbf{B}\cdot d\mathbf{r} = \int_{P_1}^{P_2} d\xi = \xi_2 - \xi_1 \tag{1.48}$$

It is clear from the above expression that the line integral about any closed path is zero and the function ξ is thereby conserved. A vector field $\mathbf{B}$ yielding the above result is thus termed *conservative*.

For $\mathbf{B}$ to be conservative, it is *necessary* that

$$\frac{\partial B_z}{\partial y} - \frac{\partial B_y}{\partial z} = 0 \qquad \frac{\partial B_x}{\partial z} - \frac{\partial B_z}{\partial x} = 0 \qquad \frac{\partial B_y}{\partial x} - \frac{\partial B_x}{\partial y} = 0 \tag{1.49}$$

These conditions may be represented by the statement that the cross product of the del operator with $\mathbf{B}$, called the *curl* of $\mathbf{B}$, must vanish:

$$\nabla \times \mathbf{B} = \mathbf{curl}\ \mathbf{B} = \begin{vmatrix} \hat{\mathbf{i}} & \hat{\mathbf{j}} & \hat{\mathbf{k}} \\ \dfrac{\partial}{\partial x} & \dfrac{\partial}{\partial y} & \dfrac{\partial}{\partial z} \\ B_x & B_y & B_z \end{vmatrix}$$

$$= \hat{\mathbf{i}}\left(\frac{\partial B_z}{\partial y} - \frac{\partial B_y}{\partial z}\right) + \hat{\mathbf{j}}\left(\frac{\partial B_x}{\partial z} - \frac{\partial B_z}{\partial x}\right) + \hat{\mathbf{k}}\left(\frac{\partial B_y}{\partial x} - \frac{\partial B_x}{\partial y}\right) = 0 \tag{1.50}$$

To ascertain the restrictions placed upon the function ξ, consider the curl of the gradient of ξ:

$$\nabla \times (\nabla\xi) = \nabla \times \left(\frac{\partial\xi}{\partial x}\hat{\mathbf{i}} + \frac{\partial\xi}{\partial y}\hat{\mathbf{j}} + \frac{\partial\xi}{\partial z}\hat{\mathbf{k}}\right)$$

$$= \begin{vmatrix} \hat{\mathbf{i}} & \hat{\mathbf{j}} & \hat{\mathbf{k}} \\ \dfrac{\partial}{\partial x} & \dfrac{\partial}{\partial y} & \dfrac{\partial}{\partial z} \\ \dfrac{\partial\xi}{\partial x} & \dfrac{\partial\xi}{\partial y} & \dfrac{\partial\xi}{\partial z} \end{vmatrix}$$

$$= \hat{\mathbf{i}}\left(\frac{\partial^2\xi}{\partial y\,\partial z} - \frac{\partial^2\xi}{\partial z\,\partial y}\right) + \hat{\mathbf{j}}\left(\frac{\partial^2\xi}{\partial z\,\partial x} - \frac{\partial^2\xi}{\partial x\,\partial z}\right) + \hat{\mathbf{k}}\left(\frac{\partial^2\xi}{\partial x\,\partial y} - \frac{\partial^2\xi}{\partial y\,\partial x}\right) \tag{1.51}$$

The order of partial differentiation above may be interchanged provided ξ and its derivatives are continuous, single-valued, and finite at all points. For a function of this type, the curl of the gradient is clearly zero, and a conservative field can be derived.

1.19 STOKES' THEOREM

Consider now one of the most important theorems of vector calculus, known as *Stokes' theorem:*

$$\oint \mathbf{D} \cdot \mathbf{dr} = \int (\mathbf{\nabla} \times \mathbf{D}) \cdot \mathbf{dA} \qquad (1.52)$$

In this expression, $\mathbf{D}$ represents any vector field, and the differential area $\mathbf{dA}$ is assigned the same direction as an outward normal at the surface. The integration of $\mathbf{\nabla} \times \mathbf{D}$ is performed over the area enclosed by the path about which $\mathbf{dr}$ is integrated.

To prove Stokes' theorem, begin with an infinitesimal square of sides dx, dy lying entirely in the XY-plane, as in Fig. 1.9. The line integral of $\mathbf{D}$ about a path beginning at the point (x, y) and proceeding to successive points $(x + dx, y)$, $(x + dx, y + dy)$, $(x, y + dy)$, returning finally to (x, y), is

$$\mathbf{D} \cdot \mathbf{dr} = D_x\left(x + \frac{dx}{2}, y\right) dx + D_y\left(x + dx, y + \frac{dy}{2}\right) dy$$

$$- D_x\left(x + \frac{dx}{2}, y + dy\right) dx - D_y\left(x, y + \frac{dy}{2}\right) dy \qquad (1.53)$$

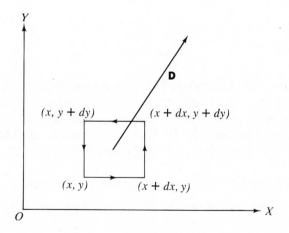

Fig. 1.9

Expanding $\mathbf{D} \cdot \mathbf{dr}$ in a two-dimensional Taylor series and retaining only terms of second order, we have

$$\mathbf{D} \cdot \mathbf{dr} = \left(\frac{\partial D_y}{\partial x} - \frac{\partial D_x}{\partial y} \right) dx \, dy \tag{1.54}$$

which may also be written

$$\mathbf{D} \cdot \mathbf{dr} = (\mathbf{\nabla} \times \mathbf{D}) \cdot \mathbf{dA}$$

where

$$\mathbf{dA} = dx \, dy \, \hat{\mathbf{k}} \tag{1.55}$$

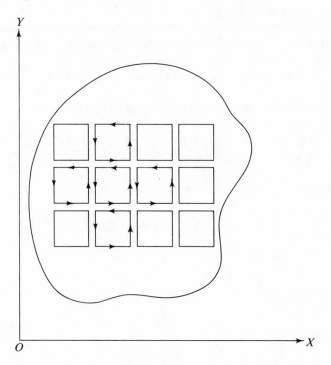

Fig. 1.10

A finite area, such as shown in Fig. 1.10, may be divided into a large number of differential squares. The result obtained above applies to each of these infinitesimal areas inasmuch as the approach taken is most general. If an integration is performed about each square, the inner paths cancel and there remains the integral over the boundary only. Stokes' theorem is thus proved.

1.20 THE DIVERGENCE OF A VECTOR—GAUSS' THEOREM

Another important theorem of vector calculus, due to Gauss, is as follows:

$$\oint \mathbf{D} \cdot \mathbf{dA} = \int (\mathbf{\nabla} \cdot \mathbf{D}) \, dV \tag{1.56}$$

where the area integration is performed over the closed surface which bounds volume V. The term $\nabla \cdot \mathbf{D}$, called the *divergence*, is defined

$$\text{div } \mathbf{D} = \nabla \cdot \mathbf{D} = \left(\frac{\partial}{\partial x} \hat{\mathbf{i}} + \frac{\partial}{\partial y} \hat{\mathbf{j}} + \frac{\partial}{\partial z} \hat{\mathbf{k}} \right) \cdot (D_x \hat{\mathbf{i}} + D_y \hat{\mathbf{j}} + D_z \hat{\mathbf{k}})$$

$$= \frac{\partial D_x}{\partial x} + \frac{\partial D_y}{\partial y} + \frac{\partial D_z}{\partial z} \tag{1.57}$$

For a proof of Gauss' theorem, consider the flux of $\mathbf{D}$ (defined as $\int \mathbf{D} \cdot d\mathbf{A}$) emanating from the various sides of the infinitesimal cube shown in Fig. 1.11. The net flux passing outward through sides I and II is

$$D_y \left(x + \frac{dx}{2}, y + dy, z + \frac{dz}{2} \right) dx\, dz - D_y \left(x + \frac{dx}{2}, y, z + \frac{dz}{2} \right) dx\, dz$$

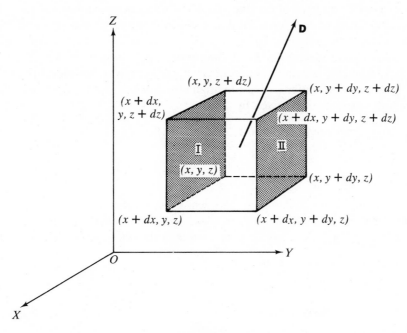

Fig. 1.11

Including all six sides, we have

$$\mathbf{D} \cdot d\mathbf{A} = D_x \left(x + dx, y + \frac{dy}{2}, z + \frac{dz}{2} \right) dy\, dz - D_x \left(x, y + \frac{dy}{2}, z + \frac{dz}{2} \right) dy\, dz$$

$$+ D_y \left(x + \frac{dx}{2}, y + dy, z + \frac{dz}{2} \right) dx\, dz - D_y \left(x + \frac{dx}{2}, y, z + \frac{dz}{2} \right) dx\, dz$$

$$+ D_z \left(x + \frac{dx}{2}, y + \frac{dy}{2}, z + dz \right) dx\, dy - D_z \left(x + \frac{dx}{2}, y + \frac{dy}{2}, z \right) dx\, dy$$

$$\tag{1.58}$$

Expanding the above expression about x, y, z and retaining terms to third order,

$$\mathbf{D} \cdot d\mathbf{A} = \frac{\partial D_x}{\partial x} \, dx \, dy \, dz + \frac{\partial D_y}{\partial y} \, dx \, dy \, dz + \frac{\partial D_z}{\partial z} \, dx \, dy \, dz$$

$$= (\nabla \cdot \mathbf{D}) \, dx \, dy \, dz = (\nabla \cdot \mathbf{D}) \, dV \qquad (1.59)$$

For a volume of finite extent, the above development is extended to include each differential cube comprising the volume. Since the net flux through the sides of the internal cubes is zero, there remains only the flux associated with the bounding surface, and the theorem is proved.

1.21 TRANSFORMATION OF A VECTOR

Consider a vector directed along one axis of a Cartesian coordinate system. What are its components in another Cartesian system sharing an origin with and rotated relative to the first system? To answer this question we begin by denoting the axes of the original system X_1', X_2', X_3' and those of the rotated system X_1, X_2, X_3. We depart from the previous notation, X, Y, Z, because it simplifies the writing of the forthcoming equations. Let the original vector be $(x_1', 0, 0)$. Its components in the unprimed system are given by

$$x_1 = x_1' \cos [\text{angle between } X_1, X_1'] = l_{11} x_1'$$

$$x_2 = x_1' \cos [\text{angle between } X_2, X_1'] = l_{21} x_1' \qquad (1.60)$$

$$x_3 = x_1' \cos [\text{angle between } X_3, X_1'] = l_{31} x_1'$$

where l_{11} is the cosine of the angle between X_1 and X_1', and so on, and where the first subscript on l refers to the unprimed axes and the second to the primed axes.

For a vector (x_1', x_2', x_3'), its components in the unprimed system are

$$x_1 = l_{11} x_1' + l_{12} x_2' + l_{13} x_3'$$

$$x_2 = l_{21} x_1' + l_{22} x_2' + l_{23} x_3' \qquad (1.61)$$

$$x_3 = l_{31} x_1' + l_{32} x_2' + l_{33} x_3'$$

where the l's again represent the direction cosines.

Six relationships can be written among the direction cosines, so that of the nine l's only three are independent:

$$l_{11}^2 + l_{21}^2 + l_{31}^2 = 1 \qquad l_{12}^2 + l_{22}^2 + l_{32}^2 = 1 \qquad l_{13}^2 + l_{23}^2 + l_{33}^2 = 1 \qquad (1.62a)$$

$$l_{11}l_{12} + l_{21}l_{22} + l_{31}l_{32} = 0 \qquad l_{12}l_{13} + l_{22}l_{23} + l_{32}l_{33} = 0$$

$$l_{13}l_{11} + l_{23}l_{21} + l_{33}l_{31} = 0 \qquad (1.62b)$$

The first three equations state the fact that the sum of the squares of the direction cosines of a line is equal to unity, and the second three express the perpendicularity of coordinate axes. The six relationships above constitute the *orthogonality condi-*

tion for a linear transformation between Cartesian coordinate systems. A quantity that transforms according to Eq. (1.61) is a vector. (A scalar is invariant under a linear coordinate transformation.)

These transformation equations, written in tensor notation, are

$$x_i = \sum_{j=1}^{3} l_{ij} x_j' \qquad i = 1, 2, 3 \tag{1.63a}$$

or

$$x_i = l_{ij} x_j' \tag{1.63b}$$

where, in Eq. (1.63b), the summation over the repeated index is implied. The orthogonality condition is derivable in tensor notation by considering the invariance of the magnitude of a vector when it is transformed from one coordinate system to another:

$$\sum_{i=1}^{3} x_i^2 = \sum_{i=1}^{3} x_i'^2 \tag{1.64}$$

or

$$\sum_{i=1}^{3} x_i^2 = \sum_{i=1}^{3} \left(\sum_{j=1}^{3} l_{ij} x_j' \right)\left(\sum_{k=1}^{3} l_{ik} x_k' \right) = \sum_{i=1}^{3} x_i'^2 \tag{1.65}$$

Rearranging the order of summation as permitted because of the commutative property of addition,

$$\sum_{i=1}^{3} l_{ij} l_{ik} x_j' x_k' = \sum_{i=1}^{3} x_i'^2 \tag{1.66}$$

From the above expression we derive the orthogonality condition in tensor form:

$$\sum_{j=1}^{3} \sum_{k=1}^{3} l_{ij} l_{ik} = \delta_{jk} \tag{1.67}$$

where δ_{jk} is termed the *Kronecker delta:*

$$\delta_{jk} = 1, \quad j = k \qquad \delta_{jk} = 0, \quad j \neq k$$

The reader will do well to reconstruct Eqs. (1.62) from Eq. (1.67).

1.22 MATRIX ALGEBRA

A *matrix* is an ordered array of scalars that obeys prescribed rules of addition and multiplication. In this text our concern is with matrices possessing real elements only, written in a rectangular array with m rows and n columns. A particular element is specified by its row number followed by its column number. Thus a_{ij} is the element of matrix A in the ith row and jth column. Alternative ways of representing matrix A are $[a_{ij}]$ or the entire array

$$A = \begin{bmatrix} a_{11} & a_{12} & \cdots & a_{1n} \\ a_{21} & a_{22} & \cdots & a_{2n} \\ \vdots & \vdots & & \\ a_{m1} & a_{m2} & \cdots & a_{mn} \end{bmatrix} \tag{1.68}$$

A vector is represented in matrix form by writing its components as either a row or column array, that is,

$$B = [b_{11} \quad b_{12} \quad b_{13}] \qquad C = \begin{bmatrix} c_{11} \\ c_{21} \\ c_{31} \end{bmatrix}$$

A vector written as a column matrix is not equivalent to one with "identical" elements, written as a row matrix.

A square matrix is characterized by an equal number of rows and columns. Obviously a vector can only be square if it possesses a single component only (and therefore exists in one-dimensional space).

The *transpose* of a matrix A, written A^T, is determined by interchanging rows and columns. Thus if A is an m by n matrix with elements a_{ij}, the elements of A^T, an n by m matrix, are a_{ji}. The transpose of a row matrix (row vector) is a column matrix. A matrix is said to be symmetric if $a_{ij} = a_{ji}$, and must be square. For this case $A^T = A$.

A square matrix in which all elements are zero except those on the main diagonal is said to be diagonalized, that is, $a_{ij} = 0$, $i \neq j$ and $a_{ij} \neq 0$, $i = j$. The unit matrix or *identity matrix*, written

$$I = \begin{bmatrix} 1 & 0 & 0 \\ 0 & 1 & 0 \\ 0 & 0 & 1 \end{bmatrix} \tag{1.69}$$

has elements such that $a_{ij} = \delta_{ij}$. The product of any matrix with an identity matrix yields the original matrix; that is, $AI = IA = A$. A matrix for which each element is zero ($a_{ij} = 0$) is termed a *null matrix*.

The determinant of square matrix A, written $\det A = |a_{ij}|$, is a *scalar*. A matrix whose determinant is zero is singular. Since a determinant is only defined for a square array, we may generalize and state that the determinants of all non-square matrices are zero and that all such matrices are singular.

Two matrices are equal only if all their corresponding elements are identical. For two m by n matrices, $A = B$ implies m by n scalar equations, $a_{ij} = b_{ij}$. Only matrices with the same numbers of rows and columns may be equated. To add two matrices it is required that they have equal numbers of rows and columns. Hence

$$A + B = C \tag{1.70a}$$

implies

$$a_{ij} + b_{ij} = c_{ij} \tag{1.70b}$$

Elements of corresponding rows and columns are thus added and retain their position in the resultant matrix. Matrix subtraction is performed as the inverse of addition. Thus

$$A - B = A + (-B) = D \tag{1.71a}$$

$$a_{ij} - b_{ij} = a_{ij} + (-b_{ij}) = d_{ij} \tag{1.71b}$$

The commutative and associative laws apply to matrix addition:

$$A + B = B + A \quad \text{and} \quad (A + B) + C = A + (B + C)$$

The multiplication of a matrix by a scalar requires that each element of the matrix be multiplied by the scalar. Therefore, if $B = cA$, $b_{ij} = ca_{ij}$. This type of multiplication is commutative, $cA = Ac$; associative, $a(bA) = (ab)A$; and distributive, $(c_1 + c_2)A = c_1A + c_2A$.

The multiplication of two matrices is performed in the same manner as the multiplication of two determinants. If

$$C = AB \tag{1.72a}$$

then

$$c_{ij} = \sum_k a_{ik}b_{kj} \tag{1.72b}$$

Consequently, to determine the ijth element of matrix C, the corresponding terms of the ith row of A and jth column of B are multiplied and the resulting products added to form c_{ij}. The number of rows of matrix A must equal the number of columns of matrix B for the multiplication to be meaningful; in this instance A and B are said to be *conformable*. For example, given the vectors

$$B = [b_{11} \quad b_{12} \quad b_{13}] \qquad C = \begin{bmatrix} c_{11} \\ c_{21} \\ c_{31} \end{bmatrix}$$

$$BC = [b_{11}c_{11} + b_{12}c_{21} + b_{13}c_{31}]$$

Note, however, that

$$CB = \begin{bmatrix} c_{11}b_{11} & c_{11}b_{12} & c_{11}b_{13} \\ c_{21}b_{11} & c_{21}b_{12} & c_{21}b_{13} \\ c_{31}b_{11} & c_{31}b_{12} & c_{31}b_{13} \end{bmatrix}$$

It is evident, then, that $CB \neq BC$, and matrix multiplication is not commutative. It is, however, associative, $A(BC) = (AB)C$, and distributive, $(A + B)C = AC + BC$.

The *inverse* of matrix D, denoted D^{-1}, is defined by the operation

$$DD^{-1} = I \tag{1.73}$$

As is now easily demonstrated, every matrix commutes with its inverse. Postmultiplying the above equation by D,

$$DD^{-1}D = D = (DD^{-1})D = D(D^{-1}D) = D$$

To construct the inverse of matrix D, consider the matrix product DD^{-1}, where

$$
D = \begin{bmatrix} d_{11} & d_{12} & \cdots & d_{1n} \\ d_{21} & d_{22} & \cdots & d_{2n} \\ \vdots & \vdots & & \vdots \\ d_{n1} & d_{n2} & & d_{nn} \end{bmatrix}
\qquad
D^{-1} = \begin{bmatrix} d'_{11} & d'_{12} & \cdots & d'_{1n} \\ d'_{21} & d_{22} & \cdots & d_{2n} \\ \vdots & \vdots & & \vdots \\ d'_{n1} & d'_{n2} & & d'_{nn} \end{bmatrix}
$$

The d_{ij} are known and the d'_{ij} are required to construct D^{-1}. The product $DD^{-1} = I$. Therefore,

$$
d_{i1}d'_{1j} + d_{i2}d'_{2j} + \cdots + d_{in}d'_{nj} = \delta_{ij} \tag{1.74a}
$$

or

$$
\sum_k d_{ik}d'_{kj} = \delta_{ij} \tag{1.74b}
$$

The above expression represents n^2 equations in d'_{ij}, which, when written out, appear as

$$
d_{11}d'_{11} + d_{12}d'_{21} + \cdots + d_{1n}d'_{1n} = 1
$$
$$
d_{21}d'_{11} + d_{22}d'_{21} + \cdots + d_{2n}d'_{n1} = 0
$$
$$
\vdots \tag{1.75}
$$
$$
d_{n1}d'_{11} + d_{n2}d'_{21} + \cdots + d_{nn}d'_{n1} = 0
$$

The solution to the above set of linear algebraic equations may be effected by applying Cramer's rule.[5]   Thus

$$
d'_{ij} = \frac{\text{cofactor}\, d_{ji}}{\det D} \tag{1.76}
$$

From the above it is clear that D must be nonsingular, lest $\det D$ be zero.

An *orthogonal* matrix is one whose inverse equals its transpose; that is, $A^{-1} = A^T$. Clearly, the determinant of an orthogonal matrix is of magnitude 1. Those with determinants $+1$ are called *proper*. For matrices with real elements, an orthogonal matrix is also termed *unitary*.

We return now to the transformation of a vector from one coordinate system to another, this time applying matrix methods. Treating X and X' as column matrices and referring to Eq. (1.61),

$$
X = AX' \tag{1.77}
$$

where

---

[5] Recall that the cofactor of element a_{ij} in a determinant is given by $(-1)^{i+j}$ multiplied by the sub-determinant remaining when the ith row and jth column are deleted.

$$A = \begin{bmatrix} l_{11} & l_{12} & l_{13} \\ l_{21} & l_{22} & l_{23} \\ l_{31} & l_{32} & l_{33} \end{bmatrix}$$

As we have noted, the transformation matrix A rotates the vector X' into X but does not change its length. For the orthogonality condition Eq. (1.67) to be valid, A must be an orthogonal matrix.

Illustrative Example 1.3
Given the matrices

$$B = \begin{bmatrix} 3 & 4 \\ -4 & 3 \end{bmatrix} \qquad C = \begin{bmatrix} 3 & 1 \\ 1 & 3 \end{bmatrix}$$

determine B^T, C^T, B^{-1}, C^{-1}.

SOLUTION
From the definition of the transpose ($a_{ij}^T = a_{ji}$),

$$B^T = \begin{bmatrix} 3 & -4 \\ 4 & 3 \end{bmatrix} \qquad C^T = \begin{bmatrix} 3 & 1 \\ 1 & 3 \end{bmatrix} = C$$

For the inverse, we apply Eq. (1.76):

$$a_{ij}^{-1} = \frac{\text{cofactor } a_{ji}}{\det A}$$

Thus

$$b_{11}^{-1} = \frac{\text{cofactor } b_{11}}{\det B} = \frac{3}{25} \qquad c_{11}^{-1} = \frac{\text{cofactor } c_{11}}{\det C} = \frac{3}{8}$$

$$b_{21}^{-1} = \frac{\text{cofactor } b_{12}}{\det B} = \frac{4}{25} \qquad c_{21}^{-1} = \frac{\text{cofactor } c_{12}}{\det C} = -\frac{1}{8}$$

$$b_{12}^{-1} = \frac{\text{cofactor } b_{21}}{\det B} = -\frac{4}{25} \qquad c_{12}^{-1} = \frac{\text{cofactor } c_{21}}{\det C} = -\frac{1}{8}$$

$$b_{22}^{-1} = \frac{\text{cofactor } b_{22}}{\det B} = \frac{3}{25} \qquad c_{22}^{-1} = \frac{\text{cofactor } c_{22}}{\det C} = \frac{3}{8}$$

and

$$B^{-1} = \frac{1}{25} \begin{bmatrix} 3 & -4 \\ 4 & 3 \end{bmatrix} = B^T \qquad C^{-1} = \frac{1}{8} \begin{bmatrix} 3 & -1 \\ -1 & 3 \end{bmatrix}$$

To check that B^{-1} and C^{-1} are indeed the correct inverses, consider

$$B^{-1}B = \frac{1}{25}\begin{bmatrix} 3 & -4 \\ 4 & 3 \end{bmatrix}\begin{bmatrix} 3 & 4 \\ -4 & 3 \end{bmatrix} = \frac{1}{25}\begin{bmatrix} 9+16 & 12-12 \\ 12-12 & 16+9 \end{bmatrix} = \begin{bmatrix} 1 & 0 \\ 0 & 1 \end{bmatrix}$$

and

$$CC^{-1} = \frac{1}{8}\begin{bmatrix} 3 & 1 \\ 1 & 3 \end{bmatrix}\begin{bmatrix} 3 & -1 \\ -1 & 3 \end{bmatrix} = \frac{1}{8}\begin{bmatrix} 9-1 & -3+3 \\ 3-3 & -1+9 \end{bmatrix} = \begin{bmatrix} 1 & 0 \\ 0 & 1 \end{bmatrix}$$

Illustrative Example 1.4

Demonstrate that the following matrix is orthogonal, determine its inverse and transpose:

$$D = \begin{bmatrix} 0.28 & 0.96 \\ -0.96 & 0.28 \end{bmatrix}$$

SOLUTION

For an orthogonal matrix $\det D = 1$ and $D^T = D^{-1}$,

$$\det D = \begin{bmatrix} 0.28 & 0.96 \\ -0.96 & 0.28 \end{bmatrix} = 0.0784 + 0.9216 = 1.000$$

The transpose of D is

$$D^T = \begin{bmatrix} 0.28 & -0.96 \\ 0.96 & 0.28 \end{bmatrix}$$

The elements of D^{-1} are $d_{ij}^{-1} = $ cofactor $d_{ji}/\det D$.   Thus

$$d_{11}^{-1} = 0.28 \qquad d_{21}^{-1} = 0.96$$
$$d_{12}^{-1} = -0.96 \qquad d_{22}^{-1} = 0.28$$

and D is an orthogonal matrix.   Thus

$$D^{-1} = \begin{bmatrix} 0.28 & -0.96 \\ 0.96 & 0.28 \end{bmatrix} = D^T$$

References

Joos, G., *Theoretical Physics*, Third Edition, Hafner Publishing Company, New York, 1956.

Kaplan, W., *Advanced Calculus*, Addison-Wesley Publishing Company, Inc., Reading, Mass., 1952.

Kreyszig, E., *Advanced Engineering Mathematics*, John Wiley & Sons, Inc., New York, 1962.

Margenau, H., and Murphy, G. M., *The Mathematics of Physics and Chemistry*, Second Edition, D. Van Nostrand Company, Inc., Princeton, N.J., 1956.

Resnick, R., and Halliday, D., *Physics*, Part 1, John Wiley & Sons, Inc., New York, 1966.

Sokolnikoff, I. S., and Redheffer, R. M., *Mathematics of Physics and Modern Engineering*, McGraw-Hill, Inc., New York, 1958.

Woods, F., *Advanced Calculus*, Ginn & Company, Boston, 1934.

EXERCISES

1.1 The constant G is the universal gravitational constant that will be discussed in Chapter 3. In the CGS system its value is 6.67×10^{-8} cm^3/g-sec^2. What is its value in the MKS and Technical English Systems?

1.2 Convert the density of water, 1 g/cm^3, to the MKS and Technical English Systems.

1.3 Demonstrate that $\mathbf{A} \cdot (\mathbf{B} \times \mathbf{C}) = (\mathbf{A} \times \mathbf{B}) \cdot \mathbf{C} = \mathbf{B} \cdot (\mathbf{C} \times \mathbf{A})$.

1.4 Demonstrate by expansion that $\mathbf{A} \times (\mathbf{B} \times \mathbf{C}) = \mathbf{B}(\mathbf{A} \cdot \mathbf{C}) - \mathbf{C}(\mathbf{A} \cdot \mathbf{B})$.

1.5 Determine the resultant, $\mathbf{A}_1 + \mathbf{A}_2 + \mathbf{A}_3$, where

$$\mathbf{A}_1 = 5\hat{\mathbf{i}} + 7\hat{\mathbf{j}} - 2\hat{\mathbf{k}}$$
$$\mathbf{A}_2 = 4\hat{\mathbf{i}} + 3\hat{\mathbf{j}} + 17\hat{\mathbf{k}}$$
$$\mathbf{A}_3 = 2\hat{\mathbf{i}} - 14\hat{\mathbf{j}} + 5\hat{\mathbf{k}}$$

1.6 Referring to Exercise 1.5, what is the angle between $\mathbf{A}_1 \times \mathbf{A}_2$ and $\mathbf{A}_1 \times \mathbf{A}_3$?

1.7 What unit vector is normal to the plane formed by the vectors $(6, 5, 4)$ and $(2, -8, 1)$?

1.8 What are the direction cosines of the line normal to the plane formed by the vectors $(0, 1, 0)$ and $(1, 1, 1)$?

1.9 Given the vectors

$$\mathbf{A}_1 = \quad 4\hat{\mathbf{i}} + 7\hat{\mathbf{j}} - 2\hat{\mathbf{k}}$$

and

$$\mathbf{A}_2 = -3\hat{\mathbf{i}} + 6\hat{\mathbf{j}} + 4\hat{\mathbf{k}}$$

a. What vector added to $\mathbf{A}_1 - \mathbf{A}_2$ makes the resultant zero?

b. What is the angle between $\mathbf{A}_1$ and $\mathbf{A}_2$?

c. What vector of magnitude 9 is normal to the plane of $\mathbf{A}_1$ and $\mathbf{A}_2$?

1.10 Apply the definition of the cross-product to prove the law of sines.

1.11 Given a vector $\mathbf{A} = (4, 8, 12)$, find a unit vector $\hat{\mathbf{e}}_n$ perpendicular to $\mathbf{A}$ and another unit vector normal to the plane of $\mathbf{A}$ and $\hat{\mathbf{e}}_n$.

1.12 Prove, using vector methods, that the sum of the square of the diagonals of a parallelogram equals the sum of the squares of the four sides.

1.13 Prove that

a. $|\mathbf{A}_1 + \mathbf{A}_2| \leq |\mathbf{A}_1| + |\mathbf{A}_2|$.

b. $|\mathbf{A}_1 \cdot \mathbf{A}_2| \leq |\mathbf{A}_1|\ |\mathbf{A}_2|$.

c. $|\mathbf{A}_1 \times \mathbf{A}_2| \leq |\mathbf{A}_1|\ |\mathbf{A}_2|$.

Under what conditions do the equalities hold?

1.14 Prove that

$$\frac{(\mathbf{A} \cdot \mathbf{B})^2 + (\mathbf{A} \times \mathbf{B}) \cdot (\mathbf{A} \times \mathbf{B})}{(\mathbf{A} \cdot \mathbf{A})(\mathbf{B} \cdot \mathbf{B})} = 1$$

1.15 Demonstrate that $\mathbf{A} \cdot (\mathbf{A} \times \mathbf{B}) = 0$.

1.16 Demonstrate that if three vectors $\mathbf{A}_1$, $\mathbf{A}_2$, $\mathbf{A}_3$ lie in a plane, any one, say $\mathbf{A}_3$, may be expressed $\mathbf{A}_3 = \alpha_1\mathbf{A}_1 + \alpha_2\mathbf{A}_2$, where α_1 and α_2 represent scalars.

1.17 Determine the direction cosines and magnitude of a vector which, when vector-multiplied by $\mathbf{A} = 5\hat{\mathbf{i}} - 6\hat{\mathbf{j}} + 4\hat{\mathbf{k}}$, produces a unit vector in the XZ-plane, 153 degrees from the X-axis.

1.18 Prove that if $\mathbf{A} + \mathbf{B} + \mathbf{C} = 0$, $\mathbf{A}$, $\mathbf{B}$, and $\mathbf{C}$ all lie in the same plane.

1.19 Prove that the cosine of the included angle between two vectors, $\mathbf{A}_1$ and $\mathbf{A}_2$, equals $l_{1x}l_{2x} + l_{1y}l_{2y} + l_{1z}l_{2z}$, where the l's represent the direction cosines of $\mathbf{A}_1$ and $\mathbf{A}_2$ with respect to a coordinate reference.

1.20 Show that $\mathbf{A} \cdot (\mathbf{B} \times \mathbf{C})$ is the volume of a parallelepiped formed by the concurrent vectors $\mathbf{A}$, $\mathbf{B}$, $\mathbf{C}$ shown in Fig. P1.20.

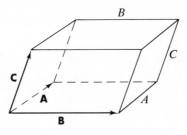

Fig. P1.20

1.21 Under what conditions is

a. $\mathbf{A} \times \dfrac{d\mathbf{A}}{dt} = 0$?

b. $\mathbf{A} \cdot \dfrac{d\mathbf{A}}{dt} = 0$?

1.22 Under what conditions is the vector product of the derivatives of two parallel vectors zero?

1.23 Given

$$\mathbf{A} = 3t^2\hat{\mathbf{i}} + 4t\hat{\mathbf{j}} + 6\hat{\mathbf{k}}$$

and

$$\mathbf{B} = \ln t\,\hat{\mathbf{i}} - e^{-t^2}\hat{\mathbf{j}} + 4t^3\hat{\mathbf{k}}$$

determine

$$\mathbf{A} \cdot \frac{d\mathbf{B}}{dt} - \mathbf{B} \cdot \frac{d\mathbf{A}}{dt}$$

1.24 Referring to Exercise 1.23, determine

$$\mathbf{A} \times \frac{d\mathbf{B}}{dt} + \mathbf{B} \times \frac{d\mathbf{A}}{dt}$$

1.25 Referring to Exercise 1.23, determine

$$\frac{d}{dt}\left[\frac{(\mathbf{A} \cdot \mathbf{B})}{|\mathbf{A}||\mathbf{B}|}\right]$$

1.26 Given

$$\mathbf{A}_1 = \sin t\hat{\mathbf{i}} + \cos t\hat{\mathbf{j}}$$

and

$$\mathbf{A}_2 = \cos t\hat{\mathbf{i}} - \sin t\hat{\mathbf{j}}$$

determine $\mathbf{A}_1 \cdot \mathbf{A}_2$ and $\mathbf{A}_1 \times \mathbf{A}_2$. Can you explain why $(d/dt)(\mathbf{A}_1 \cdot \mathbf{A}_2)$ and $(d/dt)(\mathbf{A}_1 \times \mathbf{A}_2)$ are zero without performing the scalar and vector operations?

1.27 Referring to Exercise 1.26, show that

$$\frac{d\mathbf{A}_1}{dt} = \mathbf{A}_2 \quad \text{and} \quad \frac{d\mathbf{A}_2}{dt} = -\mathbf{A}_1$$

What type of vectors are $\mathbf{A}_1$ and $\mathbf{A}_2$?

1.28 Show that the results of Illustrative Example 1.2 may be obtained by expansion. Thus

$$\frac{d}{dt}[C(\mathbf{A}\cdot\mathbf{B})] = \frac{dC}{dt}(\mathbf{A}\cdot\mathbf{B}) + C\left(\frac{d\mathbf{A}}{dt}\cdot\mathbf{B}\right) + C\left(\mathbf{A}\cdot\frac{d\mathbf{B}}{dt}\right)$$

and

$$\frac{d}{dt}[C(\mathbf{A}\times\mathbf{B})] = \frac{dC}{dt}(\mathbf{A}\times\mathbf{B}) + C\left(\mathbf{A}\times\frac{d\mathbf{B}}{dt}\right) + C\left(\frac{d\mathbf{A}}{dt}\times\mathbf{B}\right)$$

1.29 Demonstrate by expansion the following vector relationships:

$$\nabla\cdot(\phi\mathbf{A}) = (\nabla\phi)\cdot\mathbf{A} + \phi(\nabla\cdot\mathbf{A})$$

$$\nabla\times\phi(\mathbf{A}) = (\nabla\phi\times\mathbf{A}) + \phi(\nabla\times\mathbf{A})$$

$$\nabla\times(\nabla\times\mathbf{A}) = \nabla(\nabla\cdot\mathbf{A}) - (\nabla\cdot\nabla)\mathbf{A}$$

$$\nabla\cdot(\mathbf{A}\times\mathbf{B}) = \mathbf{A}\cdot(\nabla\times\mathbf{B}) - \mathbf{B}\cdot(\nabla\times\mathbf{A})$$

$$\nabla\cdot(\nabla\times\mathbf{A}) = 0$$

1.30 Demonstrate that, if $\mathbf{A} = A_x(x)\hat{\mathbf{i}} + A_y(y)\hat{\mathbf{j}} + A_z(z)\hat{\mathbf{k}}$, $\mathbf{A}$ is conservative, and that if $\mathbf{B} = \mathbf{B}(r)$, $r = (x^2 + y^2 + z^2)^{1/2}$, $\mathbf{B}$ is also conservative.

1.31 Maxwell's equations of electromagnetism, written in integral form and employing rationalized CGS units, are

$$\oint \mathbf{E}\cdot d\mathbf{A} = \int \rho\, dV \qquad \oint \mathbf{B}\cdot d\mathbf{A} = 0 \qquad \oint \mathbf{E}\cdot d\mathbf{r} + \frac{d}{dt}\int \mathbf{B}\cdot d\mathbf{A} = 0$$

$$\oint \mathbf{B}\cdot d\mathbf{r} = \int \mathbf{E}\cdot d\mathbf{A} + \frac{d}{dt}\int \mathbf{J}\cdot d\mathbf{A}$$

where **E** is the electric field intensity, **B** the magnetic induction, **J** the current density, and ρ the charge density. Express these equations in differential form by applying Stokes' and Gauss' theorems.

1.32 Consider a pair of two-dimensional coordinate systems X_1', X_2' and X_1, X_2, where X_1' and X_2' are orthogonal, as are X_1 and X_2. The angle between X_1' and X_1 is ϕ rad. Determine the transformation matrix between the primed and unprimed coordinate systems and demonstrate that this matrix satisfies the orthogonality condition. Construct the inverse of this matrix.

1.33 Solve the following set of simultaneous linear equations for x_1, x_2, x_3, employing matrices:

$$3x_1 - 2x_2 + 4x_3 = 11$$

$$2x_1 + 4x_2 - 3x_3 = 1$$

$$x_1 + 4x_2 - 2x_3 = 3$$

1.34 A similarity transformation is defined $\mathsf{B}' = \mathsf{A}^{-1}\mathsf{B}\mathsf{A}$, where B is a square matrix upon which the transformation is performed and A is an orthogonal matrix. Determine the elements of A in order that B' be diagonalized.

1.35 The trace of a matrix is defined as the sum of its diagonal elements, trace $= \sum a_{ii}$. Demonstrate that the trace of a matrix is invariant under a similarity transformation.

CHAPTER 2

KINEMATICS— THE GEOMETRY OF MOTION

2.1 INTRODUCTION

Kinematics treats the geometric description of motion without regard to forces. The quantities of usual interest are position, velocity, and acceleration. It is a constant objective of kinematics that the descriptions of motion be unambiguous and most convenient for a particular situation. In this chapter our primary concern is the kinematics of a point.

The approach taken here is to consider general particle motion in space. Special cases then follow in the illustrative examples and the exercises. Such special cases include *rectilinear motion*, in which a point undergoes motion in a straight line, and *circular motion*.

2.2 POSITION

The position of a point P with respect to the origin of an orthogonal coordinate frame is described by specifying three coordinates, as, for example, x, y, z and R, ϕ, z in Fig. 2.1. If the point is confined to move in a plane, the coordinate axes may be oriented in such a way as to require only two coordinates to locate P. The position in the XY-plane of P thus requires specification of x, y or R, ϕ. Similarly, the description of the position of a point limited to straight-line motion is accomplished by specifying a single coordinate, for example, x or y or z or R.

In vector mechanics, it is most useful to locate P by a vector running from the origin to the point. Such a vector, the *radius vector*, shown in Fig. 2.1, is designated $\mathbf{r}$. Employing the familiar Cartesian coordinate system, the radius vector is written in terms of x, y, and z, the projections of $\mathbf{r}$ upon the X-, Y-, and Z-axes, respectively, and $\hat{\mathbf{i}}$, $\hat{\mathbf{j}}$, and $\hat{\mathbf{k}}$, the unit vectors along these axes, as shown:

$$\mathbf{r} = x\hat{\mathbf{i}} + y\hat{\mathbf{j}} + z\hat{\mathbf{k}} \tag{2.1}$$

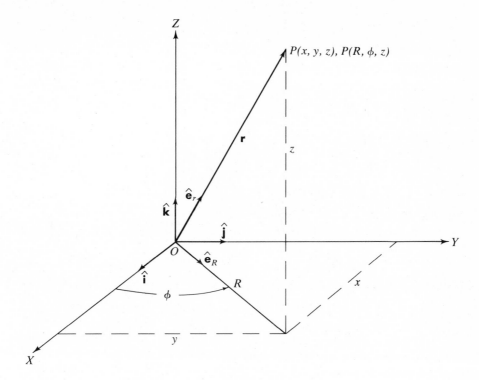

Fig. 2.1

or alternatively,

$$\mathbf{r} = r\hat{\mathbf{e}}_r \tag{2.2}$$

where $r = |\mathbf{r}| = (x^2 + y^2 + z^2)^{1/2}$ and $\hat{\mathbf{e}}_r$, the unit vector in the direction of $\mathbf{r}$, equals $\mathbf{r}/r$.

The position of the particle may, of course, be referred to any point in space. The directed line segment from such a point to the particle, the *position vector*, locates the particle relative to the chosen point. The position vector $\boldsymbol{\rho} = \mathbf{r}_2 - \mathbf{r}_1$ is, as in Fig. 2.2, for a Cartesian reference,

$$\boldsymbol{\rho} = (x_2 - x_1)\hat{\mathbf{i}} + (y_2 - y_1)\hat{\mathbf{j}} + (z_2 - z_1)\hat{\mathbf{k}} \tag{2.3}$$

where x_1, y_1, z_1 represents the point to which position is referred. If x_1, y_1, z_1 were zero (the origin), Eqs. (2.3) and (2.1) would be identical. The radius vector is therefore a position vector.

No sense of motion has thus far been evident, because the passage of time has not entered into our discussion. If points 1 and 2 represent the positions of the *same* particle at times t_1 and t_2, a *displacement vector* describes the *displacement* associated with the particle as it proceeds from one position to another, Fig. 2.3.

Z
ρ
(x_2, y_2, z_2)
at t_1
(x_1, y_1, z_1)
at t_1
$\mathbf{r}_1$
$\mathbf{r}_2$
O
Y
X

Fig. 2.2

Z
(x_2, y_2, z_2)
at t_2
Displacement vector
Path of point
(x_1, y_1, z_1) at t_1
P_1
n
Δs
Δr
P_2
r(t)
r(t + Δt)
O
Y
X

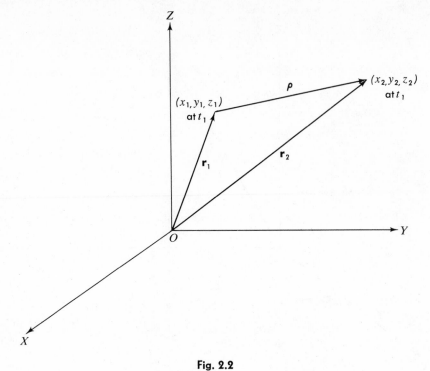

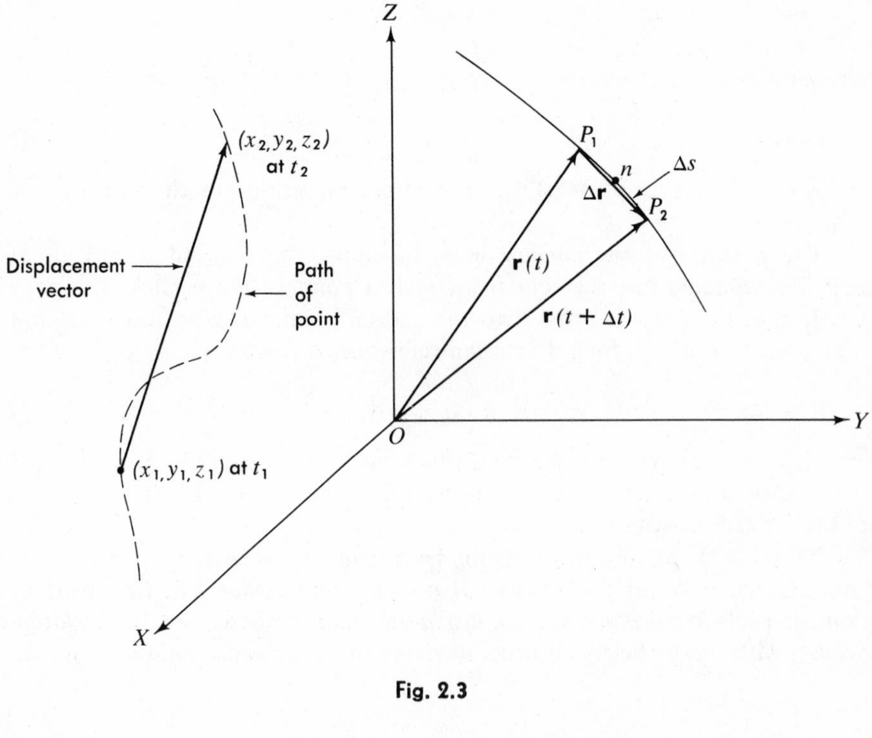

Fig. 2.3

2.3 VELOCITY

Consider the vector **r** at time t and at a later time $t + \Delta t$. The change in **r** occurring in the time interval Δt is

$$\Delta \mathbf{r} = \mathbf{r}(t + \Delta t) - \mathbf{r}(t) \tag{2.4}$$

From the definition of the derivative,

$$\frac{d\mathbf{r}}{dt} = \lim_{\Delta t \to 0} \frac{\Delta \mathbf{r}}{\Delta t} = \lim_{\Delta t \to 0} \left[\frac{\mathbf{r}(t + \Delta t) - \mathbf{r}(t)}{\Delta t} \right] \tag{2.5}$$

Equation (2.5) defines the velocity **v** of the particle with respect to the origin of the XYZ coordinate frame. Examination of Fig. 2.3 indicates that as Δt approaches zero, the vectors $\Delta \mathbf{r}$ and therefore $\Delta \mathbf{r}/\Delta t$ approach the direction of a tangent to the path drawn at point n. More formally, as Δt approaches zero, Δs, the arc length, and $|\Delta \mathbf{r}|$ become equal, so that

$$\mathbf{v} = \lim_{\Delta t \to 0} \frac{\Delta \mathbf{r}}{\Delta t} = \lim_{\Delta t \to 0} \frac{\Delta s}{\Delta t} \frac{\Delta \mathbf{r}}{\Delta s} = \frac{ds}{dt} \hat{\mathbf{e}}_t \tag{2.6}$$

where $\hat{\mathbf{e}}_t$ is a unit vector tangent to the path at every point, and $ds/dt = \dot{s}$, the velocity magnitude or speed.

From Eqs. (2.1) and (2.5) the velocity is written

$$\mathbf{v} = \frac{d}{dt}(\mathbf{r}) = \dot{\mathbf{r}} = \frac{d}{dt}(x\hat{\mathbf{i}} + y\hat{\mathbf{j}} + z\hat{\mathbf{k}}) \tag{2.7}$$

Since in this case $\hat{\mathbf{i}}$, $\hat{\mathbf{j}}$, and $\hat{\mathbf{k}}$ are constant in direction as well as magnitude, Eq. (2.7) becomes

$$\mathbf{v} = \dot{x}\hat{\mathbf{i}} + \dot{y}\hat{\mathbf{j}} + \dot{z}\hat{\mathbf{k}} \tag{2.8a}$$

or

$$\mathbf{v} = v_x\hat{\mathbf{i}} + v_y\hat{\mathbf{j}} + v_z\hat{\mathbf{k}} \tag{2.8b}$$

where $v_x = \dot{x}$, and so on. The speed is given by

$$\dot{s} = |\mathbf{v}| = (v_x^2 + v_y^2 + v_z^2)^{\frac{1}{2}} \tag{2.9}$$

Equation (2.9) may also be expressed

$$\dot{s} = (\mathbf{v} \cdot \mathbf{v})^{\frac{1}{2}} \tag{2.10}$$

The result given by Eqs. (2.8) could have been obtained directly by permitting $\Delta x/\Delta t$, $\Delta y/\Delta t$, and $\Delta z/\Delta t$ to individually undergo limiting processes as Δt approaches zero.

2.4 ACCELERATION

Acceleration bears the same relationship to velocity as velocity does to position. We have demonstrated that the velocity is tangent to the locus of the end points of the position vector as the latter varies with time. Similarly, the acceleration is tangent to the locus of the end points of the velocity vector. In this representation, $\mathbf{v}(t)$ is drawn from the origin of a frame of reference in which the axes are the orthogonal components of velocity, for example, v_x, v_y, v_z.

Acceleration, $\mathbf{a}$, is defined as the derivative with respect to time of the velocity:

$$\mathbf{a} = \lim_{\Delta t \to 0} \frac{\Delta \mathbf{v}}{\Delta t} = \frac{d\mathbf{v}}{dt} \tag{2.11}$$

Therefore,

$$\mathbf{a} = \frac{d^2\mathbf{r}}{dt^2} = \ddot{x}\hat{\mathbf{i}} + \ddot{y}\hat{\mathbf{j}} + \ddot{z}\hat{\mathbf{k}} \tag{2.12a}$$

$$\mathbf{a} = \dot{v}_x\hat{\mathbf{i}} + \dot{v}_y\hat{\mathbf{j}} + \dot{v}_z\hat{\mathbf{k}} \tag{2.12b}$$

or

$$\mathbf{a} = a_x\hat{\mathbf{i}} + a_y\hat{\mathbf{j}} + a_z\hat{\mathbf{k}} \tag{2.12c}$$

The magnitude of the acceleration does not, in general, equal the time derivative of the speed, as will be demonstrated later.

Note that although Eqs. (2.12) have been expressed in Cartesian coordinates, there is nothing sacrosanct about such a system. In other coordinate systems, discussed later, the various vectors (position, velocity, acceleration) do, to be sure, appear different. Appreciate, however, that referred to a fixed point, these vectors are unique in space; their magnitude and direction are quite independent of the manner in which they are described. The magnitude and direction of the radius vector is dependent upon the location of the origin with respect to the point under description.

2.5 PARAMETRIC REPRESENTATION OF THE POSITION VECTOR

In the foregoing development it is implicit that the various time derivatives can be determined. This is readily accomplished when the position vector $\mathbf{r}$ is expressed parametrically as a function of time:

$$\mathbf{r}(t) = x(t)\hat{\mathbf{i}} + y(t)\hat{\mathbf{j}} + z(t)\hat{\mathbf{k}} \tag{2.13}$$

Equation (2.13) implies that point P_1 on a space curve C may be defined at time t_1, since at this time the coordinates of the point $x(t_1), y(t_1), z(t_1)$ may be determined. The position of P_1 is thus given with respect to the origin:

$$\mathbf{r}(t_1) = x(t_1)\hat{\mathbf{i}} + y(t_1)\hat{\mathbf{j}} + z(t_1)\hat{\mathbf{k}} \tag{2.14}$$

Illustrative Example 2.1
Derive expressions for the velocity and acceleration of a point moving along the straight line in space: $\mathbf{r}(t) = \mathbf{A} + \mathbf{B}t$, where $\mathbf{A}$ and $\mathbf{B}$ are constant vectors.

SOLUTION

The above may be written in terms of the rectangular components of **A** and **B**:

$$\mathbf{r}(t) = (A_x + B_x t)\hat{\mathbf{i}} + (A_y + B_y t)\hat{\mathbf{j}} + (A_z + B_z t)\hat{\mathbf{k}}$$

The velocity is, therefore,

$$\mathbf{v} = \dot{\mathbf{r}} = \mathbf{B} = \text{constant} = B_x\hat{\mathbf{i}} + B_x\hat{\mathbf{j}} + B_z\hat{\mathbf{k}}$$

and the acceleration $\mathbf{a} = \mathbf{0}$.

Illustrative Example 2.2

A particle undergoes motion in a plane according to the equation

$$\mathbf{r} = 5(\sin \omega t)\hat{\mathbf{i}} + 5(\cos \omega t)\hat{\mathbf{j}} \qquad \text{where } \omega = \pi \text{ rad/sec}$$

Determine the X- and Y-components of velocity and acceleration; the magnitude and direction of the position, velocity, and acceleration vectors; and describe the trajectory.

SOLUTION

The situation described is one in which the X- and Y-motions are harmonic, that is, sine and cosine functions:

$$x = 5(\sin \pi t) \qquad y = 5(\cos \pi t)$$

The X-velocity and acceleration are, respectively,

$$\dot{x} = 5\pi(\cos \pi t) \qquad \text{and} \qquad \ddot{x} = -5\pi^2(\sin \pi t) = -\pi^2 x$$

obtained by taking the appropriate time derivatives. Similarly, the Y-velocity and acceleration are, respectively,

$$\dot{y} = -5\pi(\sin \pi t) \qquad \text{and} \qquad \ddot{y} = -5\pi^2(\cos \pi t) = -\pi^2 y$$

The magnitude of the position vector

$$r = [(5 \sin \pi t)^2 + (5 \cos \pi t)^2]^{\frac{1}{2}} = 5 = \text{constant}$$

or $x^2 + y^2 = 5^2$; the trajectory (involving spatial variables only) is a circle of radius 5.

The velocity and acceleration magnitudes are similarly determined:

$$v = [(5\pi \cos \pi t)^2 + (-5\pi \sin \pi t)^2]^{\frac{1}{2}} = 5\pi = \text{constant}$$

and

$$a = [(-5\pi^2 \sin \pi t)^2 + (-5\pi^2 \cos \pi t)^2]^{\frac{1}{2}} = 5\pi^2 = \text{constant}$$

Although the magnitudes of the foregoing quantities are constant, their directions are not. Recall that the direction of the velocity vector is everywhere tangent to the trajectory (normal to the radius in circular motion):

$$\frac{d\mathbf{r}}{dt} = \mathbf{v} = 5\pi(\cos \pi t)\hat{\mathbf{i}} - 5\pi(\sin \pi t)\hat{\mathbf{j}}$$

Since $\mathbf{r}$ and $\mathbf{v}$ are here perpendicular, their scalar product vanishes:

$$\mathbf{r} \cdot \mathbf{v} = [5(\sin \pi t)\hat{\mathbf{i}} + 5(\cos \pi t)\hat{\mathbf{j}}] \cdot [5\pi(\cos \pi t)\hat{\mathbf{i}} - 5\pi(\sin \pi t)\hat{\mathbf{j}}]$$

$$= 25\pi(\sin \pi t)(\cos \pi t) - 25\pi(\sin \pi t)(\cos \pi t) = 0$$

The acceleration $\mathbf{a} = d\mathbf{v}/dt = -5\pi^2(\sin \pi t)\hat{\mathbf{i}} - 5\pi^2(\cos \pi t)\hat{\mathbf{j}}$ is directed *toward* the origin or center of motion and is termed *centripetal*, meaning "toward the center." The scalar product of unit vectors along $\mathbf{r}$ and $\mathbf{a}$ equals -1, because they are oppositely directed:

$$\hat{\mathbf{r}} \cdot \hat{\mathbf{a}} = \left[\frac{\mathbf{r}}{r}\right] \cdot \left[\frac{\mathbf{a}}{a}\right]$$

$$= \left[\frac{5(\sin \pi t)}{5}\hat{\mathbf{i}} + \frac{5(\cos \pi t)}{5}\hat{\mathbf{j}}\right] \cdot \left[\frac{-5\pi^2(\sin \pi t)}{5\pi^2}\hat{\mathbf{i}} - \frac{5\pi^2(\cos \pi t)}{5\pi^2}\hat{\mathbf{j}}\right]$$

$$= -(\sin^2 \pi t) - (\cos^2 \pi t) = -1$$

2.6 THE KINEMATIC INTEGRALS

Beginning with $\mathbf{r}$, successive differentiations with respect to time gave us velocity and acceleration. The reverse operation is also quite useful. Given acceleration, a first-time integration yields the velocity. Another integration provides the position. Such an approach affords an efficient means of determining the instantaneous position and velocity of rockets and submarines. These systems employ inertial navigation devices which contain accelerometers to measure the components of acceleration along the orthogonal axes of a suitable coordinate reference, discussed in Chapter 3.

Equation (2.7) may be written in the form

$$\int_{\mathbf{r}_1}^{\mathbf{r}_2} d\mathbf{r} = \int_{t_1}^{t_2} \mathbf{v}\, dt \tag{2.15}$$

leading to

$$\mathbf{r}_2 - \mathbf{r}_1 = \int_{t_1}^{t_2} \mathbf{v}\, dt \tag{2.16}$$

In terms of Cartesian components, Eq. (2.16) leads to two scalar equations:

$$x_2 - x_1 = \int_{t_1}^{t_2} v_x\, dt \quad \text{etc.} \tag{2.17}$$

Clearly, the integral on the right side of Eq. (2.17) represents the area under a curve of X-velocity as a function of time. Similarly, Eq. (2.11) leads to

$$\int_{\mathbf{v}_1}^{\mathbf{v}_2} d\mathbf{v} = \int_{t_1}^{t_2} \mathbf{a}\, dt \qquad (2.18)$$

or

$$v_{x2} - v_{x1} = \int_{t_1}^{t_2} a_x\, dt \qquad \text{etc.} \qquad (2.19)$$

where the integral represents the area under an acceleration time curve, such as an accelerometer trace.

Another important integral is obtained by beginning with

$$a_x = \frac{dv_x}{dt} = \frac{dv_x}{dx}\frac{dx}{dt} = v_x \frac{dv_x}{dx} \qquad (2.20)$$

leading to

$$\int_{x_1}^{x_2} a_x\, dx = \int_{v_{x1}}^{v_{x2}} v_x\, dv_x \qquad (2.21)$$

or

$$\int_{x_1}^{x_2} a_x\, dx = \tfrac{1}{2}(v_{x2}^2 - v_{x1}^2) \qquad (2.22)$$

The integral now represents the area under the acceleration displacement curve.

2.7 RELATIVE POSITION AND DISPLACEMENT

Consider two particles, A and B, located with respect to reference XYZ by radii $\mathbf{r}_A$ and $\mathbf{r}_B$, respectively, as in Fig. 2.4. At a particular time t, the position of B relative to A is described by a relative position vector $\boldsymbol{\rho}_{BA}$:

$$\boldsymbol{\rho}_{BA} = \mathbf{r}_B - \mathbf{r}_A \qquad (2.23)$$

Equation (2.23) may be expressed

$$\mathbf{r}_B = \mathbf{r}_A + \boldsymbol{\rho}_{BA} \qquad (2.24)$$

which states that the position of B is equal to the position of A plus that of B relative to A.

As A and B undergo displacements $\Delta\mathbf{r}_A$ and $\Delta\mathbf{r}_B$ during time interval Δt, these particles are found at points A' and B'. The new relative position vector, at time $t + \Delta t$, is

$$\boldsymbol{\rho}_{B'A'} = \mathbf{r}_{B'} - \mathbf{r}_{A'} \qquad (2.25)$$

Both A and B having experienced displacements, the relative displacement of B to A during Δt is given by

$$\Delta r_{BA} = \boldsymbol{\rho}_{B'A'} - \boldsymbol{\rho}_{BA} \qquad (2.26)$$

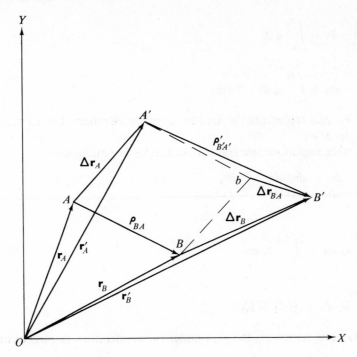

Fig. 2.4

That is, the relative displacement during interval Δt is simply the difference in the relative position vectors. Substituting $\mathbf{r}_B - \mathbf{r}_A$ for $\boldsymbol{\rho}_{BA}$ and $\mathbf{r}_{B'} - \mathbf{r}_{A'}$ for $\boldsymbol{\rho}_{B'A'}$, Eq. (2.26) becomes

$$\Delta r_{BA} = (\mathbf{r}_{B'} - \mathbf{r}_{A'}) - (\mathbf{r}_B - \mathbf{r}_A)$$

$$= (\mathbf{r}_{B'} - \mathbf{r}_B) - (\mathbf{r}_{A'} - \mathbf{r}_A)$$

$$= (\Delta \mathbf{r}_B) - (\Delta \mathbf{r}_A) \tag{2.27}$$

Thus the relative displacement of B with respect to A may be expressed as the displacement of B minus that of A. Alternatively,

$$\Delta \mathbf{r}_B = \Delta \mathbf{r}_A + \Delta \mathbf{r}_{BA} \tag{2.28}$$

Imagine that a coordinate reference is attached to and moves with particle A. B then experiences zero displacement relative to A, if during Δt it moves from B to b as shown, because the relative position vectors are identical at time t ($\boldsymbol{\rho}_{BA}$) and $t + \Delta t$ ($\mathbf{r}_b - \mathbf{r}_{A'}$). The displacement $\Delta \mathbf{r}_{BA}$ is therefore the only portion of the motion of particle B observed from a coordinate frame fixed to A, as B moves from position B to B'—hence the term "relative displacement."

2.8 RELATIVE VELOCITY AND ACCELERATION

Beginning with Eq. (2.27), the velocity of B relative to A is found by dividing by Δt and taking the limit as Δt approaches zero.   Thus

$$\mathbf{v}_{BA} = \lim_{\Delta t \to 0} \frac{\Delta \mathbf{r}_B}{\Delta t} - \lim_{\Delta t \to 0} \frac{\Delta \mathbf{r}_A}{\Delta t} \qquad (2.29a)$$

or $$\mathbf{v}_{BA} = \mathbf{v}_B - \mathbf{v}_A \qquad (2.29b)$$

where the terms on the right side of Eq. (2.29a) are recognized as the velocities of A and B with respect to XYZ.   Writing Eq. (2.29b) in alternative form,

$$\mathbf{v}_B = \mathbf{v}_A + \mathbf{v}_{BA} \qquad (2.30)$$

The concept of relative acceleration is similarly developed:

$$\mathbf{a}_{BA} = \mathbf{a}_B - \mathbf{a}_A \qquad (2.31a)$$

or $$\mathbf{a}_B = \mathbf{a}_A + \mathbf{a}_{BA} \qquad (2.31b)$$

2.9 ANGULAR VELOCITY

The quantity *angular velocity* is of particular importance in describing the motion of points located on or moving relative to rotating rigid bodies.   The motion of a point with respect to a rotating coordinate frame also requires the concept of angular velocity for its description, because such a frame is treated as a rigid body.

Consider the motion of particle P on the rigid body rotating about fixed axis OO', shown in Fig. 2.5.   In accordance with the definition of rotation, every point follows a circular path of motion.   The trajectory of each point is a circle, with each center of rotation contained on axis OO'.

Let us now define an angular velocity vector $\boldsymbol{\omega}$ of magnitude given by

$$\omega = |\boldsymbol{\omega}| = \lim_{\Delta t \to 0} \frac{\Delta \theta}{\Delta t} = \frac{d\theta}{dt} = \dot{\theta} \qquad (2.32)$$

where $\Delta \theta$ represents the angular displacement of any line on the body, such as b, perpendicular to OO'.   The direction of $\boldsymbol{\omega}$ is governed by the right-hand rule; it is identical with the direction of advance of a right-hand screw, as shown in Fig. 2.5.

It is clear that in Eq. (2.32) the incremental angular displacement $\Delta \theta$ has simply replaced the incremental linear displacement of Eqs. (2.4) and (2.5).   The matter of supplying direction to the magnitude ω is "disposed of" by the right-hand rule.   Nevertheless, we are without assurance that the angular velocity may be treated as a vector.   For example, does it behave as a vector in simple addition, which we know must yield the same result without regard to the order in which the vectors are taken?   One can be easily misled, for although finite angular displacements possess magnitude and direction (on the basis of the right-hand rule),

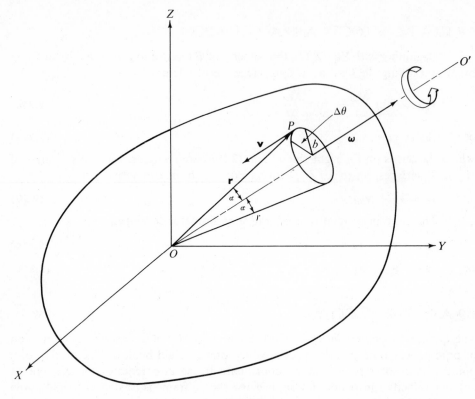

Fig. 2.5

they are nevertheless not vector quantities because they do not obey the commutative law of vector addition. In summing a number of angles, not about the same axis, the resultant angle is dependent upon the order of addition. This is demonstrated in Fig. 2.6(a) and (b), in which an object is made to undergo two counterclockwise rotations of 90 degrees, one about the X-axis, the other about the Y-axis. The order in which the angular displacements are executed is evidently crucial to the final configuration of the object, and hence *finite* angular displacements are not vector quantities. In the section that follows it is demonstrated that angular velocity is a vector, however, as are infinitesimal angular displacements.

2.10 LINEAR VELOCITY OF A POINT
   ON A ROTATING RIGID BODY

Referring again to Fig. 2.5, during time interval Δt, P experiences displacement $b\,\Delta\theta$. The magnitude of the velocity of this point, in accordance with the definition given by Eq. (2.5), is

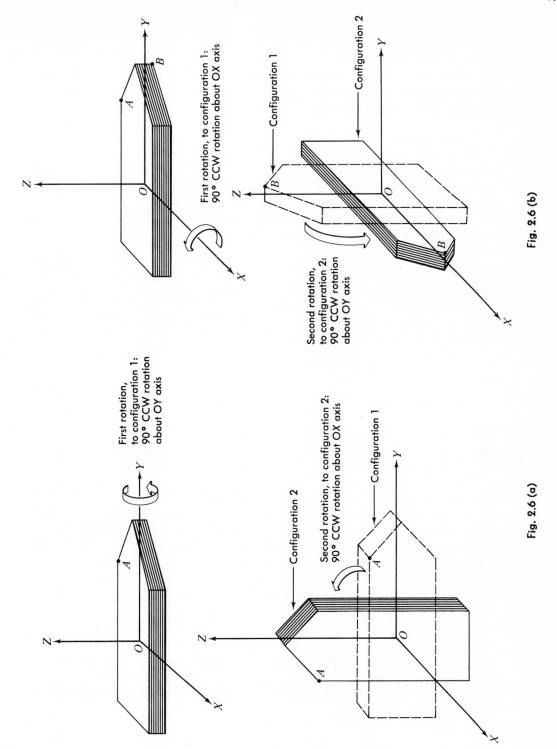

First rotation,
to configuration 1:
90° CCW rotation about OX axis

Second rotation,
to configuration 2:
90° CCW rotation
about OY axis

Configuration 1

Configuration 2

Fig. 2.6 (b)

First rotation,
to configuration 1:
90° CCW rotation
about OY axis

Second rotation, to configuration 2:
90° CCW rotation about OX axis

Configuration 2

Configuration 1

Fig. 2.6 (a)

$$|\mathbf{v}| = \lim_{\Delta t \to 0} \frac{b\Delta\theta}{\Delta t} = \omega r \sin \alpha \tag{2.33}$$

The above expression and the definition of the linear velocity as a vector quantity suggest that the velocity of a point on a rigid body experiencing pure rotation can be interpreted as the vector product

$$\mathbf{v} = \boldsymbol{\omega} \times \mathbf{r} \tag{2.34}$$

where $\mathbf{r}$ is vector running from the origin to the point P. In accordance with this equation, the velocity is a vector normal to the plane defined by $\boldsymbol{\omega}$ and $\mathbf{r}$ and in the direction determined by the right-hand rule.

If point P were to simultaneously experience linear velocities $\mathbf{v}_1$ and $\mathbf{v}_2$, its resultant velocity would be

$$\mathbf{v} = \mathbf{v}_1 + \mathbf{v}_2 = \mathbf{v}_2 + \mathbf{v}_1 \tag{2.35}$$

where the order of addition is immaterial. Assume now that the velocities $\mathbf{v}_1$ and $\mathbf{v}_2$ are attributable to simultaneous angular velocities $\boldsymbol{\omega}_1$ and $\boldsymbol{\omega}_2$. Thus $\mathbf{v}_1 = \boldsymbol{\omega}_1 \times \mathbf{r}$ and $\mathbf{v}_2 = \boldsymbol{\omega}_2 \times \mathbf{r}$, and Eq. (2.35) becomes

$$\mathbf{v} = \boldsymbol{\omega}_1 \times \mathbf{r} + \boldsymbol{\omega}_2 \times \mathbf{r} = \boldsymbol{\omega}_2 \times \mathbf{r} + \boldsymbol{\omega}_1 \times \mathbf{r} \tag{2.36a}$$

or

$$\mathbf{v} = (\boldsymbol{\omega}_1 + \boldsymbol{\omega}_2) \times \mathbf{r} = (\boldsymbol{\omega}_2 + \boldsymbol{\omega}_1) \times \mathbf{r} \tag{2.36b}$$

Therefore,

$$\boldsymbol{\omega} = \boldsymbol{\omega}_1 + \boldsymbol{\omega}_2 = \boldsymbol{\omega}_2 + \boldsymbol{\omega}_1 \tag{2.37}$$

where $\boldsymbol{\omega}_1 + \boldsymbol{\omega}_2 = \boldsymbol{\omega}_2 + \boldsymbol{\omega}_1$ has been replaced by $\boldsymbol{\omega}$, the sum of the angular velocities, clearly vector quantities, inasmuch as the results in Eqs. (2.36) are independent of the order of addition. It follows from the foregoing that *infinitesimal* angular displacements are also vectors. Consider such displacements occurring during time dt. From Eq. (2.37),

$$\boldsymbol{\omega}\, dt = \boldsymbol{\omega}_1\, dt + \boldsymbol{\omega}_2\, dt = \boldsymbol{\omega}_2\, dt + \boldsymbol{\omega}_1\, dt \tag{2.38}$$

or

$$d\boldsymbol{\theta} = d\boldsymbol{\theta}_1 + d\boldsymbol{\theta}_2 = d\boldsymbol{\theta}_2 + d\boldsymbol{\theta}_1 \tag{2.39}$$

These results should be compared with those at the conclusion of Section 2.9 regarding finite angular displacements.

The difference in velocity between two points on a rigid rotating body is found by applying Eq. (2.34),

$$v_2 - v_1 = \omega \times r_2 - \omega \times r_1 = \omega \times (r_2 - r_1) = \omega \times \rho \qquad (2.40)$$

where r_1 and r_2 are radius vectors drawn from the origin of XYZ to points P_1 and P_2, and $\rho = r_2 - r_1$. Rewriting Eq. (2.40), it is observed that

$$v_2 - v_1 = \frac{dr_2}{dt} - \frac{dr_1}{dt} = \frac{d}{dt}(r_2 - r_1) = \frac{d\rho}{dt} = \omega \times \rho \qquad (2.41)$$

so that $\omega \times \rho$ represents the time derivative of ρ, a vector fixed to a rigid body rotating at angular velocity ω. If we replace ρ by any vector fixed to a rotating rigid body, the operation $(\omega \times)$ provides a means by which the first time derivative is taken. In the above formulation, the vector is of *constant* length (because we are dealing with a rigid body) and therefore ρ experiences changing direction only.

If the foregoing principle is applied to the unit vectors $\hat{i}$, $\hat{j}$, $\hat{k}$ attached to and rotating with a rigid XYZ coordinate system experiencing angular velocity ω relative to $X'Y'Z'$, it follows that

$$\frac{d\hat{i}}{dt} = \omega \times \hat{i} = (\omega_x\hat{i} + \omega_y\hat{j} + \omega_z\hat{k}) \times \hat{i} = (\omega_z\hat{j} - \omega_y\hat{k})$$

$$\frac{d\hat{j}}{dt} = \omega \times \hat{j} = (\omega_x\hat{i} + \omega_y\hat{j} + \omega_z\hat{k}) \times \hat{j} = (\omega_x\hat{k} - \omega_z\hat{i}) \qquad (2.42)$$

$$\frac{d\hat{k}}{dt} = \omega \times \hat{k} = (\omega_x\hat{i} + \omega_y\hat{j} + \omega_z\hat{k}) \times \hat{k} = (\omega_y\hat{i} - \omega_x\hat{j})$$

These derivatives are associated with changes in *orientation* only, the magnitudes of the unit vectors remaining unchanged, as in the case of ρ previously cited.

2.11 ACCELERATION OF A POINT ON A ROTATING RIGID BODY

The acceleration of a point on a rotating rigid body is readily determined by taking the first time derivative of the velocity, Eq. (2.34):

$$a = \frac{d}{dt}(\omega \times r) = \frac{d\omega}{dt} \times r + \omega \times \frac{dr}{dt} \qquad (2.43)$$

But $dr/dt = v = \omega \times r$, and consequently

$$a = \dot{\omega} \times r + \omega \times (\omega \times r) \qquad (2.44)$$

which are the tangential and centripetal accelerations, respectively, of a point fixed on a rotating rigid body experiencing angular velocity $\boldsymbol{\omega}$ and angular acceleration $\dot{\boldsymbol{\omega}}$. It is worthwhile for the reader to relate Eq. (2.44), particularly the directions, to the accelerations studied in elementary courses in connection with circular motion.

2.12 CYLINDRICAL POLAR COORDINATES

It has already been noted that the choice of coordinate system is often dictated by the peculiarities of the physical situation under analysis. Our attention is now directed to the study of three particularly useful coordinate sets.

Referring to Fig. 2.7, the position of point P is described by

$$\mathbf{r} = r\hat{\mathbf{e}}_r = R\hat{\mathbf{e}}_R + z\hat{\mathbf{k}} \tag{2.45}$$

where $\hat{\mathbf{e}}_r$ is a unit radius vector, positive in the direction of increasing r regardless of quadrant, $\hat{\mathbf{k}}$ a unit vector in the Z-direction, $\hat{\mathbf{e}}_R$ a unit vector in the direction of increasing R, and R the projection of $\mathbf{r}$ on the XY-plane.

The unit vector parallel to the XY-plane, perpendicular to $\hat{\mathbf{e}}_R$ in the direction of increasing ϕ, is $\hat{\mathbf{e}}_\phi$. With this definition, $\hat{\mathbf{e}}_R$, $\hat{\mathbf{e}}_\phi$ and $\hat{\mathbf{k}}$ form an orthogonal right-handed set in that order ($\hat{\mathbf{e}}_R \times \hat{\mathbf{e}}_\phi = \hat{\mathbf{k}}$). The velocity is obtained by taking the first time derivative of Eq. (2.45):

$$\mathbf{v} = \dot{\mathbf{r}} = \dot{R}\hat{\mathbf{e}}_R + R\dot{\hat{\mathbf{e}}}_R + \dot{z}\hat{\mathbf{k}} \tag{2.46}$$

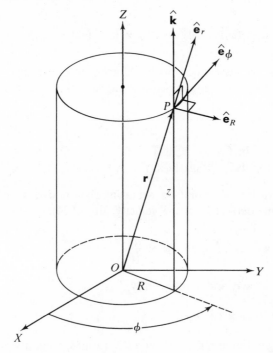

Fig. 2.7

As the particle moves along its path, R and z, in general, undergo changes of magnitude. The unit vector $\hat{\mathbf{e}}_R$ possesses a nonzero derivative, because, unlike $\hat{\mathbf{k}}$, its direction changes with time, although the magnitude of both $\hat{\mathbf{e}}_R$ and $\hat{\mathbf{k}}$, of course, remain constant. Equation (2.46) is not especially useful until $\dot{\mathbf{e}}_R$ is expressed in terms of $\hat{\mathbf{e}}_\phi$ and $\dot{\phi}$. In this connection, consider Fig. 2.8, which indicates, on the XY-plane, the change in $\hat{\mathbf{e}}_R$ which accompanies a rotation through angle $\Delta\phi$ of $\hat{\mathbf{e}}_R$.

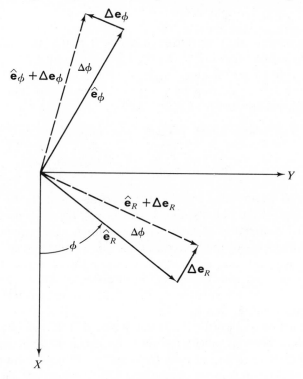

Fig. 2.8

From the figure we take $\Delta\mathbf{e}_R$ parallel to $\hat{\mathbf{e}}_\phi$ (anticipating that a limiting process is about to take place, so that the angle is necessarily small). We have, therefore,

$$\Delta\mathbf{e}_R = |\hat{\mathbf{e}}_R|\,\Delta\phi\,\hat{\mathbf{e}}_\phi = \Delta\phi\,\hat{\mathbf{e}}_\phi \tag{2.47}$$

Similarly,

$$\Delta\mathbf{e}_\phi = -|\hat{\mathbf{e}}_\phi|\,\Delta\phi\,\hat{\mathbf{e}}_R = -\Delta\phi\,\hat{\mathbf{e}}_R \tag{2.48}$$

The time derivatives of $\hat{\mathbf{e}}_R$ and $\hat{\mathbf{e}}_\phi$ are, therefore,

$$\dot{\mathbf{e}}_R = \lim_{\Delta t \to 0} \frac{\Delta\mathbf{e}_R}{\Delta t} = \lim_{\Delta t \to 0} \frac{\Delta\phi}{\Delta t}\,\mathbf{e}_\phi = \dot{\phi}\,\hat{\mathbf{e}}_\phi \tag{2.49}$$

$$\dot{\mathbf{e}}_\phi = \lim_{\Delta t \to 0} \frac{\Delta\mathbf{e}_\phi}{\Delta t} = \lim_{\Delta t \to 0} -\frac{\Delta\phi}{\Delta t}\,\hat{\mathbf{e}}_R = -\dot{\phi}\,\hat{\mathbf{e}}_R \tag{2.50}$$

or simply

$$\dot{\mathbf{e}}_R = \boldsymbol{\omega} \times \hat{\mathbf{e}}_R = (\dot{\phi}\hat{\mathbf{k}}) \times \hat{\mathbf{e}}_R = \dot{\phi}\hat{\mathbf{e}}_\phi$$

$$\dot{\mathbf{e}}_\phi = \boldsymbol{\omega} \times \hat{\mathbf{e}}_\phi = (\dot{\phi}\hat{\mathbf{k}}) \times \hat{\mathbf{e}}_\phi = -\dot{\phi}\hat{\mathbf{e}}_R$$

The velocity, Eq. (2.46), after substitution for $\dot{\mathbf{e}}_R$ becomes

$$\mathbf{v} = \dot{R}\hat{\mathbf{e}}_R + R\dot{\phi}\hat{\mathbf{e}}_\phi + \dot{z}\hat{\mathbf{k}} \tag{2.51}$$

In Eq. (2.51) $\dot{R}$ represents the time rate of change of R, that is, the radial component of velocity, $\dot{\phi}$ the angular speed of R, $R\dot{\phi}$ the transverse component of the velocity, and $\dot{z}$ the component parallel to the Z-axis.

The acceleration is derived by a procedure similar to that employed for velocity, requiring substitution for $\dot{\mathbf{e}}_\phi$ [Eq. (2.50)]:

$$\mathbf{a} = \dot{\mathbf{v}} = \ddot{R}\hat{\mathbf{e}}_R + \dot{R}\dot{\mathbf{e}}_R + (R\ddot{\phi} + \dot{R}\dot{\phi})\hat{\mathbf{e}}_\phi + R\dot{\phi}\dot{\mathbf{e}}_\phi + \ddot{z}\hat{\mathbf{k}} \tag{2.52}$$

or

$$\mathbf{a} = (\ddot{R} - R\dot{\phi}^2)\hat{\mathbf{e}}_R + (2\dot{R}\dot{\phi} + R\ddot{\phi})\hat{\mathbf{e}}_\phi + \ddot{z}\hat{\mathbf{k}} \tag{2.53}$$

There are thus two radial components of acceleration: $\ddot{R}$, attributable to the time rate of change of the radial speed $\dot{R}$, and $-R\dot{\phi}^2$, the centripetal acceleration due to the change in direction of the transverse component of velocity, $R\dot{\phi}$. There are also two components of the ϕ-acceleration: $R\ddot{\phi}$, a consequence of the change of magnitude of $R\dot{\phi}$, and $2\dot{R}\dot{\phi}$, the sum of $\dot{R}\dot{\phi}$ due to the change of the magnitude of $R\dot{\phi}$ and $R\dot{\phi}$ associated with the change in direction of $\dot{R}\hat{\mathbf{e}}_R$. These interpretations may easily be verified by referring to Eq. (2.52), noting especially the influence of the time derivatives of $\hat{\mathbf{e}}_R$ and $\hat{\mathbf{e}}_\phi$.

At this juncture it is worthwhile to emphasize that the velocity and acceleration of point P have been taken with respect to the origin of XYZ, as shown in Fig. 2.7, and that whether Cartesian or cylindrical polar coordinates are employed, the kinematic quantities result from successive time differentiations of the radius vector, running from O to P.

The differences in the appearance of the velocity and acceleration in one coordinate system as contrasted with another lie in the coordinates used to describe the kinematic quantities and in the coordinate axes upon which the velocity and acceleration are projected to determine the orthogonal components. In the case of Cartesian coordinates, the components of velocity and acceleration appear simpler in form because the coordinate triad does not move with the point, and hence the derivatives of the unit vectors $\hat{\mathbf{i}}$, $\hat{\mathbf{j}}$, $\hat{\mathbf{k}}$ are zero. That the velocity and acceleration components in the cylindrical polar system (and in the spherical and path systems to follow) appear more complex is attributable to the fact that the axes along which the components are projected move in the manner prescribed. When the vector components are added to form the *total* velocity and *total* acceleration vectors with respect to O, the results are the same, regardless of the coordinate system employed.

The application of the cylindrical polar system of coordinates can provide certain advantages when the geometry of the system manifests certain symmetries. Cylindrical polar coordinates have an appeal, for example, when one is describing helical motion or where $R = R(\phi)$ is a known function. If the particle motion is confined to a plane, only R and ϕ are pertinent, and we have the familiar plane polar coordinate system. A further simplification occurs when R is of constant length, in which case we are describing circular motion:

$$\mathbf{v} = R\dot{\phi}\hat{\mathbf{e}}_\phi = R\omega\hat{\mathbf{e}}_\phi \qquad \text{(tangential velocity)} \tag{2.54}$$

and

$$\mathbf{a} = -R\dot{\phi}^2\hat{\mathbf{e}}_R + R\ddot{\phi}\hat{\mathbf{e}}_\phi \tag{2.55a}$$

or

$$\mathbf{a} = -R\omega^2\hat{\mathbf{e}}_R + R\alpha\hat{\mathbf{e}}_\phi \tag{2.55b}$$

where $\dot{\phi} = \omega$ and $\ddot{\phi} = \dot{\omega} = \alpha$.

2.13 CYLINDRICAL-CARTESIAN TRANSFORMATIONS

When the components of a vector are known in one coordinate system, what transformation can be employed to obtain the components in another system? Consider the components of a vector $\mathbf{B}$ along the X, Y, Z directions, B_x, B_y, B_z. What are the projections of this vector in cylindrical polar coordinates? Referring to Fig. 2.9,

$$B_R = \quad (\cos \phi)B_x + (\sin \phi)B_y + (0)B_z$$

$$B_\phi = -(\sin \phi)B_x + (\cos \phi)B_y + (0)B_z \tag{2.56}$$

$$B_z = \quad (0)B_x + \quad (0)B_y + (1)B_z$$

Equations (2.56) may be written in abbreviated form employing matrices:

$$\begin{bmatrix} B_R \\ B_\phi \\ B_z \end{bmatrix} = \begin{bmatrix} \cos \phi & \sin \phi & 0 \\ -\sin \phi & \cos \phi & 0 \\ 0 & 0 & 1 \end{bmatrix} \begin{bmatrix} B_x \\ B_y \\ B_z \end{bmatrix} \tag{2.57}$$

or $\quad B_{R,\phi,z} = S_\phi B_{x,y,z} \tag{2.58}$

The vector $\mathbf{B}$ may be any vector, such as velocity and acceleration, and the square matrix S_ϕ, in Eq. (2.58) may be used for any such transformation. The square matrix is known as a transformation matrix, and the transformation described is orthogonal and unitary. Note that the transformation involves a rotation through the angle ϕ only, hence the designation S_ϕ.

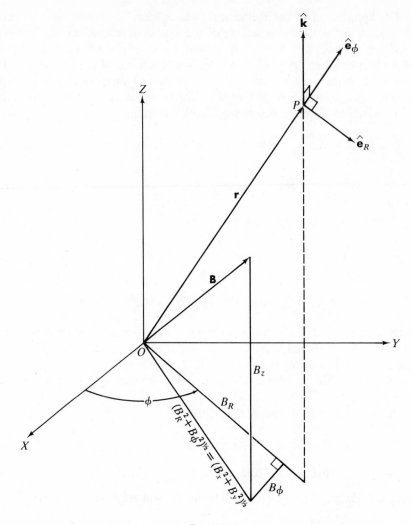

Fig. 2.9

When the components along X, Y, Z are required, they are derived from B_R, B_ϕ, B_z from the equation

$$B_{x,y,z} = S_\phi^{-1} B_{R,\phi,z} \qquad (2.59)$$

where S_ϕ^{-1} represents the inverse of S_ϕ, obtained by interchanging corresponding elements on either side of the main diagonal of the square matrix:

$$S_\phi^{-1} = \begin{bmatrix} \cos\phi & -\sin\phi & 0 \\ \sin\phi & \cos\varphi & 0 \\ 0 & 0 & 1 \end{bmatrix} \qquad (2.60)$$

Illustrative Example 2.3

The position of a particle moving in the XY-plane is given by $\mathbf{R} = 5(\cos \pi t)\hat{\mathbf{i}} + 5(\sin \pi t)\hat{\mathbf{j}}$. Determine the velocity and acceleration components in a cylindrical coordinate system employing a transformation matrix.

SOLUTION

The Cartesian components of velocity and acceleration are

$$v_x = -5\pi(\sin \pi t) \qquad v_y = 5\pi(\cos \pi t)$$

$$a_x = -5\pi^2(\cos \pi t) \qquad a_y = -5\pi^2(\sin \pi t)$$

For the velocity transformation,

$$\begin{bmatrix} v_R \\ v_\phi \\ v_z \end{bmatrix} = \begin{bmatrix} \cos \phi & \sin \phi & 0 \\ -\sin \phi & \cos \phi & 0 \\ 0 & 0 & 1 \end{bmatrix} \begin{bmatrix} v_x \\ v_y \\ v_z \end{bmatrix}$$

so that

$$v_R = -5\pi(\sin \pi t) \cos \phi + 5\pi(\cos \pi t) \sin \phi$$

$$v_\phi = -5\pi(\sin \pi t)(-\sin \phi) + 5\pi(\cos \pi t) \cos \phi$$

$$v_z = 0$$

Since $y/x = \tan \phi = \tan \pi t$, $\phi = \pi t$, and therefore $v_R = 0 \ (= \dot{R})$, $v_\phi = 5\pi = R\dot{\phi}$, $R = 5$. Also, since $R = 5$, $\dot{\phi} = \pi = $ constant. Applying the same transformation as for velocity,

$$a_R = -5\pi^2(\cos \pi t) \cos \phi - 5\pi^2(\sin \pi t) \sin \phi$$

$$a_\phi = -5\pi^2(\cos \pi t)(-\sin \phi) - 5\pi^2(\sin \pi t) \cos \phi$$

$$a_z = 0$$

Therefore,

$$a_R = -5\pi^2 \qquad (= \ddot{R} - R\dot{\phi}^2)$$

$$a_\phi = 0 \qquad (= R\ddot{\phi} + 2\dot{R}\dot{\phi})$$

2.14 SPHERICAL POLAR COORDINATES

In the spherical polar coordinate system, motion is described in terms of the variables r, θ, and ϕ, shown in Fig. 2.10, where θ is measured from the Z-axis and ϕ from the X-axis. The relationships between r, θ, ϕ and x, y, z follow:

$$x = r \sin \theta \cos \phi = R \cos \phi \qquad (2.61a)$$

$$y = r \sin \theta \sin \phi = R \sin \phi \qquad (2.61b)$$

$$z = r \cos \theta \qquad (2.61c)$$

where $r = (x^2 + y^2 + z^2)^{\frac{1}{2}}$ and $R = r \sin \theta = (x^2 + y^2)^{\frac{1}{2}}$. The angles θ and ϕ, expressed in terms of the Cartesian variables, are

$$\tan \theta = \frac{R}{z} = \frac{(x^2 + y^2)^{\frac{1}{2}}}{z} \qquad (2.62a)$$

and

$$\tan \phi = \frac{y}{x} \qquad (2.62b)$$

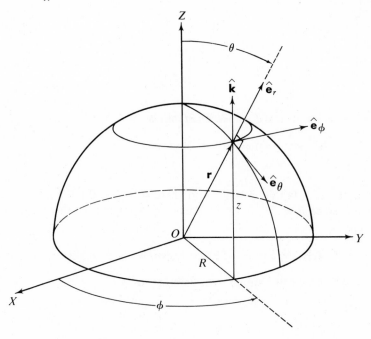

Fig. 2.10

While r, θ, and ϕ represent the variables of the coordinate system, they must be related to a set of axes; these are given by the directions of the $\hat{\mathbf{e}}_r$, $\hat{\mathbf{e}}_\theta$, $\hat{\mathbf{e}}_\phi$ unit vectors. The directions of these vectors are defined in the following manner:

$\hat{\mathbf{e}}_r$ is directed along increasing r

$\hat{\mathbf{e}}_\theta$ is perpendicular to r in the direction of increasing θ, in the plane containing r and Z

$\hat{\mathbf{e}}_\phi$ completes the right-handed triad ($\hat{\mathbf{e}}_r$ rotated into $\hat{\mathbf{e}}_\theta$ according to the right-hand rule produces the direction of $\hat{\mathbf{e}}_\phi$)

Thus $\hat{\mathbf{e}}_r$, $\hat{\mathbf{e}}_\theta$, and $\hat{\mathbf{k}}$ are all in the same vertical plane, while $\hat{\mathbf{e}}_\phi$ is parallel to the horizontal XY-plane.

It will prove useful to express the unit vectors in terms of $\hat{\mathbf{i}}$, $\hat{\mathbf{j}}$, $\hat{\mathbf{k}}$ and the angles θ and ϕ:

$$\hat{\mathbf{e}}_r = (\sin\theta\cos\phi)\hat{\mathbf{i}} + (\sin\theta\sin\phi)\hat{\mathbf{j}} + (\cos\theta)\hat{\mathbf{k}} \tag{2.63a}$$

$$\hat{\mathbf{e}}_\phi = (-\sin\phi)\hat{\mathbf{i}} + (\cos\phi)\mathbf{J} + (0)\hat{\mathbf{k}} \tag{2.63b}$$

$$\hat{\mathbf{e}}_\theta = (\cos\theta\cos\phi)\hat{\mathbf{i}} + (\cos\theta\sin\phi)\hat{\mathbf{j}} + (-\sin\theta)\hat{\mathbf{k}} \tag{2.63c}$$

The radius vector locating P is, as before,

$$\mathbf{r} = x\hat{\mathbf{i}} + y\hat{\mathbf{j}} + z\hat{\mathbf{k}} = r\hat{\mathbf{e}}_r \tag{2.64}$$

and the velocity is

$$\mathbf{v} = r\dot{\hat{\mathbf{e}}}_r + \dot{r}\hat{\mathbf{e}}_r \tag{2.65}$$

Differentiating Eq. (2.63a) for the time derivative of $\hat{\mathbf{e}}_r$:

$$\dot{\hat{\mathbf{e}}}_r = [(\cos\phi\cos\theta)\hat{\mathbf{i}} + (\cos\theta\sin\phi)\hat{\mathbf{j}} - (\sin\theta)\hat{\mathbf{k}}]\dot{\theta}$$
$$+ [(-\sin\phi)\hat{\mathbf{i}} + (\cos\phi)\hat{\mathbf{j}}](\sin\theta)\dot{\phi} \tag{2.66}$$

where the bracketed coefficients of $\dot{\theta}$ and $(\sin\theta)\dot{\phi}$ above are recognized as $\hat{\mathbf{e}}_\theta$ and $\hat{\mathbf{e}}_\phi$, respectively. We have, therefore,

$$\dot{\hat{\mathbf{e}}}_r = \dot{\theta}\hat{\mathbf{e}}_\theta + (\sin\theta)\dot{\phi}\hat{\mathbf{e}}_\phi \tag{2.67}$$

Upon substituting for $\dot{\hat{\mathbf{e}}}_r$ in Eq. (2.65),

$$\mathbf{v} = (r\dot{\theta})\hat{\mathbf{e}}_\theta + (r\sin\theta)\dot{\phi}\hat{\mathbf{e}}_\phi + \dot{r}\hat{\mathbf{e}}_r \tag{2.68}$$

The acceleration is obtained in a similar fashion:

$$\mathbf{a} = \dot{\mathbf{v}} = (r\ddot{\theta} + \dot{r}\dot{\theta})\hat{\mathbf{e}}_\theta + r\dot{\theta}\dot{\hat{\mathbf{e}}}_\theta + (r\sin\theta)\ddot{\phi}\hat{\mathbf{e}}_\phi + \dot{\phi}[(r\cos\theta)\dot{\theta} + \dot{r}\sin\theta]\hat{\mathbf{e}}_\phi$$
$$+ (r\sin\theta)\dot{\phi}\dot{\hat{\mathbf{e}}}_\phi + \ddot{r}\hat{\mathbf{e}}_r + \dot{r}\dot{\hat{\mathbf{e}}}_r \tag{2.69}$$

It is left as Exercise 2.51 to substitute for $\dot{\hat{\mathbf{e}}}_r$, $\dot{\hat{\mathbf{e}}}_\phi$, and $\dot{\hat{\mathbf{e}}}_\theta$ to obtain

$$\mathbf{a} = (\ddot{r} - r\dot{\phi}^2\sin^2\theta - r\dot{\theta}^2)\hat{\mathbf{e}}_r + (2\dot{r}\dot{\theta} - r\dot{\phi}^2\sin\theta\cos\theta + r\ddot{\theta})\hat{\mathbf{e}}_\theta$$
$$+ (2\dot{r}\dot{\phi}\sin\theta + r\ddot{\phi}\sin\theta + 2r\dot{\theta}\dot{\phi}\cos\theta)\hat{\mathbf{e}}_\phi \tag{2.70}$$

2.15 TRANSFORMATIONS INVOLVING SPHERICAL COORDINATES

The matrix for transforming from Cartesian to spherical coordinates may be read directly from Eqs. (2.63):

$$S_{\theta\phi} = \begin{bmatrix} \sin\theta\cos\phi & \sin\theta\sin\phi & \cos\theta \\ -\sin\phi & \cos\phi & 0 \\ \cos\theta\cos\phi & \cos\theta\sin\phi & -\sin\theta \end{bmatrix} \tag{2.71}$$

where the designation $S_{\theta\phi}$ indicates that both angles are involved in the transformation; that is, rotations through the angles ϕ and θ occur. When the Cartesian components of acceleration (for example) are known, the following matrix equation represents the transformation required to obtain the spherical components:

$$\begin{bmatrix} a_r \\ a_\phi \\ a_\theta \end{bmatrix} = \begin{bmatrix} \sin\theta\cos\phi & \sin\theta\sin\phi & \cos\theta \\ -\sin\phi & \cos\phi & 0 \\ \cos\theta\cos\phi & \cos\theta\sin\phi & -\sin\theta \end{bmatrix} \begin{bmatrix} a_x \\ a_y \\ a_z \end{bmatrix} \qquad (2.72)$$

The inverse of this matrix, $S_{\theta\phi}^{-1}$, is used to transform from spherical to Cartesian coordinates.

The matrix for transforming from cylindrical to spherical coordinates may be obtained in two ways:

1. By noting from Fig. 2.11 that the two systems are related by a single rotation $(\theta - 90)$.

2. By employing the transformation matrices already available. Assuming that V_r, V_ϕ, V_θ are required and V_R, V_ϕ, V_z are known, from Eq. (2.71),

$$\begin{bmatrix} V_r \\ V_\phi \\ V_\theta \end{bmatrix} = S_{\theta\phi} \begin{bmatrix} V_x \\ V_y \\ V_z \end{bmatrix} \qquad (2.73)$$

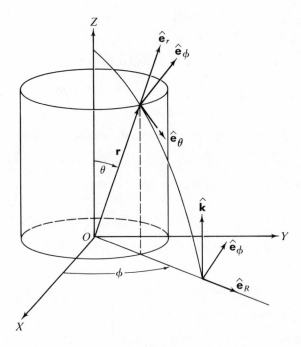

Fig. 2.11

and, from Eq. (2.57),

$$
\begin{bmatrix} V_x \\ V_y \\ V_z \end{bmatrix} = S_\phi^{-1} \begin{bmatrix} V_R \\ V_\phi \\ V_z \end{bmatrix}
$$
(2.74)

Substituting Eq. (2.74) into Eq. (2.73),

$$
\begin{bmatrix} V_r \\ V_\phi \\ V_\theta \end{bmatrix} = S_{\theta\phi} S_\phi^{-1} \begin{bmatrix} V_R \\ V_\phi \\ V_z \end{bmatrix}
$$
(2.75)

so that the product $S_{\theta\phi}S_\phi^{-1}$ is the required transformation matrix:

$$
S_{\theta\phi}S_\phi^{-1} = \begin{bmatrix} \sin\theta\cos\phi & \sin\theta\sin\phi & \cos\theta \\ -\sin\phi & \cos\phi & 0 \\ \cos\theta\cos\phi & \cos\theta\sin\phi & -\sin\theta \end{bmatrix} \begin{bmatrix} \cos\phi & -\sin\phi & 0 \\ 0 & 0 & 1 \\ \sin\phi & \cos\phi & 0 \end{bmatrix}
$$

$$
= \begin{bmatrix} \sin\theta & 0 & \cos\theta \\ 0 & 1 & 0 \\ \cos\theta & 0 & -\sin\theta \end{bmatrix}
$$
(2.76)

2.16 PATH COORDINATES

Some problems are particularly well described in terms of components tangent and normal to the curve along which the particle moves as when, for example, y is a known function of x. Consider the path of motion of a particle to be a curve in space as shown in Fig. 2.12, where the radius vector with respect to the origin of the XYZ-system changes from $\mathbf{r}(t)$ to $\mathbf{r}(t + \Delta t)$ during time interval Δt. The instantaneous position s measured along the curve is a function of time, $s = s(t)$, and the velocity at any time is

$$
\mathbf{v} = \dot{s}\hat{\mathbf{e}}_t
$$
(2.77)

where $\hat{\mathbf{e}}_t$ is a unit tangent vector moving with the particle, tangent to the curve at any time. Another time differentiation gives

$$
\mathbf{a} = \dot{\mathbf{v}} = \ddot{s}\hat{\mathbf{e}}_t + \dot{s}\dot{\hat{\mathbf{e}}}_t
$$
(2.78)

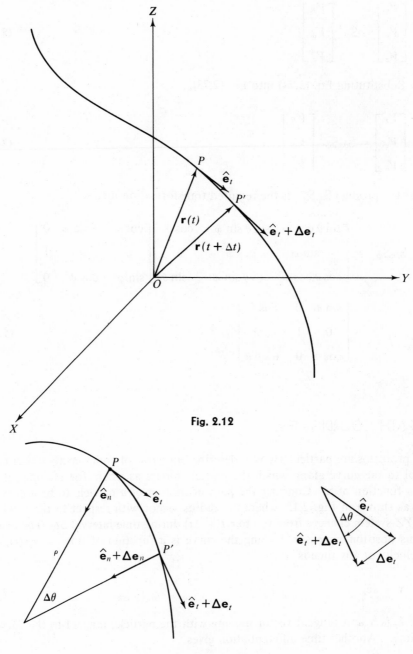

Fig. 2.12

Fig. 2.13 (a, b)

It is desirable to replace $\dot{\mathbf{e}}_t$ with a more convenient expression. Referring to Fig. 2.13, consider the magnitude of $\Delta\mathbf{e}_t$:

$$|\Delta\mathbf{e}_t| = |\hat{\mathbf{e}}_t|(\Delta\theta) = \Delta\theta \tag{2.79}$$

From Eq. (2.79) it follows that

$$\lim_{\Delta t \to 0} \left|\frac{\Delta\mathbf{e}_t}{\Delta\theta}\right| = \left|\frac{d\mathbf{e}_t}{d\theta}\right| = 1 \tag{2.80}$$

In the limit as $\Delta\theta$ approaches zero, $\Delta\mathbf{e}_t$ becomes perpendicular to $\hat{\mathbf{e}}_t$, in the direction of a unit vector normal to the path $\hat{\mathbf{e}}_n$. Therefore,

$$\frac{d\mathbf{e}_t}{d\theta} = (1)\hat{\mathbf{e}}_n = \hat{\mathbf{e}}_n \tag{2.81}$$

As $\Delta\theta$ becomes vanishingly small, $\hat{\mathbf{e}}_t$, $\Delta\hat{\mathbf{e}}_t$, and $\hat{\mathbf{e}}_t + \Delta\hat{\mathbf{e}}_t$ lie in a single plane called the *osculating plane*, which for a planar curve represents the plane of the curve.

The right-handed triad is completed by defining a third unit vector: $\hat{\mathbf{e}}_b = \hat{\mathbf{e}}_t \times \hat{\mathbf{e}}_n$. This vector, normal to the plane containing $\hat{\mathbf{e}}_t$ and $\hat{\mathbf{e}}_n$, is called the *binormal* unit vector. Referring to Eq. (2.78), an expression for $\dot{\mathbf{e}}_t$ is required:

$$\dot{\mathbf{e}}_t = \frac{d\mathbf{e}_t}{d\theta}\frac{d\theta}{dt} = \hat{\mathbf{e}}_n\frac{d\theta}{dt} = \hat{\mathbf{e}}_n\frac{d\theta}{ds}\frac{ds}{dt} \tag{2.82}$$

From Fig. 2.13(a), $ds = \rho\,d\theta$, where ρ is the radius of curvature at P and hence

$$\dot{\mathbf{e}}_t = \frac{\hat{\mathbf{e}}_n}{\rho}\dot{s} \tag{2.83}$$

The acceleration is

$$\mathbf{a} = \frac{(\dot{s})^2}{\rho}\hat{\mathbf{e}}_n + \ddot{s}\hat{\mathbf{e}}_t \tag{2.84}$$

In the familiar case of circular motion, the first term on the right side of Eq. (2.84) is usually written v^2/R or $\omega^2 R$. Whether the motion is circular or not, this component of acceleration lies in the osculating plane, is directed toward the instantaneous center of curvature (centripetal acceleration), and is normal to the path of the particle. The remaining component, the tangential acceleration, lies in the osculating plane and is tangent to the path. The magnitude of the total acceleration is given by

$$a = \left[\left(\frac{\dot{s}^2}{\rho} \right)^2 + (\ddot{s})^2 \right]^{\frac{1}{2}}$$

(2.85)

Illustrative Example 2.4

A particle moves along the plane curve $y = 2x^2$, where the X-coordinate is given by $x = 3t$. Determine the position as a function of time and the velocity and acceleration in Cartesian and path coordinates at $t = 1$.

SOLUTION

Since $y = 2x^2$ and $x = 3t$, y may be expressed as a function of time: $y = 2(3t)^2 = 18t^2$. Therefore, $\mathbf{r} = 3t\hat{\mathbf{i}} + 18t^2\hat{\mathbf{j}}$. The velocity in Cartesian coordinates is

$$\mathbf{v} = \dot{\mathbf{r}} = 3\hat{\mathbf{i}} + 36t\hat{\mathbf{j}}$$

At $t = 1$, $\mathbf{v} = 3\hat{\mathbf{i}} + 36\hat{\mathbf{j}}$ and

$$v = [(3)^2 + (36)^2]^{\frac{1}{2}} = 36.1$$

The velocity in path coordinates is $\mathbf{v} = \dot{s}\hat{\mathbf{e}}_t$, where

$$\dot{s} = [(\dot{x})^2 + (\dot{y})^2]^{\frac{1}{2}} = [(3)^2 + (36t)^2]^{\frac{1}{2}}$$

At $t = 1$, $\dot{s} = 36.1$.

The acceleration is given by

$$\mathbf{a} = \ddot{s}\hat{\mathbf{e}}_t + \frac{(\dot{s})^2}{\rho}\hat{\mathbf{e}}_n$$

where

$$\ddot{s} = \frac{d}{dt}(\dot{s}) = \frac{d}{dt}[(\dot{x})^2 + (\dot{y})^2]^{\frac{1}{2}}$$

$$= \frac{1}{2}[(\dot{x})^2 + (\dot{y})^2]^{-\frac{1}{2}}(2\dot{x}\ddot{x} + 2\dot{y}\ddot{y})$$

At $t = 1$,

$$\ddot{s} = \frac{(36)^2}{[(3)^2 + (36)^2]^{\frac{1}{2}}} = 35.9$$

This result can also be obtained by substitution of $\dot{x}(t)$ and $\dot{y}(t)$, differentiating as indicated to obtain $\ddot{s}$ and evaluating at $t = 1$.

The normal component of acceleration requires determination of the radius of curvature ρ. From differential calculus,

$$\frac{1}{\rho} = \frac{d^2y/dx^2}{[1 + (dy/dx)^2]^{3/2}}$$

where dy/dx may be expressed

$$\frac{dy}{dx} = \frac{dy}{dt}\frac{dt}{dx} = \frac{\dot{y}}{\dot{x}}$$

and

$$\frac{d^2y}{dx^2} = \frac{d}{dx}\frac{dy}{dx} = \frac{d}{dt}\frac{dy}{dx}\frac{dt}{dx} = \left[\frac{d}{dt}\frac{\dot{y}}{\dot{x}}\right]\frac{1}{\dot{x}}$$

Differentiating as indicated and substituting into the expression for $1/\rho$:

$$\frac{1}{\rho} = \frac{\dot{x}\ddot{y} - \dot{y}\ddot{x}}{[(\dot{x})^2 + (\dot{y})^2]^{3/2}}$$

At $t = 1$,

$$\frac{1}{\rho} = \frac{(3)(36)}{[(3)^2 + (36)^2]^{3/2}}$$

The normal acceleration is

$$\frac{(\dot{s})^2}{\rho} = \frac{(36.1)^2(3)(36)}{(36.1)^3} = 2.98$$

and therefore the acceleration magnitude is

$$a = [(35.9)^2 + (2.98)^2]^{1/2} = 36$$

The slope of the curve at the point in question is

$$\left(\frac{dy}{dx}\right)_{t=1} = \left(\frac{\dot{y}}{\dot{x}}\right)_{t=1} = \frac{36}{3} = 12$$

Since $\tan\theta = 12$, $\cos\theta = 0.083$, and $\sin\theta = 0.995$.

Taking the X-components of the tangential and normal accelerations,

$$a_x = (35.9)(\cos\theta) - (2.98)(\sin\theta) = 0$$

as previously determined.

2.17 MOTION REFERRED TO MOVING COORDINATE AXES

Velocities and accelerations are often measured with respect to a coordinate frame of reference which experiences rotation and translation with respect to another system, often fixed. Kinematic measurements made with respect to a frame fixed to the earth's surface represent such an instance. It is important to relate velocity and acceleration thus determined to the values they assume with respect to another coordinate frame. The special significance of one such frame, called an inertial frame of reference, is discussed in Chapter 3.

Consider the position, velocity, and acceleration of point P with respect to the origin of coordinate frame $X'Y'Z'$. We designate these quantities $\mathbf{r}, \dot{\mathbf{r}}, \ddot{\mathbf{r}}$. Now consider the motion of P described with respect to another reference XYZ, which experiences angular velocity $\boldsymbol{\omega}$ and translation $\dot{\mathbf{R}}$ relative to $X'Y'Z'$, as in Fig. 2.14.

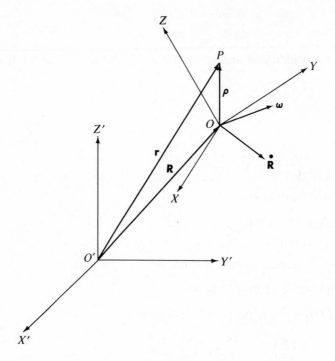

Fig. 2.14

In this figure,

$\mathbf{r}$ = radius vector running from the origin of $X'Y'Z'$ directly to P

$\mathbf{R}$ = vector running from the origin of $X'Y'Z'$ to the origin of XYZ, as shown

$\boldsymbol{\rho}$ = radius vector from the origin of XYZ to P

The position and velocity of P with respect to XYZ are

$$\boldsymbol{\rho} = x\hat{\mathbf{i}} + y\hat{\mathbf{j}} + z\hat{\mathbf{k}} \tag{2.86}$$

and

$$\dot{\boldsymbol{\rho}} = \dot{x}\hat{\mathbf{i}} + \dot{y}\hat{\mathbf{j}} + \dot{z}\hat{\mathbf{k}} \tag{2.87}$$

where it is noted that $\hat{\mathbf{i}}, \hat{\mathbf{j}}, \hat{\mathbf{k}}$ are unit vectors fixed to X, Y, Z. The projections of $\boldsymbol{\rho}$ along the X-, Y-, and Z-axes are x, y, z, and the time derivative of $\boldsymbol{\rho}$ is taken with respect to the XYZ-frame.

The first derivative of $\boldsymbol{\rho}$ with respect to $X'Y'Z'$ is

$$\dot{\boldsymbol{\rho}}' = \frac{d}{dt}(x\hat{\mathbf{i}} + y\hat{\mathbf{j}} + z\hat{\mathbf{k}}) = (\dot{x}\hat{\mathbf{i}} + \dot{y}\hat{\mathbf{j}} + \dot{z}\hat{\mathbf{k}}) + x\dot{\hat{\mathbf{i}}} + y\dot{\hat{\mathbf{j}}} + z\dot{\hat{\mathbf{k}}} \qquad (2.88)$$

where the terms $x\dot{\hat{\mathbf{i}}}$, and so on, are differentiated as products, in which the unit vectors are *not* constant, because they change direction when viewed from the primed frame. Recalling Eqs. (2.42),

$$\dot{\hat{\mathbf{i}}} = \boldsymbol{\omega} \times \hat{\mathbf{i}}$$

$$\dot{\hat{\mathbf{j}}} = \boldsymbol{\omega} \times \hat{\mathbf{j}}$$

and $\qquad\qquad\qquad\qquad\qquad\qquad\qquad\qquad\qquad\qquad\qquad$ (2.89)

$$\dot{\hat{\mathbf{k}}} = \boldsymbol{\omega} \times \hat{\mathbf{k}}$$

After substitution and collection of terms, Eq. (2.88) becomes

$$\dot{\boldsymbol{\rho}}' = (\dot{x}\hat{\mathbf{i}} + \dot{y}\hat{\mathbf{j}} + \dot{z}\hat{\mathbf{k}}) + \boldsymbol{\omega} \times (x\hat{\mathbf{i}} + y\hat{\mathbf{j}} + z\hat{\mathbf{k}}) \qquad (2.90)$$

or

$$\dot{\boldsymbol{\rho}}' = \dot{\boldsymbol{\rho}} + \boldsymbol{\omega} \times \boldsymbol{\rho} \qquad (2.91)$$

Thus the time derivative of the radius vector as viewed from $X'Y'Z'$ is $\dot{\boldsymbol{\rho}}$ plus an additional term, $\boldsymbol{\omega} \times \boldsymbol{\rho}$, due to the relative rotation of XYZ and $X'Y'Z'$.

Equation (2.91) may be generalized to represent the time derivative with respect to $X'Y'Z'$ of any vector $\mathbf{A}$:

$$\left(\frac{d\mathbf{A}}{dt}\right)' = \frac{d\mathbf{A}}{dt} + \boldsymbol{\omega} \times \mathbf{A} \qquad (2.92)$$

It is now our task to determine the relationship between the motion of P referred to XYZ and $X'Y'Z'$.

The position of P with respect to O' may be written

$$\mathbf{r} = \mathbf{R} + \boldsymbol{\rho} \qquad (2.93)$$

and the velocity of P with respect to $X'Y'Z'$ is obtained by determining the time derivative of Eq. (2.93) as viewed from $X'Y'Z'$,

$$\mathbf{v}' = \left(\frac{d\mathbf{r}}{dt}\right)' = \left(\frac{d\mathbf{R}}{dt}\right)' + \left(\frac{d\boldsymbol{\rho}}{dt}\right)' \qquad (2.94)$$

where

$$\left(\frac{d\mathbf{R}}{dt}\right)' = \dot{\mathbf{R}}$$

and

$$\left(\frac{d\rho}{dt}\right)' = \frac{d\rho}{dt} + \omega \times \rho$$

so that

$$\mathbf{v}' = \dot{\mathbf{R}} + \dot{\rho} + \omega \times \rho \tag{2.95}$$

$\dot{\rho}$ is often designated $\dot{\rho}_{rel}$ or $\mathbf{v}_{rel}$, because it represents the relative motion of P with respect to XYZ.

The acceleration of P with respect to $X'Y'Z'$ requires the time derivative of $\mathbf{v}'$ viewed from the primed axes,

$$\mathbf{a}' = \left[\frac{d}{dt}(\mathbf{v})'\right]' = \left(\frac{d^2\mathbf{r}}{dt^2}\right)' \tag{2.96a}$$

or

$$\mathbf{a}' = \left[\frac{d}{dt}(\dot{\mathbf{R}})\right]' + \left[\frac{d}{dt}(\dot{\rho})\right]' + \left[\frac{d}{dt}(\omega \times \rho)\right]' \tag{2.96b}$$

where the primes outside the brackets again indicate derivatives viewed from the primed coordinate reference.

The first term on the right side of Eq. (2.96b) represents the acceleration of O with respect to O'. The second term is the time derivative of the relative velocity ($\mathbf{v}_{rel}$) of P as viewed in $X'Y'Z'$. This differentiation is treated in accordance with Eq. (2.92),

$$\left[\frac{d}{dt}(\dot{\rho})\right]' = \frac{d}{dt}(\dot{\rho}) + \omega \times \dot{\rho} = \ddot{\rho} + \omega \times \dot{\rho} \tag{2.97}$$

where $\ddot{\rho} = \mathbf{a}_{rel}$ or $\dot{\mathbf{v}}_{rel}$.

The final term on the right side of Eq. (2.96b) is

$$\left[\frac{d}{dt}(\omega \times \rho)\right]' = \omega \times \left(\frac{d\rho}{dt}\right)' + \left(\frac{d\omega}{dt}\right)' \times \rho$$

$$= \omega \times \left[\left(\frac{d\rho}{dt}\right) + \omega \times \rho\right] + \dot{\omega} \times \rho$$

$$= \omega \times \mathbf{v}_{rel} + \omega \times (\omega \times \rho) + \dot{\omega} \times \rho \tag{2.98}$$

Therefore,

$$\mathbf{a}' = \ddot{\mathbf{R}} + \mathbf{a}_{rel} + \omega \times \mathbf{v}_{rel} + \omega \times \mathbf{v}_{rel} + \omega \times (\omega \times \rho) + \dot{\omega} \times \rho \tag{2.99}$$

where $\ddot{\mathbf{R}}$ is the acceleration of O relative to O'. Collecting terms,

$$\mathbf{a}' = \ddot{\mathbf{R}} + \mathbf{a}_{rel} + 2\omega \times \mathbf{v}_{rel} + \omega \times (\omega \times \rho) + \dot{\omega} \times \rho \tag{2.100}$$

In Eq. (2.100), the term $2\omega \times \mathbf{v}_{rel}$ is called the *Coriolis acceleration*,[1] and its significance is treated at length in Chapter 9.

---

[1] Coriolis, G. G., *Traité de mécanique*, 1743.

2.18 MOTION REFERRED TO AXES UNDERGOING TRANSLATION ONLY

Coordinate frame XYZ is in pure translation with respect to $X'Y'Z'$ if each axis of XYZ maintains a fixed orientation relative to $X'Y'Z'$. This is readily seen for the the case in which X is parallel to X', and so on, as shown in Fig. 2.15. Since $\boldsymbol{\omega} = \mathbf{0}$ and $\dot{\boldsymbol{\omega}} = \mathbf{0}$, the kinematic equations [Eqs. (2.95) and (2.100)] become

$$\mathbf{v}' = \dot{\mathbf{R}} + \mathbf{v}_{rel} \qquad (2.101)$$

and

$$\mathbf{a}' = \ddot{\mathbf{R}} + \mathbf{a}_{rel} \qquad (2.102)$$

For the special case in which $\mathbf{a}_{rel} = \mathbf{0}$, particle acceleration in both systems is identical.

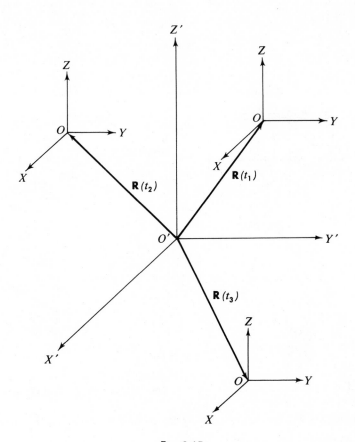

Fig. 2.15

Illustrative Example 2.5

A helicopter traveling with a constant forward velocity of 50 ft/sec hoists an object at a speed of 2 ft/sec. Determine the velocity and acceleration of the object with respect to the ground at the instant the hoist line makes an angle of 30 degrees with the vertical, and experiences an angular velocity and angular acceleration of 1 rad/sec clockwise and 5 rad/sec^2 counterclockwise. At the instant in question, 20 ft of line are out.

SOLUTION

Two solutions are given below to demonstrate that although the selection of a coordinate system influences the individual terms of Eqs. (2.95) and (2.100), it does not change the final result.

In the left-hand column, the primed system is, at the instant under consideration, coincident with the body axes of the helicopter, although fixed with respect to the earth, as shown in Fig. 2.16. The XYZ coordinate frame is fixed to the helicopter, as shown.

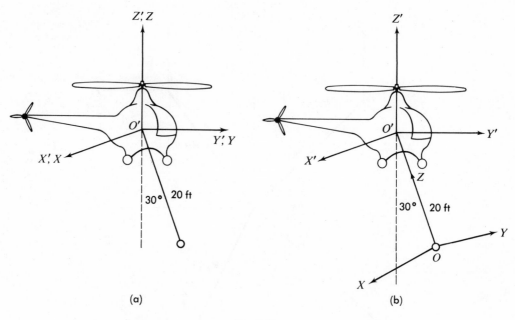

Fig. 2.16

In the right-hand column, the primed system is as described above, but XYZ is fixed to the object being hoisted.

All vectors will be expressed in terms of unit vectors $\hat{\mathbf{i}}'$, $\hat{\mathbf{j}}'$, $\hat{\mathbf{k}}'$ in the $X'Y'Z'$-system:

$\boldsymbol{\rho} = 20(0.500\hat{\mathbf{j}}' - 0.866\hat{\mathbf{k}}')$ $= 10\hat{\mathbf{j}}' - 17.32\hat{\mathbf{k}}'$	$\boldsymbol{\rho} = \mathbf{0}$ (since object is located at origin of XYZ)
$\boldsymbol{\omega} = \mathbf{0}$ (since XYZ translates with respect to $X'Y'Z'$)	$\boldsymbol{\omega} = -\hat{\mathbf{i}}'$ rad/sec
$\dot{\boldsymbol{\omega}} = \mathbf{0}$	$\dot{\boldsymbol{\omega}} = 5\hat{\mathbf{i}}'$ rad/sec^2
$\dot{\mathbf{R}} = 50\hat{\mathbf{j}}'$ ft/sec	$\dot{\mathbf{R}} = 50\hat{\mathbf{j}}' + 2(-0.500\hat{\mathbf{j}}'$ $+ 0.866\hat{\mathbf{k}}') + (-\hat{\mathbf{i}}')$ $\times\ 20(0.500\hat{\mathbf{j}}'$ $- 0.866\hat{\mathbf{k}}')$ $= 31.68\hat{\mathbf{j}}' - 8.27\hat{\mathbf{k}}'$ $\dot{\mathbf{R}}$ is the sum of the velocity components of O with respect to O': the hoist velocity with respect to the helicopter plus that of the helicopter relative to the primed axes
$\ddot{\mathbf{R}} = \mathbf{0}$	$\ddot{\mathbf{R}} = (-\hat{\mathbf{i}}') \times [(-\hat{\mathbf{i}}')$ $\times\ (10\hat{\mathbf{j}}' - 17.32\hat{\mathbf{k}}')]$ $+ (5\hat{\mathbf{i}}') \times (10\hat{\mathbf{j}}'$ $- 17.32\hat{\mathbf{k}}')$ $\ddot{\mathbf{R}}$ is the sum of the acceleration components of O with respect to O': These are the centripetal and tangential accelerations of O with respect to the helicopter plus the acceleration of the helicopter relative to O' (zero in this case). $\ddot{\mathbf{R}} = 76.6\hat{\mathbf{j}}' + 67.32\hat{\mathbf{k}}'$ ft/sec^2
$\mathbf{v}_{\text{rel}} = 2(-0.500\hat{\mathbf{j}}' + 0.866\hat{\mathbf{k}}')$ $+ (-\hat{\mathbf{i}}') \times (10\hat{\mathbf{j}}'$ $- 17.32\hat{\mathbf{k}}')$ $= -18.32\hat{\mathbf{j}}' - 8.27\hat{\mathbf{k}}'$	$\mathbf{v}_{\text{rel}} = \mathbf{0}$
$\mathbf{a}_{\text{rel}} = (-\hat{\mathbf{i}}') \times [(-\hat{\mathbf{i}}')$ $\times\ (10\hat{\mathbf{j}}' - 17.32\hat{\mathbf{k}}')]$ $+ (5\hat{\mathbf{i}}') \times (10\hat{\mathbf{j}}'$ $- 17.32\hat{\mathbf{k}}')$ $= 76.6\hat{\mathbf{j}}' + 67.32\hat{\mathbf{k}}'$	$\mathbf{a}_{\text{rel}} = \mathbf{0}$
$\boldsymbol{\omega} \times (\boldsymbol{\omega} \times \boldsymbol{\rho}) = \mathbf{0}$ $(\boldsymbol{\omega} = \mathbf{0})$	$\boldsymbol{\omega} \times (\boldsymbol{\omega} \times \boldsymbol{\rho}) = \mathbf{0}$ $(\boldsymbol{\rho} = \mathbf{0})$
$\dot{\boldsymbol{\omega}} \times \boldsymbol{\rho} = \mathbf{0}$ $(\dot{\boldsymbol{\omega}} = \mathbf{0})$	$\dot{\boldsymbol{\omega}} \times \boldsymbol{\rho} = \mathbf{0}$ $(\boldsymbol{\rho} = \mathbf{0})$
$2\boldsymbol{\omega} \times \mathbf{v}_{\text{rel}} = \mathbf{0}$ $(\boldsymbol{\omega} = \mathbf{0})$	$2\boldsymbol{\omega} \times \mathbf{v}_{\text{rel}} = \mathbf{0}$ $(\mathbf{v}_{\text{rel}} = \mathbf{0})$
$\mathbf{v}' = (-18.32\hat{\mathbf{j}}' - 8.27\hat{\mathbf{k}}')$ $+ (50\hat{\mathbf{j}}')$ $= 31.68\hat{\mathbf{j}}' - 8.27\hat{\mathbf{k}}'$ ft/sec	$\mathbf{v}' = \mathbf{0} + (31.68\hat{\mathbf{j}}' - 8.27\hat{\mathbf{k}}')$ $= 31.68\hat{\mathbf{j}}' - 8.27\hat{\mathbf{k}}'$ ft/sec
$\mathbf{a}' = 76.6\hat{\mathbf{j}}' + 67.32\hat{\mathbf{k}}'$ ft/sec^2	$\mathbf{a}' = 76.6\hat{\mathbf{j}}' + 67.32\hat{\mathbf{k}}'$ ft/sec^2

References

Blass, G. A., *Theoretical Physics*, Appleton-Century-Crofts, New York, 1962.

Goodman, L. E., and Warner, W. H., *Dynamics*, Wadsworth Publishing Co., Inc., Belmont, Calif., 1963.

Lindsay, R. B., *Physical Mechanics*, Third Edition, D. Van Nostrand Company, Inc., Princeton, N.J., 1961.

Osgood, W., *Mechanics*, The Macmillan Company, New York, 1946.

Shames, I. H., *Engineering Mechanics*, Second Edition, Prentice-Hall, Inc., Englewood Cliffs, N.J., 1967.

Synge, J. L., and Griffith, B. A., *Principles of Mechanics*, Third Edition, McGraw-Hill, Inc., New York, 1959.

Whittaker, E. T., *Analytical Dynamics*, Fifth Edition, Dover Publications, Inc., New York, 1944.

Zajac, A., *Basic Principles and Laws of Mechanics*, D. C. Heath and Company, Boston, 1966.

EXERCISES

2.1 The position of a particle relative to a Cartesian coordinate frame is given by the equation

$$\mathbf{r} = x(t)\hat{\mathbf{i}} + y(t)\hat{\mathbf{j}} + z(t)\hat{\mathbf{k}}$$

where

$$x(t) = 5 + 6t^2 \text{ m}$$

$$y(t) = 3e^{-2t} \text{ m}$$

$$z(t) = 6 \text{ m}$$

and t is in seconds.

    a. Determine the X-, Y-, and Z-displacements between $t = 0$ and $t = 1$.
    b. What are the components of velocity at $t = 1$?
    c. What are the components of acceleration at $t = 1$?
    d. From the equations derived to determine (b) and (c), sketch the curves of velocity and acceleration as functions of time for each component and describe their characteristics.

2.2 The equations describing the helical motion of a point are

$$x = A(\cos bt)$$

$$y = A(\sin bt)$$

$$z = Bt$$

Determine the velocity and acceleration, and prove that the speed and the magnitude of the acceleration are constant.

2.3 A particle moves along a path $x = 3t^2$, $y = 2t^2 + 3t$, $z = 5t$. Determine the velocity and acceleration components in the direction of a vector

$$\mathbf{A} = 3\hat{\mathbf{i}} + 2\hat{\mathbf{j}} - 6\hat{\mathbf{k}}$$

at $t = 2$ sec.

2.4 A particle undergoes two-dimensional simple harmonic motion given by

$$x = A \sin (\omega t)$$

$$y = B \cos (\omega t + \alpha)$$

Determine the *trajectory* of the particle, that is, the path described in terms of spatial coordinates only. Sketch the trajectory. Under what circumstances will the motion be a straight line and a circle?

2.5 A *hodograph* is the locus of the velocity vector end point drawn from the origin of a given coordinate system. Determine the hodograph for the motions described in Exercise 2.4.

2.6 The acceleration of a particle is given by

$$\mathbf{a} = (7 \sin 2t + 3e^t)\hat{\mathbf{i}}$$

If the initial displacement and velocity are zero, determine their values at any time t.

2.7 A particle undergoes motion in a plane. Its X- and Y-positions, described as functions of time, are

$$x = 2t + 6$$
$$y = 3t^2$$

At $t = 2$, determine the magnitude and direction cosines of the velocity and acceleration.

2.8 What is the relative velocity between an aircraft flying due east at 600 mph and an automobile traveling at 50 mph in a northerly direction? If the aircraft is initially 20,000 ft directly above the automobile, how far apart will they be in 1 hr?

2.9 The position of a particle moving in a plane is given by

$$x = \sin(\omega t)$$
$$y = \cos(\alpha \omega t)$$

Show that the trajectory repeats itself periodically only if α is a rational number.

2.10 A rigid body rotates about an axis passing through the origin of a Cartesian reference system. The direction cosines of the axis are 0.707, 0.500, 0.500.

    a. For $\omega = 8$ rad/sec, determine the velocity of a point on the body located by $\mathbf{r} = 5\hat{\mathbf{i}} + 2\hat{\mathbf{j}} - 5\hat{\mathbf{k}}$ ft.

    b. What is the relative velocity of two points on the body having the coordinates (3, 2, 2) and (−5, 6, 7) ft?

2.11 Aircraft A, traveling at 300 mph in a southerly direction, is, at $t = 0$, 200 miles northwest of aircraft B. What heading should B take to intercept A in the shortest time, assuming B travels at 500 mph? How long will it take?

2.12 Particles A and B of a rigid body are located with respect to the origin by

$$\mathbf{r}_A = 5\hat{\mathbf{i}} + 6\hat{\mathbf{j}} + 2\hat{\mathbf{k}} \text{ ft}$$
$$\mathbf{r}_B = 2\hat{\mathbf{i}} - 3\hat{\mathbf{j}} - 6\hat{\mathbf{k}} \text{ ft}$$

If $\boldsymbol{\omega} = 12\hat{\mathbf{j}}$ rad/sec, determine the velocity of A relative to B.

2.13 The Cartesian components of velocity are (5, 8, 3) when a particle is at position (1, −5, 2). Find the components of velocity in cylindrical polar coordinates.

2.14 Repeat Exercise 2.13 for a spherical polar coordinate system.

2.15 For particle position described by the following equations, determine the velocity, speed, and acceleration in polar coordinates:

$$\mathbf{r} = R\hat{\mathbf{e}}_R + z\hat{\mathbf{k}} \qquad \text{and} \qquad \phi = Bt$$

where

$$R(t) = Ae^t$$
$$z(t) = 0$$

Give a geometric interpretation to your results.

2.16 A particle moves in a plane in accordance with the equations

$$R = 12te^{-t}$$
$$\phi = 6t$$

Determine the X- and Y-components of velocity and acceleration.

2.17 The position of a particle undergoing motion in a plane is described by

$$x = A \sinh (\omega t) - Bt$$
$$y = A[\cosh (\omega t) - 1]$$

Determine the radial and tangential components of velocity and acceleration.

2.18 Earth satellites S_1 and S_2 are traveling in coplanar circular orbits of radii r_1 and r_2 with periods T_1 and T_2. Describe the relative position, velocity, and acceleration of S_1 with respect to S_2.

2.19 A satellite S moves about the earth in a circular equatorial orbit of radius R and period T. A rocket is fired from the equator with velocity $v_0\hat{\mathbf{k}}$ and acceleration $-g\hat{\mathbf{k}}$. Where must the satellite be when the rocket is fired for an intercept to occur?

2.20 Demonstrate that a particle moving on the surface of a sphere possesses tangential velocity only but, in general, nonzero radial and tangential accelerations.

2.21 The position of a particle is given by

$$r = A \cos \theta \quad \text{and} \quad \theta = 6t^2$$

Determine the velocity and acceleration in polar coordinates.

2.22 The components of acceleration of a particle in spherical polar coordinates are $(0, 2, -3)$. What are the Cartesian components if the particle is at position $(-1, 1, -1)$?

2.23 An aircraft at an altitude of 10,000 ft and a velocity of 550 mph climbs at an angle of 15 degrees with the horizontal. A radar installation detects the aircraft when it is 30 miles away, measured horizontally. Find the components of the aircraft velocity using spherical polar coordinates. Assume that the aircraft velocity vector and the line of sight of the antenna are in a single vertical plane.

2.24 Repeat Exercise 2.23 assuming that the aircraft velocity vector is directed 15 degrees above the horizontal and 30 degrees out of plane.

2.25 What are the Cartesian components of velocity if $\theta = 30$ degrees, $\phi = 60$ degrees, and $\mathbf{v} = 30\hat{\mathbf{e}}_r + 20\hat{\mathbf{e}}_\phi + 15\hat{\mathbf{e}}_\theta$ cm/sec?

2.26 A particle undergoes circular motion in the YZ-plane. The center of the circle is at the origin and its radius is 10 ft. If the velocity of the particle at $(0, 0, 10)$ is $-5\hat{\mathbf{j}}$ ft/sec, and the acceleration tangent to the path at this point is 12 ft/sec^2, determine for this point:

    a. the components of velocity and acceleration in Cartesian coordinates.
    b. the components of velocity and acceleration in path coordinates.

2.27 A point moves along a circular path of radius ρ according to the relation

$$s = At^2 + Bt + C$$

where s represents the distance traveled measured from a prescribed starting point. What are the velocity and acceleration at $t = t_1$?

2.28 A particle moves along the curve

$$x = 4t$$
$$y = 3t^2 + 8$$
$$z = 12t^2 + 2t$$

Determine the direction cosines of the unit tangent vector $\hat{\mathbf{e}}_t$, the unit principal normal vector $\hat{\mathbf{e}}_n$, and the unit binormal vector $\hat{\mathbf{e}}_b$ at $t = 10$ sec.

2.29 Determine the normal and the tangential components of acceleration at $t = 1$ sec for a particle, the position of which is given by

$$\mathbf{r} = 3t\hat{\mathbf{i}} + 4t^2\hat{\mathbf{j}} - 10\hat{\mathbf{k}} \text{ ft}$$

2.30 The position of a particle is described by

$$\mathbf{r} = 5 \sin (3t)\hat{\mathbf{i}} + 3t\hat{\mathbf{j}}$$

At $t = 0.5$ sec, what are the velocity, direction cosines of $\hat{\mathbf{e}}_t$, radius of curvature, and the normal and tangential components of acceleration?

2.31 The equations of motion for a projectile, subject to certain simplifying assumptions (see Chapter 7), are

$$x = v_{0x}t$$

and $\quad y = v_{0y}t - \frac{1}{2}gt^2$

where v_{0x} and v_{0y} represent the initial X- and Y-components of the velocity and g is the familiar gravitational acceleration. Determine the components of acceleration in path coordinates.

2.32 A particle experiences constant-speed motion at 5 ft/sec along the path shown in Fig. P2.32 until point 3, where the speed increases uniformly at a rate of 1 ft/sec/ft of path. Determine the acceleration at points 1, 2, 3, 4, and 5.

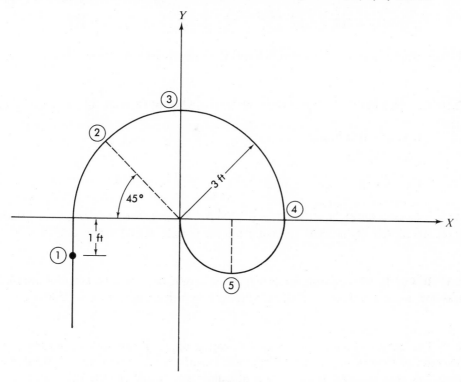

Fig. P2.32

2.33 A particle, initially at rest, experiences straight-line motion with velocity given by

$$v = At^3 + Bt^2 + Ct$$

Determine the displacement and acceleration at $t = t_f$ if the initial displacement is zero.

2.34 A particle having initial velocity v_0, moving along the X-axis, has an acceleration given by

$$a_x = -Cv_x^2$$

where C is a positive constant and v_x represents the X-velocity. How long does it take for the particle to come to rest and how far has it traveled?

2.35 A particle is projected downward through a liquid offering resistance proportional to the velocity. The acceleration is expressed by

$$a_z = 32.2 - 10v_z \text{ ft/sec}^2$$

If the initial velocity is 20 ft/sec, what is the velocity as a function of time?

2.36 The readings of a set of accelerometers may be approximated by the equations

$$a_x = 2 + 0.5t \text{ ft/sec}^2$$
$$a_y = -3t$$
$$a_z = -32.2$$

At $t = 0$, $x = 700$ ft, $y = 300$ ft, $z = 2500$ ft. What is the position at $t = 10$ sec if the initial velocity is zero? How far has the vehicle moved in this time?

2.37 In Fig. P2.37 is plotted aircraft forward acceleration as a function of distance traveled along a runway. What velocity will the vehicle have at $x = 3500$ ft?

2.38 The output of an accelerometer mounted on an automobile is plotted as a function of time in Fig. P2.38. Draw a curve of velocity versus time. What distance has been traveled at $t = 30$ sec, assuming zero initial conditions?

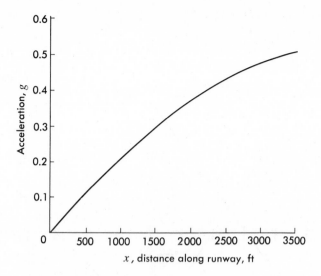

Fig. P2.37

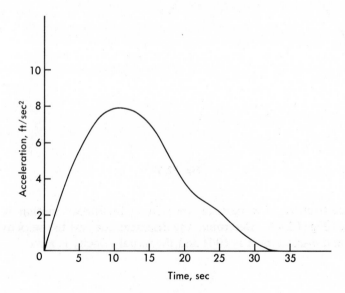

Fig. P2.38

2.39 At time zero, a particle located at $x = 3$ ft and $y = -2$ ft possesses X- and Y-velocities of 20 and 30 ft/sec, respectively. Determine the position coordinates at $t = 12$ sec, as well as the X- and Y-velocities for the acceleration given in Fig. P2.39.

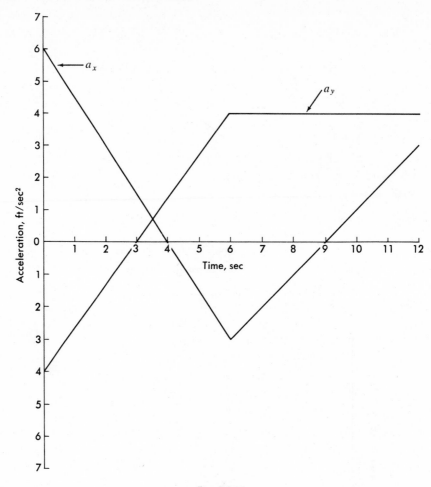

Fig. P2.39

2.40 The acceleration of a particle undergoing rectilinear motion is described graphically in Fig. P2.40. Determine the displacement and the velocity at $t = 10$ sec if the initial displacement is 10 ft and the initial velocity is zero.

2.41 Consider the arrangement shown in Fig. P2.41, in which the platform rotates at constant speed $\omega_1 = 50$ rad/sec about axis A-A while that axis rotates at constant speed $\omega_2 = 40$ rad/sec about axis B-B. In terms of the space-fixed XYZ-axes, determine:

    a. the total angular velocity of the platform.
    b. the angular acceleration of the platform.

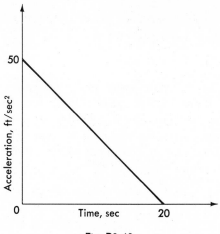

Fig. P2.40

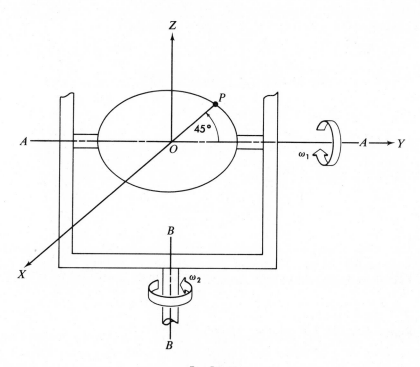

Fig. P2.41

2.42 Referring to Exercise 2.41, determine the velocity and acceleration of point P when the platform is vertical, as shown. The distance from the origin to P is 3 in.

2.43 Again referring to Exercise 2.41, what velocity and acceleration will the particle experience if $\dot{\omega}_1 = 1$ rad/sec^2 and $\dot{\omega}_2 = 3$ rad/sec^2?

2.44 A slide S moves at a constant speed of 1 ft/sec relative to rotating arm R as shown in Fig. P2.44. If the angular speed of R is constant at 2 rad/sec counterclockwise, determine the velocity and acceleration of point P when R makes an angle of 45 degrees with the Y-axis and $O'P$ is 1 ft. Rotation is entirely in the YZ-plane. (*Note:* Select a moving coordinate system fixed to the arm at P.)

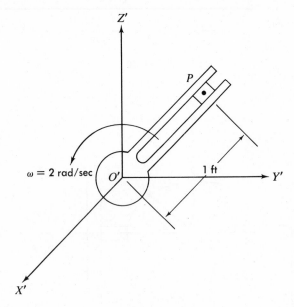

Fig. P2.44

2.45 Repeat Exercise 2.44 but use a moving coordinate system fixed to the slide.

2.46 Repeat Exercise 2.44 with $\dot{\omega} = 3$ rad/sec^2 clockwise.

2.47 Consider the three concentric disks of radii R_1, R_2, and R_3 shown in Fig. P2.47. If ω_1 is the constant angular velocity of disk 1 relative to disk 2, ω_2 is the constant angular velocity of 2 relative to 3, and disk 3 rotates at constant angular velocity ω_3 with respect to a fixed reference X', Y', Z', determine the velocity and

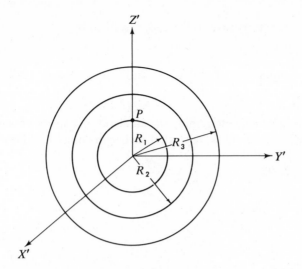

Fig. P2.47

acceleration of P relative to the fixed system. (*Note:* Employ a rotating coordinate system fixed to disk 3.)

2.48 Repeat Exercise 2.47 using the rotating-coordinate-system approach twice. First fix such a system to disk 2, then to disk 3.

2.49 Consider the disk shown in Fig. P2.49 rotating about the fixed Z'-axis according to the expression

$$\phi = A \sin (bt)$$

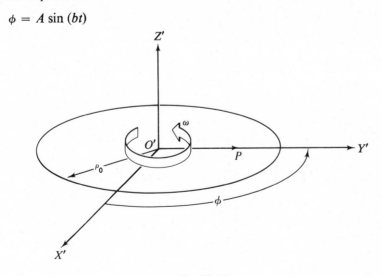

Fig. P2.49

A particle P has motion relative to the disk described by

$$\rho = \rho_0 \sin (bt)$$

Determine the components of the acceleration of P relative to $X'Y'Z'$ in polar coordinates.

2.50 Repeat Exercise 2.49, determining the acceleration components using a coordinate system attached to the disk, such that its origin coincides with that of $X'Y'Z'$.

2.51 Derive Eq. (2.70).

CHAPTER 3

NEWTON'S LAWS OF MOTION— THE PARTICLE

3.1 INTRODUCTION

Having dealt with the manner in which the motion of particles may be described, we now consider the matter of causality: the relationship between the factors influencing the motion and the results produced. Broadly speaking, all of classical mechanics, including fluid mechanics, sound, and elasticity, can be logically traced to a set of laws of motion formulated by Sir Isaac Newton. In this chapter our concern is with these laws and how they relate to the particle or point mass. In subsequent chapters the dynamics of a particle is extended to various systems of particles.

3.2 THE POINT MASS

The idealization of a body of finite extent as a point mass offers many advantages and has great utility in the analysis of many real situations. Consider the following instances:

1. Each particle of a finite body undergoing pure translation, such as a block sliding down an incline, experiences identical velocity and acceleration. Therefore, the motion of one particle is the motion of every particle, and the block may be regarded as a point mass.

2. Each particle of a body experiencing simultaneous translation and rotation does not experience identical velocity and acceleration. In many instances, however, all we are really interested in is the motion of a single point, called the *center of mass*. For example, if the trajectory of a planet or artificial earth satellite is required, it is often satisfactory to deal only with the motion of the mass center,

despite the fact that rotation about this point may occur. The trajectory is, in fact, the path of the mass center.

The point-mass idealization is thus seen to relate itself not so much to the size of the body, but to the type of information sought concerning its motion.

3.3 MASS AND NEWTON'S LAWS

Imagine two particles to be entirely isolated from all effects in the universe, except those which they may cause to act upon one another. Following the approach of Mach,[1] assume that it is determined experimentally that each particle develops an acceleration because of some mutual interaction, and that these accelerations are oppositely directed along the straight line connecting the particles. The interaction cited may, for example, be due to impact. Labeling the particles 1 and 2,

$$a_{12} = -K_{21}a_{21} \tag{3.1}$$

where

a_{12} = acceleration of particle 1 because of the influence of particle 2

a_{21} = acceleration of particle 2 because of the influence of particle 1

K_{21} = a constant of proportionality; a positive scalar quantity valid for all interactions involving particles 1 and 2

An experiment following the same format may be performed involving particles 1 and 3, and then 2 and 3, with the following results:

$$a_{13} = -K_{31}a_{31} \tag{3.2}$$

and

$$a_{23} = -K_{32}a_{32} \tag{3.3}$$

To this juncture our "imaginary" experiment (which incidentally can actually be performed) leads to the following rather important proportionalities between the various accelerations:

$$K_{21} = -\frac{a_{12}}{a_{21}}$$

$$K_{31} = -\frac{a_{13}}{a_{31}}$$

and

$$K_{32} = -\frac{a_{23}}{a_{32}}$$

Continuing the experiment, it is found that

$$\frac{K_{31}}{K_{21}} = K_{32} \tag{3.4}$$

---

[1] Mach, E., *Science of Mechanics*, Fifth Edition, Translated by T. J. McCormack, Paquin Printers, Chicago, 1942.

which gives [after substitution into Eq. (3.3)]

$$K_{21}a_{23} = -K_{31}a_{32} \tag{3.5}$$

What we have related has been repeated in countless experiments, and the conclusions drawn may be accepted as quite general provided the velocities are small relative to the velocity of light. From Eq. (3.5) it is clear that we may select particle 1 as the standard for whatever property is responsible for the observed proportionality of accelerations. By this time the reader has undoubtedly recognized K_{31} to be the mass of particle 3 relative to the standard or test mass. Since we recognize that a standard mass must exist, it is only necessary to speak of the mass of particle 3. Equation (3.5) may now be written

$$m_2a_{23} = -m_3a_{32} \tag{3.6}$$

In Eq. (3.5) the term K_{31} or mass 3 is a property not only unrelated to mass 2 with which 3 interacts but is presumably unchanged by the presence of any other masses. Thus mass 3 is to be treated not as a property of the interaction but rather of the matter in particle 3, a quantitative measure of inertia.

The foregoing "experiments" relate to *inertial* mass, defined in terms of the responses of different masses to identical dynamical influences.

There are alternative approaches in defining mass. Consider, for example, two different volumes of a given homogenous material suspended by identical springs in a gravitational field. The spring supporting the larger object will deform a greater amount by virtue of its interaction with the earth. Mass may thus be defined in terms of the relative deformation of a spring in a constant gravitational field. Such experiments, defining the *gravitational* mass, are essentially based upon the equality of *masses* displaying equal *weights*. It is of the greatest consequence that experiment proves the gravitational and inertial masses equal—that the magnitude derived from an experiment completely independent of gravity is identical with one found in an experiment which is entirely dependent upon gravity. What we have described is a consequence of the *principle of equivalence*,[2] a postulate of Einstein's general theory of relativity.

We recognize that it is the force that is equal and opposite on either side of Eq. (3.6). By "interaction" of particle 3 upon 2, we mean the force that 3 causes to act upon 2, and we write, therefore,

$$F_{23} = m_2a_{23} \tag{3.7}$$

and

$$F_{32} = m_3a_{32} = -F_{23} = -m_2a_{23} \tag{3.8}$$

The force that one mass causes to act upon another is observed to be exactly equal and opposite that which the second exerts upon the first. The reader probably recognizes in Eq. (3.8) a statement of Newton's law of action and reaction, and at

---

[2] Witten, L., *Gravitation: An Introduction to Current Research*, John Wiley & Sons, Inc., New York, 1962.

this point he may very well be growing impatient, in that Newton's laws of motion have not as yet been stated explicitly.

From Eq. (3.7) it is clear that it is force (often referred to as a push or pull) which causes a mass to experience acceleration. *Newton's second law of motion*[3] may be stated: *Change in motion is proportional to the net force and takes place in the direction of the straight line in which the force acts.*

Newton was thus the first to make clear that the role of force is to alter, rather than maintain, a given state of motion.

By motion, Newton referred to the product of mass and velocity, the linear momentum **p**, and we therefore write

$$\mathbf{F} = \frac{d}{dt}(m\mathbf{v}) = \frac{d}{dt}(\mathbf{p}) = \dot{\mathbf{p}} \tag{3.9}$$

The momentum form of the second law, given above, is most general, and is in fact applicable to those situations in which momentum, but not necessarily mass, can be ascribed to the "particle." Such is the case of the photon of electromagnetic energy.

Differentiating Eq. (3.9) we obtain

$$\mathbf{F} = m\frac{d\mathbf{v}}{dt} + \mathbf{v}\frac{dm}{dt} \tag{3.10a}$$

which for the case of *constant* mass reduces to the familiar form

$$\mathbf{F} = m\frac{d\mathbf{v}}{dt} = m\mathbf{a} \tag{3.10b}$$

Since velocity and linear momentum are vectors, one must be alert to the fact that changes in these quantities may manifest themselves as changes in magnitude or direction or both.

Equation (3.10b) may be stated in terms of components parallel to a convenient set of axes as follows:

$$\mathbf{F} = F_x\hat{\mathbf{i}} + F_y\hat{\mathbf{j}} + F_z\hat{\mathbf{k}} = m(a_x\hat{\mathbf{i}} + a_y\hat{\mathbf{j}} + a_z\hat{\mathbf{k}}) \tag{3.11a}$$

$$\mathbf{F} = m(\dot{v}_x\hat{\mathbf{i}} + \dot{v}_y\hat{\mathbf{j}} + \dot{v}_z\hat{\mathbf{k}}) \tag{3.11b}$$

or

$$\mathbf{F} = m(\ddot{x}\hat{\mathbf{i}} + \ddot{y}\hat{\mathbf{j}} + \ddot{z}\hat{\mathbf{k}}) \tag{3.11c}$$

In the foregoing development of the concept of mass, only two masses interacted in a given experiment, to the exclusion of all other effects. It is empirically verified that should more than two masses interact, the accelerations produced may be added vectorially. Thus

$$\mathbf{a}_{1,2,3,4} = \mathbf{a}_{12} + \mathbf{a}_{13} + \mathbf{a}_{14} \tag{3.12}$$

---

[3] *Principia Mathematica Philosophiae Naturalis* of Sir I. Newton, Translated by A. Motte, Revised by F. Cajori, originally published 1687, University of California Press, Berkeley, Calif., 1947.

where the left side represents the acceleration of particle 1 due to the combined influences of masses 2, 3, and 4. Here we have an example of the principle of superposition in physics, for it is observed that the individual effects (the accelerations) are merely added to determine the total effect. Equation (3.12) leads one to believe, and correctly so, that the forces responsible for the total acceleration of particle 1 may also be added vectorially,

$$\mathbf{F}_{12} + \mathbf{F}_{13} + \mathbf{F}_{14} = \sum_{n=1}^{4} \mathbf{F}_{1n} = m\mathbf{a}_1 \qquad (\mathbf{F}_{11} = \mathbf{0}) \qquad (3.13a)$$

or, in general,

$$\sum_{i=1}^{n} \mathbf{F}_i = m\mathbf{a} \qquad (3.13b)$$

From Newton's second law, the *first law* follows quite logically: *Everybody continues in its state of rest or uniform motion in a straight line, unless it is compelled by forces to change that state.*

When no net force acts (that is, the sum of all forces is zero), Eq. (3.9) leads to the result that the momentum must remain constant. If the body is initially at rest, it remains so. If it possesses nonzero momentum, the momentum does not change. This condition implies that the motion continues in a straight line (that is, unchanging direction), because by change we must include deviation from a fixed direction. The first law of Newton thus describes a property of inertia, common to all material objects: the tendency of matter to retain a given state of motion unless acted upon by forces. Although the property of inertia is common to all matter, we appreciate that matter is seldom, if ever, left alone to continue in straight-line motion indefinitely, or in an indefinite state of rest, even in atomic systems. As long as there are two masses in the universe, gravitational forces must act; as long as masses touch, contact forces exist.

Newton's *third law* of action and reaction[3] has already been introduced: *To every action, there is always an equal and opposite reaction; or the mutual actions of any two bodies are always equal and oppositely directed along the same straight line* (the *strong form* of this law). The *weak form* does not require that the action-reaction pair be along the same straight line.

The most common error in interpretation made in connection with the third law is to assume that both the action and reaction forces act on the same body. If this pair of forces is equal in magnitude and oppositely directed, how then can any mass ever move? The answer is, of course, that the action-reaction pair acts on different masses. The third law refers to the mutual actions of any *two* bodies. Examples of such forces are those which act between contacting bodies and gravity-type forces. The force that causes an elevator to accelerate is equal and opposite to the force causing the supporting cable to experience strain. The elevator accelerates because of an unbalanced force acting upon it; the *reaction* force acts on the cable.

3.4 STATICS OF A PARTICLE

The statics or equilibrium of a particle is described by stating

$$\sum_{i=1}^{n} \mathbf{F}_i = \mathbf{0} \tag{3.14}$$

We need not discuss, at this juncture, equilibrium with respect to rotation, inasmuch as rotation of a point has no meaning.

To state that the vector sum of forces is zero leads to any number of scalar equations:

$$\sum F_x = 0 \qquad \sum F_y = 0 \qquad \sum F_z = 0$$

$$\sum F'_x = 0 \qquad \sum F'_y = 0 \qquad \sum F'_z = 0 \qquad \text{etc.} \tag{3.15}$$

Equation (3.14) means that the sum of forces is zero in all directions. One should be careful to make a judicious choice of coordinate direction, not to assure a correct result, but to reduce the complexity of the calculations. For a two- or three-dimensional system, the directions along which forces are summed need not be orthogonal. To assure independence of the equations of statics, the directions selected for a two-dimensional system should not be parallel; for a three-dimensional system, they should not be coplanar.

Illustrative Example 3.1
What force is required to maintain the mass shown (Fig. 3.1) in a state of equilibrium?

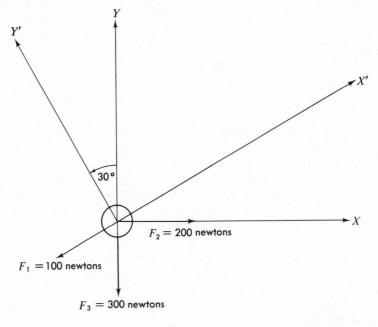

Fig. 3.1

SOLUTION

In the unprimed system,

$$\sum_{i=1}^{n} \mathbf{F}_i = \mathbf{0} = \mathbf{F}_e - 300\hat{\mathbf{j}} - 100(\cos 30)\hat{\mathbf{i}} - 100(\sin 30)\hat{\mathbf{j}} + 200\hat{\mathbf{i}}$$

where $\mathbf{F}_e$ represents the equilibrant (force required for equilibrium):

$$\mathbf{F}_e = 113.3\hat{\mathbf{i}} - 350\hat{\mathbf{j}} \ \text{N}$$

The magnitude of $\mathbf{F}_e$ is $[(113.3)^2 + (350)^2]^{1/2} = 367$ N. In the primed system,

$$\sum_{i=1}^{n} \mathbf{F}_i = \mathbf{0} = \mathbf{F}'_e - 100\hat{\mathbf{i}}' - 300(\cos 30)\hat{\mathbf{j}}' - 300(\sin 30)\hat{\mathbf{i}}'$$

$$+ 200(\cos 30)\hat{\mathbf{i}}' - 200(\sin 30)\hat{\mathbf{j}}'$$

$$\mathbf{F}'_e = -77\hat{\mathbf{i}}' - 360\hat{\mathbf{j}}'$$

$$F'_e = [(77)^2 + (360)^2]^{1/2} = 367 \ \text{N} = F_e$$

3.5 INERTIAL FRAMES OF REFERENCE

In Chapter 2 it was observed that to discuss motion without regard to a reference has little meaning. Clearly then there is a common question posed by Newton's laws of motion: To what coordinate frame is motion referred when one speaks of states of rest or uniform straight-line motion, or of changes in motion? Newton's laws actually imply that there exists one or more coordinate frames for which they are valid. Such coordinate systems are termed *inertial frames of reference*, because their axes are either fixed or experience constant-velocity motion. It is as though they exist in a state of inertia as described by Newton's first law. This law may be regarded, therefore, as providing the basis for inertial frames of reference. A *primary inertial system* is one that is at rest relative to the mean position of the "fixed stars." We now demonstrate that $\mathbf{F} = m\mathbf{a}$ is equally valid in all *secondary inertial systems:* rigid coordinate frames experiencing constant velocity and zero rotation with respect to a primary system. Two such systems are shown in Fig. 3.2.

Consider first the velocity of a mass m measured in each system, bearing in mind that the secondary system experiences a constant velocity $\mathbf{v}_r$ with respect to the primary system. For the primary system

$$\mathbf{v}^* = \dot{\mathbf{r}} \tag{3.16}$$

and for the secondary system

$$\mathbf{v}' = \dot{\mathbf{r}}' \tag{3.17}$$

The relationship between these two velocities is

$$\mathbf{v}^* = \mathbf{v}' + \mathbf{v}_r \tag{3.18}$$

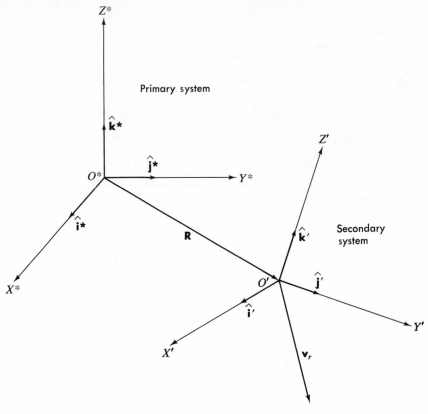

Fig. 3.2

Differentiating Eq. (3.18) with respect to time,

$$\mathbf{a}^* = \mathbf{a}' \tag{3.19}$$

since $\dot{\mathbf{v}}_r = \mathbf{0}$. The magnitude of the force vector and its orientation in space are independent of coordinate reference, and the mass is identical in either reference. (The results of special relativity modify the foregoing analysis and are discussed in Chapter 10.)

We may, therefore, conclude that

$$\sum \mathbf{F} = m\mathbf{a}^* = m\mathbf{a}' \tag{3.20}$$

Thus a primary or secondary system may be used in Newtonian formulations. The reader should take particular note of the fact that force and acceleration vectors may have different direction cosines in the primary and secondary systems, because the respective coordinate axes need not be parallel. These vectors are, however, uniquely oriented in space, independently of how they are viewed. In a given vector equation, however, only one coordinate frame should be used. We may write either

$$\sum \mathbf{F}^* = \sum F_x^* \hat{\mathbf{i}}^* + \sum F_y^* \hat{\mathbf{j}}^* + \sum F_z^* \hat{\mathbf{k}}^*$$
$$= m(a_x^* \hat{\mathbf{i}}^* + a_y^* \hat{\mathbf{j}}^* + a_z^* \hat{\mathbf{k}}^*) \tag{3.21}$$

or

$$\sum \mathbf{F}' = \sum F_x' \hat{\mathbf{i}}' + \sum F_y' \hat{\mathbf{j}}' + \sum F_z' \hat{\mathbf{k}}'$$
$$= m(a_x' \hat{\mathbf{i}}' + a_y' \hat{\mathbf{j}}' + a_z' \hat{\mathbf{k}}') \tag{3.22}$$

Finally, it is reasonable to wonder whether one must always refer motion to inertial frames of reference in the "practical" applications of dynamics. The answer rests with the amount of error acceptable in a given situation, and requires that a judgment be made. In many cases, a reference frame fixed to the earth (and obviously not experiencing constant velocity and zero rotation with respect to the fixed stars) will yield satisfactory results.

The accelerations in terms of cylindrical polar, spherical polar, and path coordinates (Chapter 2) pertain to a point in motion relative to the origin of a Cartesian frame of reference. If this reference is an intertial frame of reference, the accelerations derived in Chapter 2 may be applied to Eq. (3.10b). In cylindrical polar coordinates,

$$F_R = m(\ddot{R} - R\dot{\phi}^2) \tag{3.23a}$$

$$F_\phi = m(R\ddot{\phi} + 2\dot{R}\dot{\phi}) \tag{3.23b}$$

$$F_z = m\ddot{z} \tag{3.23c}$$

In spherical polar coordinates,

$$F_r = m(\ddot{r} - r\dot{\theta}^2 - r\dot{\phi}^2 \sin^2 \theta) \tag{3.24a}$$

$$F_\theta = m(r\ddot{\theta} + 2\dot{r}\dot{\theta} - r\dot{\phi}^2 \sin \theta \cos \theta) \tag{3.24b}$$

$$F_\phi = m[(r\ddot{\phi} + 2\dot{r}\dot{\phi}) \sin \theta + 2r\dot{\phi}\dot{\theta} \cos \theta] \tag{3.24c}$$

In path coordinates,

$$F_t = m\ddot{s} \tag{3.25a}$$

$$F_n = \frac{m\dot{s}^2}{\rho} \tag{3.25b}$$

The reader should especially note that the above components of acceleration are *not* measured relative to the origins of the various *non-Cartesian* systems. They are rather the accelerations relative to the origin of XYZ. In each case, the sum of the individual components is simply $\ddot{\mathbf{r}}$, the second time derivative of the vector from O_{xyz} to P. The results are clearly different than would obtain, for example, by considering the $\hat{\mathbf{e}}_R$, $\hat{\mathbf{e}}_\phi$, $\hat{\mathbf{k}}$ axes as forming a secondary inertial frame. Because the origin of $\hat{\mathbf{e}}_R$, $\hat{\mathbf{e}}_\phi$, $\hat{\mathbf{k}}$ accelerates (with P), it cannot be an inertial frame of reference.

3.6 STRENGTHS AND WEAKNESSES OF NEWTON'S LAWS

We should understand at the outset that criticisms of the laws of motion as postulated by Newton are not intended to deny their validity, within certain limits of applicability. That these fundamentals have so ably withstood the test of time, despite a somewhat tenuous beginning, is certainly a tribute not only to Newton's analytical ability, but to the intuitive genius that was surely his.

There are two major weaknesses of Newton's laws. These are the failure to prescribe the coordinate references for which the laws hold, and the failure to define force. Newton himself realized that a suitable coordinate frame was lacking, for indeed, not only does the second law suffer for lack of such a frame, but the first law as well, as we have pointed out.

Although the definitions of velocity and acceleration are quite straightforward, and a definition of mass can be arrived at, perhaps with somewhat more difficulty, there remains the all-important question of force. Without an adequate definition of force, Newton's second law might be treated as a defining expression for force rather than a fundamental law of physics. Two factors mitigate against this viewpoint. To begin with, Newton did indeed make substantial contributions to our understanding of the nature of force in his formulation of the law of gravitation as well as the law of action and reaction. Second, we can measure force quite independently of motion, as, for example, in noting the deformation of a spring supporting a mass in a gravitational field.

That Newton did not anticipate the dependence of mass upon velocity would appear of minor importance, especially in view of the momentum formulation of the second law, which does not require mass to remain constant. Even if Newton had preferred the $F = ma$ form, the constancy of mass is an excellent first approximation.

The overriding strengths of Newton's laws are the immense body of accomplishment to which they have directly and indirectly led. The scientific and technical achievements in elasticity, orbital mechanics, fluid mechanics, and many other subdisciplines of mechanics speak eloquently for Newton.

3.7 APPLICATIONS OF THE SECOND LAW

Consider a particle of constant mass acted upon simultaneously by a number of different forces, Fig. 3.3(a). On the basis of Newton's second law, these forces cause an acceleration in the direction of the force resultant F_R as shown in Fig. 3.3(b). The individual forces each make a contribution to the acceleration, so that an acceleration polygon [Fig. 3.3(c)] is geometrically similar to a force polygon when the components are taken in the same order. As required of all vector addition, the resultant is independent of the order.

The application of Newton's second law to a real situation is, in principle, quite simple. One examines the physical system, considers all the forces acting,

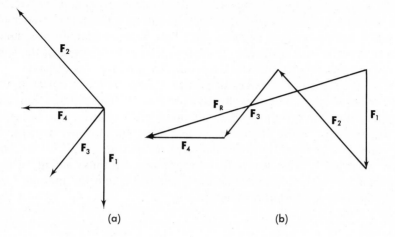

(a)   (b)

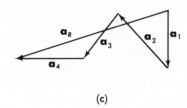

(c)

Fig. 3.3

and proceeds to describe them in some analytic form. The resultant force and acceleration vectors are uniquely oriented in space regardless of the inertial frame to which motion is referred. Experience soon teaches that time spent in making a judicious selection of coordinate reference eliminates much unnecessary labor. In practice, the process of analysis is not executed as smoothly as we have indicated, for two important reasons:

1. The analytic descriptions of many of the forces encountered are often incomplete or unknown. The nature of friction, for example, is still the subject of considerable scientific research and analysis, as is the nuclear binding force.

2. Because the mathematical description of a given force may be quite complex, the resulting equation of motion ($\mathbf{F} = m\ddot{\mathbf{r}}$) may defy solution except by numerical or computer techniques.

These difficulties are compounded when one considers that rarely does a single force act upon a mass. It is usual for several forces to act at once, each possessing its own peculiarities.

We are continually making idealizations and approximations which must serve as more or less adequate descriptions of nature. The formal statements of the "laws" of motion, gravity, friction, and so on, have been largely ascertained on the basis of experimental evidence. These laws lead to predictions of the behavior of physical systems, predictions which can only be judged by experiment.

What are some of the forces of particular importance in our study of mechanics? Most of man's everyday efforts, as well as his longer-range technological activities, are directed toward both overcoming and exploiting the influences of gravity and friction. These effects are inseparable from our environment, and as such are taken for granted, in much the same manner as the air we breathe and the water we drink. Other forces that we examine are those acting on charged particles in the presence of electric and magnetic fields and the internal forces in materials which cause resistance to deformation.

3.8 THE FREE-BODY DIAGRAM

The free-body diagram represents the most important step in the analysis of many practical situations because it makes a significant contribution to the organization of thought prior to the mathematical formulation of a problem.

In preparing a free-body diagram of a particle, all the forces acting *on* the body under study are indicated. Special care must be taken not to include forces (reactions) that the body exerts on surrounding or contacting objects. The free-body diagram represents an interpretation of the actual condition of loading. The quantitative results of analysis predicated upon this representation can be no better than the assumptions made in its preparation.

3.9 THE FORCE OF GRAVITY

The reader recalls that a body in free fall travels toward the center of the earth with an acceleration g, approximately 980 cm/sec^2 or 32.2 ft/sec^2. This acceleration is due to the force of attraction between the earth and an object of mass m:

$$W = mg \qquad (3.26)$$

In Eq. (3.26) g is the so-called acceleration of gravity and W is the weight of the body. Let us examine the origin of this equation more closely.

In 1666 Newton hypothesized in *Principia Mathematica* that any two bodies in the universe are attracted to one another by a force that depends directly upon the product of their masses and inversely as the square of the distance separating them. The Newtonian theory became a "universal law of gravitation" when Cavendish[4] in 1797 demonstrated by direct observation that two laboratory masses actually do attract one another in accordance with Newton's theory.

Consider now point masses m_i and m_j separated by a distance r_{ij}. Since

---

[4] Cavendish, H., *Philosophical Transactions 17* (1798).

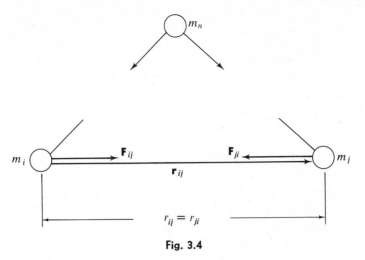

Fig. 3.4

mass i attracts mass j, the force of m_i on m_j is directed from j to i, as shown in Fig. 3.4, and designated $\mathbf{F}_{ji}$. Similarly, the force acting on m_i due to the gravitational influence of m_j is $\mathbf{F}_{ij}$. In general, other gravitational masses are also present, and their influences must be taken into account. In vector form, Newton's law of gravity is written

$$\mathbf{F}_{ij} = \frac{Gm_im_j}{r_{ij}^2}\left(\frac{\mathbf{r}_{ij}}{r_{ij}}\right) \tag{3.27}$$

where G, the universal gravitational constant, equals $6.670 \pm 0.0015 \times 10^{-8}$ dyn-cm^2/g^2; $\mathbf{r}_{ij}$ is a vector directed from i to j along the straight line connecting the masses. The magnitude of G is determined under carefully controlled conditions, by measuring the force of attraction between two laboratory masses separated by a known distance. The term in parentheses in Eq. (3.27) is equal to unit vector $\hat{\mathbf{r}}_{ij}$ along $\mathbf{r}_{ij}$, thus giving direction to the force. Alternative forms of Eq. (3.27) are

$$\mathbf{F}_{ij} = \frac{Gm_im_j}{r_{ij}^3}\mathbf{r}_{ij} = \frac{Gm_im_j}{r_{ij}^2}\hat{\mathbf{r}}_{ij} \tag{3.28}$$

We emphasize that the law of gravity describes a mutual force of attraction; thus $\mathbf{F}_{ij}$ and $\mathbf{F}_{ji}$ are equal in magnitude but oppositely directed ($\mathbf{r}_{ij} = -\mathbf{r}_{ji}$). The above treatment of Newton's law of gravity relates to the gravitational attraction of point masses, and although it may be satisfactory under certain circumstances to consider an object on the surface of the earth a point mass, the legitimacy of so treating the entire earth must be justified.

Consider Fig. 3.5(a), which shows a spherical earth consisting of concentric shells, each exerting a gravitational force upon a point mass m_p. From the combined effect of each point mass comprising a single shell, the total influence upon

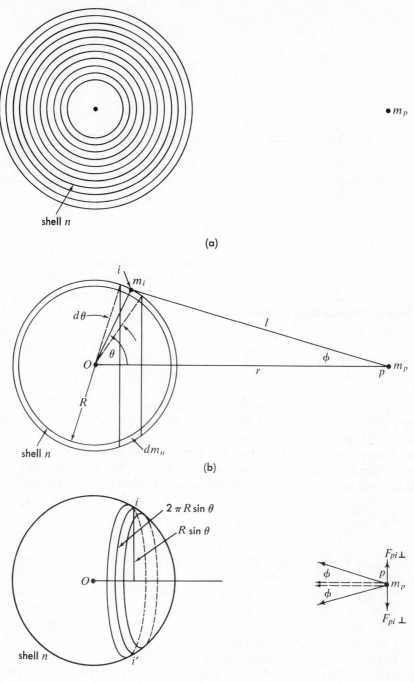

(a)

(b)

(c)
Fig. 3.5

m_p of the entire spherical mass is deduced. To begin with, a single point mass m_i located in shell n exerts a force upon m_p of magnitude [Fig. 3.5(b)]

$$F_{pi} = \frac{Gm_i m_j}{l^2} \qquad (3.29)$$

Since it is unwieldy to consider each mass point in a shell separately, assume that the nth shell is characterized by a uniform mass density per unit surface area ρ_n. The mass of band dm_n shown in Fig. 3.5(b) is

$$dm_n = \rho_n (2\pi R \sin \theta) R \, d\theta \qquad (3.30)$$

where

$$R = \text{radius of the sphere}$$

$2\pi R \sin \theta = $ circumference of band as shown in Fig. 3.5(c)

$R \, d\theta = $ width of band; therefore, $(2\pi R \sin \theta) R \, d\theta$ represents the area of a band of the nth shell

Each point mass of which dm_n is comprised has exerted upon it a force acting along a line l. This force may be resolved into components normal to and colinear with the line Op. For each mass m_i of shell n, there exists an identical mass m_i' directly across from it on the shell as shown. Clearly the components $F_{pi\perp}$ and $F_{pi\perp}'$ are equal and opposite, and hereafter we need only concern ourselves with the parallel components of F_{pi}. The total force on m_p due to a single band is, therefore,

$$dF = \frac{G[\rho_n (2\pi R \sin \theta) R \, d\theta] \, m_p \cos \varphi}{l^2} \qquad (3.31)$$

The influence of an entire shell is determined by integration. Since the variables l, θ, and ϕ are related by geometry, Eq. (3.31) may be easily expressed in terms of a single variable l. Applying the law of cosines twice for the triangle Opi:

$$R^2 = l^2 + r^2 - 2lr \cos \phi \qquad (3.32)$$

and

$$l^2 = R^2 + r^2 - 2Rr \cos \theta \qquad (3.33)$$

Taking the differential of Eq. (3.33) ($2l \, dl = 2Rr \sin \theta \, d\theta$) we obtain

$$\sin \theta \, d\theta = \frac{l \, dl}{Rr} \qquad (3.34)$$

Solving for $\cos \phi$ in Eq. (3.32),

$$\cos \phi = \frac{l^2 + r^2 - R^2}{2lr} \qquad (3.35)$$

Substituting for $\sin \theta \, d\theta$ and $\cos \phi$ in Eq. (3.31) and integrating over the entire shell,

$$F_n = \frac{Gm_p(\rho_n \pi R^2)}{Rr^2} \int_{l=r-R}^{l=R+r} \left(1 + \frac{r^2 - R^2}{l^2}\right) dl \tag{3.36}$$

The mass of shell n, $m_n = \rho_n(4\pi R^2)$. Making this substitution in Eq. (3.36) and integrating,

$$F_n = \frac{Gm_p m_n(4R)}{4Rr^2} = \frac{Gm_p m_n}{r^2} \tag{3.37}$$

where the value of the integral is $4R$. Since Eq. (3.37) is identical in form with Eq. (3.29), we should now be convinced of what was suspected all along: A spherical shell of mass m_n exerts a force of attraction upon a point mass external to the shell as though the mass of the shell were concentrated at its center. The same result applies to each of the shells comprising the sphere, so that the total force upon m_p due to m, the mass of the sphere, is

$$F_{pm} = \frac{Gm_p m}{r^2} \tag{3.38}$$

On the basis of the foregoing development, the force of gravity acting on a point mass is directed from the mass to the center of the earth. The relationship between mg and the law of gravity is now clear. For a mass subjected to the influence of the earth's mass,

$$m_p g = \frac{Gm_p m_e}{r^2} \tag{3.39}$$

or

$$g = \frac{Gm_e}{r^2} \tag{3.40}$$

where the effect of rotation has been neglected. In Eqs. (3.39) and (3.40) m_e represents the mass of the earth and r is the distance from the center of the earth to the mass.

Since laboratory experiments provide an accurate value of G, and g and r at the earth's surface are known, the mass of the earth may be calculated from Eq. (3.40). Taking $r = r_e = 6.37 \times 10^8$ cm, $g = 980$ cm/sec^2, and $G = 6.67 \times 10^{-8}$ dyn-cm^2/g^2, m_e is approximately 6.0×10^{27} g.

The influence of r upon g is given by

$$g = \frac{K}{r^2} \tag{3.41}$$

where

$$K = Gm_e \cong 4.0 \times 10^{20} \ (\text{cm/sec}^2)\text{cm}^2$$

The change in g with r is obtained by differentiating:

$$\frac{dg}{dr} = -\frac{2K}{r^3} \tag{3.42}$$

Making the appropriate substitutions in Eq. (3.42),

$$\frac{dg}{dr} = -3.1 \times 10^{-6} \ (\text{cm/sec}^2) \ /\text{cm above } r_e$$

Since the earth is not a perfect sphere, and not even a perfect oblate spheroid, it follows that g actually depends upon latitude and longitude in addition to altitude. In fact, a map of the gravitational field is quite complex, inasmuch as there are also influences such as mineral deposits and mountain ranges.

We now examine briefly the dependence of g upon latitude, saving for Chapter 9 a more elaborate explanation. Imagine a mass m to be suspended from a spring scale at the equator (latitude $\lambda = 0$ degrees). The difference between the weight (the upward supporting force of the spring) and the gravitational force is the unbalanced vertical force, which equals the product of mass and acceleration. The mass experiences centripetal acceleration because of the difference between F_{grav} and W:

$$F_{\text{grav}} - W = \frac{mv^2}{r_e} \tag{3.43}$$

where $F_{\text{grav}} = Gm_e m/r_e^2$, $W = mg$, and v represents the tangential velocity of the earth at the equator. Substituting for F_{grav} and W, and solving for g, we obtain

$$g_{\text{equator}} = \frac{Gm_e}{r_e^2} - \frac{v^2}{r_e} \tag{3.44}$$

At the poles, the second term on the right side of Eq. (3.44) is zero. The difference between g at the equator and at the poles is approximately 4.6 cm/sec² or 0.15 ft/sec². The following formula has been determined empirically:

$$g = 32.089(1 + 0.00524 \sin^2 \lambda)(1 - 0.000000096z) \ \text{ft/sec}^2 \tag{3.45}$$

where λ represents the latitude angle and z the altitude above sea level.

3.10 INTEGRATION OF NEWTON'S SECOND LAW

The least troublesome force resultant is one that is constant, regardless of the position of the mass or its derivatives. For example, consider a body in free fall, subjected to a gravitational force, assumed constant. If the retarding forces imposed by the air are disregarded, gravity is the only force acting. Let us determine the displacement, velocity, and acceleration of the body dropped z_0 above the earth with initial velocity v_{0z}. A convenient coordinate system is one in which vertical displacement is measured upward and zero corresponds to ground level. Applying Newton's second law,

$$\sum_{i=1}^{n} \mathbf{F}_i = -mg\hat{\mathbf{k}} \tag{3.46}$$

Therefore,

$$-mg\hat{\mathbf{k}} = m\frac{d^2}{dt^2}(\mathbf{r})$$

$$= m\frac{d^2}{dt^2}(x\hat{\mathbf{i}} + y\hat{\mathbf{j}} + z\hat{\mathbf{k}}) \tag{3.47}$$

Equating coefficients of $\hat{\mathbf{i}}$, $\hat{\mathbf{j}}$, $\hat{\mathbf{k}}$,

$$m\ddot{x} = 0 \tag{3.48a}$$

$$m\ddot{y} = 0 \tag{3.48b}$$

$$m\ddot{z} = -mg \tag{3.48c}$$

Since the X- and Y-accelerations are zero, and the initial X- and Y-velocities are zero, these velocities remain zero throughout. The final X- and Y-coordinates of the particle are what they were initially.

Integration of Eq. (3.48c) twice yields

$$\int \ddot{z}\, dt = \dot{z} = -gt + C_1 \tag{3.49}$$

and

$$\int \dot{z}\, dt = z = -\tfrac{1}{2}gt^2 + C_1 t + C_2 \tag{3.50}$$

Since we have begun with Newton's second law, a second-order differential equation in position, it follows that two integrations result in an equal number of constants, evaluated on the basis of known conditions. The constant C_1 is determined from the initial condition: at $t = 0$, $\dot{z} = v_{0z}$, $C_1 = v_{0z}$. The familiar result is thus

$$v_z = \dot{z} = -gt + v_{0z} \tag{3.51}$$

Similarly, $C_2 = z_0$, the initial Z-position with respect to the origin of the coordinate system. Thus

$$z = z_0 + v_{0z}t - \tfrac{1}{2}gt^2 \tag{3.52}$$

The foregoing is an example of *rectilinear* motion.

3.11 POSITION-DEPENDENT FORCES

A better approximation of the differential equation describing the free fall of a particle represents the force of gravity as varying inversely as the distance squared from the center of the earth. This is an example of a position-dependent force. The counterpart of Eq. (3.48c) is

$$m\ddot{z} = -\frac{Gm_e m}{r^2} \tag{3.53}$$

where z is the distance measured above the earth's surface (Fig. 3.6), and r is the distance measured from center of earth to the falling mass. Since z and r differ by a constant, r_e, the derivatives of z and r are identical and r may replace z in Eq. (3.53). In order to integrate the resulting differential equation of motion,

$$\frac{d^2r}{dt^2} = -\frac{Gm_e}{r^2} \tag{3.54}$$

consider the following identity:

$$\frac{d}{dt} = \frac{dr}{dt}\frac{d}{dr} = v\frac{d}{dr}$$

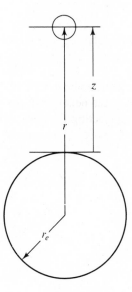

Fig. 3.6

Then

$$\frac{d^2r}{dt^2} = v\frac{dv}{dr} \tag{3.55}$$

and, therefore,

$$\int_{v_0}^{v} v\,dv = -\int_{r_0}^{r} \frac{Gm_e}{r^2}\,dr \tag{3.56}$$

where v_0 is the initial velocity, corresponding to r_0, the initial position of the mass relative to the center of the earth. Integrating Eq. (3.56) we have

$$\tfrac{1}{2}(v^2 - v_0^2) = Gm_e\left(\frac{1}{r} - \frac{1}{r_0}\right) \tag{3.57}$$

The problem of free fall is, in principle, identical with that of a projectile fired vertically upward. Equation (3.57) can, therefore, be employed to calculate an approximate velocity of escape from the influence of the earth's gravitational field. The initial velocity v_0 (occurring at $r = r_e$) is in this instance the velocity of escape. The boundary condition requires that as r grows without limit, the velocity becomes negligible. Substitution of this condition into Eq. (3.57) yields the minimum escape velocity,

$$v_{escape} = \left(\frac{2Gm_e}{r_e}\right)^{1/2} \tag{3.58}$$

or, since $g \cong Gm_e/r_e^2$,

$$v_{escape} = (2gr_e)^{1/2} \tag{3.59}$$

Employing the constants already noted, $v_{escape} = 11.2 \times 10^5$ cm/sec or 36,800 ft/sec.

Illustrative Example 3.2

A particle is projected up an inclined plane with an initial velocity v_0 of 100 cm/sec and an initial angle θ_0 of 135 degrees between the velocity and the line of maximum slope, as shown in Fig. 3.7. Neglecting friction, what is the particle velocity when θ has the values 90, 45, and 0 degrees? Use path coordinates.

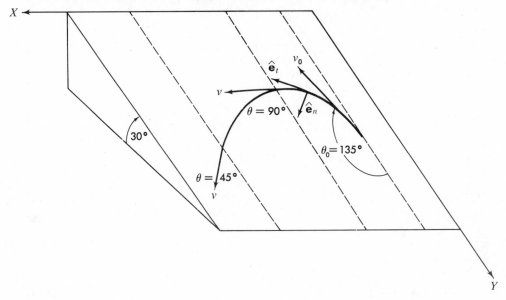

Fig. 3.7

SOLUTION

Motion occurs entirely in a single plane. The forces acting on the particle (gravity and the normal reaction) must be resolved into components F_t and F_n, tangent and normal to the path, in the plane of motion. No acceleration occurs normal to the plane of the incline, because the particle is in equilibrium in this direction. The motion tangent to the path is given by

$$F_t = mg \sin \beta \cos \theta = m\ddot{s} = m \frac{dv}{dt}$$

or

$$\frac{dv}{dt} = g \sin \beta \cos \theta$$

where β is the angle of the incline.

The motion normal to the path is given by

$$F_n = mg \sin \beta \sin \theta = \frac{m(\dot{s})^2}{\rho} = \frac{mv^2}{\rho}$$

or

$$\frac{v^2}{\rho} = g \sin \beta \sin \theta$$

Substituting

$$\frac{dv}{ds} \frac{ds}{dt} = v \frac{dv}{ds}$$

for dv/dt and $-d\theta/ds$ for $1/\rho$ (note the relative direction of change of s and θ to account for the minus sign), the equations of motion become

$$v \frac{dv}{ds} = g \sin \beta \cos \theta$$

$$v^2 \frac{d\theta}{ds} = -g \sin \beta \sin \theta$$

Dividing the first expression above by the second and separating variables,

$$\frac{dv}{v} = -\cot \theta \, d\theta$$

Integrating,

$$\ln v = -\ln \sin \theta + \ln C$$

or

$$v = \frac{C}{\sin \theta}$$

The constant of integration is determined by substituting the initial condition $\theta = \theta_0 = 135$ degrees, $v = v_0 = 100$ cm/sec. Thus $C = 70.7$ cm/sec, and the velocity at any θ is given by

$$v = \frac{70.7}{\sin \theta}$$

At $\theta = 90$ degrees, $v = 70.7$ cm/sec; at $\theta = 45$ degrees, $v = 100$ cm/sec. As θ approaches zero, corresponding to large distances of travel down the incline, the velocity grows without limit.

It is interesting to observe that the velocity component parallel to the top edge of the inclined plane remains constant, because no forces act in that direction:

$$v_x = \left(\frac{70.7}{\sin \theta}\right) \sin \theta = 70.7 \text{ cm/sec}$$

The Cartesian component along a line of maximum slope is

$$v_y = \left(\frac{70.7}{\sin \theta}\right) \cos \theta = \frac{70.7}{\tan \theta}$$

At $\theta = 90$ degrees, v_y is clearly zero, this angle representing the condition of maximum upward motion.

3.12 FORCE BETWEEN CHARGED PARTICLES

It is interesting to explore another law of attraction (or repulsion) which, like gravitation, depends inversely upon the square of a distance. Coulomb [5] deduced experimentally (1785), and it has been verified on firmer experimental grounds since, that a mutual force exists between charged objects (Fig. 3.8). If the intervening distance is large with respect to their size, the force in vacuo is expressed

$$\mathbf{F}_{12} = \frac{Kq_1 q_2}{r_{12}^2} \hat{\mathbf{r}}_{21} \tag{3.60}$$

where K is a constant that depends upon physical units and q_1 and q_2 are the charges.

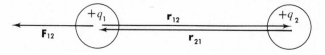

Fig. 3.8

If q_1 and q_2 are of like sign, the mutual force is repulsive, as indicated by the nomenclature $\mathbf{F}_{12}$ and $\hat{\mathbf{r}}_{21}$ in Eq. (3.60). In the rationalized MKS system,

$$K = \frac{1}{4\pi \epsilon_0}$$

A consistent set of units is one in which F is expressed in newtons, q in coulombs, and r in meters. The permittivity of free space ϵ_0 is 8.85415×10^{-12} C^2/N-m^2. In the unrationalized CGS system, $K = 1$ (dimensionless).

---

[5] Coulomb, C. A., *Mémoires de l'Académie Royales des Sciences* (1788).

A charged particle placed in an electric field is acted on by a force that depends upon the charge and the electric field strength,

$$\mathbf{F} = q\mathbf{E} \tag{3.61}$$

where, in electrostatics, $\mathbf{E}$ is a function of position. This equation is applied in Chapter 7 in studying the motion of electrons subjected to the influence of electric fields in a cathode-ray tube.

3.13 RESTORING FORCES

The gravitational and electrostatic forces have already been cited as examples of forces that depend upon position or distance. Another case is the force attributable to the action of a spring, as shown in Fig. 3.9. If the spring is "linear," the force

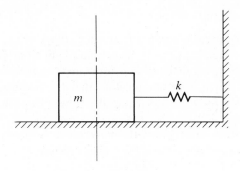

Fig. 3.9

required to extend or compress it varies linearly with the amount by which it is deformed relative to a neutral or undeformed position. The force $\mathbf{F}_s$ required to deform the spring must act in the direction of the spring displacement. If the X-axis is selected as the spring axis,

$$\mathbf{F}_s = kx\hat{\mathbf{i}} \tag{3.62}$$

where k is known as the spring constant (units of dyn/cm, N/m, lb/ft, and so on), and x is the total displacement of the spring and may be positive or negative. The force which the spring exerts upon the mass must, according to Newton's third law, be equal and opposite to that deforming the spring. Thus

$$\mathbf{F} = -\mathbf{F}_s = -kx\hat{\mathbf{i}} \tag{3.63}$$

This expression may be easily rationalized on the basis of experience. For displacement of the mass to the right, the spring is compressed. The force on the mass is directed toward the left, opposite the direction of the displacement. Similarly, a displacement to the left causes a right-directed force to act on the mass.

In the Y-direction, the normal and gravitational forces cancel, and there is a zero resultant. In the absence of friction, the spring force is the only X-force acting. The differential equation of motion is thus

$$-kx\hat{\mathbf{i}} = m\ddot{x}\hat{\mathbf{i}} \tag{3.64}$$

or

$$\ddot{x} + \left(\frac{k}{m}\right)x = 0 \tag{3.65}$$

Equation (3.65), as we explore in Chapter 6, has a solution:

$$x = x_0 \cos \omega_0 t + \left(\frac{v_0}{\omega_0}\right)\sin \omega_0 t \tag{3.66}$$

The velocity is therefore

$$\dot{x} = -x_0\omega_0 \sin \omega_0 t + v_0 \cos \omega_0 t \tag{3.67}$$

where $\omega_0 = (k/m)^{\frac{1}{2}}$, the natural circular frequency (rad/sec), x_0 represents the displacement, and v_0 the velocity at $t = 0$. Equation (3.66) describes a case of simple harmonic motion. The spring tends to return the mass to the equilibrium position, $x = 0$ in this instance, for it is at this position that the force is zero. The spring thus provides a *restoring force*, and, because it is linear in position, it is termed a *linear restoring force*.

Once displaced from equilibrium, the mass oscillates indefinitely about $x = 0$, as there is no dissipation of energy in a frictionless system.

3.14 TIME-DEPENDENT FORCES

Applications exist in which it is possible to express force as an explicit function of time. Often, as in the case of impulsive forces, time-dependent representations are more convenient than far more complex alternatives.

For the one-dimensional case, $F_x = F_x(t)$. The acceleration is, therefore,

$$\ddot{x} = a_x(t) = \frac{F_x(t)}{m}$$

The velocity and position are obtained by successive integrations with respect to time:

$$\dot{x} = v_x(t) = \int_{t_0}^{t} \frac{F_x(t')\, dt'}{m} + C_1 \tag{3.68}$$

or

$$v_x(t) = \int_{t_0}^{t} a_x(t')\, dt' + C_1 \tag{3.69}$$

and

$$x(t) = \int_{t_0}^{t} \left[\int_{t_0}^{t'} \frac{F_x(t'') \, dt''}{m} + C_1 \right] dt' + C_2 \qquad (3.70)$$

or

$$x(t) = \int_{t_0}^{t} v_x(t') \, dt' + C_2 \qquad (3.71)$$

where t' and t'' are dummy variables, retaining t as the remaining variable of the desired equations.

Illustrative Example 3.3

A particle of mass m is acted upon by a one-dimensional force $F_0 \sin (\omega t - \phi)$, where F_0, ω, and ϕ represent constants. If the initial position and velocity are zero, derive equations for position and velocity for all positive t.

SOLUTION

The differential equation of motion, in the absence of other forces, is

$$F_0 \sin (\omega t - \phi) = m \ddot{x}$$

so that the velocity may be obtained through a first time integration:

$$\dot{x} = v_x(t) = \int_{t_0}^{t} \frac{F_0}{m} \sin (\omega t' - \phi) \, dt' + C_1$$

$$= -\frac{F_0}{m\omega} [\cos (\omega t - \phi) - \cos (\omega t_0 - \phi)] + C_1$$

Taking the initial time, t_0, to be zero,

$$v_x(t) = -\frac{F_0}{m\omega} [\cos (\omega t - \phi) - \cos \phi] + C_1$$

where we have used $\cos (-\phi) = \cos \phi$.

The constant of integration, C_1, is determined by substituting the condition that at $t = 0$, $v_x = 0$. This yields $C_1 = 0$, and the final expression for velocity is thus

$$v_x(t) = \frac{F_0}{m\omega} [\cos \phi - \cos (\omega t - \phi)]$$

A second integration gives the displacement,

$$x(t) = \int_{0}^{t} \frac{F_0}{m\omega} [\cos \phi - \cos (\omega t' - \phi)] \, dt' + C_2$$

$$= \frac{F_0}{m\omega^2} [\omega t \cos \phi - \sin (\omega t - \phi) - \sin \phi] + C_2$$

where we have used $\sin (-\phi) = -\sin \phi$. Introducing the initial condition, at $t = 0$, $x = 0$: $C_2 = 0$.

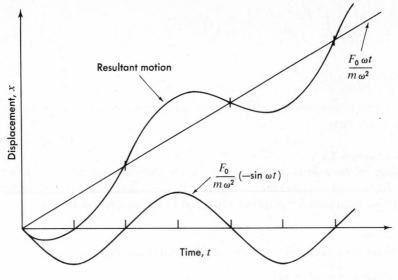

Fig. 3.10

Thus $x(t)$ contains a term linear in time in addition to the sinusoidal and constant terms. Because of the term linear in t, motion does not repeat itself about a fixed point, as shown in Fig. 3.10 for $\phi = 0$.

3.15 IMPULSIVE FORCES

It is sometimes convenient to represent an impulsive force, that is, one of short duration, as entirely dependent upon time. Such a force may be produced by a quick blow upon a mass by a hammer or by an explosive charge. In determining the form of the describing equation, bear in mind that our purpose will be best served by a smooth function displaying a rapid rise, a distinct maximum, and a rapid decay. Consider, for example, the force given by

$$F(t) = ate^{-bt} \qquad (t \geq 0) \tag{3.72}$$

The maximum force is obtained by setting $dF/dt = 0$,

$$\frac{dF}{dt} = 0 = -abte^{-bt} + ae^{-bt} \tag{3.73}$$

giving $t_{max} = 1/b$ and $F_{max} = a/be$.

The reader will note that as t grows, Eq. (3.72) does not predict a zero force but rather an asymptotic approach to zero. For large b it is possible to have a rapid decay after F_{max}. Equation (3.72) may therefore be employed to *approximately* match a given impulsive force.

It is interesting to explore the velocity characteristics resulting from an impulsive force thus described. The differential equation of motion is

$$ate^{-bt} = m\ddot{x} \tag{3.74}$$

and the velocity is therefore

$$v(t) = \int_0^t \frac{at'e^{-bt'}}{m} dt' + C_1 \tag{3.75}$$

or

$$v(t) = \frac{ae^{-bt}}{mb^2}(-bt - 1) + \frac{a}{mb^2} + C_1 \tag{3.76}$$

As t grows without bound, the limiting velocity becomes

$$v_{\lim} = \frac{a}{mb^2} + C_1 \tag{3.77}$$

where C_1 represents the initial velocity.

3.16 FRICTION

Frictional forces act at surfaces of contact to impede or prevent relative motion and are therefore opposite the direction of motion or the direction in which motion tends. There are many factors, chemical as well as physical, influencing friction and related phenomena such as wear.[6]

No surface is perfectly smooth but is characterized by waviness as well as the presence of asperities. When two materials are brought together, therefore, the significant contact area is not the gross surface common to the bodies but the total of all the local areas of contact. When the surfaces are pressed together, it is found that the areas of contact increase until the total force causing contact is sustained. Thus friction is related to the force required to overcome the local interlocking and adhesion of the surface irregularities. In order to initiate motion, a force greater than the *static* friction force must be supplied.

Once motion occurs, we are concerned with *kinetic* friction, which is related to the ease of sliding. With sliding, there are high local temperatures produced at the surfaces of contact. As might be expected, material softening and welding may occur, even for moderate sliding velocities. The temperatures attained are related to the rate of heat removal and hence the thermal conductivity of the materials.

In the sliding of metals, therefore, friction is a function of the force required to interrupt or shear the local junctions of adhesion and welding. Once again, the

[6] Bowden, F. P., and Tabor, D., *The Friction and Lubrication of Solids*, Oxford University Press, New York, 1950.

total junction area or true area is approximately independent of the gross area and more or less proportional to the contact force.

Other variables include the relative thickness of any lubricants that may be present, as well as adsorbed films of gases or oxides that may exist. If the lubricating films are thick, friction is principally related to hydrodynamic or fluid mechanical phenomena. With thin films of lubricants present, there can be penetration by the asperities, and localized adhesion can occur. Friction obviously constitutes an area of challenging research activity.

Despite the lack of satisfaction that one may feel regarding the state of knowledge, there are at least three compelling reasons to employ very simple postulates governing friction:

a. The role of friction is often not so significant as to warrant great complexities in mathematical formulation.
b. Considerable uncertainty exists in the experimental data.
c. Complex expressions for frictional forces usually lead to differential equations of motion which resist simple solution.

Consider a block of weight W resting on a "flat" surface as shown in the free-body diagram of Fig. 3.11. If a small horizontal force P is applied, we know from everyday experience that the block does not move; a frictional force f develops

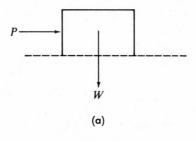

W

(a)

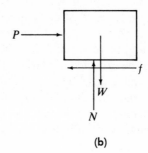

(b)

Fig. 3.11

as a consequence of the effort to move the block. Increasing the magnitude of P still causes no motion and we conclude that the friction force increases by an amount sufficient to maintain a state of equilibrium. Eventually a value of P is reached which causes motion of the block, and we discover that the magnitude of P required to maintain a low velocity is less than that required to initiate motion.

The "laws" governing friction between dry surfaces (Coulomb friction[7]) apply approximately to the situation described above:

a. The frictional force does not depend upon the gross contact area.
b. The frictional force is proportional to the force N normal to the contacting surface.
c. The frictional force does not depend upon the relative velocity between the surfaces (sliding velocity) provided the velocity is low.
d. For low velocities, the friction is lower than for a state of impending motion.
e. The frictional force exists because of P and therefore cannot exceed a value required to maintain the body in a state of equilibrium.

The second statement above may be written

$$f = \mu N \tag{3.78}$$

or, in terms of a definition,

$$\mu = \frac{f}{N} \tag{3.79}$$

where μ is the *coefficient of friction*. Whenever the friction and normal forces are known, the coefficient of friction (which may exceed unity) may be calculated. A subscript s applied to μ pertains to static friction and refers to a state of impending motion. The frictional force calculated on the basis of μ_s and N is the maximum frictional force before motion is initiated. A subscript k or d pertains to sliding, kinetic, or dynamic friction. The coefficient of sliding friction is often treated as a constant over a limited range of velocity.

Illustrative Example 3.4

The coefficient of kinetic friction is determined to be a function of velocity given by Fig. 3.12. Determine, as a function of time, the velocity of a 10-g mass acted upon by a horizontal force of 2.94×10^3 dyn.

---

[7] Coulomb, C. A., "Théorie des machines simples," *Mémoires de l'Académie Royales des Sciences* (1785).

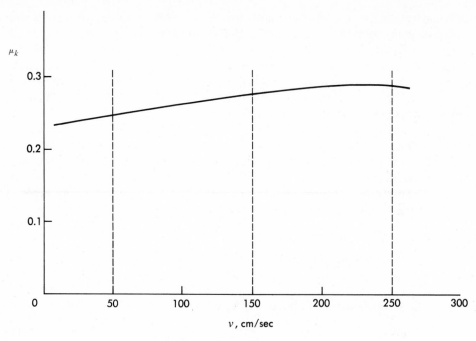

Fig. 3.12

SOLUTION

Since the only applied force is horizontal, $N = mg$. The X-equation is, therefore,

$$F - \mu_k N = ma_x$$

Substituting $N = mg$ and solving for a_x,

$$a_x = \frac{F}{m} - \mu_k g = 294 - \mu_k(980)$$

Since we lack a mathematical expression, $\mu_k = f(v)$, we shall assume that μ_k remains constant over a limited range of velocity. If the selected ranges prove too large, they may be reduced. For the ranges

$$0 < v_x < 100 \text{ cm/sec}; \quad \mu_k = 0.25 \text{ and } a_x = 49 \text{ cm/sec}^2$$

$$100 < v_x < 200 \text{ cm/sec}; \quad \mu_k = 0.28 \text{ and } a_x = 20 \text{ cm/sec}^2$$

$$200 < v_x < 300 \text{ cm/sec}; \quad \mu_k = 0.29 \text{ and } a_x = 10 \text{ cm/sec}^2$$

The time elapsed for the first velocity interval is obtained from $v_x = v_{0x} + a_x t$. Since $v_{0x} = 0$, $t = 100/49 = 2.04$ sec. The time elapsed during the second velocity interval is given by

$$t = \frac{v_x - v_{0x}}{a_x} = \frac{200 - 100}{20} = 5.0 \text{ sec}$$

where $v_{0x} = 100$ cm/sec represents the initial velocity of the second interval. Similarly, for the third interval, $t = 10.0$ sec. Plotting the results obtained (Fig. 3.13), we observe that because of the increasing frictional resistance with increasing velocity, the velocity–time curve appears to approach a limiting value.

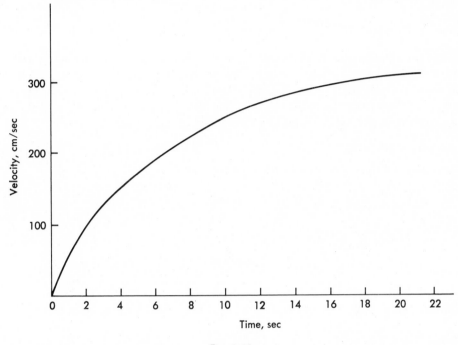

Fig. 3.13

When this *terminal velocity* is reached, the external force tending to increase the velocity is exactly balanced by the frictional force, so that the net X-force is zero. A similar phenomenon occurs when an object is acted upon simultaneously by a gravitational force and a force due to fluid friction. We are thus led to yet another type of frictional effect. The retarding force can be shown experimentally and analytically to depend upon many variables, including fluid properties (for example, density, viscosity) and the geometry and velocity of the object. For very low velocities and relatively constant fluid properties, the force of fluid resistance takes the form

$$\mathbf{F} = -b\mathbf{v} \tag{3.80}$$

where the negative sign indicates that the force acts opposite the velocity vector. At higher velocities,

$$\mathbf{F} = -cv^2\hat{\mathbf{v}} \tag{3.81}$$

where $\hat{\mathbf{v}}$ is a unit vector in the direction of $\mathbf{v}$. In some cases, the resistance may vary as the velocity to some power between 1 and 2. The above expressions are obviously approximate, but generally any desired degree of accuracy may be obtained provided one is prepared to solve, numerically or otherwise, the complex differential equations of motion which result. In Chapter 7, such motions are explored further.

3.17 FORCE DUE TO MAGNETIC FIELD

In the previous section we discussed the velocity-dependent force acting on a mass traveling through a resisting medium. Such retarding forces were observed to be colinear (and opposite) to the velocity. The force on a charged particle of velocity **v** passing through a magnetic induction field **B** is perpendicular to the velocity and is governed (in the MKS system) by the following expression:[8]

$$\mathbf{F} = q(\mathbf{v} \times \mathbf{B}) \tag{3.82}$$

This matter is treated further in Chapter 7.

References

Becker, R. A., *Introduction to Theoretical Mechanics*, McGraw-Hill, Inc., New York, 1954.

Lindsay, R. B., *Physical Mechanics*, Third Edition, D. Van Nostrand Company, Inc., Princeton, N.J., 1961.

Resnick, R., and Halliday, D., *Physics*, Part 1, John Wiley & Sons, Inc., New York, 1966.

Sommerfeld, A., *Mechanics*, Translated by M. O. Stern, Academic Press, Inc., New York, 1957.

Synge, J. L., and Griffith, B. A., *Principles of Mechanics*, Third Edition, McGraw-Hill, Inc., New York, 1959.

Taylor, E. F., *Introductory Mechanics*, John Wiley & Sons, Inc., New York, 1964.

EXERCISES

3.1 A mass m is imparted an initial velocity v_0 up a frictionless inclined plane making an angle α with the horizontal. Derive an expression for the time required for the mass to stop. Begin with a free-body diagram.

---

[8] Lorentz, H. A., *Archives Néerlandaises XXI* (1887).

3.2 Amend the derivation for Exercise 3.1 to include a constant coefficient of friction μ_k.

3.3 An object is thrown vertically upward with an initial velocity of 2 m/sec. Determine:

    a. the time required to reach the maximum height.
    b. the maximum height.
    c. the velocity midway up and midway down.

3.4 A man who normally weighs 200 lb enters an elevator normally weighing 1000 lb. Determine the force exerted by the man on the floor of the elevator and the tension in the cable supporting the elevator during the following phases of upward motion:

    a. acceleration of 5 ft/sec^2.
    b. constant velocity of 10 ft/sec.
    c. deceleration of 3 ft/sec^2.

3.5 Using your same selection for positive displacement, repeat Exercise 3.4 for the downward phase:

    a. acceleration of 5 ft/sec^2.
    b. constant velocity of 10 ft/sec.
    c. deceleration of 3 ft/sec^2.

3.6 Given a constant coefficient of friction μ between a rubber tire and a roadway, determine the maximum acceleration an automobile may experience.

3.7 What is the shortest stopping distance of an automobile traveling at 60 mph if the coefficient of friction between the tires and roadway is 0.7 and the reaction time of the driver is 0.5 sec. What factors would you take into account if a more accurate calculation were required?

3.8 A roadway contains a bend of radius R. Determine the minimum coefficient of friction required between the tires and the pavement if a turn is to be negotiated without skidding at v_0. Take the roadway to be unbanked.

3.9 Referring to Exercise 3.8, if there were essentially no friction, what bank angle θ would be required to prevent skidding?

3.10 If an automobile maintains a constant speed of 80 ft/sec around a banked curve of radius 200 ft, at what angle must the roadway be banked to avoid skidding if the coefficient of friction $\mu_s = 0.5$?

3.11 A small object is in equilibrium as shown in Fig. P3.11. If its weight is 10 lb, calculate the force in the cord and the normal force exerted by the incline on the mass. Neglect friction.

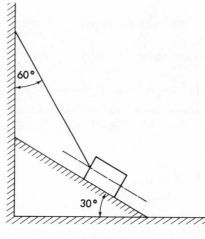

Fig. P3.11

3.12 What coefficient of static friction between the object and the inclined plane causes the tension in the cord to equal one half the value found in Exericse 3.11?

3.13 Two spheres each weighing 10 lb are stacked as shown in Fig. P3.13. For the geometry and weights given, determine all the contact forces. Neglect friction.

3.14 Compute the value of the gravitational acceleration on the moon, assuming it to be composed of material similar to that of the earth.

3.15 Determine the escape velocity from the moon.

3.16 A mass m is supported by two wires of length l (Fig. P3.16). The wires are connected to rings that ride over a horizontal rod. If the coefficient of static friction between the rod and ring is μ_s determine the maximum separation S of the rings before they begin to slip.

3.17 A 10-g mass moves at constant speed $\dot{s} = 100$ cm/sec along a path shown in Fig. P3.17, the motion occurring in a plane. If the radius of curvature is 1000 cm, determine the value of θ at which the normal force of the path upon the object vanishes. For $\theta = 80$ degrees, what are the components of force tangent and normal to the path required to maintain the motion as described?

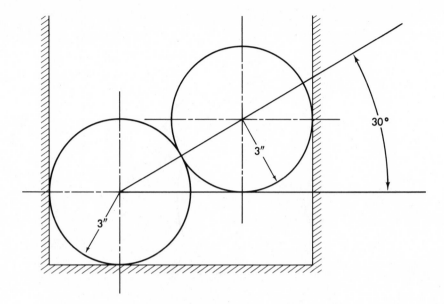

Fig. P3.13

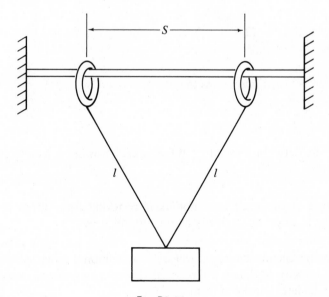

Fig. P3.16

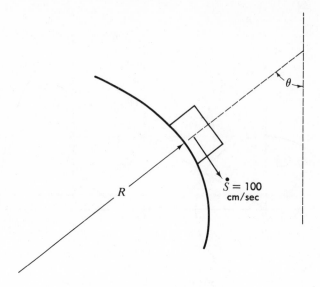

Fig. P3.17

3.18 A 12-g mass is acted upon by a constant force given by $\mathbf{F} = 12\hat{\mathbf{i}} + 22\hat{\mathbf{j}} - 6\hat{\mathbf{k}}$ dyn. If the initial position is $\mathbf{r} = 5\hat{\mathbf{i}} - 2\hat{\mathbf{j}}$ cm and the initial velocity is $2\hat{\mathbf{j}}$ cm/sec, determine the components of acceleration, velocity, and displacement at $t = 5$ sec.

3.19 Using Eq. (3.57), demonstrate that the velocity with which an object strikes the earth is given by $v = (2gz)^{\frac{1}{2}}$ if the initial velocity is zero and z is the height from which it falls.

3.20 The force between two atoms may be expressed by

$$F = -\frac{A}{r^7} + \frac{B}{r^n}$$

where n may lie between 9 and 15. What is the position as a function of time of an atom of mass m for $n = 13$?

3.21 A particle is subjected to a nonlinear restoring force given by $F = -\alpha x^3$. Determine the position and velocity as function of time.

3.22 A position-dependent force $F = F_0 e^{-\beta x}$ acts upon a mass m. Determine the position and velocity at time τ if the initial values are x_0 and v_0, respectively. The solution may be left in integral form.

3.23 Fig. P3.23 describes a time-dependent force acting upon a 2-g mass. Determine the position and velocity at $t = 10$ sec if the initial position and velocity are zero.

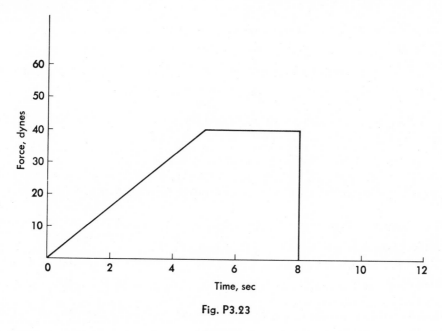

Fig. P3.23

3.24 Repeat Exercise 3.23 given an initial position of 100 cm and an initial velocity of 30 cm/sec.

3.25 Prove that the asymptotic velocity of a particle acted on by an impulsive force $F = ate^{-bt}$ is also the maximum velocity.

3.26 Derive an expression for particle displacement resulting from the impulsive force given in Exercise 3.25. Assume zero initial conditions.

3.27 A block of mass m, possessing velocity v_0, is acted upon by a one-dimensional time-dependent force, $F(t) = F_0(1 + \beta t)$. Determine the distance traveled and the velocity at $t = \tau$.

3.28 Observations made of a particle moving through a viscous medium indicate that the velocity is given by $v(t) = v_0 e^{-\alpha t}$, where v_0 represents the initial velocity. Determine the distance traveled at $t = \tau$. Prove that the velocity is linearly dependent upon distance. What is the maximum distance traveled?

3.29 An impulsive force is represented by the equation

$$F(t) = F_0(e^{-a_1 t} - e^{-a_2 t})$$

where a_1 and a_2 are positive constants. Determine:

    a. the maximum force.
    b. the velocity of a mass m if the initial velocity is v_0.
    c. the displacement of the mass for an initial displacement x_0.
    d. the average force acting over the time period τ.

Compare the displacement and velocity resulting from the average force of part (d) with that caused by the actual force acting over the same time interval.

3.30 A 2-kg mass experiences an impulsive force given by $F(t) = 500te^{-100t}$ N, where t is the time in seconds. Determine:

    a. the time at which the maximum force occurs.
    b. the velocity at F_{max} if the objects start from rest.
    c. the distance traveled at 90 percent of the limiting velocity.

3.31 A particle of mass m is acted upon by a time-dependent force $F(t) = Ate^{-\alpha t^2}$.

    a. Determine $v(t)$.
    b. For an initial velocity $v_0 = -A/2m\alpha$, determine the maximum particle displacement.

3.32 An electron of charge e moves in a horizontal plane with velocity v. A vertical magnetic induction field B is applied perpendicular to the plane of motion. Demonstrate that the electron will experience constant-speed circular motion of radius $R = mv/eB$.

3.33 Repeat Illustrative Example 3.2 assuming a constant coefficient of sliding friction of 0.2.

CHAPTER 4

IMPULSE AND MOMENTUM— THE PARTICLE

4.1 INTRODUCTION

In this chapter treatment is given to the time integral of Newton's second law. We are thus led from a vector relationship between force and acceleration to another vector equation connecting the time integral of force to changes in velocity. Even when the force is not a known function of time, much can often be learned of its average magnitude or duration on the basis of the associated changes in momentum.

4.2 LINEAR IMPULSE AND LINEAR MOMENTUM

In Chapter 3 the momentum form of the second law was written as a first-order ordinary differential equation

$$\mathbf{F} = \frac{d}{dt}(m\mathbf{v}) = \frac{d\mathbf{p}}{dt} \tag{4.1}$$

where $\mathbf{F}$ represents the sum of all forces acting upon a point mass. Multiplying Eq. (4.1) by dt,

$$\mathbf{F}\,dt = d(m\mathbf{v}) = d\mathbf{p} \tag{4.2}$$

Thus a force $\mathbf{F}$ of duration dt produces a change in linear momentum, $d(m\mathbf{v})$. Integrating Eq. (4.2) from initial time t_0 (when the velocity is $\mathbf{v}_0$) to final time t,

$$\int_{t_0}^{t} \mathbf{F}\,dt' = m\mathbf{v} - m\mathbf{v}_0 \tag{4.3}$$

where the definite integral is termed the *linear impulse*, and the right side of Eq. (4.3) is the change of linear momentum. Equation (4.3), referred to as the equation of impulse and momentum, clearly demonstrates that it is the *change* in linear momentum, which is in the same direction as the impulse.

The impulse–momentum equation may be written in terms of three scalar components.   For a Cartesian frame of reference,

$$\int_{t_0}^{t} F_x \, dt' = mv_x - mv_{0x} \qquad \text{etc.} \tag{4.4}$$

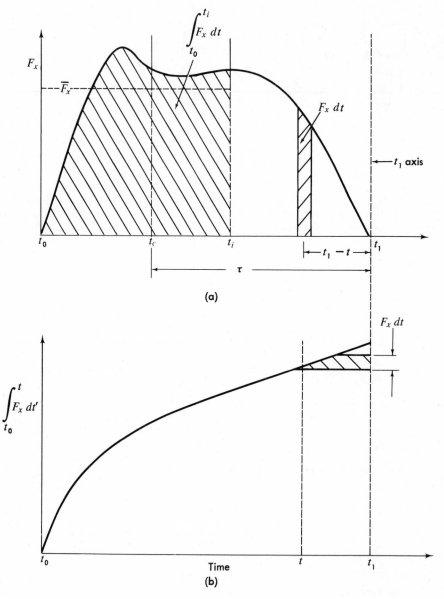

Fig. 4.1

One such integral is represented by the shaded area beneath the force–time curve as shown in Fig. 4.1(a).

The integral may also be expressed as the product of an average force and the time interval Δt over which it acts. For example,

$$\int_{t_0}^{t_i} F_x \, dt = \bar{F}_x \, \Delta t \tag{4.5}$$

where $\Delta t = t_i - t_0$ and $\bar{F}_x$ is the average X-force.

The impulse–momentum approach to problems in dynamics has its most important application where the force is a known function of time, or the average force for the time interval is known, where it is desirable to connect the force acting directly with the particle velocity, and where, even though $F(t)$ is unknown, the impulse may nevertheless be determined.

Illustrative Example 4.1

The force–time curve for a chemical rocket is given in Fig. 4.2, with the rectangular areas used to approximate the actual area. Assuming the total mass to remain constant, determine the velocity of the rocket at burnout (t_b) if its initial velocity is zero and the rocket is directed vertically upward.

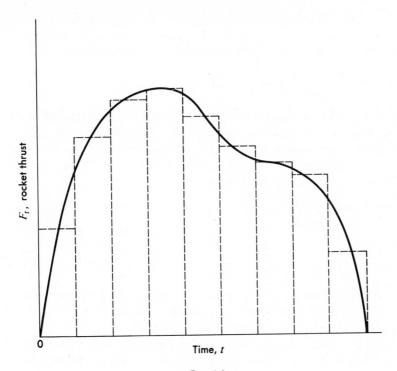

Fig. 4.2

SOLUTION

There are two principal contributors to the total impulse, neglecting air drag: the propulsive force and the gravitational attraction of the earth. The equation of impulse and momentum including both influences is

$$\int_0^{t_b} F_t \, dt + \int_0^{t_b} (-mg) \, dt = mv_b - 0$$

where g has been assumed constant and the first integral on the left side above is the area under the curve of Fig. 4.2. Solving for v_b,

$$v_b = \frac{\int_0^{t_b} F_t \, dt - mgt_b}{m}$$

In the case of variable mass, Eq. (4.3) can be employed for each time interval in Fig. 4.2 by assuming F_t and m constant for a given interval. The velocity thus determined at the end of one interval represents the initial velocity for the next. If the thrust and mass are expressible as reasonably simple mathematical functions of time, the impulse may often be analytically determined, as is done in Chapter 11.

4.3 IMPULSE DUE TO IMPULSIVE FORCE

In Chapter 3 the time-dependent force $F = ate^{-bt}$ was offered as one that could serve in many instances to describe impulsive (short-duration) loading. The linear impulse resulting from such a force acting from $t = 0$ to some time t is

$$\int_0^t at'e^{-bt'} \, dt' = \frac{ae^{-bt}}{b^2}(-bt - 1) + \frac{a}{b^2} \tag{4.6}$$

Equation (4.6) is equal to the change in linear momentum of the particle, $mv - mv_0$, where v_0 represents the initial velocity. It is characteristic of the impulse thus described to approach a limiting value. As the reader will recall, this impulsive force leads to a limiting velocity for large t. From Eq. (4.6), as t grows without limit, the change of linear momentum is

$$(mv - mv_0)_{\text{lim}} = \frac{a}{b^2} \tag{4.7a}$$

or

$$v_{\text{lim}} = \frac{a}{mb^2} + v_0 \tag{4.7b}$$

which is identical with Eq. (3.77). That limiting values of impulse and velocity occur follows directly from the fact that $F(t)$ asymptotically approaches zero, offering less and less force to accelerate the particle.

The form of Eq. (4.6) is such that the impulse is zero when the interval of time over which $F(t)$ acts is zero. This formulation, although perhaps adequate to describe certain reasonably short processes, does not suffice for those impulsive forces in which the force is extremely high and the time interval extremely short. Here we have the interesting case of the impulse integral possessing a finite magnitude, despite the fact that the associated force may approach infinite values and the time interval zero. This is truly an impulsive force. For the impulse to be evaluated, one need not know the force as a function of time but the change in linear momentum of the particle. Should information be available (from experimental data) as to the duration of contact between the mass and that which is causing the force, it is clear from Eq. (4.5) that an average force can be calculated. Similarly, if the average force can be deduced, as, for example, from examination of the deformation of the mass (clearly not a particle concept), some estimate of the time of contact may be made. From the foregoing discussion it is evident that although the principle of impulse and momentum has great utility when force is a known function of time, it is by no means limited to such applications.

As a corollary to the discussion above concerning the action of large forces over zero time intervals, we note that the associated velocity must, if this is the model selected, undergo a step change and therefore represent a discontinuous function of time. The change of position that occurs during the impulse is nevertheless zero, and, as a consequence, the position–time curve is a continuous function.

In summary, from a knowledge of the total impulse, any "mathematical" force may be constructed which is negligible, except during a very short time interval. One should, of course, make every effort to select force functions which are easily treated.

4.4 CONSERVATION OF LINEAR MOMENTUM

If the force acting on a particle is zero, the linear impulse is also zero. Consequently the linear momentum remains constant throughout the time interval in question. We are thus led to one of the important conservation laws of physics: *The linear momentum of a particle is conserved (remains constant) if zero net force acts upon it.*

In equation form, when

$$\sum \mathbf{F} = 0, \; \mathbf{p} = \text{constant vector} \tag{4.8}$$

Linear momentum may be conserved in one direction and not another. For example, if $F_x = F_y = 0$ and $F_z \neq 0$,

$$p_x = p_{0x} \qquad p_y = p_{0y} \qquad p_z \neq p_{0z}$$

Illustrative Example 4.2
It is found that it takes approximately 0.2 sec for a 4 kg mass, given an initial velocity up an incline of 2 m/sec, to stop. If the angle of the incline is 30 degrees with the horizontal, determine the average coefficient of sliding friction. Refer to Fig. 4.3.

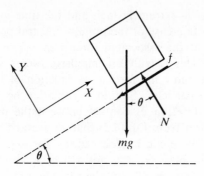

Fig. 4.3

SOLUTION

Applying the equation of impulse and momentum in the direction parallel to the incline as shown,

$$\int_0^t (-\mu_k mg \cos \theta - mg \sin \theta) \, dt' = mv_x - mv_{0x}$$

Since an average μ_k is sought, we shall regard it as a constant. The integral thus becomes

$$m(-\mu_k g \cos \theta - g \sin \theta) \, \Delta t = m(v_x - v_{0x})$$

Solving for μ_k,

$$\mu_k = \frac{v_{0x} - v_x}{g(\cos \theta) \, \Delta t} - \tan \theta$$

For the data given $\mu_k \cong 0.6$.

4.5 PARTICLE DISPLACEMENT

Let us consider the relationship between impulse and particle displacement. Solving Eq. (4.4) for velocity, treating motion in one direction only, we have

$$v_x - v_{0x} = \frac{1}{m} \int_{t_0}^t F_x \, dt' \tag{4.9}$$

In order to determine the X-position, Eq. (4.9) is multiplied by dt and a second integration performed:

$$\int_{t_0}^{t_1} v_x \, dt - v_{0x} \int_{t_0}^{t_1} dt = \int_{t_0}^{t_1} dt \left(\frac{1}{m} \int_{t_0}^t F_x \, dt' \right) \tag{4.10}$$

Thus

$$x_1 - x_0 = v_{0x}(t_1 - t_0) + \int_{t_0}^{t_1} dt \left(\frac{1}{m} \int_{t_0}^t F_x \, dt' \right) \tag{4.11}$$

where x_0 represents the particle position at $t = t_0$.

The geometric interpretation of the impulse integral is simply the area beneath the force–time curve, as already noted. Similarly, the double integral

$$\int_{t_0}^{t_1} dt \left(\int_{t_0}^{t} F_x \, dt' \right)$$

may be represented as the area beneath a curve of impulse as a function of time, is shown in Fig. 4.1(b). On this plot, the small shaded area shown is $(F_x \, dt)(t_1 - t)$, so that the total area is given by

$$\int_{t_0}^{t_1} F_x(t_1 - t) \, dt$$

But this is, of course, equal to the double integral discussed above, evaluated between the same limits. The quantity $(F_x \, dt)(t_1 - t)$ may also be interpreted as the moment of the shaded area $F_x \, dt$ in Fig. 4.1(a) taken about the t_1-axis. The double integral may thus be viewed as the summation of all the moments of $F_x \, dt$ about t_1. If, therefore, the location of the centroid of the area $\int_{t_0}^{t_1} F_x \, dt$ is known, the double integral may be easily evaluated,

$$\int_{t_0}^{t_1} dt \left(\int_{t_0}^{t} F_x \, dt' \right) = \tau \int_{t_0}^{t_1} F_x \, dt \tag{4.12}$$

where $\tau = t_1 - t_0$, as shown in Fig. 4.1(a). Here t_c represents the position of the centroid of the impulse.

Illustrative Example 4.3
Figure 4.4 describes an approximate force–time characteristic of a rocket. Determine the position of the projectile at t_1 if the initial velocity and position are zero.

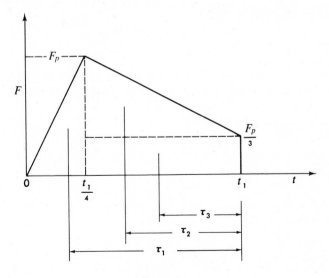

Fig. 4.4

SOLUTION

We apply Eq. (4.11) with v_{0x} and x_0 both zero. The term $\tau \int_0^{t_1} F_x \, dt$ [Eq. (4.12)] may be regarded as composed of three parts corresponding to the three simple geometric shapes shown. Thus

$$\int_0^{t_1} dt \left(\int_0^t F_x \, dt' \right) = \tau_1 \int_0^{t_1/4} F \, dt + \tau_2 \int_0^{t_1} F \, dt + \tau_3 \int_0^{t_1} F \, dt$$

where

$$\tau_1 = \frac{3}{4} t_1 + \frac{1}{3} \frac{t_1}{4} = \frac{5 t_1}{6} \text{ and the corresponding area is } \frac{F_p t_1}{8}$$

$$\tau_2 = \frac{2}{3} \left(\frac{3}{4} t_1 \right) = \frac{t_1}{2} \text{ and the corresponding area is } \frac{3}{8} t_1 \left(\frac{2}{3} F_p \right)$$

$$\tau_3 = \frac{1}{2} \left(\frac{3}{4} t_1 \right) = \frac{3 t_1}{8} \text{ and the corresponding area is } \frac{3}{4} t_1 \left(\frac{F_p}{3} \right)$$

Therefore,

$$x = \frac{31}{96} \frac{F_p t_1^2}{m}$$

Illustrative Example 4.4

A constant force of 2 N acts on a mass of 1 kg beginning at time $t_0 = 3$ sec, when the mass has a velocity of 12 m/sec in the direction of the force. Determine the position of the particle at $t = 10$ sec, given an initial position $x_0 = 5$ m.

SOLUTION

We have a direct application of Eq. (4.11):

$$x = 5 + (12)(10 - 3) + \int_3^{10} dt \left(\frac{1}{1} \int_3^t 2 \, dt' \right) = 138 \text{ m}$$

Note that time need not begin at zero. Had the derivative form of the second law been employed, the solution would have begun with $F = m\ddot{x}$, from which

$$\ddot{x} = \frac{F}{m} = 2 \text{ m/sec}^2$$

Integrating,

$$v = \dot{x} = 2t + C_1$$

The initial condition, $v_0 = 12$ at $t_0 = 3$, yields $C_1 = 6$ m/sec. A second integration gives the position,

$$x = t^2 + 6t + C_2$$

where the initial condition on displacement is, at $t_0 = 3$, $x_0 = 5$ m, giving $C_2 = -22$ m. Thus

$$x = t^2 + 6t - 22$$

and at $t = 10$, $x = 138$ m.

In either approach t_0 could have been set equal to zero, in which case the time of interest $t = 7$ sec.

4.6 DISCONTINUOUS TIME-DEPENDENT FORCE

It is possible that $F(t)$ may be represented as a discontinuous function of time or described by more than one mathematical function of time, as, for example, in the series of bursts shown in Fig. 4.5. In this case, Eq. (4.11) may be written

$$x_3 - x_0 = v_{0x}(t_3 - t_0) + \int_{t_0}^{t_1} dt \left(\frac{1}{m} \int_{t_0}^{t} F_x \, dt' \right)$$

$$+ \int_{t_1}^{t_2} dt \left(\frac{1}{m} \int_{t_0}^{t} F_x \, dt' \right) + \int_{t_2}^{t_3} dt \left(\frac{1}{m} \int_{t_0}^{t} F_x \, dt' \right) \qquad \text{(4.13)}$$

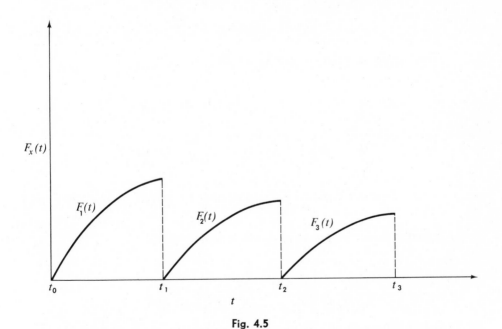

Fig. 4.5

where the integral of the impulse has been conveniently divided in accordance with force–time curve of Fig. 4.5, and where the expressions for the integrals in parentheses are readily determined from Fig. 4.6, a plot of impulse as a function of time.

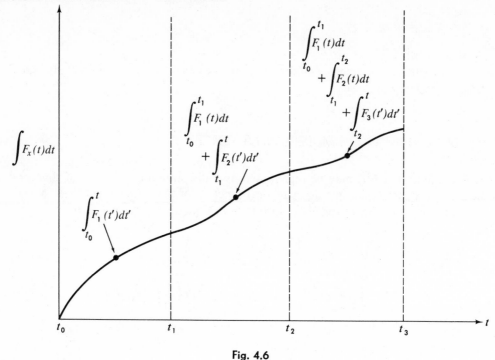

Fig. 4.6

The displacement for the case under discussion is

$$x_3 - x_0 = v_{0x}(t_3 - t_0) + \int_{t_0}^{t_1} dt \left[\frac{1}{m} \int_{t_0}^{t} F_1(t') \, dt' \right]$$

$$+ \int_{t_1}^{t_2} dt \left[\frac{1}{m} \int_{t_0}^{t_1} F_1(t) \, dt + \frac{1}{m} \int_{t_0}^{t} F_2(t') \, dt' \right]$$

$$+ \int_{t_2}^{t_3} dt \left[\frac{1}{m} \int_{t_0}^{t_1} F_1(t) \, dt + \frac{1}{m} \int_{t_1}^{t_2} F_2(t) \, dt + \frac{1}{m} \int_{t_2}^{t} F_3(t') \, dt' \right]$$

(4.14)

Any required extension of Eq. (4.14) is similarly determined.

4.7 MOMENT OF LINEAR MOMENTUM (ANGULAR MOMENTUM)

In deriving the equation of impulse and momentum, we observed a useful concept generated by operating upon Newton's second law. In that instance the operation was a first time integration. Consider now a development resulting from the cross product with $\mathbf{r}_P$ of each side of $\mathbf{F} = \dot{\mathbf{p}}$,

$$\mathbf{r}_P \times \mathbf{F} = \mathbf{r}_P \times \dot{\mathbf{p}}$$

(4.15)

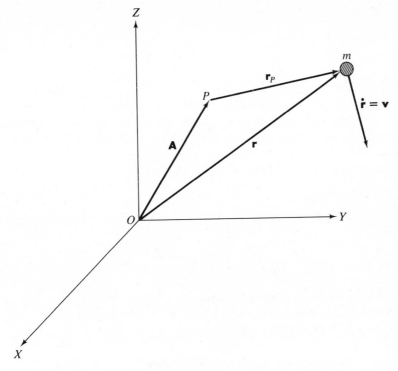

Fig. 4.7

where $\mathbf{r}_P$ represents a vector from a point fixed with respect to an inertial frame of reference, to the particle (Fig. 4.7). Point P may, of course, be the origin. The linear momentum and, more particularly, the velocity $\mathbf{v}$ are taken relative to the same reference frame. The expression $\mathbf{r}_P \times \mathbf{F}$ is the moment of force $\mathbf{F}$ taken about P, $\mathbf{M}_P$.

We shall find it useful to prove that

$$\mathbf{r}_P \times \dot{\mathbf{p}} = \frac{d}{dt}(\mathbf{r}_P \times \mathbf{p}) \tag{4.16}$$

To do this, examine the derivative:

$$\frac{d}{dt}(\mathbf{r}_P \times \mathbf{p}) = \mathbf{r}_P \times \dot{\mathbf{p}} + \dot{\mathbf{r}}_P \times \mathbf{p} \tag{4.17}$$

For Eq. (4.16) to be valid, $\dot{\mathbf{r}}_P \times \mathbf{p}$ must be zero. Since $\mathbf{r}_P$ may be written

$$\mathbf{r}_P = \mathbf{r} - \mathbf{A} \tag{4.18}$$

the first time derivative is

$$\dot{\mathbf{r}}_P = \dot{\mathbf{r}} \tag{4.19}$$

inasmuch as $\mathbf{A}$ is a constant vector running from the origin to P. The term $\dot{\mathbf{r}}_P \times \mathbf{p}$ may now be written $\dot{\mathbf{r}} \times \mathbf{p}$, or in view of the definition of $\mathbf{p}$, $\dot{\mathbf{r}} \times m\dot{\mathbf{r}}$. Since $\dot{\mathbf{r}}$ and $m\dot{\mathbf{r}}$ are parallel, their vector product is zero, and Eq. (4.16) is proved.

Substituting Eq. (4.16) in Eq. (4.15) we obtain

$$\mathbf{r}_P \times \mathbf{F} = \frac{d}{dt}(\mathbf{r}_P \times \mathbf{p}) \tag{4.20a}$$

or

$$\mathbf{M}_P = \dot{\mathbf{H}}_P \tag{4.20b}$$

where $\mathbf{H}_P = \mathbf{r}_P \times \mathbf{p}$ is the moment of linear momentum about point P. This physical quantity is often referred to as the angular momentum, for reasons that will become apparent when we study the mechanics of rigid bodies and Lagrange's equations.

An alternative representation of Eq. (4.20b),

$$\mathbf{M}_P \, dt = d\mathbf{H}_P \tag{4.21}$$

leads to the principle of angular impulse and angular momentum, which parallels Eq. (4.3):

$$\int_1^2 \mathbf{M}_P \, dt = \mathbf{H}_{P2} - \mathbf{H}_{P1} \tag{4.22}$$

where the integral is the angular impulse.

Substituting $\mathbf{M}_P = M_x\hat{\mathbf{i}} + M_y\hat{\mathbf{j}} + M_z\hat{\mathbf{k}}$, $\mathbf{r}_P = r_x\hat{\mathbf{i}} + r_y\hat{\mathbf{j}} + r_z\hat{\mathbf{k}}$, and $\mathbf{p} = p_x\hat{\mathbf{i}} + p_y\hat{\mathbf{j}} + p_z\hat{\mathbf{k}}$ in Eq. (4.20) we have

$$M_x\hat{\mathbf{i}} + M_y\hat{\mathbf{j}} + M_z\hat{\mathbf{k}} = \frac{d}{dt}(H_x\hat{\mathbf{i}} + H_y\hat{\mathbf{j}} + H_z\hat{\mathbf{k}})$$

$$= \frac{d}{dt}[(r_x\hat{\mathbf{i}} + r_y\hat{\mathbf{j}} + r_z\hat{\mathbf{k}}) \times (p_x\hat{\mathbf{i}} + p_y\hat{\mathbf{j}} + p_z\hat{\mathbf{k}})]$$

$$= \frac{d}{dt}[(r_yp_z - r_zp_y)\hat{\mathbf{i}} + (r_zp_x - r_xp_z)\hat{\mathbf{j}} + (r_xp_y - r_yp_x)\hat{\mathbf{k}}] \tag{4.23}$$

In component form,

$$M_x = \frac{d}{dt}(r_yp_z - r_zp_y) \tag{4.24a}$$

$$M_y = \frac{d}{dt}(r_zp_x - r_xp_z) \tag{4.24b}$$

and

$$M_z = \frac{d}{dt}(r_xp_y - r_yp_x) \tag{4.24c}$$

Expanding the moment of force,

$$\mathbf{M}_P = \mathbf{r}_P \times \mathbf{F} = (r_x\hat{\mathbf{i}} + r_y\hat{\mathbf{j}} + r_z\hat{\mathbf{k}}) \times (F_x\hat{\mathbf{i}} + F_y\hat{\mathbf{j}} + F_z\hat{\mathbf{k}})$$

$$M_x\hat{\mathbf{i}} + M_y\hat{\mathbf{j}} + M_z\hat{\mathbf{k}} = (r_yF_z - r_zF_y)\hat{\mathbf{i}} + (r_zF_x - r_xF_z)\hat{\mathbf{j}} + (r_xF_y - r_yF_x)\hat{\mathbf{k}} \tag{4.25}$$

Consequently,

$$M_x = r_y F_z - r_z F_y = \frac{d}{dt}(r_y p_z - r_z p_y) \tag{4.26a}$$

$$M_y = r_z F_x - r_x F_z = \frac{d}{dt}(r_z p_x - r_x p_z) \tag{4.26b}$$

$$M_z = r_x F_y - r_y F_x = \frac{d}{dt}(r_x p_y - r_y p_x) \tag{4.26c}$$

The reader should convince himself that $r_y F_z - r_z F_y$ is indeed the moment of force about the X-axis, and so on. Carrying out the differentiation in Eq. (4.26a), for example, one quickly verifies the result:

$$r_y F_z - r_z F_y = r_y \dot{p}_z + \dot{r}_y p_z - r_z \dot{p}_y - \dot{r}_z p_y$$

But $\dot{p}_z = F_z$, $\dot{p}_y = F_y$, $\dot{r}_y p_z = m \dot{y} \dot{z}$, and $\dot{r}_z p_y = m \dot{z} \dot{y}$, which, when substituted above, gives the desired equality.

4.8 CONSERVATION OF ANGULAR MOMENTUM

When the moment of force is zero, which occurs when $\mathbf{r}_P = 0$, $\mathbf{F} = 0$, or when $\mathbf{r}_P$ and $\mathbf{F}$ are parallel,

$$\frac{d}{dt}(\mathbf{r}_P \times \mathbf{p}) = 0 \tag{4.27}$$

and $\mathbf{H}_P = \mathbf{r}_P \times \mathbf{p} = $ constant vector.

Under these circumstances the moment of momentum vector is constant in magnitude and direction, and we speak of "conservation of angular momentum." We thus have two closely related concepts dealing with linear and angular momentum:

For $\mathbf{F} = 0$, $\mathbf{p} = $ constant; linear momentum is conserved.

For $\mathbf{M}_P = 0$, $\mathbf{H}_P = \mathbf{r}_P \times \mathbf{p} = $ constant; moment of linear momentum (angular momentum) is conserved.

A number of forces of special importance in physics are described as *central forces*, because they are directed toward a fixed point in space throughout the motion. The force of gravity exerted on the earth by the sun is an obvious example of a central force. Taking moments about the fixed point, it is clear that $\mathbf{r}_P$ and $\mathbf{F}$ are colinear and hence possess a zero vector product. Consequently, the angular momentum in central force motion remains constant. Since $\mathbf{r} \times m\mathbf{v}$ represents, for this case, a constant vector, its direction (normal to the plane of $\mathbf{r}$ and $m\mathbf{v}$) must be fixed, and we conclude that the motion governed by a central force occurs in a fixed plane.

That the time derivative of the angular momentum is zero if $\mathbf{M}_P = \mathbf{0}$ is readily confirmed by the following:

$$\frac{d}{dt}(\mathbf{r}_P \times m\mathbf{v}) = \frac{d\mathbf{r}_P}{dt} \times m\mathbf{v} + \mathbf{r}_P \times \frac{d}{dt}(m\mathbf{v}) = \mathbf{v} \times m\mathbf{v} + \mathbf{r}_P \times \mathbf{F} = \mathbf{0}$$

where $\mathbf{v}$ and $m\mathbf{v}$, and $\mathbf{r}_P$ and $\mathbf{F}$ are parallel pairs of vectors.

It is of interest to explore the above representation in polar coordinates:

$$\mathbf{H}_P = \mathbf{r}_P \times m\mathbf{v} = \mathbf{r}_P \times m(v_R\hat{\mathbf{e}}_R + v_\phi\hat{\mathbf{e}}_\phi) = \text{constant vector}$$

Since $\hat{\mathbf{e}}_R$ and $\mathbf{r}_P$ are colinear, $\mathbf{r}_P \times \hat{\mathbf{e}}_R = \mathbf{0}$, and

$$\mathbf{H}_P = \mathbf{r}_P \times mv_\phi\hat{\mathbf{e}}_\phi = mr_Pv_\phi\hat{\mathbf{k}} = \text{constant vector}$$

where $\hat{\mathbf{k}}$ is normal to the plane of $\hat{\mathbf{e}}_R$ and $\hat{\mathbf{e}}_\phi$ (refer to Chapter 2). But $v_\phi = r_P\dot{\phi}$, and consequently

$$H_P = mr_P^2\dot{\phi} = \text{constant}$$

At any points in the trajectory, the angular momenta are equal:

$$mr_1^2\dot{\phi}_1 = mr_2^2\dot{\phi}_2$$

Chapter 8 is devoted to a more extensive study of central force motion.

References

Feynman, R. P., Leighton, R. B., and Sands, M., *The Feynman Lectures on Physics*, Volume 1, Addison-Wesley Publishing Company, Inc., Reading, Mass., 1963.

Fowles, G. R., *Analytical Mechanics*, Holt, Rinehart and Winston, Inc., New York, 1962.

Joos, G., *Theoretical Physics*, Third Edition, Hafner Publishing Company, New York, 1956.

Nara, H. R., *Vector Mechanics for Engineers*, Parts 1 and 2, John Wiley & Sons, Inc., New York, 1962.

Sommerfeld, A., *Mechanics*, Translated by M. O. Stern, Academic Press, Inc., New York, 1957.

Timoshenko, S. P., and Young, D. H., *Advanced Dynamics*, McGraw-Hill, Inc., New York, 1948.

EXERCISES

4.1 A mass of 50 g is acted upon by a constant force of 1100 dyn. If the direction cosines of the force are (0.707, 0, 0.707), determine the components of velocity after 10 sec if the initial velocity is $v_0 = 20\hat{j}$ cm/sec.

4.2 A one-dimensional force–time curve is given in Fig. P4.2. Determine the velocity of a mass of 1 kg after 20 sec if the initial velocity is 20 m/sec.

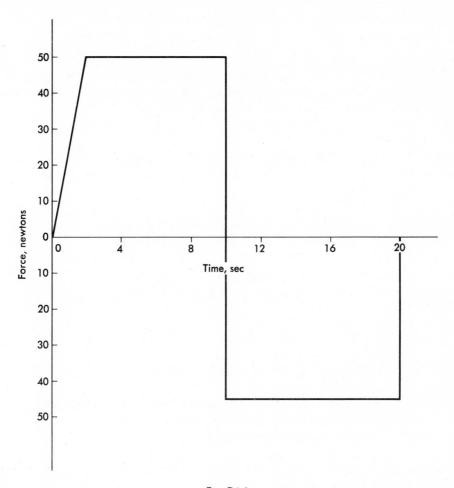

Fig. P4.2

4.3 It is required that a mass of 26 kg attain a velocity of 100 m/sec in 10 sec. If initially at rest, what value must the constant b assume for a force $F = bt$?

4.4 A mass of 10 g is given an initial velocity of 120 cm/sec up a frictionless inclined plane making an angle of 30 degrees with the horizontal. How long will it take for the mass to stop? How long a time is required for the mass to attain a velocity of 60 cm/sec down the incline?

4.5 Repeat Exercise 4.4 assuming a constant coefficient of friction of 0.1.

4.6 Assuming a constant coefficient of friction of 0.15 between a 2 kg mass and a plane inclined at 20 degrees, determine the velocity of the mass after 10 sec if initially at rest. The mass is pulled up the plane by a cable parallel to the plane, where the tension in the cable varies with time according to the relation

$$F = 50(1 + 3t) \text{ N}$$

4.7 It is found that a 300 g ball initially possessing velocity $\mathbf{v} = 50\hat{\mathbf{i}}$ cm/sec has, after a collision with another ball, a velocity $\mathbf{v} = 100\hat{\mathbf{i}} + 50\hat{\mathbf{j}}$ cm/sec. What average force acted, if the time of impact is found to be 0.001 sec?

4.8 A particle of mass m and horizontal velocity v_i collides with a smooth inclined plane making an angle θ with the horizontal. The particle then moves up the surface. Derive an expression for the impulse on the mass. What final velocity will the particle have?

4.9 What are the components of the impulse caused by a force of magnitude 40 lb having direction cosines 0, 0.500, 0.866 and of 3 sec duration?

4.10 A mass m is imparted an initial velocity v_{0y} in a constant gravitational field. Employing the impulse–momentum equation, prove that the velocity v_y at any time is given by

$$v_y = v_{0y} - gt$$

4.11 Repeat Illustrative Example 4.3 employing Eq. (4.11).

4.12 If the reaction time of a driver is 0.5 sec, calculate the shortest time required to stop a 4000 lb automobile traveling at 60 mph. Assume a constant coefficient of friction between tires and pavement of 0.6.

4.13 A rocket is subjected to bursts of propulsive force described in Fig. P4.13. At time $t = 0$, the velocity is v_0 and the displacement is x_0. During the first interval, the mass may be assumed constant at m_1. For the subsequent bursts the mass is m_2 and m_3. Determine the velocity and displacement at t_3.

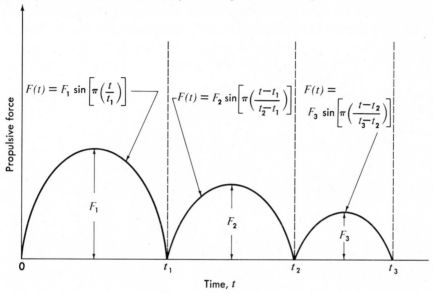

Fig. P4.13

4.14 A 5 g particle moves inward at a constant speed of 10 cm/sec while rotating on a frictionless horizontal surface as shown in Fig. P4.14. If the angular velocity is 12 rad/sec when $r = 20$ cm, determine the radial and tangential components of particle velocity when $r = 10$ cm.

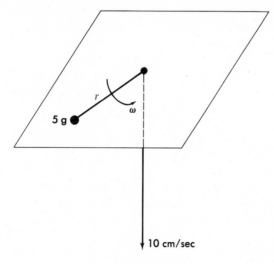

Fig. P4.14

4.15 The particle shown in Fig. P4.15 is moving outward along arm A at a constant speed of 0.3 m/sec while the arm rotates about axis O-O. If the angular velocity is 1 rad/sec when $r = 2$ m, and then a time-varying torque $M = 25t^2$ N-m is applied, what angular velocity will the arm experience when $r = 2.5$ m?

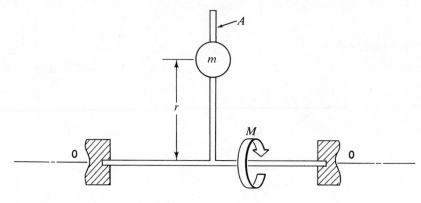

Fig. P4.15

CHAPTER 5

WORK AND ENERGY — THE PARTICLE

5.1 INTRODUCTION

The utility of Newton's second law was extended in Chapter 4 by taking the first time integral to obtain the impulse–momentum relationship. The cross product of position with $\mathbf{F} = \dot{\mathbf{p}}$ led to concepts involving the angular momentum. These techniques were shown to result in a number of additional vector quantities.

In this chapter, beginning with the dot product of force with displacement, the far-reaching scalar equations of work and energy are developed.

5.2 WORK AND KINETIC ENERGY

Beginning with Newton's second law for a particle and taking the scalar product of each side with an infinitesimal displacement $\mathbf{dr}$,

$$\mathbf{F} \cdot \mathbf{dr} = m \left(\frac{d^2\mathbf{r}}{dt^2} \right) \cdot \mathbf{dr} \tag{5.1}$$

where $\mathbf{F}$ represents the force resultant. The left side of Eq. (5.1) defines the work dW done by $\mathbf{F}$ as the particle proceeds through an infinitesimal displacement. Work, the scalar product of force and displacement, may be expressed in terms of the definition of the scalar product:

$$dW = |\mathbf{F}||\mathbf{dr}| \cos{(\mathbf{F}, \mathbf{dr})} = F\, dr \cos\theta \tag{5.2}$$

It is apparent that only those components of force tangent (parallel or antiparallel) to the displacement contribute to the work, as shown in Fig. 5.1.

Integrating Eq. (5.1) between the limits of position a and b, corresponding to times t_a and t_b,

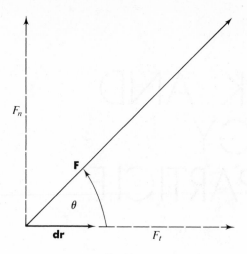

Fig. 5.1

$$W = \int_a^b \mathbf{F} \cdot d\mathbf{r} = \int_a^b \left(m \frac{d^2\mathbf{r}}{dt^2} \right) \cdot d\mathbf{r} = \int_{t_a}^{t_b} m \frac{d^2\mathbf{r}}{dt^2} \cdot \frac{d\mathbf{r}}{dt} dt = \frac{1}{2} m \int_{t_a}^{t_b} \frac{d}{dt} (\dot{\mathbf{r}} \cdot \dot{\mathbf{r}}) \, dt$$

$$= \frac{1}{2} m \int_{t_a}^{t_b} \frac{d}{dt} (v^2) \, dt = \frac{1}{2} mv_b^2 - \frac{1}{2} mv_a^2 = T_b - T_a = \Delta T \tag{5.3}$$

where T is the kinetic energy of the particle; $T \geq 0$. Equation (5.3) is referred to as the work-energy equation.

Since $\mathbf{F}$ represents the combined effect of all the forces acting, W must be the sum of the individual work terms associated with each force. Equation (5.3) may thus be written

$$W = \sum_{i=1}^n W_i = \int_a^b \left(\sum_{i=1}^n \mathbf{F}_i \right) \cdot d\mathbf{r} = \Delta T \tag{5.4}$$

where W_i is the work done by the ith force $\mathbf{F}_i$ in executing the path $a - b$.

The integrals in Eqs. (5.3) and (5.4) may be written in an alternative form to underscore the importance of the tangential component of force,

$$W = \int_a^b \mathbf{F} \cdot d\mathbf{r} = \int_a^b \mathbf{F} \cdot \frac{d\mathbf{r}}{ds} \, ds = \int_a^b F_t \, ds = \Delta T \tag{5.5}$$

where F_t is the component of force tangent at each point along the path and ds is an infinitesimal of length along a-b.

Expressing Eq. (5.3) in Cartesian coordinates and expanding,

$$W = \int_{x_a}^{x_b} F_x\,dx + \int_{y_a}^{y_b} F_y\,dy + \int_{z_a}^{z_b} F_z\,dz = \Delta T \qquad (5.6)$$

where F_x, F_y, F_z are the X-, Y-, and Z-components of F along the path.

Examination of the work-energy equation leads one to conclude that work and energy are expressible in the same physical units and that net work equals the change in kinetic energy. The kinetic energy decreases when the net force is anti-parallel to the displacement. Frictional forces, for example, oppose motion and contribute negative work. If a block is pushed along a plane by a constant parallel force F, a distance L, the work done by this force is FL. A friction force f contributes the work $-fL$. The net work is therefore $FL - fL$, and it is this work which is equated to the change in kinetic energy.

It is clear that work causes a change in kinetic energy. Conversely, the motion of a particle makes it capable of causing a force to act through a displacement, resulting in a total or partial conversion of kinetic energy into work. The latter viewpoint leads to the statement often made that energy represents a capacity to do work.

The work-energy equation was derived entirely from Newton's second law and we should expect it to yield no essentially new information. What it does offer, however, is an alternative approach in analysis. This formulation is particularly useful when the forces are expressed as functions of position and when it is desirable to study the relationship between particle velocity and displacement. The integral is a line integral, in general requiring for evaluation a knowledge of force as a function of position everywhere along the path.

In conservative force systems, as we shall soon establish, the work-energy equation's utility is enhanced because the work integral is path-independent.

5.3 WORK AND POWER

Another scalar of particular importance results from the dot product of force and velocity:

$$P = \mathbf{F} \cdot \mathbf{v} \qquad (5.7)$$

where P represents the power.

In terms of the forces comprising $\mathbf{F}$,

$$P = \sum_{i=1}^{n} P_i = \left(\sum_{i=1}^{n} \mathbf{F}_i \right) \cdot \mathbf{v} = \sum_{i=1}^{n} (\mathbf{F} \cdot \mathbf{v}) \qquad (5.8)$$

Substituting the product of mass and acceleration for force in Eq. (5.7),

$$P = (m\mathbf{a}) \cdot \mathbf{v} = (m\dot{\mathbf{v}}) \cdot \mathbf{v} = \frac{d}{dt} \left(\tfrac{1}{2} m \mathbf{v} \cdot \mathbf{v}\right) = \frac{d}{dt} \left(\tfrac{1}{2} m v^2\right) \qquad (5.9)$$

Thus power is the time rate of change of kinetic energy or equivalently the rate at which work is done. From Eq. (5.9),

$$dT = P \, dt \qquad (5.10)$$

Integrating,

$$T_b - T_a = \int_{t_a}^{t_b} P \, dt = W \qquad (5.11)$$

Equation (5.11) offers an alternative means of determining the work when the power is known as a function of time.

Illustrative Example 5.1

Determine the net work done on a block weighing 50 lb as it moves 10 ft down an inclined plane as shown in Fig. 5.2(a). If the block starts from rest, what is its final velocity? The coefficient of kinetic friction is 0.1.

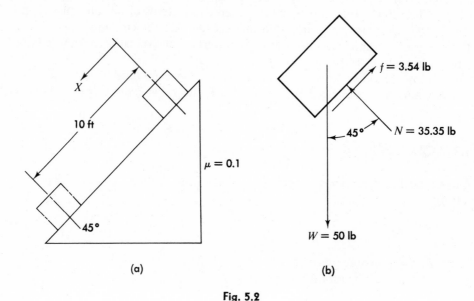

(a) (b)

Fig. 5.2

SOLUTION

From the free-body diagram of Fig. 5.2(b), it is clear that only the force of friction and the component of weight parallel to the plane (50 sin 45) result in nonzero contributions to the net work:

$$W = \int_0^{10} \left(\sum_{i=1}^n \mathbf{F}_i \right) \cdot d\mathbf{r} = \underbrace{(35.35)(10)}_{W_{\text{gravity}}} - \underbrace{(3.54)(10)}_{W_{\text{friction}}} = 318 \text{ lb-ft}$$

$$= \Delta T = \frac{1}{2} mv_{10}^2 - \frac{1}{2} mv_0^2 = \frac{1}{2} \left(\frac{50}{32.2} \right) v_{10}^2$$

$$v_{10} = 20.2 \text{ ft/sec}$$

If friction had been neglected,

$$W = 353.5 \text{ lb-ft} \qquad \text{and} \qquad v_{10} = 21.3 \text{ ft/sec}$$

Illustrative Example 5.2
What work is done by the force

$$\mathbf{F} = (x^2 + z^2)\hat{\mathbf{i}} + (2z)\hat{\mathbf{j}} + 3\hat{\mathbf{k}} \text{ dyn}$$

along the path $x = y = 2z$ between the limits $(2, 2, 1)$ and $(8, 8, 4)$ cm?

SOLUTION

$$\int_a^b \mathbf{F} \cdot d\mathbf{r} = \int_{x_a}^{x_b} F_x \, dx + \int_{y_a}^{y_b} F_y \, dy + \int_{z_a}^{z_b} F_z \, dz$$

$$\int_2^8 F_x \, dx = \int_2^8 (x^2 + z^2) \, dx = \int_2^8 \left(x^2 + \frac{x^2}{4} \right) dx = 210 \text{ dyn-cm}$$

$$\int_2^8 F_y \, dy = \int_2^8 2z \, dy = \int_2^8 y \, dy = 30 \text{ dyn-cm}$$

$$\int_1^4 F_z \, dz = \int_1^4 3 \, dz = 9 \text{ dyn-cm}$$

Therefore, $W = 249$ dyn-cm.

5.4 CONSERVATIVE SYSTEMS AND THE POTENTIAL

In the work-energy equation, the only positions represented on the right side [for example, Eqs. (5.4) and (5.6)] are the initial and final points on the path of motion. The work integral, however, is in general a function of the entire path.

Suppose that it is legitimate to represent the force in Eq. (5.3) as a function of spatial coordinates only: $\mathbf{F} = \mathbf{F}(x, y, z)$. This force function describes a time-independent field of force. The work done in moving a particle subjected to this force field from position a to b along an arbitrary path I, as shown in Fig. 5.3, is

$$W_I = \int_a^b \mathbf{F} \cdot \mathbf{dr} \qquad (5.12)$$

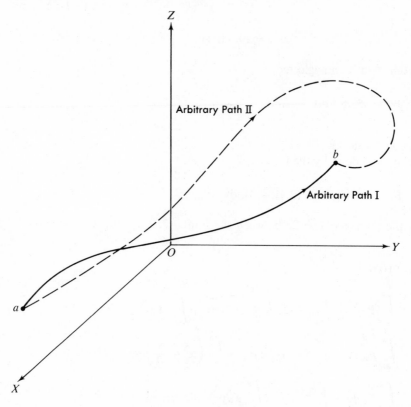

Fig. 5.3

If another arbitrary path, II, is selected to proceed from a to b, the work is

$$W_{II} = \int_a^b \mathbf{F} \cdot \mathbf{dr} \qquad (5.13)$$

Since the paths chosen are arbitrary, they are representative of all paths between a and b. If it is found, therefore, that $W_I = W_{II}$, the work cannot depend upon path but only upon the limits of integration. Forces satisfying the condition of path independence are termed *conservative forces* and form conservative fields of force (see Section 1.18). As demonstrated in Section 5.7, the term "conservative" refers to the fact that mechanical energy remains constant.

Consider now the relationship between the work done by the force in moving from a to b along direct path II and the work done by the force for the reverse path along II from b to a. Referring to Fig. 5.4, we observe that each infinitesimal particle displacement of the direct path is the negative of each corresponding displacement along the reverse path. The force at any arbitrary point P along the path must be the same regardless of the direction of motion, because the force is a function of position only and not of the direction of motion. The preceding leads to the result

$$\int_{II}{}_{a}^{b} \mathbf{F} \cdot d\mathbf{r} = \int_{II}{}_{b}^{a} \mathbf{F} \cdot (-d\mathbf{r}) = -\int_{II}{}_{b}^{a} \mathbf{F} \cdot d\mathbf{r} \tag{5.14}$$

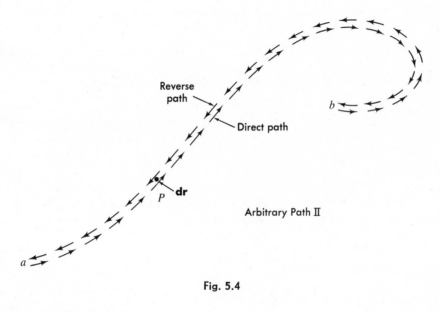

Reverse path

Direct path

P $d\mathbf{r}$

Arbitrary Path II

Fig. 5.4

For conservative forces,

$$\int_{I}{}_{a}^{b} \mathbf{F} \cdot d\mathbf{r} = \int_{II}{}_{a}^{b} \mathbf{F} \cdot d\mathbf{r} \tag{5.15}$$

and, therefore, after substituting Eq. (5.14) we have

$$\int_{a}^{b} \mathbf{F} \cdot d\mathbf{r} + \int_{b}^{a} \mathbf{F} \cdot d\mathbf{r} = 0 \tag{5.16}$$

regardless of path followed. Written more succinctly, the work done along any closed path by a conservative force must be zero:[1]

$$W = \oint \mathbf{F} \cdot d\mathbf{r} = 0 \tag{5.17}$$

---

[1] The region enclosed must not contain a singularity.

If the line integral $\int_b^a \mathbf{F} \cdot \mathbf{dr}$ depends only upon the limits of integration, it follows that the integrand is an exact differential of some function Φ, termed a *force potential:*

$$d\Phi = \mathbf{F} \cdot \mathbf{dr} = F_x \, dx + F_y \, dy + F_z \, dz \tag{5.18}$$

From Eq. (5.18),

$$\int_a^b \mathbf{F} \cdot \mathbf{dr} = \int_a^b d\Phi = \Phi_b - \Phi_a \tag{5.19}$$

Since the force depends solely upon spatial variables, so must Φ:

$$d\Phi = \frac{\partial \Phi}{\partial x} \, dx + \frac{\partial \Phi}{\partial y} \, dy + \frac{\partial \Phi}{\partial z} \, dz \tag{5.20}$$

Equating coefficients of like differentials in Eqs. (5.18) and (5.20),

$$F_x = \frac{\partial \Phi}{\partial x} \qquad F_y = \frac{\partial \Phi}{\partial y} \qquad F_z = \frac{\partial \Phi}{\partial z} \tag{5.21}$$

or

$$\mathbf{F} = \frac{\partial \Phi}{\partial x} \hat{\mathbf{i}} + \frac{\partial \Phi}{\partial y} \hat{\mathbf{j}} + \frac{\partial \Phi}{\partial z} \hat{\mathbf{k}} \tag{5.22}$$

The coordinate system employed to describe the force field may be selected on any basis, and in general $\mathbf{F}$ is represented by

$$\mathbf{F} = \text{grad } \Phi = \nabla \Phi \tag{5.23}$$

where, for Cartesian coordinates,

$$\nabla \Phi = \frac{\partial \Phi}{\partial x} \hat{\mathbf{i}} + \frac{\partial \Phi}{\partial y} \hat{\mathbf{j}} + \frac{\partial \Phi}{\partial z} \hat{\mathbf{k}} \tag{5.24}$$

For the cylindrical polar system,

$$\nabla \Phi = \frac{\partial \Phi}{\partial R} \hat{\mathbf{e}}_R + \frac{1}{R} \frac{\partial \Phi}{\partial \phi} \hat{\mathbf{e}}_\phi + \frac{\partial \Phi}{\partial z} \hat{\mathbf{k}} \tag{5.25}$$

In spherical polar coordinates,

$$\nabla \Phi = \frac{\partial \Phi}{\partial r} \hat{\mathbf{e}}_r + \frac{1}{r} \frac{\partial \Phi}{\partial \theta} \hat{\mathbf{e}}_\theta + \frac{1}{r \sin \theta} \frac{\partial \Phi}{\partial \phi} \hat{\mathbf{e}}_\phi \tag{5.26}$$

On the basis of the foregoing development, it may be stated that a vector force field is conservative in a given region if there exists a scalar function of position Φ, the partial derivative of which with respect to any coordinate is equal to the component of force in the direction of the coordinate.

Given that a force is a function of coordinates only, what test can be applied to determine whether a scalar potential Φ exists? From the theory of differential equations, recall that $\mathbf{F}\cdot\mathbf{dr} = F_x\,dx + F_y\,dy + F_z\,dz$ is the total differential of the function Φ only if the following conditions are satisfied:

$$\frac{\partial F_x}{\partial y} = \frac{\partial F_y}{\partial x} \qquad \frac{\partial F_x}{\partial z} = \frac{\partial F_z}{\partial x} \qquad \frac{\partial F_y}{\partial z} = \frac{\partial F_z}{\partial y} \tag{5.27}$$

The conditions given in Eq. (5.27) are equivalent to stating that the curl of $\mathbf{F}$ or $\boldsymbol{\nabla} \times \mathbf{F}$ is equal to zero. In Cartesian coordinates,

$$\boldsymbol{\nabla} \times \mathbf{F} = \begin{vmatrix} \hat{\mathbf{i}} & \hat{\mathbf{j}} & \hat{\mathbf{k}} \\ \dfrac{\partial}{\partial x} & \dfrac{\partial}{\partial y} & \dfrac{\partial}{\partial z} \\ F_x & F_y & F_z \end{vmatrix} = \left(\frac{\partial F_z}{\partial y} - \frac{\partial F_y}{\partial z}\right)\hat{\mathbf{i}} + \left(\frac{\partial F_x}{\partial z} - \frac{\partial F_z}{\partial x}\right)\hat{\mathbf{j}}$$

$$+ \left(\frac{\partial F_y}{\partial x} - \frac{\partial F_x}{\partial y}\right)\hat{\mathbf{k}} \tag{5.28}$$

Thus in the case of a constant gravitational force field,

$$\mathbf{F} = -mg\hat{\mathbf{k}}$$

$$F_x = F_y = 0 \qquad F_z = -mg$$

Applying the conditions of Eq. (5.27),

$$\frac{\partial F_x}{\partial y} = \frac{\partial F_y}{\partial x} = \frac{\partial F_z}{\partial y} = \frac{\partial F_y}{\partial z} = \frac{\partial F_z}{\partial x} = \frac{\partial F_x}{\partial z} = 0 \qquad \text{curl } \mathbf{F} = \mathbf{0}$$

and therefore a scalar potential Φ exists. From $\partial\Phi/\partial z = F_z = d\Phi/dz$,

$$\Phi = \int F_z\,dz = -mgz + C$$

In the case of the force required to extend or compress a linear spring,

$$\mathbf{F} = kx\hat{\mathbf{i}}$$

$$F_x = kx \qquad F_y = F_z = 0$$

where k is the spring constant. Applying the test given by Eq. (5.27), the partial derivatives are all equal to zero. The potential

$$\Phi = \tfrac{1}{2}kx^2 + C$$

5.5 POTENTIAL ENERGY

Unlike kinetic energy, which is a function only of velocity, potential energy V depends solely upon the position of a particle within a conservative force field. The difference of potential energy of a particle in moving from a to b, $V_b - V_a$, is defined as the negative of the work done by a conservative force in proceeding from a to b:

$$\Delta V = V_b - V_a = -\int_a^b \mathbf{F} \cdot \mathbf{dr} \tag{5.29}$$

If the potential energy at some datum or reference point is set equal to zero, the potential energy at some other point P with respect to the datum is

$$V_P = -\int_{datum}^P \mathbf{F} \cdot \mathbf{dr} = \int_P^{datum} \mathbf{F} \cdot \mathbf{dr} \tag{5.30}$$

The datum of potential energy is arbitrary and usually selected on the basis of convenience. For example, the zero of potential energy of an electron attracted to a proton (hydrogen atom) corresponds to infinite separation of the two. Since only differences in potential energy have meaning, any datum will result in identical ΔV.

The convention given above is such that the potential energy increases when the force acts opposite to the direction of the integration.

From Eq. (5.29) it is clear that for a conservative force field

$$-dV = \mathbf{F} \cdot \mathbf{dr} \tag{5.31}$$

Since it has already been established that for such a force

$$\mathbf{F} \cdot \mathbf{dr} = d\Phi \tag{5.32}$$

it follows that $d\Phi = -dV$ and

$$\mathbf{F} = -\operatorname{grad} V = -\boldsymbol{\nabla} V \tag{5.33}$$

For a Cartesian coordinate system, therefore,

$$F_x = -\frac{\partial V}{\partial x} \qquad F_y = -\frac{\partial V}{\partial y} \qquad F_z = -\frac{\partial V}{\partial z} \tag{5.34}$$

As Eq. (5.33) is quite general, once the potential energy V is known, the force component in any arbitrary direction may be found from

$$F_\eta = -\frac{\partial V}{\partial \eta} \tag{5.35}$$

where η represents displacement in the direction of unit vector $\hat{\mathbf{e}}_\eta$.

Note that the force components derived on the basis of Eq. (5.34) would be identical regardless of any additive constant associated with the potential energy V.

Substituting Eq. (5.34) into Eq. (5.27),

$$\frac{\partial^2 V}{\partial x \, \partial y} = \frac{\partial^2 V}{\partial y \, \partial x} \qquad \frac{\partial^2 V}{\partial y \, \partial z} = \frac{\partial^2 V}{\partial z \, \partial y} \qquad \frac{\partial^2 V}{\partial z \, \partial x} = \frac{\partial^2 V}{\partial x \, \partial z} \tag{5.36}$$

Illustrative Example 5.3

A frictionless simple pendulum is released from rest as shown in Fig. 5.5. Determine the work done and the change of potential energy as the pendulum bob rotates through $\pi/2$ rad to the lowest potential energy.

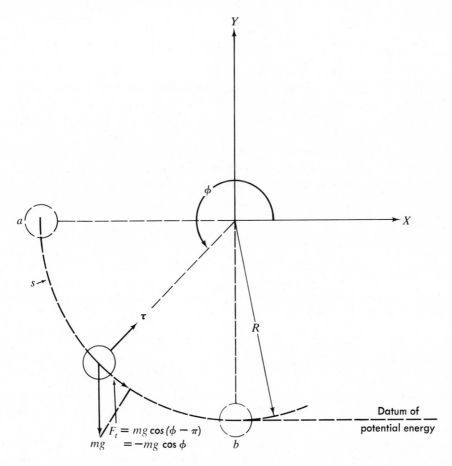

Fig. 5.5

SOLUTION

The work integral is evaluated over the path of travel of the pendulum bob, the arc corresponding to $\pi/2$ rad,

$$W = \int_a^b \mathbf{F} \cdot \mathbf{dr} = \int_a^b F_t \, ds$$

where F_t is the component of force tangent to the arc:

$$F_t = -mg \cos \phi$$

Since $ds = R\,d\phi$, the work integral may be expressed in terms of ϕ:

$$W = \int_{\pi}^{3\pi/2} (-mg \cos \phi) R\,d\phi = -mgR \left[\sin \phi \right]_{\pi}^{3\pi/2} = mgR$$

The change of potential energy is the negative of the work done by a conservative force:

$$\Delta V = -W = -mgR$$

The work, being independent of the path of integration, may also have been determined on the basis of the change in vertical position, R.

5.6 EQUILIBRIUM AND STABILITY

Since the conditions of equilibrium for a particle are $F_x = F_y = F_z = 0$, an alternative means of representing the equilibrium of conservative force systems is available in terms of Eq. (5.34):

$$-\frac{\partial V}{\partial x} = -\frac{\partial V}{\partial y} = -\frac{\partial V}{\partial z} = 0 \tag{5.37}$$

The variation in potential energy offers more than a test for equilibrium. It can also provide insight into the type of equilibrium characterizing the system within a given region.

Examine, for example, the simple pendulum shown in Fig. 5.6(a) and the accompanying plot of potential energy as a function of displacement s. From Fig. 5.6(b), $dV/ds = 0$ at $s = 0$, and consequently, at this point, the pendulum is in equilibrium in a downward-acting gravitational field. If the bob is displaced a small amount to the right (positive s-direction), the force acting on the pendulum is negative, because $F = -dV/ds$ and dV/ds is positive in this region. Thus the force resulting from a slight disturbing displacement to the right tends to restore the pendulum to a state of equilibrium. Similarly, a displacement to the left will cause a positive restoring force to act. Such a physical situation illustrates a state of stable equilibrium; the potential energy is a minimum. From the plot of dV/ds shown in Fig. 5.6(b), we observe that for stable equilibrium $d^2V/ds^2 > 0$.

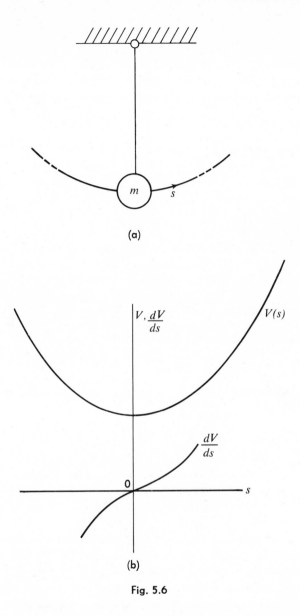

(a)

(b)

Fig. 5.6

Contrast the above case with the inverted pendulum shown in Fig. 5.7(a). Once again the conditions of equilibrium are satisfied at $s = 0$, where $dV/ds = 0$. Equilibrium is unstable, however, because any small displacement of the bob to the right or left of the equilibrium position will cause the pendulum to move still further from equilibrium. This is clear from the direction of the force obtained

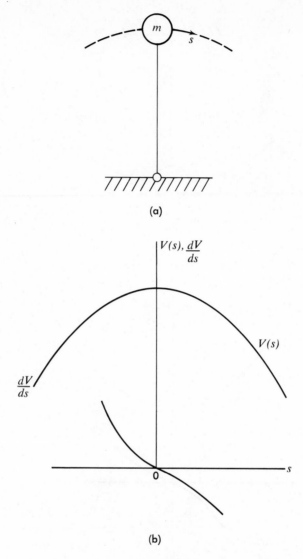

(a)

(b)

Fig. 5.7

from consideration of $-dV/ds$ for positions on either side of $s = 0$. Examination of dV/ds as a function of s [Fig. 5.7(b)] indicates that $d^2V/ds^2 < 0$. Thus a state of equilibrium coupled with maximum potential energy is unstable.

Finally, we consider a case of neutral equilibrium for the cylinder shown in Fig. 5.8(a). Here $dV/ds = 0$, not only at $s = 0$ but at positions to the right and left. The cylinder is in a state of equilibrium at all points on the plane. Furthermore, displacements from $s = 0$ do not cause restoring forces to act; the displacement is neither amplified nor diminished, and the cylinder remains at a new equi-

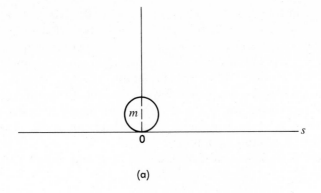

(a)

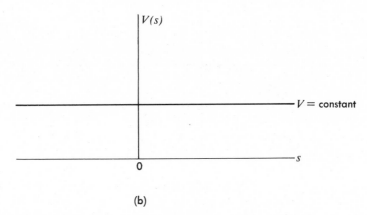

(b)

Fig. 5.8

librium position following a disturbance. As shown in Fig. 5.8(b), there is no minimum or maximum in potential energy.

Consider the interesting case of a mass subjected to the one-dimensional force:

$$F(x) = -a^2x^2$$

leading to a potential energy

$$V = \frac{a^2x^3}{3} + C$$

since

$$V = -\int F(x)\,dx$$

The particle is in equilibrium at $x = 0$ since $F(0) = 0$. Taking $(d^2V/dx^2)_0$ as a test for stability, no conclusion can be drawn as to the type of stability, because at $x = 0$, $d^2V/dx^2 = 0$. From $d^2V/dx^2 = 2a^2x$, it is observed that for $x > 0$, $d^2V/dx^2 > 0$, and for $x < 0$, $d^2V/dx^2 < 0$. The equilibrium appears stable for positive displacements and unstable for negative displacements. Any arbitrary positive displacement, however, supplies energy to the system causing the particle to possess a negative velocity upon its return to $x = 0$. The resulting negative displacement is then amplified because of the negative force which acts for $x < 0$. The system is therefore in a state of unstable equilibrium.

Examine now Figs. 5.9(a) and 5.9(b), curves of V and dV/dx as functions

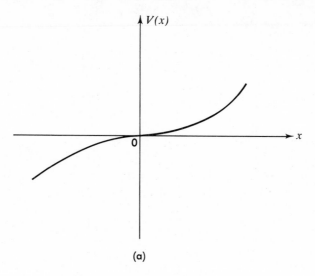

(a)

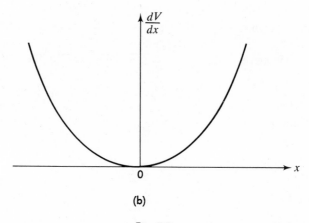

(b)

Fig. 5.9

of x. Since $dV/dx > 0$ for $x > 0$, the force acting is negative tending to return the system to $x = 0$. For $x > 0$, dV/dx is likewise negative, resulting in a departure from the equilibrium state.

Illustrative Example 5.4

A mass m is supported by two linear springs (k_1, k_2) located between rigid supports, Fig. 5.10. Determine the equilibrium position of the mass in terms of the spring constants, the undeformed spring lengths l_1 and l_2, and the distance L between supports.

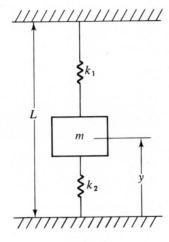

Fig. 5.10

SOLUTION

The deformation δ of each spring may be described in terms of the variable y, the vertical position of the mass above the lower support:

$$\delta = \text{(final length)} - \text{(undeformed length)}$$

$$\delta_1 = (L - y) - l_1$$

$$\delta_2 = y - l_2$$

The system potential energy is determined by adding the gravitational and elastic components:

$$V(y) = mgy + \tfrac{1}{2}k_1\delta_1^2 + \tfrac{1}{2}k_2\delta_2^2$$

$$= mgy + \tfrac{1}{2}k_1(L - y - l_1)^2 + \tfrac{1}{2}k_2(y - l_2)^2$$

Next we set $dV/dy = 0$ to obtain the equilibrium position:

$$\frac{dV}{dy} = 0 = mg - k_1(L - y - l_1) + k_2(y - l_2)$$

Solving for y (which is now y_{eq}) the equilibrium position is

$$y_{eq} = \frac{k_1(L - l_1) + k_2 l_2 - mg}{k_1 + k_2}$$

Since $d^2V/dy^2 = k_1 + k_2$, a positive number for all y, the equilibrium is stable at y_{eq}.

Equilibrium and its relationship to a minimum in the potential energy are worthy of further elaboration. Suppose a potential-energy function $V(x)$ possesses a minimum at x_0, as shown in Fig. 5.11. Since at x_0, $dV/dx = 0$, we recognize

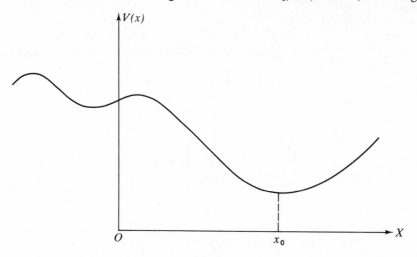

Fig. 5.11

this point as one for which the system is in equilibrium. Furthermore, for the potential illustrated, $(d^2V/dx^2)_{x_0} > 0$, so that the equilibrium is stable. Expanding $V(x)$ in a Taylor series about x_0 we have

$$V(x) = V(x_0) + \left(\frac{dV}{dx}\right)_{x_0} (x - x_0) + \frac{1}{2!}\left(\frac{d^2V}{dx^2}\right)_{x_0} (x - x_0)^2$$

$$+ \frac{1}{3!}\left(\frac{d^3V}{dx^3}\right)_{x_0} (x - x_0)^3 + \cdots \tag{5.38}$$

It is a characteristic of stable equilibrium that small disturbances about the equilibrium point set up restoring influences. Since our attention will be confined to small displacements near x_0, only the lowest power of $x - x_0$ need be retained in Eq. (5.38). When the substitution $(dV/dx)_{x_0} = 0$ is made, Eq. (5.38) reduces to

$$V(x) \cong V(x_0) + \frac{1}{2}\left(\frac{d^2V}{dx^2}\right)_{x_0} (x - x_0)^2 \tag{5.39}$$

and the force associated with this potential energy in the neighborhood of x_0 is, therefore,

$$F = -\frac{dV}{dx} = -\left(\frac{d^2V}{dx^2}\right)_{x_0} (x - x_0') \tag{5.40}$$

The force is thus linearly dependent upon particle displacement from equilibrium just as the force of a linear spring. This result should not be too surprising inasmuch as the potential energy for a linear spring is identical with Eq. (5.39), where the spring constant k replaces $(d^2V/dx^2)_{x_0}$.

The origin may be shifted to the right an amount x_0, making the potential and force functions of x' ($x' = x - x_0$). Of greater significance is the fact that whether the potential minimum is associated with a spring or an analogous physical system, the important variable is the displacement from equilibrium, wherever that point may lie.

The spring–mass system has already been shown to be a harmonic oscillator, and it is clear that in a small neighborhood about x_0, any system possessing a minimum in potential energy will display similar characteristics. This is true regardless of the nature of the potential function beyond the region of small displacements. As $x - x_0$ increases, however, terms such as $(x - x_0)^3$ may no longer be safely discarded, and more and more terms in Eq. (5.38) must be retained to assure a reasonable approximation to the actual potential energy. The motion will then be more complex than simple harmonic.

5.7 CONSERVATION OF MECHANICAL ENERGY

By combining the work-energy equation, Eq. (5.3), with the defining expression for difference of potential energy, the principle of conservation of mechanical energy results:

$$V_b - V_a = -\int_a^b \mathbf{F} \cdot d\mathbf{r} = T_a - T_b \tag{5.41}$$

or

$$V_a + T_a = V_b + T_b = E = \text{constant} \tag{5.42}$$

This important principle states that in a conservative system, the *mechanical energy* E, the sum of the kinetic and potential energies, remains constant. Mechanical energy is thus conserved. Equations (5.41) and (5.42) must be restricted to conservative systems because they include the defining relationship for potential energy, applicable only to such systems.

Consider now the motion of a particle in a one-dimensional conservative force field. Applying Eq. (5.42),

$$T + V = E = \tfrac{1}{2}m\dot{x}^2 + V(x) \tag{5.43}$$

Solving for the velocity,

$$v = \dot{x} = \pm \left\{\frac{2}{m}[E - V(x)]\right\}^{1/2} \tag{5.44}$$

Thus

$$v \sim (E - V)^{\frac{1}{2}}$$

Time may be obtained as a function of displacement through an analytical or numerical integration of Eq. (5.44):

$$t_b - t_a = \pm \int_a^b \frac{dx}{\{2/m[E - V(x)]\}^{\frac{1}{2}}} \tag{5.45}$$

Even without mathematical analysis, much can be learned from careful examination of a plot of $V(x)$ as a function of x as in Fig. 5.12(a). Bearing in

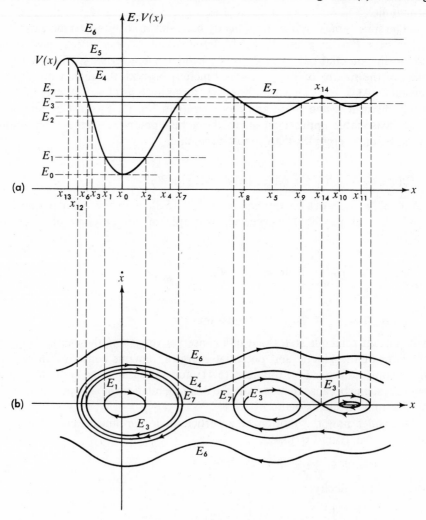

Fig. 5.12

mind the general restriction that the potential energy can at most be equal to the total mechanical energy (at which position the kinetic energy must be zero), Table 5.1 summarizes the types of motion possible for various values of E. Since the difference between the total and potential energies is the kinetic energy, the deeper the potential valleys, the higher the maximum particle velocities attained. For example, a particle possessing total energy E_6 will accelerate and decelerate as it passes over the hills and valleys that describe the potential energy.

Table 5.1

TOTAL MECHANICAL ENERGY	REGION	COMMENT
E_0	Particle at x_0.	E_0 is the lowest possible system energy compatible with $V(x)$. Stable equilibrium at x_0.
E_1	Particle may oscillate between x_1 and x_2.	At x_1 and x_2 (called turning points), the particle reverses direction, because these are points of zero kinetic energy.
E_2	Particle may oscillate between x_3 and x_4 or be in stable equilibrium at x_5.	There are two turning points. x_5 is a *metastable* state, because it does not represent the lowest stable energy state.
E_3	Particle may oscillate between x_6 and x_7, x_8 and x_9, or x_{10} and x_{11}.	There are six turning points.
E_4	See comment.	Turning point at x_{12}. Particle proceeding to this position from the right, stops and reverses direction.
E_5	See comment.	Particle in unstable equilibrium at x_{13}.
E_6	See comment.	No turning points.

The nature of the possible oscillations can be predicted on the basis of the shape of the potential-energy curve between the limits of oscillation. For example, if the particle were confined between x_8 and x_9 possessing energy E_3, its motion would be essentially simple harmonic, because $V(x)$ is nearly parabolic. Since $V(x)$ is symmetrical, the restoring forces are also symmetrical about mean position x_5. This would not be the case for motion between x_6 and x_7, however, where the motion is not even approximately simple harmonic. The particle moving to the left toward x_6 experiences a much more severe deceleration than it does in moving toward x_7, because of the relative steepness of the potential-energy curve near x_6. Motion between x_7 and x_8, x_9 and x_{10}, and so on, is prohibited, because the kinetic energy would be negative and the velocity imaginary in such regions.

The form of Eq. (5.43) suggests that motion may be represented by employing $\dot{x}$ and x as coordinates. These variables uniquely describe the state of the system at any time in what is termed the *phase plane*, with the locus of points called the *phase trajectory*. Such a depiction need not be limited to one-dimensional motion. For *n*-dimensional motion, 2*n* coordinates are required to describe the trajectory in *phase space*.

It may be deduced from Eq. (5.43), in which E is constant, that closed curves in the phase plane represent conservative systems experiencing periodic motion; that is, $x(t) = x(t + \tau)$ and $v(t) = v(t + \tau)$, τ signifying the period of repetition.

In the specific case of the undamped linear oscillator, for which $V(x) = \frac{1}{2}kx^2$, Eq. (5.43) becomes

$$\tfrac{1}{2}m\dot{x}^2 + \tfrac{1}{2}kx^2 = E$$

or

$$\frac{\dot{x}^2}{2E/m} + \frac{x^2}{2E/k} = 1$$

which is an ellipse with semiaxes $(2E/k)^{1/2}$ and $(2E/m)^{1/2}$. Each value of energy thus represents a unique ellipse.

It is interesting and informative to examine the phase-plane depiction of the particle motion corresponding to the potential-energy curve of Fig. 5.12(a). At energy E_1, the motion is periodic between the limits x_1 and x_2, but the closed curve in the phase plane of Fig. 5.12(b) departs from that of an ellipse because the oscillator is not linear. The arrowheads in this figure indicate the direction in which motion proceeds, that is, the direction of increasing time. That the motion must be clockwise as shown should be verified by the reader.

There are three closed curves associated with energy E_3, indicating three possible points about which motion may occur, depending upon the initial position of the particle. For energy E_4, the motion is no longer periodic; the particle is observed to change direction at x_{12} only. No reversal of direction occurs at energy E_6. Consequently, a particle continues to move in its initial direction. This accounts for the two curves labeled E_6 in the phase plane. Finally, position x_{14} is a point of instability inasmuch as the potential energy is at a local maximum. This instability is readily discerned in the phase plane by noting that at x_{14}, motion may proceed to the right or left. Whichever direction the particle moves, it does not continue in this direction indefinitely, as it soon reaches a turning point.

The phase trajectories of nonconservative systems spiral inward or outward depending upon whether energy is dissipated (as by friction) or supplied by an external source.

Illustrative Example 5.5

A frictionless one-dimensional spring–mass system consists of a mass of 8 g attached by means of a linear spring ($k = 20$ dyn/cm) to a rigid wall. Determine the velocity-displacement characteristics for the system at three values of total mechanical energy: 1000, 490, and 160 dyn-cm.

SOLUTION

Since the system is conservative, $T + V = E$, where $V = \frac{1}{2}kx^2$ and $T = \frac{1}{2}m\dot{x}^2$. Solving for $\dot{x}$,

$$\dot{x} = \pm\left[\frac{2}{m}(E - \tfrac{1}{2}kx^2)\right]^{1/2} = [\tfrac{2}{8}(E - 10x^2)]^{1/2}$$

For $E = 1000$ dyn-cm, x ranges between ±10 cm and $\dot{x}$ between ±16.6 cm/sec. Figure 5.13 shows the ellipses of $\dot{x}$ as a function of x for various total energies.

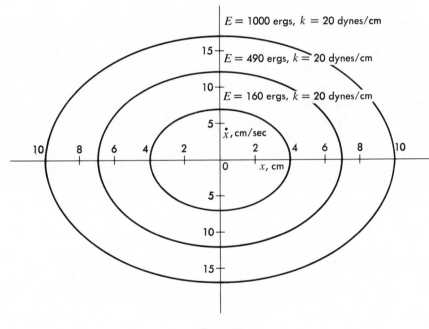

Fig. 5.13

5.8 ENERGY CONSIDERATIONS IN CONSTRAINED MOTION

In solving the equations of motion, the matter of *constraint* has been implicit regardless of the particular approach taken. By constraint is meant some limitation placed upon the possible motions of a given system. These limitations are usually represented in geometric terms. Thus in the problems of Chapter 3 dealing with masses sliding down inclined planes, there were one-sided constraints limiting motions in such a way that the plane itself was not penetrated. An example of complete constraint is the case of a hollow bead sliding along a wire. Another familiar constrained system is the simple pendulum, in which the mass is constrained to move in such a way as to remain a constant distance from a fixed point. An example of a nongeometric constraint is provided by a governor that prevents the speed of a machine from exceeding a prescribed value. In Chapter 14, we treat the matter of constraints in a formal fashion and, in so doing, discuss their classification.

The effect of constraint may be to limit the degrees of freedom of the system. For example, consider a mass constrained to move on the surface of a sphere. Whereas an unconstrained mass has three degrees of freedom, that is, may move independently in the X-, Y-, and Z-directions, the mass constrained as described is subject to the following relationship among variables:

$$x^2 + y^2 + z^2 = r^2 = r_0^2 \tag{5.46}$$

where r_0 is the radius of the sphere, and the origin of X, Y, Z and the center of the sphere coincide. Since r is fixed, it is clear that when two of the three variables are determined, the third is also determined. The system thus has only two degrees of freedom. The reduction in the number of degrees of freedom may be generalized by stating that in a system described by n coordinates and m independent equations of constraint, the number of degrees of freedom is $n - m$. Employing a different coordinate system does not change $n - m$. This may be observed in the example under discussion by simply selecting spherical coordinates rather than Cartesian. Since r is fixed, the constraint equation is $r = r_0 = $ constant, and the remaining coordinates are θ and ϕ. Thus $n - m$ remains 2. The number of degrees of freedom is a function not of the coordinate system but rather of the physical system.

Consider now the case of the bead constrained to move on a fixed wire as shown in Fig. 5.14(a). Referring to the free-body diagram [Fig. 5.14(b)], the

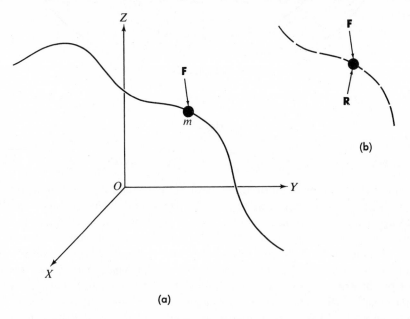

(b)

(a)

Fig. 5.14

forces acting on the bead are **F**, the sum of all applied forces, and **R**, the reaction of the wire on the mass. **R** is thus the constraining force. The equation of motion is

$$\mathbf{F} + \mathbf{R} = m(\ddot{x}\hat{\mathbf{i}} + \ddot{y}\hat{\mathbf{j}} + \ddot{z}\hat{\mathbf{k}}) = m\dot{\mathbf{v}} \tag{5.47}$$

The scalar product with **v** of Eq. (5.47) gives

$$\mathbf{F} \cdot \mathbf{v} + \mathbf{R} \cdot \mathbf{v} = m\dot{\mathbf{v}} \cdot \mathbf{v} \tag{5.48}$$

For frictionless constraint, **R** is everywhere perpendicular to the path. Since **v** is tangent to the path it follows that $\mathbf{R} \cdot \mathbf{v} = 0$, and Eq. (5.48) may therefore be expressed

$$\mathbf{F} \cdot \mathbf{v} = m\dot{\mathbf{v}} \cdot \mathbf{v} = \frac{d}{dt}(\tfrac{1}{2}m\mathbf{v} \cdot \mathbf{v}) \tag{5.49}$$

or

$$F_x\dot{x} + F_y\dot{y} + F_z\dot{z} = \frac{d}{dt}(\tfrac{1}{2}mv^2) \tag{5.50}$$

leading to the integral form

$$\int (F_x\,dx + F_y\,dy + F_z\,dz) = \tfrac{1}{2}mv^2 + C' \tag{5.51}$$

If **F** represents a conservative force, Eq. (5.34) applies and may be substituted into Eq. (5.51), yielding

$$-\int \left(\frac{\partial V}{\partial x}\,dx + \frac{\partial V}{\partial y}\,dy + \frac{\partial V}{\partial z}\,dz\right) = T + C' \tag{5.52}$$

or simply

$$T + V = C$$

where C represents the total system energy. We have thus obtained the anticipated result: Mechanical energy is conserved as long as the particle does not violate the constraints imposed upon it and provided the force field is conservative. In the case at hand, it is convenient to express Eq. (5.52) in terms of the displacement variable measured along the wire:

$$\tfrac{1}{2}m\dot{s}^2 + V(s) = C \tag{5.53}$$

5.9 NONCONSERVATIVE FORCES

Forces that cannot be expressed in terms of a potential are termed *nonconservative*. Frictional forces are always directed opposite to the motion and are not single-valued functions of position alone, hence nonconservative. Time-dependent forces can likewise not be derived from a position-dependent scalar potential.

The influence of frictional forces is very much dependent upon the path length, the work done in overcoming friction increasing with length of path. Here we have a direct contrast between conservative and nonconservative force systems. In the discussion of force systems composed of both conservative and nonconservative forces which follows, the dissipative nature of frictional forces is demonstrated. Not all nonconservative forces *remove* mechanical energy from a system, however. It is important to note that the motion of a mechanical system is initiated by *adding* energy through the action of nonconservative forces.

5.10 FORCE SYSTEMS CONTAINING CONSERVATIVE AND NONCONSERVATIVE FORCES

An object falling freely through a vacuum under the influence of a gravitational field is an example of a conservative force system. If the mass instead falls through a resistive medium such as air, it is not always satisfactory to neglect the effects of the dissipative forces. This is amply demonstrated in Chapter 7. Many real situations in physics and engineering lend themselves to practical interpretation and solution only when they are represented by force systems containing both conservative and nonconservative forces. The work-energy relationship may be employed in such cases, because it is general:

$$\int_a^b \mathbf{F} \cdot \mathbf{dr} = \int_a^b \mathbf{F}_c \cdot \mathbf{dr} + \int_a^b \mathbf{F}_{nc} \cdot \mathbf{dr} = T_b - T_a \tag{5.54}$$

where the net force $\mathbf{F}$ has been separated into conservative $(\mathbf{F}_c)$ and nonconservative $(\mathbf{F}_{nc})$ parts. Recalling that the difference of potential energy is defined in terms of only the conservative forces,

$$\int_a^b \mathbf{F}_c \cdot \mathbf{dr} = V_a - V_b \tag{5.55}$$

Combining Eqs. (5.54) and (5.55),

$$V_a + T_a + \int_a^b \mathbf{F}_{nc} \cdot \mathbf{dr} = V_b + T_b \tag{5.56}$$

The work done in overcoming friction is always negative, because $\mathbf{F}_{nc}$ is opposite the displacement. Equation (5.56) demonstrates, therefore, that the influence of friction is dissipative, decreasing the total mechanical energy of the system. Stated another way,

$$(V_b - V_a) + (T_b - T_a) = \int_a^b \mathbf{F}_{nc} \cdot \mathbf{dr} \tag{5.57}$$

or

$$\Delta(V + T) = \Delta E = \int_a^b \mathbf{F}_{nc} \cdot \mathbf{dr} \tag{5.58}$$

In the case of sliding, the mechanical energy dissipated as a result of friction manifests itself as a rise in temperature of the contacting bodies, indicating their increased internal energy. It is of interest to recall that processes in which work is converted into internal energy as a result of friction are thermodynamically irreversible. Thus the original state of the system and its surroundings cannot be restored by simply reversing the original process. Such an occurrence would be in direct violation of the second law of thermodynamics. Stated in different (but equivalent) terms, in processes involving friction there occurs the transfer of heat through finite temperature differences. Such processes are irreversible.

Illustrative Example 5.6

A force system composed of a conservative force (due to a linear spring) and a nonconservative force (due to Coulomb friction) is shown in Fig. 5.15. If the mass is initially displaced $\delta = 20$ cm to the left of the undeformed spring position and released, determine the manner in which the total mechanical energy of the system varies with position until motion ceases.

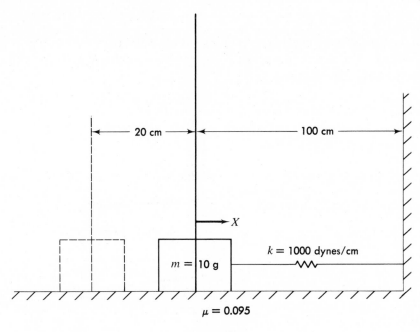

Fig. 5.15

SOLUTION

The energy of the system is being continuously dissipated because of the influence of friction. Initially the linear spring possesses the entire system energy: $E = \frac{1}{2}k\delta^2 = \frac{1}{2}(1000)(400) = 200{,}000$ ergs. When the mass returns to $x = 0$ from $x = -20$, its kinetic energy is calculated as follows:

$$T_{x=0} + V_{x=0} = T_{x=-20} + V_{x=-20} + W_f$$

where W_f, the friction work, equals μmgx, of sign opposite that of the motion. Since $T_{x=-20} = 0$ and $V_{x=0} = 0$,

$$T_{x=0} = 200{,}000 - (0.095)(10)(980)(20) = 181{,}200 \text{ ergs}$$

How far to the right does the mass now proceed? The total system energy at $x = 0$ of 181,200 ergs is decreased by an amount μmgx, where x now represents the total displacement to the right of $x = 0$. Since the kinetic energy is zero at the limit of travel to the right,

$$181{,}200 - (0.095)(10)(980)x = \tfrac{1}{2}(1000)x^2$$

$$x = 18.1 \text{ cm}$$

The remaining root of x is extraneous because it corresponds to a higher energy state than the initial total energy.

The foregoing analysis is repeated until, for the system at rest,

$$-kx_f + \mu mg = 0$$

where x_f represents the final mass position. At this point the system energy is $\tfrac{1}{2}kx_f^2$.

Figure 5.16 clearly indicates the dissipation of mechanical energy with each oscillation until motion ceases.

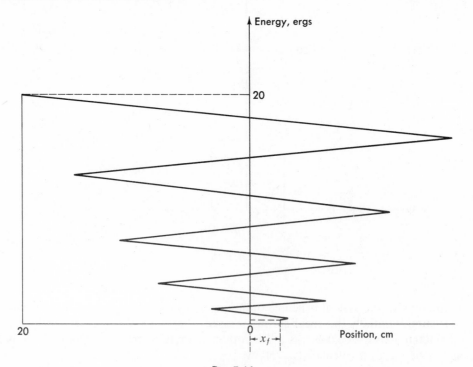

Fig. 5.16

References

Goodman, L. E., and Warner, W. H., *Dynamics*, Wadsworth Publishing Co., Inc., Belmont, Calif., 1963.

Greenwood, D. T., *Principles of Dynamics*, Prentice-Hall, Inc., Englewood Cliffs, N.J., 1965.

Marion, J. B., *Classical Dynamics*, Academic Press, Inc., New York, 1965.

Wangsness, R. K., *Introduction to Theoretical Physics*, John Wiley & Sons, Inc., New York, 1963.

EXERCISES

5.1 What work is required to raise a 1 kg mass 10,000 km above the surface of the earth? What is the potential energy of the mass at this position?

5.2 A mass of 0.1 kg is attached by means of a linear spring ($k = 2000$ N/m) to a rigid wall. If the mass is initially displaced 0.1 m to the left of the equilibrium position and released, determine the maximum velocity of the mass. What is the total system energy? Assume frictionless contact between the mass and supporting surface.

5.3 A block weighing 10 lb is attached to a vertical spring ($k = 20$ lb/in.). If the spring is compressed 2 in. above the undeformed position and the mass is then released with a downward velocity of 5 ft/sec, calculate:

    a. the total potential energy of the system (relative to a suitable datum) at the maximum upward position.

    b. the velocity of the mass when the spring is again undeformed.

    c. the potential energy of the system at the lowest position.

    d. the lowest position of the system.

    e. the equilibrium position of the mass.

5.4 An object of mass 100 kg is pulled 10 m up a frictionless incline making a 20-degree angle with the horizontal. If the speed of the mass is constant and the rope pulling the object makes an angle of 5 degrees above the incline, calculate:

    a. the work done by gravity.

    b. the work done by the rope tension.

5.5 Repeat Exercise 5.4 assuming a constant coefficient of friction of 0.4 between the mass and the inclined plane. What work is done by the force of friction?

5.6 An object of mass 20 kg is dropped a distance of 100 m, striking a linear spring (20,000 N/m). Determine the maximum compression of the spring assuming that no energy is lost during contact between mass and spring.

5.7 Repeat Exercise 5.6 assuming the spring force acting on the object to equal $-100x^3$ N, where x is in meters.

5.8 A 10 g bullet is fired directly through a wall of 2 cm thickness. During this process the velocity is reduced from 30,000 to 20,000 cm/sec. Calculate the effective force of friction, assuming it to remain constant throughout the penetration.

5.9 A 100 g sphere, dropped onto a surface from a height of 10 m, is found to possess a rebound velocity of 1000 cm/sec. What fraction of the total initial energy is lost in the collision?

5.10 A block weighing 50 lb is projected down a 30 degree inclined plane with a velocity of 12 ft/sec. If the coefficient of friction is assumed constant at 0.3, what distance measured along the incline is required for the mass to halve its velocity? What change in potential energy occurs, and what work is done by friction and the normal reaction of the incline on the mass?

5.11 An object of mass m, initially at rest, is dropped from a height h onto a linear spring k. Calling the constant frictional force associated with the spring f, derive an expression to predict the maximum height to which the object rebounds.

5.12 A 5 g particle experiences a displacement from $(0, 0, 1)$ to $(5, -3, 2)$ cm, during which a constant force $\mathbf{F} = 3\hat{\mathbf{i}} + 2\hat{\mathbf{j}} - 6\hat{\mathbf{k}}$ dyn acts. Determine the change in kinetic energy. Is the force conservative? What is the potential?

In Exercises 5.13 through 5.17, determine the potential energy for those force fields which are conservative.

5.13 $\mathbf{F} = (2x + 2y)\hat{\mathbf{i}} + 2x\hat{\mathbf{j}}$.

5.14 $\mathbf{F} = (4xy - z^3)\hat{\mathbf{i}} + 2x^2\hat{\mathbf{j}} - 3xz^2\hat{\mathbf{k}}$.

5.15 $\mathbf{F} = 2x\hat{\mathbf{i}} + y\hat{\mathbf{j}} - z\hat{\mathbf{k}}$.

5.16 $\mathbf{F} = 3R^3 \sin \phi\hat{\mathbf{e}}_R + 3R^2 \cos \phi\hat{\mathbf{e}}_\phi + 3z^2\hat{\mathbf{k}}$.

5.17 $\mathbf{F} = 2x^2\hat{\mathbf{i}} + (4y - 3z)\hat{\mathbf{j}} + (2x + 3y)\hat{\mathbf{k}}$.

5.18 The potential energy of a particle varies inversely as its distance from the origin, that is, $V = A/r$, where A is a constant. Determine the cylindrical polar components of force and verify that the force is conservative.

5.19 Using the potential energy of Exercise 5.18 determine the Cartesian components of force and verify that the force is conservative.

5.20 A 50 lb weight is supported by two identical springs ($k = 100$ lb/in.) as shown in Fig. P5.20. If the angle between each spring and the horizontal is 45 degrees when the springs are undeformed, determine the value of θ and y at equilibrium. The free length of each spring is 1 ft.

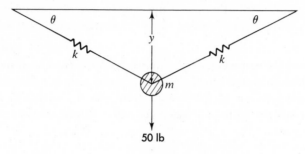

Fig. P5.20

5.21 Referring to Fig. P5.21, point O represents the center of a field in which a repulsive force $F = 2 \times 10^{-6}/r^2$ dyn/g acts. Here r represents the distance from any point to point O. If a mass of 1 g possesses a velocity of 300 cm/sec at infinity and comes to rest at r_1, what impulse acts on the mass as it travels a distance of 10^{-6} cm from r_2 to r_1? What are the kinetic energy and momentum at r_2?

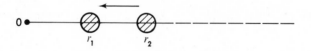

Fig. P5.21

5.22 Demonstrate that Eq. (5.53) is equivalent to $F_s = m\ddot{s}$, where F_s represents the component of external force parallel to the path of motion.

5.23 The force of attraction between a charged body and a point charge located at the origin is given by

$$\mathbf{F} = -\frac{A}{r^2}\hat{\mathbf{e}}_r$$

where A represents a constant and r is the distance from the origin to the body. Determine the potential function if one exists.

5.24 If in Exercise 5.23, $A = 1000$ dyn-cm^2, calculate the work done in moving the body from $r = 5$ cm to $r = 10$ cm. What work is done in moving it from $r = 8$ to $r = 6$ cm? Determine the potential energy at $r = 10$ cm, taking infinite r to be a point of zero potential.

5.25 A particle of mass 10 g is supported by a massless string and held so that the string makes an angle θ of 30 degrees with the vertical. Selecting a datum, determine the potential energy just as the mass is released. Establish the potential-energy function and verify the lowest point to be the equilibrium position. What are the velocity and potential energy at $\theta = 15$ degrees and $\theta = 0$ degrees?

5.26 A pendulum is constructed of a massless spring (k) and a mass (m). If the system is held at an initial angle θ_0 with the vertical and released, determine the equilibrium configuration of the system. Assume the spring to be undeformed (length l_0) just before the mass is released. Also determine the maximum kinetic energy of the system.

5.27 Given the following field of force, determine the force potential:

$$\mathbf{F} = 2 \sin x \hat{\mathbf{i}} + y^2 \hat{\mathbf{j}} - z \hat{\mathbf{k}}$$

5.28 Given the force of Exercise 5.23, write the equation of conservation of energy in cylindrical polar coordinates.

5.29 Given the potential energy $V = 3x^2 + 2y - z$, determine the corresponding force field.

5.30 Determine the force field corresponding to the function $V = r^2 \cos 2\phi$.

5.31 Demonstrate that the following force field is conservative. What point must be omitted from the force field? Find the potential energy. $\mathbf{F} = (A/r)\hat{\mathbf{e}}_r$.

5.32 What is the equilibrium position of the mass shown in Fig. P5.32 in which the linear springs (k_1, k_2) have undeformed lengths l_1 and l_2? The system is assumed frictionless.

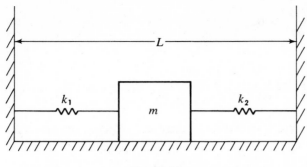

Fig. P5.32

5.33 Repeat Exercise 5.32 for a system similar to that described except that the spring force varies as the square of the deformation.

5.34 Determine the relationships among the constants for which the following force is conservative:

$$\mathbf{F} = (ax^2 + bx + cy)\hat{\mathbf{i}} + (2dx + ey^3)\hat{\mathbf{j}} + fz\hat{\mathbf{k}}$$

5.35 Find the region in which the mass of Exercise 5.32 is in equilibrium if the coefficient of static friction is μ_s.

5.36 The Yukawa potential for a pion (a nuclear particle unstable outside the nucleus) inside a nucleus is given by

$$V(r) = \frac{-G_0 e^{-\beta r}}{r}$$

where G_0 and β are positive constants and r represents the distance from the center of the nucleus to the pion. Determine the force as well as the equilibrium position of the pion.

5.37 Find the points of stable and unstable equilibrium and the turning points for a particle in the following potential field:

$$V(r) = V_0 \left(\frac{r}{a} - \frac{a}{r}\right) e^{-r/a}$$

where $a > 0$.

5.38 Demonstrate that if $V = V(x, y)$, the criteria for stable equilibrium in addition to $\partial V/\partial x = \partial V/\partial y = 0$ are

$$\frac{\partial^2 V}{\partial x^2} + \frac{\partial^2 V}{\partial y^2} > 0$$

and

$$\left(\frac{\partial^2 V}{\partial x \, \partial y}\right)^2 - \left(\frac{\partial^2 V}{\partial x^2} \frac{\partial^2 V}{\partial y^2}\right) < 0$$

5.39 Prove that curl $\mathbf{F} = \mathbf{0}$ is a *necessary* and *sufficient* condition for the force to be conservative.

5.40 Under what conditions is a force, which is solely dependent upon r, conservative?

5.41 One component of force acting upon a particle is $ar^2\hat{\mathbf{e}}_\theta$. What other components must be present to make $\mathbf{F}$ conservative?

5.42 An anharmonic oscillator of mass m is subject to the force $F(x) = -\beta x^3$. Determine the maximum velocity if the initial conditions are x_0, v_0.

CHAPTER 6

OSCILLATIONS OF A PARTICLE

6.1 INTRODUCTION

A full treatment of vibration in general and the linear oscillator in particular might have been included in Chapters 3 and 5 as examples of integration of Newton's second law and application of the energy approach. It is hoped that the special attention afforded here will give emphasis to the characteristics shared by so many physical systems. Although the importance of mathematics as an invaluable tool of analysis cannot be denied, the reader should, as early as possible, make an effort to gain insight into the physics underlying a given situation. It can be challenging and rewarding to examine a mechanical or electrical system, allocate particular roles to the cast of constituents, and predict in a general way the behavior under various conditions even before equations are written down. This approach is quite fruitful in the area of linear oscillations because of the analogous behavior of many real systems.

6.2 FREE OSCILLATION — NEWTON'S SECOND LAW

Consider the case of a mass m attached by means of a spring to a rigid wall as shown in Fig. 6.1(a). The actual system under analysis may bear little visual similarity to this illustration. To obtain a tractable equation of motion, a real situation may be reduced to this form only after considerable compromise with reality.

We are confronted with the usual problem: Determine the position and the velocity of the mass as a function of time. A number of idealizations prove extremely helpful:

    a. The spring is subject to neither internal nor external frictional forces.
    b. The spring is without mass.
    c. The force–displacement characteristic of the spring is linear.

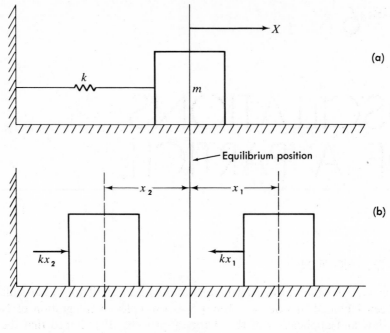

Fig. 6.1

d. The mass is subject to no frictional forces.

e. The wall is rigid, so the wall recoil due to motion of the mass may be neglected.

Despite the fact that we begin with many simplifications and approximations, the solutions are often surprisingly accurate, because the essential aspects of the problem have not been compromised.

In our analysis we assume further that the mass is constrained to move in a straight line, parallel to the supporting surface. Consequently only one coordinate is required to specify the position of the mass at any time.

Consistent with the above assumptions, the only net force acting on the mass is that of the spring. Let us select the position of the mass when the spring is undeformed as the origin of the X-coordinate system. If the spring obeys Hooke's law [1]—*ut tensio, sic vis*—the force acting on the mass is negatively proportional to the displacement of the mass from its equilibrium position, as in Fig. 6.1(b):

$$F = -kx \tag{6.1}$$

The constant of proportionality, k, is a measure of the stiffness of the spring and is usually called the *spring constant* or *stiffness constant*. It is measured in units of force per unit displacement, for example, newtons per meter, pounds per foot.

---

[1] Hooke, R., *De Pontencia Restitutiva*, 1678.

It can be calculated from material and geometric properties of the spring, or measured. The spring displacement must be small enough so that no permanent deformation is introduced. Situations in which the force–displacement characteristics of the spring are nonlinear often lead to complex differential equations of motion. Even in such cases, however, the linear spring approximation may often yield satisfactory results, provided motion is restricted to small displacements from equilibrium.

Applying Newton's second law of motion for the constant-mass case,

$$\sum F_x = -kx = m\ddot{x} \tag{6.2a}$$

or

$$m\ddot{x} + kx = 0 \tag{6.2b}$$

Before we solve this equation, mention should be made that similar equations occur in many entirely different areas of physics. Consider, for example, an electrical circuit containing no sources of potential, consisting simply of an ideal capacitor of capacitance C and an ideal inductor of inductance L, connected as in Fig. 6.2. The voltage drop across the capacitor, Q/C, is equal to the emf induced in the inductor, $-L\ddot{Q}$, according to Kirchhoff's second law,

$$\frac{Q}{C} = -L\ddot{Q} \tag{6.3}$$

where Q is the charge. Clearly Q is analogous to x, as L and $1/C$ are analogous to m and k, respectively.

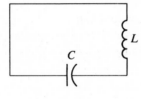

Fig. 6.2

Once we have solved Eq. (6.2), we have, by analogy, also solved Eq. (6.3). Specialized techniques available to solve any of these equations will also apply to any other analogous equation.

The best approach to the solution of Eq. (6.2b), a linear second-order homogeneous differential equation, is to make a shrewd guess of the correct solution, $x = x(t)$. This technique implies having seen the equation or a similar one before. In the absence of such foresight, some logic is required. The equation is linear in the dependent variable x and in its time derivatives. In addition, the coefficients of x and $\ddot{x}$ are constant. Since the derivative of an exponential is a constant multiplied by the exponential, a solution of the form

$$x(t) = Be^{st} \tag{6.4}$$

is justified, where B and s are constants determined by substitution into Eq. (6.2b). Differentiating Eq. (6.4),

$$\dot{x} = sBe^{st} \tag{6.5}$$

and

$$\ddot{x} = s^2Be^{st} \tag{6.6}$$

Substituting for x and $\ddot{x}$ in Eq. (6.2b),

$$mBs^2e^{st} + kBe^{st} = 0 \tag{6.7}$$

Dividing both sides of Eq. (6.7) by Be^{st},

$$s^2 = -\frac{k}{m} \tag{6.8}$$

Thus s has two values, both complex, $\pm i(k/m)^{1/2}$, where $i = (-1)^{1/2}$. A common designation for $(k/m)^{1/2}$ is ω_0, the natural circular frequency of vibration. Either value of s will make $X = Be^{st}$ a solution of Eq. (6.2b). Since at present we have no information concerning B, a solution of the form

$$x = B_1e^{i\omega_0 t} + B_2e^{-i\omega_0 t} \tag{6.9}$$

is general and also satisfies Eq. (6.2b) (where B_1 and B_2 are constants yet to be determined). We live in a real world; therefore, x must be real, but B_1 and B_2 may be complex. Thus the solution of the equation of motion contains two constants of integration which can be evaluated only by applying some subsidiary information such as the position and velocity at a specified time. We conveniently measure our time from some instant $t = 0$ when the position is x_0 and the velocity v_0, both considered known. Substituting $t = 0$ into Eq. (6.9),

$$x_0 = B_1 + B_2 \tag{6.10}$$

Differentiating Eq. (6.9),

$$\dot{x} = i\omega_0 B_1 e^{i\omega_0 t} - i\omega_0 B_2 e^{-i\omega_0 t} \tag{6.11}$$

which yields, at $t = 0$,

$$v_0 = i\omega_0 B_1 - i\omega_0 B_2 \tag{6.12}$$

Solving Eqs. (6.10) and (6.12) for B_1 and B_2,

$$B_1 = \frac{1}{2}\left(x_0 + \frac{v_0}{i\omega_0}\right)$$

$$B_2 = \frac{1}{2}\left(x_0 - \frac{v_0}{i\omega_0}\right) \tag{6.13}$$

Substitution for B_1 and B_2 in Eq. (6.9) gives the solution of Eq. (6.2):

$$x = \frac{1}{2}\left(x_0 + \frac{v_0}{i\omega_0}\right)e^{i\omega_0 t} + \frac{1}{2}\left(x_0 - \frac{v_0}{i\omega_0}\right)e^{-i\omega_0 t} \tag{6.14}$$

We now employ Euler's identity to obtain a more convenient expression for $x(t)$:

$$e^{i\omega_0 t} = \cos \omega_0 t + i \sin \omega_0 t$$

$$e^{-i\omega_0 t} = \cos \omega_0 t - i \sin \omega_0 t$$

Substitution into Eq. (6.14) yields

$$x = \frac{1}{2}\left(x_0 + \frac{v_0}{i\omega_0}\right)(\cos \omega_0 t + i \sin \omega_0 t) + \frac{1}{2}\left(x_0 - \frac{v_0}{i\omega_0}\right)(\cos \omega_0 t - i \sin \omega_0 t) \tag{6.15}$$

Carrying out the arithmetic operations and collecting terms, we have

$$x = x_0 \cos \omega_0 t + \frac{v_0}{\omega_0} \sin \omega_0 t \tag{6.16}$$

The displacement $x(t)$ in Eq. (6.16) is clearly real, and may be still more succinctly written

$$x = A \cos (\omega_0 t + \theta_0) \tag{6.17}$$

where A and θ_0 are the constants of integration appropriate to this form. Expansion of $\cos (\omega_0 t + \theta_0)$ yields

$$x = A \cos \omega_0 t \cos \theta_0 - A \sin \omega_0 t \sin \theta_0 \tag{6.18}$$

Equating like coefficients of $\sin \omega_0 t$ and $\cos \omega_0 t$ in Eqs. (6.16) and (6.18),

$$A \cos \theta_0 = x_0 \tag{6.19a}$$

and

$$-A \sin \theta_0 = \frac{v_0}{\omega_0} \tag{6.19b}$$

Dividing Eq. (6.19b) by Eq. (6.19a),

$$\tan \theta_0 = -\frac{v_0}{\omega_0 x_0} \tag{6.20a}$$

or

$$\theta_0 = \tan^{-1}\left(-\frac{v_0}{\omega_0 x_0}\right) \tag{6.20b}$$

Squaring Eqs. (6.19a) and (6.19b) and adding,

$$A^2 \cos^2 \theta_0 + A^2 \sin^2 \theta_0 = x_0^2 + \left(\frac{v_0}{\omega_0}\right)^2$$

but

$$\cos^2 \theta_0 + \sin^2 \theta_0 = 1$$

Therefore,

$$A^2 = x_0^2 + \left(\frac{v_0}{\omega_0}\right)^2 \tag{6.21}$$

Finally,

$$x = \left[x_0^2 + \left(\frac{v_0}{\omega_0}\right)^2\right]^{\frac{1}{2}} \cos(\omega_0 t + \theta_0) \tag{6.22}$$

$$\dot{x} = -\omega_0 \left[x_0^2 + \left(\frac{v_0}{\omega_0}\right)^2\right]^{\frac{1}{2}} \sin(\omega_0 t + \theta_0) \tag{6.23}$$

$$\ddot{x} = -\omega_0^2 \left[x_0^2 + \left(\frac{v_0}{\omega_0}\right)^2\right]^{\frac{1}{2}} \cos(\omega_0 t + \theta_0) \tag{6.24}$$

Equations (6.14), (6.16), and (6.22) are equivalent representations of the mass position, subject to a single linear restoring force negatively proportional to displacement.

Note that the displacement repeats itself at time intervals of $2\pi/\omega_0$. Thus the position, velocity, and acceleration at any time have the same respective values at intervals of $2\pi/\omega_0$, as shown in Fig. 6.3. This time is called the *period*, T. The number of times per second that the repetition occurs is the reciprocal of the period termed the *natural frequency*, f_0. Thus

$$T = \frac{2\pi}{\omega_0} = \frac{1}{f_0} \tag{6.25}$$

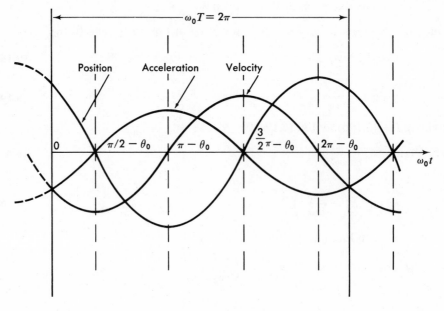

Fig. 6.3

Since the argument of the cosine function in Eq. (6.22) is expressed in radians, ω_0 has the units radians per second. The reader is urged to check the units of ω_0 from its defining expression.

The maximum distance the mass moves from equilibrium is the *amplitude*, equal to

$$\left[x_0^2 + \left(\frac{v_0}{\omega_0} \right)^2 \right]^{\frac{1}{2}}$$

in Eq. (6.22).

Equations (6.22), (6.23), and (6.24) describe *simple harmonic motion*. *Harmonic motion* is described by the sum of sine and cosine functions which repeat themselves at integral *multiples* of a single frequency. In simple harmonic motion, a *single* frequency describes the periodicity.

Another mechanical system, the simple pendulum, will now be explored to illustrate further the linear oscillator. A mass m, considered a point mass, is suspended from a massless cord of length L, as shown in Fig. 6.4(a). The cord is connected to a rigid support by means of a frictionless pinned arrangement. We shall assume the mass constrained so as to move in a plane. Consequently, only a single coordinate, ψ, is required to describe completely the position of the mass at any time. Once again, we are dealing with a single degree of freedom system.

A free-body diagram of the mass, Fig. 6.4(b), reveals two forces acting: the force of gravity pulling the mass toward the center of the earth, and the cord tension pulling the mass toward the pin. A component of the gravitational force serves

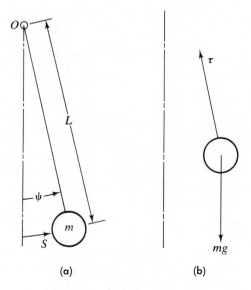

(a)   (b)

Fig. 6.4

as the restoring force, acting to return the system to its lowest potential energy at $\psi = 0$, playing much the same role as the spring discussed earlier.

One approach is to take moments of force about the pin and equate the sum of the moments to the product of moment of inertia I and angular acceleration $\ddot{\psi}$. The moment of inertia for rotating systems is analogous to the property of mass for translating systems. For a point mass, $I = mL^2$, where L is the distance from the point about which moments are taken, to the mass. The moment contribution of the weight is $-mgL \sin \psi$, where the minus sign is consistent with a moment of force which tends to return the system to $\psi = 0$. Recall the analogous influence of the spring. The moment of the cord tension is zero, because its line of action passes through point O. The differential equation of motion is, therefore,

$$-mgL \sin \psi = mL^2 \ddot{\psi} \qquad (6.26a)$$

or

$$\ddot{\psi} + \frac{g}{L} \sin \psi = 0 \qquad (6.26b)$$

Before we solve Eq. (6.26) let us arrive at the same equation in another manner. Writing Newton's second law for the S-direction, we have

$$-mg \sin \psi = m\ddot{s} \qquad (6.27)$$

Since $s = L\psi$, $\ddot{s} = L\ddot{\psi}$, substitution of which into Eq. (6.27) again gives Eq. (6.26).

As Eq. (6.26) now stands, it does not represent a linear oscillator because the restoring force is not linearly proportional to the position variable. If, however, ψ is limited to small angles (less than $1/10$ rad) $\sin \psi$ may be approximated by ψ, and Eq. (6.26) may be rewritten

$$\ddot{\psi} + \frac{g}{L} \psi = 0 \qquad (6.28)$$

The angle ψ corresponds to the displacement x and g/L is analogous to k/m. The solution of Eq. (6.28) is, therefore,

$$\psi = \psi_{max} \cos (\omega_0 t + \theta_0) \qquad (6.29)$$

where $\omega_0 = (g/L)^{1/2}$ and ψ_{max} is the maximum angle or amplitude of oscillation. It is important to realize that $\dot{\psi}$ is the angular velocity of the pendulum, with the same dimensions as ω_0. The former refers to the time rate of change of the angle measured from the vertical and is not constant. The latter refers to the angular rate at which the angle ψ repeats itself and is a constant.

6.3 AN ENERGY APPROACH

The principle of conservation of mechanical energy offers an alternative means of analyzing conservative systems. We are, of course, restricted to forces derivable entirely from scalar potentials. For the frictionless spring–mass system discussed earlier,

$$T + V = \tfrac{1}{2}m(\dot{x})^2 + \tfrac{1}{2}kx^2 = E \tag{6.30}$$

where E is the total energy, a constant.   Rearranging Eq. (6.30),

$$\frac{\dot{x}^2}{[(2E/m)^{1/2}]^2} + \frac{x^2}{[(2E/k)^{1/2}]^2} = 1 \tag{6.31}$$

Considering x and $\dot{x}$ as variables, Eq. (6.31) describes an ellipse for which $(2E/m)^{1/2}$ and $(2E/k)^{1/2}$ are the semimajor and semiminor axes.   In this connection, the reader is reminded of Illustrative Example 5.5.

The energy may be evaluated from a knowledge of position and velocity at any time.   Let us assume that at $t = 0$, $x = x_0$ and $v = \dot{x} = v_0$.   Therefore,

$$E = \tfrac{1}{2}mv_0^2 + \tfrac{1}{2}kx_0^2 = \tfrac{1}{2}m(\dot{x})^2 + \tfrac{1}{2}kx^2 \tag{6.32}$$

Rearranging Eq. (6.32) and solving for velocity,

$$\dot{x} = \left[\frac{2}{m}(E - \tfrac{1}{2}kx^2) \right]^{1/2} \tag{6.33}$$

If $\dot{x}$ is regarded as the quotient of differentials dx and dt and the integration proceeds from x_0 to x and 0 to t, Eq. (6.33) leads to

$$\int_{x_0}^{x} \frac{dx'}{[(2/m)(E - \tfrac{1}{2}kx'^2)]^{1/2}} = \int_{0}^{t} dt' \tag{6.34}$$

For an integral of the form $\int dz/(1 - z^2)^{1/2}$, the solution is $-\arccos z$.   Equation (6.34) can be converted to this form by letting

$$z = \left(\frac{k}{2E} \right)^{1/2} x'$$

and

$$dx' = \left(\frac{2E}{k} \right)^{1/2} dz$$

Upon substitution, Eq. (6.34) becomes

$$t = \int_{x'=x_0}^{x'=x} \frac{(2E/k)^{1/2}\, dz}{[(2E/m) - (2E/m)z^2]^{1/2}} = \int_{x'=x_0}^{x'=x} \left(\frac{m}{k} \right)^{1/2} \frac{dz}{(1 - z^2)^{1/2}} \tag{6.35}$$

so that

$$\left(\frac{k}{m} \right)^{1/2} t = [-\arccos z]_{x'=x_0}^{x'=x} = \arccos \left(\frac{k}{2E} \right)^{1/2} x_0 - \arccos \left(\frac{k}{2E} \right)^{1/2} x$$

or

$$x = \left(\frac{2E}{k}\right)^{\frac{1}{2}} \cos\left[\left(\frac{k}{m}\right)^{\frac{1}{2}} t - \theta_1\right] \qquad (6.36)$$

where

$$\theta_1 = \text{arc cos} \left(\frac{k}{2E}\right)^{\frac{1}{2}} x_0 \qquad (6.37)$$

We must now show that (Eq. 6.36) is the same as Eq. (6.22), with identical constants. The equation for θ_1 may be written

$$\theta_1 = \text{arc cos} \left(\frac{kx_0^2}{2E}\right)^{\frac{1}{2}}$$

Substituting the value for E from Eq. (6.32),

$$\theta_1 = \text{arc cos} \left(\frac{\frac{1}{2}kx_0^2}{\frac{1}{2}kx_0^2 + \frac{1}{2}mv_0^2}\right)^{\frac{1}{2}}$$

On the basis of the foregoing, Fig. 6.5 may be drawn, from which it is clear that

$$\tan \theta_1 = \left(\frac{\frac{1}{2}mv_0^2}{\frac{1}{2}kx_0^2}\right)^{\frac{1}{2}}$$

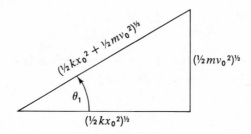

Fig. 6.5

Recalling that $m/k = 1/\omega_0^2$, $\theta_1 = \tan^{-1}(v_0/\omega_0 x_0) = -\theta_0$ in Eq. (6.22). In addition,

$$\left(\frac{2E}{k}\right)^{\frac{1}{2}} = \left(\frac{mv_0^2 + kx_0^2}{k}\right)^{\frac{1}{2}} = \left(\frac{v_0^2}{\omega_0^2} + x_0^2\right)^{\frac{1}{2}} = A$$

and therefore Eq. (6.36) is written $x = A \cos(\omega_0 t + \theta_0)$.

Illustrative Example 6.1

Write the differential equation of motion and determine the natural frequency of oscillation for the pendulum shown in Fig. 6.6. Consider the spring to remain horizontal for small deformations.

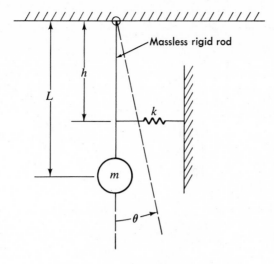

Fig. 6.6

SOLUTION

There are two restoring forces: gravitational attraction and the spring force. Summing moments of force about point O, a frictionless pivot, we have

$$-mgL \sin \theta - k(h \sin \theta)(h \cos \theta) = mL^2 \ddot{\theta}$$

where $I = mL^2$. If oscillations are limited to small angles, $\sin \theta \cong \theta$, $\cos \theta \cong 1$, and the simplified equation is

$$-mgL\theta - kh^2\theta = mL^2\ddot{\theta}$$

The above may be written in the more easily recognized form,

$$\ddot{\theta} + \frac{mgL + kh^2}{mL^2}\theta = 0$$

from which

$$\omega_0 = \left(\frac{mgL + kh^2}{mL^2}\right)^{1/2}$$

Illustrative Example 6.2

A cylindrical object of density ρ_c floats in a liquid of density ρ_l as shown in Fig. 6.7. Write the differential equation of motion and determine the natural frequency of oscillation when the cylinder is depressed by an external force and released.

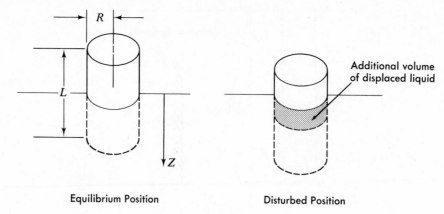

Equilibrium Position   Disturbed Position

Fig. 6.7

SOLUTION

A coordinate system is selected in which the undisturbed position of the cylinder corresponds to $z = 0$, as shown. When the object is depressed, a buoyant force is established equal to the weight of water displaced by the additional submersion beyond that required to maintain static equilibrium. For any z, the unbalanced buoyant force is

$$F_b = -\pi R^2 z \rho_l g$$

where the minus sign indicates that F_b acts to return the object to equilibrium. Writing Newton's second law,

$$-\pi R^2 z \rho_l g = \rho_c \pi R^2 L \ddot{z} \qquad \ddot{z} + \frac{g \rho_l}{L \rho_c} z = 0$$

from which the natural frequency is

$$\omega_0 = \left(\frac{g \rho_l}{L \rho_c}\right)^{\frac{1}{2}}$$

The reader should compare these results with those obtained for the simple pendulum.

Illustrative Example 6.3

Determine the differential equation of motion and the natural frequency for the manometer liquid shown in Fig. 6.8. Employ two approaches: energy and force.

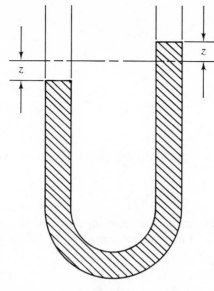

Fig. 6.8

SOLUTION BY ENERGY METHOD
Since each particle of liquid possesses the same average speed at any instant, the kinetic energy of the liquid

$$T = \tfrac{1}{2}mv^2 = \tfrac{1}{2}\rho A L \dot{z}^2$$

where ρ is the liquid density, L the total liquid column length, and A the cross-sectional area of the manometer tube. For an arbitrary liquid displacement as shown, the potential-energy gain of the liquid may be determined by considering the work done against the gravitational field in moving a column of liquid of length z from the left side to the right side of the tube: $V = (\rho g A z)z$. Alternatively, the linear spring relationship may be employed, $V = \tfrac{1}{2}kz^2$, where k is the force required to obtain a unit displacement of the column. The force required to hold the column at the displaced position z is $\rho g A(2z)$. Consequently, $k = 2\rho g A$ and again $V = \rho g A z^2$.

The total system energy is constant:

$$\tfrac{1}{2}\rho A L \dot{z}^2 + \rho g A z^2 = E = \text{constant}$$

Since the time rate of change of energy is zero,

$$\frac{d}{dt}(T + V) = 0 = \rho A L \dot{z}\ddot{z} + 2\rho g A \dot{z}z$$

or

$$\ddot{z} + \left(\frac{2g}{L}\right)^{1/2} z = 0$$

from which

$$\omega_0 = \left(\frac{2g}{L}\right)^{1/2}$$

Again it is interesting to note the similarity between this system and that of the simple pendulum.

FORCE SOLUTION

For any arbitrary displacement, the restoring force is the unbalanced force acting on the liquid, $-(\rho g A)2z$. For a system mass of $\rho A L$, the equation of motion is

$$-2A\rho g z = \rho A L \ddot{z}$$

which is the same result as previously obtained.

Illustrative Example 6.4

A particle of mass m is subject to the influence of the potential energy,

$$V(x) = (1 - \alpha x)e^{-\alpha x}$$

in the region $x \geq 0$, where α represents a positive constant. Determine the location of the equilibrium point(s), the nature of the equilibrium, and the natural frequency of oscillation for motion restricted to small displacements in the neighborhood of equilibrium.

SOLUTION

At equilibrium, $dV/dx = 0$. For the potential energy given,

$$\frac{dV}{dx} = e^{-\alpha x}(\alpha^2 x - 2\alpha)$$

and there is one equilibrium point at $x = 2/\alpha$. The second derivative evaluated at equilibrium provides a test for the nature of equilibrium:

$$\frac{d^2V}{dx^2} = e^{-\alpha x}(3\alpha^2 - \alpha^3 x)$$

At $x = 2/\alpha$, $d^2V/dx^2 = \alpha^2 e^{-2}$. Since $d^2V/dx^2 > 0$, equilibrium is stable.

We have already demonstrated that d^2V/dx^2 evaluated at an equilibrium point plays the role of the spring constant, provided motion is restricted to small values. The natural circular frequency is, therefore,

$$\omega_0 = \left(\frac{k}{m}\right)^{1/2} = \left(\frac{\alpha^2 e^{-2}}{m}\right)^{1/2} = \frac{\alpha}{e}(m)^{1/2}$$

Figure 6.9, a plot of $V(x)$ as a function of x, indicates that for large displacements about $x = 2/\alpha$, the motion is *anharmonic*.

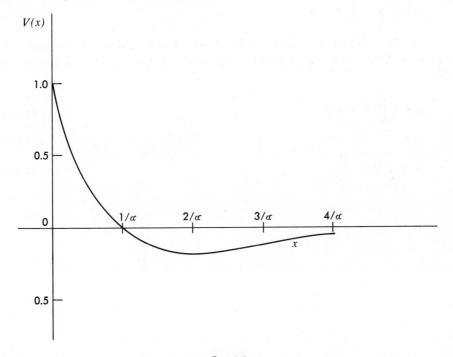

Fig. 6.9

6.4 FREE OSCILLATION WITH DAMPING

We have observed in Chapter 5 that Coulomb friction causes the amplitude of an oscillating system to diminish with time. When a fluid film exists between the mass and the surface over which it slides, it is often justified by fluid mechanical theory and experiment to assume the friction to be velocity-dependent. When the relative velocity of sliding is not "too high," the damping force, as this friction is termed, is linearly proportional to (and opposite) the velocity. For the one-dimensional spring–mass system shown in Fig. 6.1,

$$F_d = -b\dot{x} \tag{6.38a}$$

where b represents the damping coefficient, possessing dimensions of force per unit velocity (for example, newton-seconds per meter, pound-seconds per foot),

$$b \equiv \frac{-F_d}{\dot{x}} \tag{6.38b}$$

From Newton's second law:

$$-kx - b\dot{x} = m\ddot{x} \tag{6.39a}$$

or

$$\ddot{x} + \frac{b}{m}\dot{x} + \frac{k}{m}x = 0 \tag{6.39b}$$

For the electric circuit shown in Fig. 6.10, the voltage drop across resistor R, according to Ohm's law, is $R\dot{Q}$, where Q represents electric charge. For the circuit, shown,

$$R\dot{Q} + \frac{Q}{C} = -L\ddot{Q} \tag{6.40a}$$

or

$$\ddot{Q} + \frac{R}{L}\dot{Q} + \frac{Q}{LC} = 0 \tag{6.40b}$$

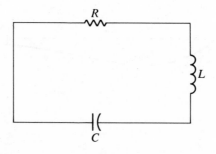

Fig. 6.10

Since Eqs. (6.39) and (6.40) are identical in form, the solution of one is, by analogy, the solution of the other.

In a more exact treatment of the simple pendulum, it may be desirable to consider the air resistance. According to Stokes' law governing the free fall of a sphere through a viscous medium, the frictional force is proportional to the velocity of the sphere. We shall consider the pendulous mass as a small sphere in order to account for the frictional torque developed, and the thickness of the cord small enough to result in negligible friction.

The instantaneous linear velocity of the mass $L\dot{\psi}$ is perpendicular to the cord. The viscous force causes a frictional moment about O equal to $-LF_d$. Denoting the Stokes' law constant of proportionality as C, and substituting for F_d, the frictional moment is written $-L^2C\dot{\psi}$. The resulting modified version of Eq. (6.26a) for small angles becomes

$$-mgL\psi - L^2C\dot{\psi} = mL^2\ddot{\psi} \tag{6.41}$$

and another equation of the form of Eq. (6.39a) results.

As in the case of undamped motion, a solution of the form $x = Be^{st}$ is assumed. Substituting for x, $\dot{x}$, and $\ddot{x}$ in Eq. (6.39b),

$$s^2 Be^{st} + \frac{b}{m} sBe^{st} + \frac{k}{m} Be^{st} = 0$$

Dividing by Be^{st} yields the quadratic algebraic equation

$$s^2 + \frac{b}{m} s + \frac{k}{m} = 0 \tag{6.42}$$

the roots of which are

$$s_1 = -\frac{b}{2m} - \left[\left(\frac{b}{2m} \right)^2 - \frac{k}{m} \right]^{1/2} \tag{6.43a}$$

$$s_2 = -\frac{b}{2m} + \left[\left(\frac{b}{2m} \right)^2 - \frac{k}{m} \right]^{1/2} \tag{6.43b}$$

There are three distinct possibilities for s_1 and s_2, each leading to a different class of solution to Eq. (6.39) and to correspondingly different types of motion:

$$\left(\frac{b}{2m} \right)^2 > \frac{k}{m} \qquad s_1 \text{ and } s_2 \text{ are real and negative}$$

$$\left(\frac{b}{2m} \right)^2 < \frac{k}{m} \qquad s_1 \text{ and } s_2 \text{ are complex conjugates}$$

$$\left(\frac{b}{2m} \right)^2 = \frac{k}{m} \qquad s_1 \text{ and } s_2 \text{ are real and equal}$$

The three cases are referred to as *overdamped*, *underdamped*, and *critically damped*, respectively. For each, two linearly independent solutions are required for a complete solution.

For overdamped motion, the displacement is

$$x = B_1 e^{s_1 t} + B_2 e^{s_2 t} \tag{6.44}$$

Applying the initial conditions, $x = x_0$ and $\dot{x} = v_0$ at $t = 0$:

$$x_0 = B_1 + B_2$$

and

$$v_0 = B_1 s_1 + B_2 s_2$$

leading to

$$B_1 = \frac{x_0 s_2 - v_0}{s_2 - s_1}$$

and

$$B_2 = \frac{v_0 - x_0 s_1}{s_2 - s_1}$$

The displacement, subject to these initial conditions, is, therefore,

$$x = \frac{x_0 s_2 - v_0}{s_2 - s_1} e^{s_1 t} + \frac{v_0 - x_0 s_1}{s_2 - s_1} e^{s_2 t} \tag{6.45}$$

If $v_0 = 0$, the displacement approaches the equilibrium position without overshoot, as shown in Fig. 6.11, since s_1 and s_2 are negative.

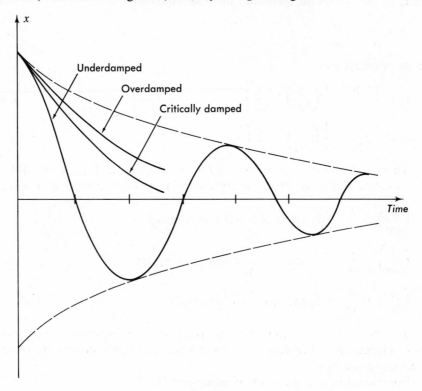

Fig. 6.11

For underdamped motion, s_1 and s_2 are conveniently written

$$s_1 = -\frac{b}{2m} - i\left[\frac{k}{m} - \left(\frac{b}{2m}\right)^2\right]^{1/2} = -\frac{b}{2m} - i\Omega \tag{6.46a}$$

and

$$s_2 = -\frac{b}{2m} + i\left[\frac{k}{m} - \left(\frac{b}{2m}\right)^2\right]^{1/2} = -\frac{b}{2m} + i\Omega \tag{6.46b}$$

where $\Omega = [(k/m) - (b/2m)^2]^{1/2}$ is real. The displacement now becomes

$$x = B_1 e^{-(b/2m)t - i\Omega t} + B_2 e^{-(b/2m)t + i\Omega t}$$
$$= e^{-(b/2m)t}(B_1 e^{-i\Omega t} + B_2 e^{i\Omega t}) \tag{6.47}$$

Since the solution of Eq. (6.39) is real, we follow a procedure similar to that for Eq. (6.14). Underdamped motion is thus described by

$$x = e^{-(b/2m)t}(B_1' \cos \Omega t + B_2' \sin \Omega t) \qquad (6.48)$$

where B_1' and B_2' are real, whereas B_1 and B_2 in Eq. (6.47) are complex. For the case of zero initial velocity, and initial displacement x_0, $B_1' = x_0$ and $B_2' = x_0 b/2m\Omega$. Thus

$$x = e^{-(b/2m)t}\left(x_0 \cos \Omega t + \frac{x_0 b}{2m\Omega} \sin \Omega t\right) \qquad (6.49)$$

or

$$x = Ae^{-(b/2m)t}[\cos (\Omega t + \alpha)] \qquad (6.50)$$

where

$$A = \left[x_0^2 + \left(\frac{x_0 b}{2m\Omega}\right)^2\right]^{\frac{1}{2}}$$

and

$$\alpha = -\tan^{-1}\left(\frac{2m\Omega}{b}\right)$$

Referring to Eq. (6.50) it is clear that $x = 0$ when $\cos (\Omega t + \alpha) = 0$; that is, $\Omega t + \alpha = \pi/2, 3\pi/2, 5\pi/2$, and so on. There are two zeros of displacement for each complete oscillation. To unambiguously define the period, reference must therefore be made to the interval of time between the zeros of displacement corresponding to repeated algebraic signs of the velocity. For the case above, $\Omega T = 2\pi$, or

$$T = \frac{2\pi}{[(k/m) - (b/2m)^2]^{\frac{1}{2}}} \qquad (6.51)$$

and

$$f = \frac{1}{2\pi}\left[\frac{k}{m} - \left(\frac{b}{2m}\right)^2\right]^{\frac{1}{2}} \qquad (6.52)$$

(It should be noted that a Fourier expansion of x yields Ω as the fundamental or lowest frequency.)

Referring again to Eq. (6.50), we observe that the maxima and minima in x occur when $\cos (\Omega t + \alpha) = \pm 1$ and, therefore, the envelope of this equation is given by

$$x = \pm Ae^{-(b/2m)t} \qquad (6.53)$$

as shown in Fig. 6.11.

Underdamped motion is thus characterized by oscillation of decreasing amplitude and of period longer than that for undamped motion. The maxima and minima of the oscillations do not occur midway between the zeros, as is demonstrated in the following example.

Illustrative Example 6.5

Determine the locations of the maximum and minimum displacements for damped harmonic motion described by

$$x(t) = Ae^{-\lambda t} \sin \Omega t$$

SOLUTION

The maximum and minimum displacements occur when

$$\dot{x} = 0 = Ae^{-\lambda t} \Omega \cos \Omega t - A\lambda e^{-\lambda t} \sin \Omega t$$

Dividing by $A\lambda e^{-\lambda t} \Omega \cos \Omega t$, the transcendental equation is of somewhat simpler form:

$$\frac{\Omega}{\lambda} = \tan \Omega t$$

The required values occur at the points of intersection of the functions on the right and left sides of the above equation, as shown in Fig. 6.12. As λ approaches

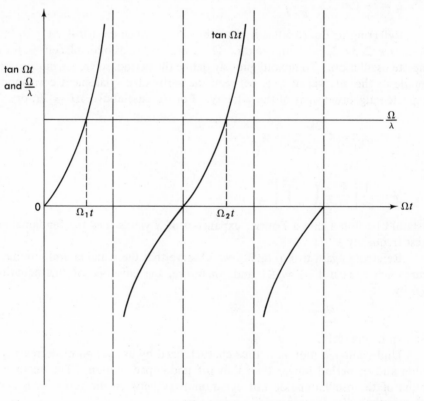

Fig. 6.12

zero, $\tan \Omega t$ grows without limit, and consequently the intersections occur at $\pi/2$, $3\pi/2$, $5\pi/2$, and so on. This is, of course, the result for undamped motion.

Critical damping occurs when the frequency of the oscillatory term in Eqs. (6.49) and (6.50) just equals zero. To determine experimentally the value of the *critical damping coefficient*, oscillations are initiated by releasing a displaced mass with zero initial velocity. Damping is increased slowly until the frequency of damped oscillation approaches zero. At critical damping, there is no overshoot beyond the equilibrium position and the mass returns toward equilibrium sooner than for overdamped motion.

Mathematically, a damped oscillator never comes to rest in a finite time. Actually the motion of a real oscillator ceases because of the action of frictional influences which are not represented by simple velocity damping.

When the values of s are identical, as is the case of critical damping, $B_1 e^{-(b/2m)t}$ and $B_2 e^{-(b/2m)t}$ do not represent linearly independent solutions of Eq. (6.39), and a solution of the form

$$x = (B_1' + B_2't)e^{-(b/2m)t} \tag{6.54}$$

is required. This may be verified by substituting into Eq. (6.39).

Since for critical damping, $(b/2m)^2 = k/m$, we employ the relation

$$b = b_{\text{cr}} = 2(km)^{1/2} \tag{6.55}$$

in verifying Eq. (6.54), where b_{cr} is referred to as the critical damping coefficient. Differentiating Eq. (6.54),

$$\dot{x} = (B_1' + B_2't)\left(\frac{-b}{2m}\right)e^{-(b/2m)t} + B_2'e^{-(b/2m)t}$$

and

$$\ddot{x} = (B_1' + B_2't)\left(\frac{b}{2m}\right)^2 e^{-(b/2m)t} + 2B_2'\left(\frac{-b}{2m}\right)e^{-(b/2m)t}$$

Substituting for x, $\dot{x}$, and $\ddot{x}$ in Eq. (6.39),

$$m\left[(B_1' + B_2't)\left(\frac{b}{2m}\right)^2 e^{-(b/2m)t} + 2B_2'e^{-(b/2m)t}\left(\frac{-b}{2m}\right)\right]$$

$$+ b\left[(B_1' + B_2't)\left(\frac{-b}{2m}\right)e^{-(b/2m)t} + B_2'e^{-(b/2m)t}\right]$$

$$+ k[(B_1' + B_2't)e^{-(b/2m)t}] = 0$$

Dividing by $e^{-(b/2m)t}$ and arranging in powers of t^0 and t^1,

$$\left(\frac{B_1'b^2}{4m} - B_2'b - \frac{b^2B_1'}{2m} + B_2'b + kB_1'\right)t^0 + \left(\frac{b^2B_2'}{4m} - \frac{b^2B_1'}{2m} + kB_2'\right)t^1 = 0$$

Each of the terms in parentheses must vanish simultaneously and independently for Eq. (6.54) to be the general solution for any time. From the first term, we have

$$B_1'\left(-\frac{b^2}{4m} + k\right) = 0$$

Although, in general, B_1' need not be zero, $-b^2/4m + k$ is zero for critical damping. From the second term in parentheses,

$$B_2'\left(-\frac{b^2}{4m} + k\right) = 0$$

giving the same result as above. Thus Eq. (6.54) is the solution we seek.

In examining the curves of Fig. 6.11, including that for critical damping, bear in mind that the mass in each instance is displaced from equilibrium an amount x_0 but possesses no velocity when released.

Illustrative Example 6.6
An electric dipole, consisting of a pair of oppositely charged electrons attached to a massless rod of length l, is placed in a uniform electric field of magnitude E. Find the natural frequency of oscillation if the dipole, shown in Fig. 6.13, is slightly rotated from its equilibrium position and released. If a damping torque proportional to angular velocity of the dipole is present, determine the frequency of oscillation.

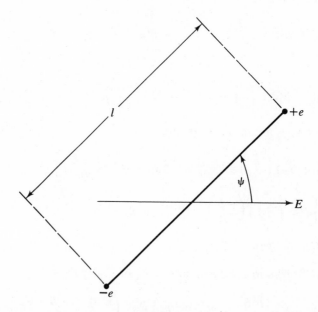

Fig. 6.13

SOLUTION

Although the net force on the dipole is zero ($eE - eE$), the net torque is $-eEl \sin \psi$. The torque is equated to the product of system moment of inertia (about the center of mass) and the angular acceleration. For the undamped case,

$$-eEl \sin \psi = I\ddot{\psi} = \tfrac{1}{2}ml^2\ddot{\psi}$$

where $I = 2m(l/2)^2 = \tfrac{1}{2}ml^2$.

For small angles,

$$\ddot{\psi} + \frac{2eE}{ml}\psi = 0$$

The undamped natural circular frequency is, therefore,

$$\omega_0 = \left(\frac{2eE}{ml}\right)^{\frac{1}{2}}$$

With damping present,

$$-eEl\psi - C\dot{\psi} = \tfrac{1}{2}ml^2\ddot{\psi}$$

where C represents the torsional damping coefficient (torque per unit angular velocity). The above equation may be written

$$\ddot{\psi} + \frac{2C}{ml^2}\dot{\psi} + \frac{2eE}{ml}\psi = 0$$

Assuming a solution of the form $\psi = e^{st}$, taking the appropriate derivatives and substituting into the equation of motion,

$$s^2 + \left(\frac{2C}{ml^2}\right)s + \frac{2eE}{ml} = 0$$

and

$$s = -\frac{C}{ml^2} \pm \left[\left(\frac{C}{ml^2}\right)^2 - \frac{2eE}{ml}\right]^{\frac{1}{2}}$$

The condition for oscillation of the dipole is

$$\frac{2eE}{ml} > \left(\frac{C}{ml^2}\right)^2$$

and the circular frequency of oscillation is

$$\Omega = \left[\frac{2eE}{ml} - \left(\frac{C}{ml^2}\right)^2\right]^{\frac{1}{2}}$$

As the damping coefficient increases, Ω diminishes, until at $C = (2eEml^3)^{\frac{1}{2}}$, oscillation ceases.

6.5 FORCED OSCILLATION

In the case of free oscillation, the forces discussed depend linearly upon position and velocity. When these forces are of a dissipative nature, the system is obliged to run down because no additional energy is supplied by an external source. Many systems exist in which an external time-dependent force, $F(t)$, acts to supply energy, resulting in motion termed *forced oscillation*.

The general equation of motion for one-dimensional forced oscillation of a linear spring–mass system with velocity damping, subject to a purely time-dependent external driving force, is

$$-kx - b\dot{x} + F(t) = m\ddot{x} \tag{6.56a}$$

or

$$\ddot{x} + \frac{b}{m}\dot{x} + \frac{k}{m}x = \frac{F(t)}{m} \tag{6.56b}$$

It remains our problem to find the position of the mass as a function of time. We have already solved this equation for the case where $F(t) = 0$. When $F(t)$ does not equal zero we need find one additional or particular solution to Eq. (6.56b) and add to it the complementary solutions of Eq. (6.39b) to obtain the *general* solution. As with the previous equations, there are only two arbitrary constants to be evaluated from a pair of specified initial conditions.

Consider first a simple case: a constant external force. An example of such a system is given by a mass suspended vertically as shown in Fig. 6.14. Let the mass be subjected to an upward external force equal to its weight for all time prior to $t = 0$. The net force acting on the spring is zero and hence the spring is undeformed; that is, the mass is at equilibrium and at rest. At $t = 0$, the external force

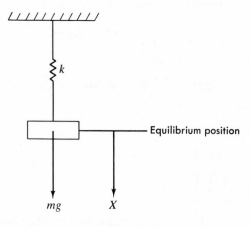

Fig. 6.14

is removed. The differential equation of motion for this forced system with damping is

$$-kx - b\dot{x} + mg = m\ddot{x} \tag{6.57}$$

Rewriting Eq. (6.57) we have

$$\ddot{x} + \frac{b}{m}\dot{x} + \frac{k}{m}x = g \tag{6.58}$$

Depending upon the values of b, k, and m, the motion is undamped, overdamped, or critically damped. Since $F(t)$ is constant, let us assume that the particular solution x_p is also a constant, say D. Substituting $x_p = D$ into Eq. (6.58), $D = mg/k$. Thus

$$x = \frac{mg}{k} + B_1 e^{s_1 t} + B_2 e^{s_2 t} \tag{6.59}$$

where the complementary and particular solutions have been added. Here s_1 and s_2 are the roots of Eq. (6.42) if $s_1 \neq s_2$. [If $s_1 = s_2$,

$$x = \frac{mg}{k} + (B_1' + B_2' t)e^{st}] \tag{6.60}$$

Applying the initial conditions $x = 0$, $\dot{x} = 0$ at $t = 0$,

$$x(0) = 0 = \frac{mg}{k} + B_1 + B_2 \tag{6.61a}$$

$$\dot{x}(0) = 0 = B_1 s_1 + B_2 s_2 \tag{6.61b}$$

Solving Eqs. (6.61a) and (6.61b) simultaneously,

$$B = \frac{-s_2(mg/k)}{s_2 - s_1} \quad \text{and} \quad B_2 = \frac{s_1(mg/k)}{s_2 - s_1}$$

so that the solution becomes

$$x = \frac{mg/k}{s_2 - s_1}(-s_2 e^{s_1 t} + s_1 e^{s_2 t}) + \frac{mg}{k} \tag{6.62}$$

Equation (6.62) contains a time-dependent term and a constant term. If we define a new variable $x' = x - mg/k$, it is clear that the motion of the mass is damped harmonic about $x' = 0$. The external constant force has the effect of shifting the equilibrium position from $x = 0$ to $x = mg/k$. The problem might have been formulated in an alternative manner by choosing $x' = 0$ as the equilibrium position about which *unforced* harmonic motion occurs.

In the discussion of forced vibration which follows, emphasis is given the particular solution which when added to the complementary solution yields the complete solution. The reader is cautioned that the constants of integration evaluated from initial conditions must be applied to the *complete solution* and not to the complementary solution alone. The complementary solution describes the transient character of the motion. We may consider that after a sufficiently long time, this component undergoes decay and is no longer significant. The steady-state nature of the motion is, therefore, described by the particular solution, and for this reason it assumes special importance.

Fourier's theorem states that any arbitrary periodic function $F(t)$, having only a finite number of discontinuities and defined over a time interval T, can be expanded in a series of sinusoidal and cosinusoidal functions at harmonic frequencies. The series, called a Fourier series, may have an infinite number of terms. It is discussed more fully in Chapter 13. In the analysis following we consider a case of a single-frequency external force which is cosinusoidal. Were this force more complicated, it would be necessary to decompose it by Fourier analysis into its constituent frequencies. The resultant motion is then the sum of the contributions at each frequency. To support the validity of this approach, an application of the *principle of superposition*, it is incumbent upon us to show that the "spring" continues to obey Hooke's law.

Consider the force $F(t)$ represented by

$$F(t) = F_{max} \cos \omega t \tag{6.63}$$

where F_{max} is the amplitude of the force, its maximum value, and ω is the circular frequency of the forcing function, not to be confused with ω_0.

The equation of motion now becomes

$$-kx - b\dot{x} + F_{max} \cos \omega t = m\ddot{x} \tag{6.64}$$

Substituting Euler's identity,

$$\cos \omega t = \text{real part } e^{i\omega t} \tag{6.65}$$

into Eq. (6.64):

$$m\ddot{x} + b\dot{x} + kx = F_{max} (\text{real part } e^{i\omega t}) \tag{6.66}$$

Equation (6.66) is unwieldy because of the term "real part $e^{i\omega t}$." If $F_{max}e^{i\omega t}$ replaces F_{max} real part $e^{i\omega t}$, the effect is to replace a real force by a complex force. When this is done we have

$$m\ddot{x} + b\dot{x} + kx = F_{max}e^{i\omega t} \tag{6.67}$$

As a consequence of this modification, the displacement, velocity, and acceleration derived from Eq. (6.67) will have both real and imaginary components. The latter result from the inclusion of the imaginary part of the forcing function but do not influence the real part of $x(t)$, $\dot{x}(t)$, and $\ddot{x}(t)$. This is a consequence of the principle of superposition, valid for linear systems.

Since the left side of Eq. (6.67) is linear in x and its derivatives, let us assume a particular solution of the form

$$x = x_m e^{st} \tag{6.68}$$

where x_m and s are to be determined. Taking the necessary time derivatives of $x(t)$ and substituting into Eq. (6.67),

$$mx_m s^2 e^{st} + bx_m s e^{st} + kx_m e^{st} = F_{max} e^{i\omega t} \tag{6.69a}$$

or

$$x_m e^{st}(ms^2 + bs + k) = F_{max} e^{i\omega t} \tag{6.69b}$$

The time-dependent factors on either side of Eq. (6.69b) must be equal and, therefore, $s = i\omega$.

Dividing Eq. (6.69b) by e^{st},

$$x_m(ms^2 + bs + k) = F_{max} \tag{6.70}$$

from which

$$x_m = \frac{F_{max}}{ms^2 + bs + k} = \frac{F_{max}}{k - \omega^2 m + i\omega b} \tag{6.71}$$

Dividing the numerator and denominator of Eq. (6.71) by k and making the substitution $\omega_0 = k/m$, we obtain

$$x_m = \frac{F_{max}/k}{1 - (\omega/\omega_0)^2 + (i\omega b/k)} \tag{6.72}$$

The denominator of Eq. (6.72) is complex. Recall that a complex number may be represented as follows:

$$\alpha + i\beta = (\alpha^2 + \beta^2)^{\frac{1}{2}} e^{i\theta}$$

where $\theta = \tan^{-1}(\beta/\alpha)$. The denominator becomes, therefore,

$$\left\{ \left[1 - \left(\frac{\omega}{\omega_0}\right)^2 \right]^2 + \left(\frac{\omega b}{k}\right)^2 \right\}^{\frac{1}{2}} e^{i\theta}$$

and Eq. (6.72) is written

$$x_m = \frac{(F_{max}/k)e^{-i\theta}}{\{[1 - (\omega/\omega_0)^2]^2 + (\omega b/k)^2\}^{\frac{1}{2}}} \tag{6.73}$$

where

$$\theta = \tan^{-1} \frac{\omega b/k}{[1 - (\omega/\omega_0)^2]} \tag{6.74}$$

Substituting x_m into $x = x_m e^{i\omega t}$,

$$x = \frac{(F_{max}/k)e^{i(\omega t - \theta)}}{\{[1 - (\omega/\omega_0)^2]^2 + (\omega b/k^2)\}^{\frac{1}{2}}} \tag{6.75}$$

Since we desire only the real part of x,

$$x = \frac{(F_{max}/k) \cos(\omega t - \theta)}{\{[1 - (\omega/\omega_0)^2]^2 + (\omega b/k)^2\}^{\frac{1}{2}}} = x_{max} \cos(\omega t - \theta) \qquad (6.76)$$

where

$$x_{max} = \frac{F_{max}/k}{\{[1 - (\omega/\omega_0)^2]^2 + (\omega b/k)^2\}^{\frac{1}{2}}}$$

We observe that the displacement varies at the same frequency as the applied force but it is not "in phase" with it, lagging in phase by θ. As in the case of the damped unforced oscillator, the phase angle is simply the argument of the cosine function. The difference in phase of two quantities *at the same frequency* is the difference in their phase angles.

Another quantity of interest is the velocity,

$$v = \dot{x} = \frac{(F_{max}/k) i\omega e^{i(\omega t - \theta)}}{\{[1 - (\omega/\omega_0)^2]^2 + (\omega b/k)^2\}^{\frac{1}{2}}} \qquad (6.77)$$

but since $i = e^{i\pi/2}$

$$v = \frac{(F_{max}/k)\omega e^{i(\omega t - \theta + \pi/2)}}{\{[1 - (\omega/\omega_0)^2]^2 + (\omega b/k)^2\}^{\frac{1}{2}}}$$

$$= v_{max} e^{i(\omega t - \theta + \pi/2)} \qquad (6.78)$$

where

v_{max} = maximum velocity

$$= \frac{(F_{max}/k)\omega}{\{[1 - (\omega/\omega_0)^2]^2 + (\omega b/k)^2\}^{\frac{1}{2}}}$$

Equation (6.78) represents the velocity in complex polar form. The actual velocity is the real part of this equation:

$$v = v_{max} \cos(\omega t - \theta + \pi/2)$$

$$= -v_{max} \sin(\omega t - \theta) \qquad (6.79)$$

Following a similar procedure, the acceleration is

$$a = \dot{v} = \frac{-(F_{max}/k)\omega^2 e^{i(\omega t - \theta)}}{\{[1 - (\omega/\omega_0)^2]^2 + (\omega b/k)^2\}^{\frac{1}{2}}} = -a_{max} e^{i(\omega t - \theta)} \qquad (6.80)$$

where

$$a_{max} = \frac{(F_{max}/k)\omega^2}{\{[1 - (\omega/\omega_0)^2]^2 + (\omega b/k)^2\}^{\frac{1}{2}}} = x_{max}\omega^2$$

The reader should note that the acceleration is directed opposite the displacement, and hence the two quantities differ in phase by π rad.

It is convenient to represent quantities such as force and displacement as rotating vectors in the complex plane called *phasors*. As seen in Fig. 6.15, these quantities are of constant magnitude and varying direction. The magnitude is given by the coefficient of the complex exponential, and the angle measured counter-

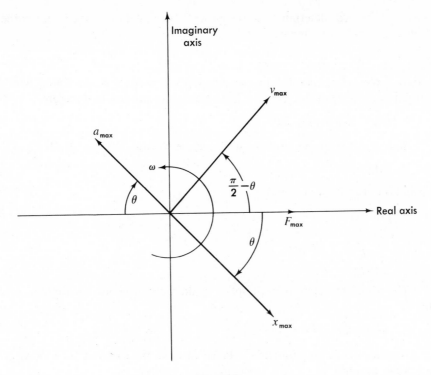

Fig. 6.15

clockwise from the real axis in the complex plane is given by $1/i$ multiplied by the exponent. Phasors rotate counterclockwise with angular velocity ω, the forcing circular frequency. Thus the projections of the phasors on the real axis supply the instantaneous values of physical interest.

Illustrative Example 6.7

A mass m is connected by means of a linear spring k to a wall which experiences *velocity shock*. At $t = 0$, the wall shown in Fig. 6.16 instantaneously attains a constant velocity u to the right. If the mass is at rest at $t = 0$, and if the spring is

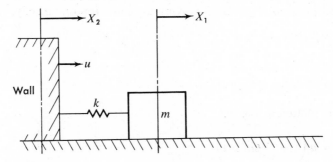

Fig. 6.16

initially undeformed, determine the response of the system and the spring force as a function of time. Assume frictional influences to be negligible.

SOLUTION

The only force acting parallel to the direction of motion is associated with the deformation $x_1 - x_2$ of the spring. The equation of motion is therefore

$$-k(x_1 - x_2) = m\ddot{x}_1$$

Since the displacement of the wall may be written

$$x_2 = ut$$

the equation of motion becomes

$$m\ddot{x}_1 + kx_1 = kut$$

or

$$\ddot{x}_1 + \omega_0^2 x_1 = \omega_0^2 ut$$

where $\omega_0^2 = k/m$.

The transient solution $x_{1,t}$ applies to the homogeneous equation $\ddot{x}_1 + \omega_0^2 x_1 = 0$:

$$x_{1,t} = C_1 \sin \omega_0 t + C_2 \cos \omega_0 t$$

Because the forcing function is linear in time, the assumed form of $x_{1,ss}$, the steady-state solution, is

$$x_{1,ss} = At + B$$

where A and B are constants to be evaluated by substitution into the differential equation of motion. Since $\dot{x}_{1,ss} = A$ and $\ddot{x}_{1,ss} = 0$,

$$\omega_0^2 (At + B) = \omega_0^2 ut$$

or $A = u$, $B = 0$.

The complete solution is therefore

$$x_1(t) = C_1 \sin \omega_0 t + C_2 \cos \omega_0 t + ut$$

to which the conditions $x_1 = 0$, $\dot{x}_1 = 0$ at $t = 0$ must be applied. From the condition on initial position, $C_2 = 0$. Applying the condition on velocity,

$$\dot{x}_1 = \omega_0 C_1 \cos \omega_0 t + u$$

At $t = 0$,

$$0 = \omega_0 C_1 + u \qquad C_1 = -\frac{u}{\omega_0}$$

The response, plotted in Fig. 6.17, is

$$x_1(t) = u\left(t - \frac{1}{\omega_0} \sin \omega_0 t\right)$$

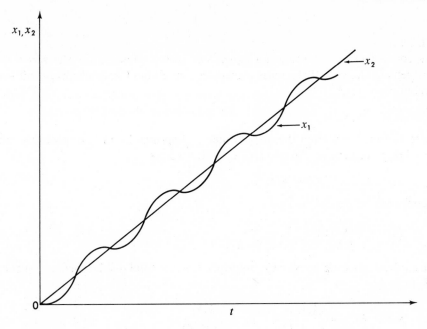

Fig. 6.17

The spring force

$$F = -k(x_1 - x_2) = \frac{ku}{\omega_0} \sin \omega_0 t$$

Illustrative Example 6.8

A mass is connected to a rigid wall by means of a dashpot and two springs as shown in Fig. 6.18. Derive the differential equation of motion if the mass is subjected to a sinusoidal forcing function $F = F_{max} \sin \omega t$. Neglect frictional effects between the mass and its support.

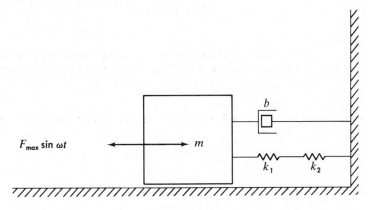

Fig. 6.18

SOLUTION

It is convenient to determine an equivalent spring constant for the series spring arrangement shown. If the mass is displaced an amount x, each spring will experience an equal force,

$$k_1 x_1 = k_2 x_2$$

where x_1 and x_2 represent the displacements of springs 1 and 2, respectively, due to a total displacement x. Since $x_1 + x_2 = x$, we have

$$k_1 x_1 = k(x - x_2) = k_2 x_2$$

Therefore,

$$x_2 = \frac{k_1 x}{k_1 + k_2} \qquad x_1 = \frac{k_2 x}{k_1 + k_2}$$

The equivalent spring constant k_e multiplied by the total displacement x must equal $k_1 x_1$ or $k_2 x_2$:

$$k_e x = k_1 x_1 = \frac{k_1 k_2 x}{k_1 + k_2} \qquad k_e = \frac{k_1 k_2}{k_1 + k_2}$$

Alternatively,

$$k_e x = k_2 x_2 = \frac{k_1 k_2 x}{k_1 + k_2} \qquad k_e = \frac{k_1 k_2}{k_1 + k_2}$$

For a dashpot force of the form $F_d = -b\dot{x}$, the equation of motion is

$$-k_e x - b\dot{x} + F_{max} \sin \omega t = m\ddot{x}$$

or

$$m\ddot{x} + b\dot{x} + k_e x = F_{max} \sin \omega t = \text{imag part } (F_{max} e^{i\omega t})$$
$$= \text{real part } (F_{max} e^{i(\omega t - \pi/2)})$$

Illustrative Example 6.9

Determine the root-mean-square (rms) values of displacement, velocity, and acceleration for a damped forced harmonic oscillator operating at steady state.

SOLUTION

Equations (6.76), (6.79), and (6.80) give the steady-state values of x, $\dot{x}$, and $\ddot{x}$, respectively:

$$x = x_{max} \cos (\omega t - \theta)$$

$$v = v_{max} \cos (\omega t - \theta + \pi/2)$$

$$a = -x_{max} \omega^2 \cos (\omega t - \theta) = -a_{max} \cos (\omega t - \theta)$$

The rms values are found by employing the defining expression

$$g_{\text{rms}} = \left\{ \frac{\int_0^T g^2 \, dt}{\int_0^T dt} \right\}^{\frac{1}{2}}$$

where g is an arbitrary periodic function of time with period T. Therefore,

$$x_{\text{rms}} = \left\{ \frac{\int_0^T x^2 \, dt}{\int_0^T dt} \right\}^{\frac{1}{2}} = \left\{ \frac{\int_0^{2\pi/\omega} [x_{\text{max}} \cos (\omega t - \theta)]^2 \, dt}{\int_0^{2\pi/\omega} dt} \right\}^{\frac{1}{2}}$$

$$= \frac{x_{\text{max}}}{(2\pi/\omega)^{\frac{1}{2}}} \left\{ \int_0^{2\pi/\omega} \cos^2 (\omega t - \theta) \, dt \right\}^{\frac{1}{2}}$$

$$= \frac{x_{\text{max}}}{(2\pi/\omega)^{\frac{1}{2}}} \left\{ \int_0^{2\pi/\omega} [\tfrac{1}{2} + \tfrac{1}{2} \cos 2(\omega t - \theta)] \, dt \right\}^{\frac{1}{2}} = \frac{x_{\text{max}}}{\sqrt{2}}$$

Similarly,

$$v_{\text{rms}} = \left\{ \frac{\int_0^{2\pi/\omega} [v_{\text{max}} \cos (\omega t - \theta + \pi/2)]^2 \, dt}{\int_0^{2\pi/\omega} dt} \right\}^{\frac{1}{2}}$$

$$= \frac{v_{\text{max}}}{(2\pi/\omega)^{\frac{1}{2}}} \left\{ \int_0^{2\pi/\omega} [\tfrac{1}{2} + \tfrac{1}{2} \cos 2(\omega t - \theta + \pi/2)] \, dt \right\}^{\frac{1}{2}}$$

$$= \frac{v_{\text{max}}}{\sqrt{2}} = \frac{\omega x_{\text{max}}}{\sqrt{2}}$$

It is left for the reader to confirm that

$$a_{\text{rms}} = \frac{a_{\text{max}}}{\sqrt{2}} = \frac{\omega^2 x_{\text{max}}}{\sqrt{2}}$$

6.6 AMPLIFICATION FACTOR

Equation (6.76) for the displacement can be written

$$x = \frac{(F_{\text{max}}/k) \cos (\omega t - \theta)}{\{[1 - (\omega/\omega_0)^2]^2 + [2(b/b_{\text{cr}})(\omega/\omega_0)]^2\}^{\frac{1}{2}}} \tag{6.81}$$

where $b_{\text{cr}} = 2(km)^{\frac{1}{2}}$, the same damping factor as for the critically damped free oscillator.

The amplification factor α, a dimensionless quantity, is defined

$$\alpha = \frac{x_{\text{max}}}{F_{\text{max}}/k} = \frac{1}{\{[1 - (\omega/\omega_0)^2]^2 + [2(b/b_{\text{cr}})(\omega/\omega_0)]^2\}^{\frac{1}{2}}} \tag{6.82}$$

The term F_{max}/k represents the static displacement resulting from a constant force F_{max}. The amplification factor is thus the ratio of the displacement under dynamic conditions to the static displacement ($\omega = 0$).

Figure 6.19 is a graphical representation of α as a function of the frequency ratio ω/ω_0 for different b/b_{cr}.

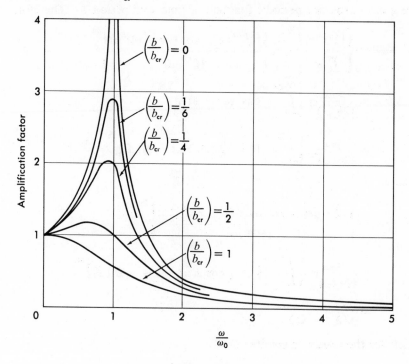

Fig. 6.19

For a given system, α depends solely upon ω, the frequency of the impressed force. Only in an "ideal" system where the damping factor is zero does the amplification approach infinity as ω approaches ω_0.

Physically, unbounded motion does not occur, because real quantities such as deformation and electrical current lead to system failures even at large finite amplitudes. Moreover, damping occurs in all but atomic and molecular systems and tends to limit the amplification. In *any* system, however, in which a linear restoring term is assumed, the linearity is limited to relatively small displacements and therefore the equations predicting large amplitudes no longer accurately represent the physics. Nevertheless, they remain as a guide to the nature of system behavior.

With damping present, the amplification will be a maximum when $\omega \cong \omega_0$. The exact expression for the resonant frequency ω_r (corresponding to maximum amplification factor α) may be found by setting $d\alpha/d(\omega/\omega_0) = 0$:

$$\frac{\omega_r}{\omega_0} = \left[1 - 2\left(\frac{b}{b_{cr}}\right)^2 \right]^{\frac{1}{2}} \tag{6.83}$$

6.7 VELOCITY RESONANCE

It is clear from Eq. (6.78) that the amplitude of the velocity, v_{max}, varies with impressed frequency ω. When v_{max} is a maximum, velocity resonance occurs. The impressed frequency for velocity resonance differs from the frequency for displacement resonance ω_r, except for the undamped case. It may be determined by setting $dv_{max}/d(\omega/\omega_0) = 0$. It is left as an exercise for the reader to show that the frequencies for velocity and displacement resonance differ from one another, but both approach ω_0 as b approaches zero. Velocity resonance corresponds to the resonance of current rather than charge in an electrical circuit, because current is analogous to velocity.

6.8 VIBRATION ISOLATION

There are many instances in which it is important to isolate an apparatus from the harsh effects of its vibrational environment. How, for example, can delicate instrumentation be mounted to the wall of a vibrating chassis in such a way as to minimize the vibration of its contents? One arrangement to effect such isolation of vibration is shown in Fig. 6.20, where a mass is mounted to a wall by means of a linear spring and a dashpot.

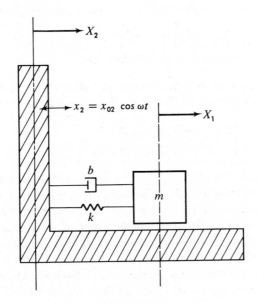

Fig. 6.20

The wall is assumed to experience vibration according to the relation $x_2 = x_{02} \cos \omega t$, although more complex analyses can be carried out in which random vibration simulates the actual environment. For the model selected, the spring and damping forces depend, respectively, upon the relative displacement and velocity of the mass and the wall. The equation of motion for the mass is, therefore,

$$-k(x_1 - x_2) - b(\dot{x}_1 - \dot{x}_2) = m\ddot{x}_1 \tag{6.84}$$

which may be rewritten

$$m\ddot{x}_1 + b\dot{x}_1 + kx_1 = b\dot{x}_2 + kx_2$$

$$= -bx_{02}\omega \sin \omega t + kx_{02} \cos \omega t \tag{6.85}$$

Dividing by m,

$$\ddot{x} + \frac{b}{m}\dot{x}_1 + \frac{k}{m}x_1 = \frac{k}{m}x_{02} \cos \omega t - \frac{b}{m}x_{02}\omega \sin \omega t$$

$$= \frac{x_{02}}{m}[k^2 + (\omega b)^2]^{\frac{1}{2}} \cos(\omega t + \beta) \tag{6.86}$$

where $\beta = \tan^{-1}(\omega b/k)$. Since Eq. (6.86) is of the same form as Eq. (6.64) with

$$\frac{x_{02}}{m}[k^2 + (\omega b)^2]^{\frac{1}{2}}$$

substituted for F_{max}/m, the solutions are also identical in form. The amplitude of the steady-state response, on the basis of Eq. (6.81), is therefore

$$x_{1,max} = \frac{(x_{02}/k)[k^2 + (\omega b)^2]^{\frac{1}{2}}}{\{[1 - (\omega/\omega_0)^2]^2 + [2(b/b_{cr})(\omega/\omega_0)^2]\}^{\frac{1}{2}}} \tag{6.87}$$

As an indication of the effectiveness of isolation, we calculate the ratio of the amplitude of the response to that of the input vibration:

$$\frac{x_{1,max}}{x_{02}} = \frac{\{1 + [2(b/b_{cr})(\omega/\omega_0)]^2\}^{\frac{1}{2}}}{\{[1 - (\omega/\omega_0)^2]^2 + [2(b/b_{cr})(\omega/\omega_0]^2\}^{\frac{1}{2}}} \tag{6.88}$$

Figure 6.21 is a plot of $x_{1,max}/x_{02}$ as a function of ω/ω_0 for various values of b/b_{cr}. Since the objective is system isolation, it is clear that we seek values of $x_{1,max}/x_{02} < 1$. These occur at $\omega/\omega_0 > \sqrt{2}$, for all values of b/b_{cr}. The curves indicate that in this region damping has the effect of increasing the amplitude of the vibration of the mass relative to that of the undamped case. When a system is brought up to its operating speed from rest, it must pass through its natural frequency, $\omega/\omega_0 = 1$, if the operating ω exceeds ω_0. Even though damping increases the steady-state ratio $x_{1,max}/x_{02}$, it serves to limit the amplitude during such transients.

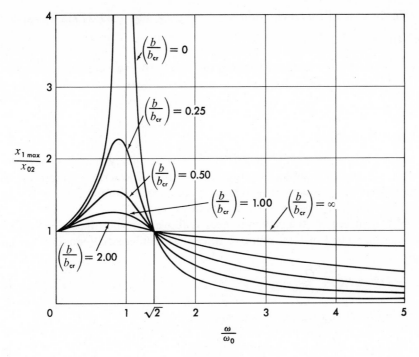

$\left(\dfrac{b}{b_{cr}}\right) = 0$

$\left(\dfrac{b}{b_{cr}}\right) = 0.25$

$\left(\dfrac{b}{b_{cr}}\right) = 0.50$

$\left(\dfrac{b}{b_{cr}}\right) = 1.00$ $\left(\dfrac{b}{b_{cr}}\right) = \infty$

$\left(\dfrac{b}{b_{cr}}\right) = 2.00$

$\dfrac{x_{1\,max}}{x_{02}}$

$\dfrac{\omega}{\omega_0}$

Fig. 6.21

6.9 FURTHER ENERGY CONSIDERATIONS

Additional insight may be gained by again applying the energy approach to the linear oscillator. Multiplying Eq. (6.2) by $\dot{x}$,

$$m\ddot{x}\dot{x} + kx\dot{x} = 0 \qquad (6.89)$$

where

$$m\ddot{x}\dot{x} = \frac{d}{dt}\left[\frac{m}{2}\dot{x}^2\right]$$

is the time rate of change of kinetic energy, $\dot{T}$, and

$$kx\dot{x} = \frac{d}{dt}\frac{k}{2}x^2$$

is the time derivative of the potential energy, $\dot{V}$. We may write Eq. (6.89) as

$$\dot{T} + \dot{V} = 0 = \dot{E} \qquad (6.90)$$

where E represents total system energy.

In the case of free oscillation with damping, Eq. (6.39) is multiplied by $\dot{x}$:

$$m\ddot{x}\dot{x} + b\dot{x}^2 + kx\dot{x} = 0 \tag{6.91}$$

The substitutions used in the derivation of Eq. (6.90) now give

$$\dot{T} + \dot{V} = -b\dot{x}^2 \tag{6.92a}$$

or

$$\dot{E} = -b\dot{x}^2 \tag{6.92b}$$

Since b is positive, it is clear that the rate of energy change is not zero but negative, and therefore system energy is depleted. The term $-b\dot{x}$ represents the frictional force: Consequently, $-b\dot{x}\dot{x}$, the product of frictional force and velocity, is the power dissipated as the system runs down.

As an analogy of the system discussed above, consider the series RLC circuit containing no source of electromotive force. The differential equation for charge Q is

$$-L\ddot{Q} = R\dot{Q} + \frac{Q}{C} \tag{6.93}$$

Multiplying by $\dot{Q}$ and rearranging terms,

$$\frac{d}{dt}\tfrac{1}{2}L\dot{Q}^2 + \frac{d}{dt}\frac{Q^2}{2C} = -R\dot{Q}^2 \tag{6.94}$$

Recall that the current $I = \dot{Q}$, so that Eq. (6.94) may be written

$$\frac{d}{dt}\tfrac{1}{2}LI^2 + \frac{d}{dt}\frac{Q^2}{2C} = -I^2R \tag{6.95}$$

The first term on the left side of Eq. (6.95) is the rate at which energy is stored in the inductance; the second term is the rate of energy storage in the capacitor. The sum of these rates equals the time rate of change of system energy or, in other words, the rate of energy dissipated in the resistance.

The equation of motion of a system undergoing forced vibration, multiplied by $\dot{x}$, yields

$$m\ddot{x}\dot{x} + b\dot{x}^2 + kx\dot{x} = F(t)\dot{x} \tag{6.96}$$

which may be rewritten

$$\dot{T} + \dot{V} = -bx^2 + F(t)\dot{x} \tag{6.97}$$

The entire left side is again the rate of change of system energy, and the first term on the right represents the rate at which mechanical energy is dissipated (converted ultimately into heat). The term $F(t)\dot{x}$ is the externally supplied power, the rate at which the external force does work on the system.

After the transient phase is caused by damping to decay to essentially zero, a steady-state motion persists as long as the forcing is maintained. The average rate of energy supplied as a consequence of $F(t)$ must equal the average dissipation rate, if a steady state, in which average system energy remains constant, is to develop.

6.10 JERK

The dynamical quantity *jerk*, defined as the time derivative of acceleration, can lead to new perspectives in the analysis of complex mechanical systems. Consider, for example, the problem of ascertaining the response of a large rocket subjected to a suddenly applied force at one end, as is the case during ignition. The force is not transmitted instantaneously to the other end, but travels through the rocket with an appropriately averaged sonic velocity v_s. After a time τ, on the order of l/v_s (where l is a characteristic length in the direction of the force), it is assumed that the axial force distribution is once again uniform. During the critical period $0 < t < \tau$, however, time-dependent forces may exist nonuniformly throughout the rocket, of such magnitude as to cause structural failure almost instantaneously after the motors are turned on.

It has been hypothesized that an acceleration wave passing through the rocket material is responsible for the high force which must develop during startup. To simplify analysis, a single-mass linear oscillator subjected to a force linearly dependent upon the jerk is selected as an equivalent system. The equation of motion for the period τ is thus

$$h\dddot{x} + m\ddot{x} + b\dot{x} + kx = F(t) \tag{6.98}$$

where h is a proportionality constant having units of force per unit jerk (as dyn-sec^3/cm).

The applied force $F(t)$ is, in general, aperiodic, containing all frequencies (often called a white spectrum of frequencies). Although this force cannot be described by a mathematically simple function, it can be measured. In an effort to obtain the response, consider a single component of $F(t)$, $F_{max} \cos \omega t$. Again employing complex representation,

$$F = F_{max}e^{i\omega t} \qquad x = x_m e^{i\omega t} \tag{6.99}$$

where, in general, x_m is complex and F has a complex part $(\sin \omega t)F_{max}$. Differentiating and substituting into Eq. (6.98) we obtain

$$hx_m(-i\omega)^3 e^{i\omega t} - mx_m\omega^2 e^{i\omega t} + bi\omega x_m e^{i\omega t} + kx_m e^{i\omega t} = F_{max}\,e^{i\omega t} \tag{6.100}$$

Dividing by $e^{i\omega t}$ and solving for x_m,

$$x_m = \frac{F_{max}}{(k - m\omega^2) + i(\omega b - h\omega^3)} \tag{6.101}$$

and the amplitude of vibration is

$$x_{max} = \frac{F_{max}}{[(k - m\omega^2)^2 + (\omega b - h\omega^3)^2]^{1/2}} \tag{6.102}$$

Examination of Eq. (6.102) reveals that the jerk contribution $h\omega^3$ reduces the magnitude of the second term of the denominator, thereby increasing the amplitude. This is a consequence of the jerk and damping forces being 180 degrees out of phase and tending to cancel.

6.11 HARMONIC MOTION IN TWO DIMENSIONS

Until now, a single variable (for example, position, angle, charge) and its time derivatives sufficed to describe the system dynamics. Consider at this time a single mass connected to a rigid wall by means of two linear springs, as shown in Fig. 6.22. Clearly, two position coordinates are required to identify the position of the mass.

When the mass is given a displacement x, y, both springs exert a restoring force, *each* with components in the X- and Y-directions. If the equilibrium lengths of the springs are L_x and L_y, the forces on the mass, in component form, are

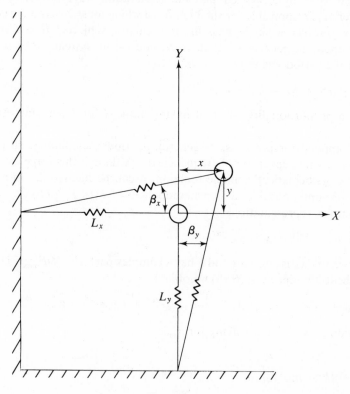

Fig. 6.22

$$F_x = \frac{-k_x\{[(L_x + x)^2 + y^2]^{\frac{1}{2}} - L_x\}(L_x + x)}{[(L_x + x)^2 + y^2]^{\frac{1}{2}}}$$

$$- \frac{k_y\{[x^2 + (L_y + y)^2]^{\frac{1}{2}} - L_y\}x}{[x^2 + (L_y + y)^2]^{\frac{1}{2}}} \qquad (6.103a)$$

$$F_y = \frac{-k_x\{[(L_x + x)^2 + y^2]^{\frac{1}{2}} - L_x\}y}{[(L_x + x)^2 + y^2]^{\frac{1}{2}}}$$

$$- \frac{k_y\{[x^2 + (L_y + y)^2]^{\frac{1}{2}} - L_y\}(L_y + y)}{[x^2 + (L_y + y)^2]^{\frac{1}{2}}} \qquad (6.103b)$$

When x and y are small compared with L_x and L_y, a first-order approximation of the forces yields

$$F_x = -k_x x \qquad (6.104a)$$

$$F_y = -k_y y \qquad (6.104b)$$

In so doing, the X- and Y-motions are no longer coupled, so the equation for the X-direction involves only x and its derivatives. The same applies to the Y-motion. The equations of motion are, therefore,

$$F_x = -k_x x = m\ddot{x} \qquad (6.105a)$$

$$F_y = -k_y y = m\ddot{y} \qquad (6.105b)$$

and the solutions for x and y are

$$x = x_{max} \cos\left[\left(\frac{k_x}{m}\right)^{\frac{1}{2}} t + \alpha_x\right] = x_{max} \cos(\omega_{0x}t + \alpha_x) \qquad (6.106a)$$

$$y = y_{max} \cos\left[\left(\frac{k_y}{m}\right)^{\frac{1}{2}} t + \alpha_y\right] = y_{max} \cos(\omega_{0y}t + \alpha_y) \qquad (6.106b)$$

The constants x_{max}, y_{max}, α_x, and α_y represent constants of integration which are evaluated from appropriate initial conditions. The trajectory of the mass (y as a function of x) may be rather complicated. The pattern described is called a *Lissajous figure*, Fig. 6.23. If the ratio ω_{0y}/ω_{0x} is a rational number, the motion is closed and periodic; otherwise the curve fills up the rectangle $2x_{max}$, $2y_{max}$ without ever repeating itself.

Some interesting cases occur when $\omega_{0x} = \omega_{0y}$. If $\alpha_x = \alpha_y$, then the ratio of y/x is a constant and the motion is a straight line. If α_x does not equal α_y, the motion is elliptic. If the motions differ in phase by 90 degrees ($\alpha_x = \alpha_y + \pi/2$) and the amplitudes are the same ($x_{max} = y_{max} = A$), then circular motion occurs:

$$x^2 + y^2 = A^2 \cos^2 \omega_0 t + A^2 \cos^2\left(\omega_0 t + \frac{\pi}{2}\right) = A^2 \qquad (6.107)$$

Lissajous patterns may be conveniently demonstrated by means of an oscillo-

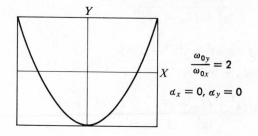

$$\frac{\omega_{0y}}{\omega_{0x}} = 2$$

$$\alpha_x = 0, \, \alpha_y = 0$$

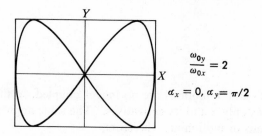

$$\frac{\omega_{0y}}{\omega_{0x}} = 2$$

$$\alpha_x = 0, \, \alpha_y = \pi/2$$

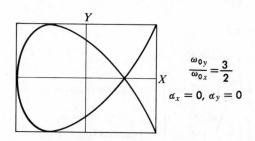

$$\frac{\omega_{0y}}{\omega_{0x}} = \frac{3}{2}$$

$$\alpha_x = 0, \, \alpha_y = 0$$

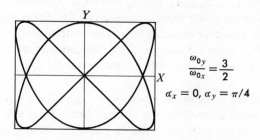

$$\frac{\omega_{0y}}{\omega_{0x}} = \frac{3}{2}$$

$$\alpha_x = 0, \, \alpha_y = \pi/4$$

Fig. 6.23

scope as discussed in Chapter 7. Signal-generator outputs possessing the appropriate frequency and phase relationships are connected to the X- and Y-deflection plates of a cathode-ray tube.

For the three-dimensional case, the mathematical analysis is the same. For uncoupled three-dimensional motion,

$$-k_x x = m\ddot{x} \qquad -k_y y = m\ddot{y} \qquad -k_z z = m\ddot{z} \qquad \text{(6.108)}$$

and the natural frequencies of oscillation are

$$\omega_{0x} = \left(\frac{k_x}{m}\right)^{\frac{1}{2}} \qquad \omega_{0y} = \left(\frac{k_y}{m}\right)^{\frac{1}{2}} \qquad \omega_{0z} = \left(\frac{k_z}{m}\right)^{\frac{1}{2}} \qquad \text{(6.109)}$$

When the frequencies are equal, three-dimensional ellipsoids occur. Degenerate cases include the sphere and straight line.

6.12 THE DUHAMEL INTEGRAL [2]

The impulse–momentum equation leads to an interesting and useful result when applied to a spring–mass system subject to an arbitrary force $F(t)$ as shown in Fig. 6.24.

Recall that the displacement of an undamped spring–mass system with initial displacement x_0 and velocity v_0 is given by

$$x = x_0 \cos \omega_0 t + \frac{v_0}{\omega_0} \sin \omega_0 t \qquad \text{(6.110)}$$

Consider now the influence of only a small segment of $F(t)$ shown by the shaded

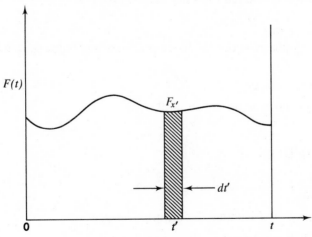

Fig. 6.24

[2] Von Karman, T., and Biot, M., *Mathematical Methods in Engineering*, McGraw-Hill, Inc., New York, 1940.

area in Fig. 6.24. Its effect will be to impart a small additional velocity $\dot{x}'$ found by applying the impulse–momentum relationship,

$$F_{x'}\, dt' = m\dot{x}' \tag{6.111a}$$

or

$$\dot{x}' = \frac{F_{x'}\, dt'}{m} \tag{6.111b}$$

What is the displacement at t due only to the impulse $F_{x'}\, dt'$ occurring at t'? The time of interest is not t but $t - t'$, the elapsed time between the impulse and the displacement we seek. The contribution to the displacement is determined by treating $\dot{x}'$ as though it were an initial velocity in Eq. (6.110):

$$\frac{F_{x'}\, dt'}{m\omega_0} \sin \omega_0(t - t') \tag{6.112}$$

and the influence of $F(t)$, made up of many $F_{x'}\, dt'$ impulses, is

$$\frac{1}{m\omega_0} \int_0^t F_{x'}(t') \sin \omega_0(t - t')\, dt' \tag{6.113}$$

The integral is a direct result of the principle of superposition which, because the system is linear, permits us to treat the additional effects as though no prior velocity or displacement existed. The total displacement must incorporate Eq. (6.113) in addition to the contribution due to conditions at $t = 0$, Eq. (6.110):

$$x = x_0 \cos \omega_0 t + \frac{v_0}{\omega_0} \sin \omega_0 t + \frac{1}{m\omega_0} \int_0^t F_{x'}(t') \sin \omega_0(t - t')\, dt' \tag{6.114}$$

The integral above is referred to as the *Duhamel integral*.

Illustrative Example 6.10

Determine the displacement of an undamped linear oscillator subject to the forcing function

$$F(t) = F_{max} \cos \omega t$$

with the initial conditions $x_0 = 0$, $v_0 = 0$.

SOLUTION

Applying Eq. (6.114),

$$x = \frac{1}{m\omega_0} \int_0^t F_{max} \cos \omega t' \sin \omega_0(t - t')\, dt'$$

Employing the trigonometric identity

$$\cos A \sin B = \tfrac{1}{2}[\sin (A + B) - \sin (A - B)]$$

the integral becomes

$$x = \frac{F_{max}}{2m\omega_0} \int_0^t \{\sin [(\omega - \omega_0)t' + \omega_0 t] - \sin [(\omega + \omega_0)t' - \omega_0 t]\}\, dt'$$

Integrating,

$$x = \frac{F_{max}}{2m\omega_0} \left\{ \frac{-\cos\left[(\omega - \omega_0)t' + \omega_0 t\right]}{\omega - \omega_0} + \frac{\cos\left[(\omega + \omega_0)t' - \omega_0 t\right]}{\omega + \omega_0} \right\}_0^t$$

$$= \frac{F_{max}}{2m\omega_0} \left[\frac{(-\cos \omega t)(\omega + \omega_0) + (\cos \omega t)(\omega - \omega_0)}{\omega^2 - \omega_0^2} \right.$$

$$\left. + \frac{(\cos \omega_0 t)(\omega + \omega_0) - (\cos \omega_0 t)(\omega - \omega_0)}{\omega^2 - \omega_0^2} \right]$$

Dividing numerator and denominator by k and simplifying, the displacement becomes

$$x = \frac{(F_{max}/k)(\cos \omega t - \cos \omega_0 t)}{1 - (\omega/\omega_0)^2}$$

This result agrees with Eq. (6.76) for undamped forced oscillation. The additional term above corresponds to the transient solution of Eq. (6.64).

6.13 NONLINEAR OSCILLATIONS

We have already seen that if the force applied to a mass is negatively proportional to displacement from the equilibrium position, harmonic motion occurs. This has been shown to be approximately true for many position-dependent forces which cause only small displacement.

Consider a force $F(x)$, expanded in a Taylor series about x_0, an equilibrium point:

$$F(x) = F(x_0) + \left(\frac{dF}{dx}\right)_{x_0} (x - x_0) + \frac{1}{2!} \left(\frac{d^2F}{dx^2}\right)_{x_0} (x - x_0)^2$$

$$+ \frac{1}{3!} \left(\frac{d^3F}{dx^3}\right)_{x_0} (x - x_0)^3 + \cdots \qquad (6.115)$$

Since x_0 is a point of equilibrium, $F(x_0) = 0$, and for small displacements about x_0,

$$F(x) \cong \left(\frac{dF}{dx}\right)_{x_0} (x - x_0)$$

It was pointed out in Chapter 5, that as the displacement increases, it may no longer suffice to neglect higher-order terms in $x - x_0$. Making the substitution $x - x_0 = y$, and setting

$$\left(\frac{dF}{dx}\right)_{x_0} = k_1 \qquad \left(\frac{d^2F}{dx^2}\right)_{x_0} = k_2 \qquad \text{etc.}$$

Eq. (6.115) is written

$$F(x) = k_1 y + k_2 y^2 + k_3 y^3 + \cdots \qquad (6.116)$$

For stable equilibrium, k_1 must be negative. Setting $k_1 = -k$ (where $k > 0$), the equation of motion is

$$-ky + k_2 y^2 + k_3 y^3 + \cdots = m\ddot{y} \tag{6.117}$$

We shall limit ourselves to considerations of only those forces which can lead to stable equilibrium, thereby setting the coefficients $k_2, k_4, \cdots$, equal to zero. Equation (6.117) now becomes

$$m\ddot{y} + ky = k_3 y^3 + \cdots \tag{6.118}$$

It can be readily demonstrated that the equation for the simple pendulum with finite oscillations may be reduced to an equation of the same form as Eq. (6.118). Recall that for the simple pendulum,

$$-mg \sin \psi = mL\ddot{\psi} \tag{6.119}$$

Suppose ψ_{max} is not small enough to warrant approximating $\sin \psi$ by ψ but is small enough that we need only consider one term beyond ψ in an expansion of $\sin \psi$ ($\sin \psi = \psi - \psi^3/3!$). Thus

$$-mg\psi + \frac{mg\psi^3}{6} = mL\ddot{\psi} \tag{6.120a}$$

or

$$\ddot{\psi} + \omega_0^2 \psi = \frac{\omega_0^2 \psi^3}{6} \tag{6.120b}$$

where $\omega_0 = (g/L)^{1/2}$.

As an approximation to the solution, let $\psi = a \cos \omega t$, where ω does not necessarily equal ω_0. The term ψ^3 then generates a term involving $\cos \omega t$ and $\cos^3 \omega t$. [Recall from trigonometry that

$$\cos^3 \omega t = \cos \omega t \cos^2 \omega t = \cos \omega t \left(\tfrac{1}{2} + \tfrac{1}{2} \cos 2\omega t\right)$$

$$= \tfrac{1}{2} \cos \omega t + \tfrac{1}{4} \cos 3\omega t + \tfrac{1}{4} \cos \omega t$$

$$= \tfrac{3}{4} \cos \omega t + \tfrac{1}{4} \cos 3\omega t]$$

For a somewhat better approximation, therefore, a trial function

$$\psi = a_1 \cos \omega t + a_3 \cos 3\omega t \tag{6.121}$$

is selected. Sinusoidal functions are not required if the initial conditions are $\psi = \psi_0$, $\dot{\psi} = 0$.

Differentiating Eq. (6.121) and substituting into Eq. (6.120b),

$$-a_1 \omega^2 \cos \omega t - 9a_3 \omega^2 \cos 3\omega t + \omega_0^2 a_1 \cos \omega t + \omega_0^2 a_3 \cos 3\omega t$$

$$= \frac{\omega_0^2}{6} a_1^3 \frac{3}{4} \cos \omega t + \frac{\omega_0^2}{6} a_1^3 \frac{1}{4} \cos 3\omega t + \frac{\omega_0^2}{2} a_1^2 a_3 \cos 3\omega t \cos^2 \omega t$$

$$+ \frac{\omega_0^2}{2} a_1 a_3^2 \cos \omega t \cos^2 3\omega t + \frac{\omega_0^2}{6} a_3^3 \cos^3 3\omega t \tag{6.122}$$

Combining like terms in $\cos \omega t$ and $\cos 3\omega t$ and neglecting the terms leading to even harmonics of $\cos \omega t$, since $a_3 \ll a_1$,

$$(-a_1\omega^2 + a_1\omega_0^2 - \frac{1}{8}\omega_0^2 a_1^3)\cos \omega t + \left(-9a_3\omega^2 + a_3\omega_0^2 - \frac{a_1^3\omega_0^2}{24}\right)\cos 3\omega t \cong 0$$

(6.123)

The coefficients of $\cos \omega t$ and $\cos 3\omega t$ must equal zero independently in order that Eq. (6.123) be valid for all t:

$$-a_1\omega^2 + a_1\omega_0^2 - \frac{a_1^3}{8}\omega_0^2 = 0$$

(6.124a)

$$-9a_3\omega^2 + a_3\omega_0^2 - \frac{a_1^3\omega_0^2}{24} = 0$$

(6.124b)

Solving for ω^2 in terms of ω_0 and a_1 [Eq. (6.124a)],

$$\omega^2 = \omega_0^2\left(1 - \frac{a_1^2}{8}\right)$$

(6.125)

The other solution is $a_1 = 0$, a trivial case. Solving for a_3 in terms of a_1, Eq. (6.124b),

$$a_3 = -\frac{a_1^3}{3(64 + 9a_1^2)}$$

(6.126)

From the initial condition on displacement,

$$\psi_0 = a_1 + a_3$$

If the solution is limited to angles of less than 20 degrees, the approximation $64 \gg 9a_1^2$ is valid, leading to

$$a_3 \cong -\frac{a_1^3}{192}$$

(6.127)

$$\psi_0 \cong a_1 - \frac{a_1^3}{192} \cong a_1$$

(6.128)

Thus the motion is

$$\psi \cong \psi_0 \cos \omega t - \frac{\psi_0^3}{192}\cos 3\omega t$$

(6.129)

where

$$\omega = \omega_0\left(1 - \frac{\psi_0^2}{8}\right)^{1/2} \cong \omega_0\left(1 - \frac{\psi_0^2}{16} + \cdots\right)$$

(6.130)

and the period

$$T = \frac{2\pi}{\omega} = \frac{2\pi}{\omega_0}\left(1 - \frac{\psi_0^2}{8}\right)^{-1/2} \cong \frac{2\pi}{\omega_0}\left(1 + \frac{\psi_0^2}{16} + \cdots\right)$$

(6.131)

The frequency is now a function of amplitude; this was not the case for the linear restoring force. The motion is harmonic since the cosine terms contain multiples of a fundamental frequency. Nevertheless, it is not simple harmonic because more than one frequency occurs.

We investigate next an exact analysis of the simple pendulum. Consider once again the equation of motion,

$$-mgL \sin \psi = mL^2 \ddot{\psi} \tag{6.132}$$

Multiplying by $\dot{\psi}/mL^2$ and letting $g/L = \omega_0^2$,

$$\ddot{\psi}\dot{\psi} = -\omega_0^2 \dot{\psi} \sin \psi \tag{6.133}$$

Multiplying by dt and integrating,

$$\tfrac{1}{2}\dot{\psi}^2 = \omega_0^2(\cos \psi - \cos \psi_0) \tag{6.134}$$

where $-\omega_0^2 \cos \psi_0$ is the constant of integration. Solving for dt and integrating t' from 0 to t and ψ' from ψ_0 to ψ,

$$\int_0^t dt' = t = \int_{\psi_0}^{\psi} \frac{d\psi'}{\omega_0[2(\cos \psi' - \cos \psi_0)]^{\frac{1}{2}}} \tag{6.135}$$

Equation (6.135) is an incomplete elliptic integral of the first kind.[3] To convert it to standard representation, we use the trigonometric identity $\cos \psi = 1 - 2\sin^2(\psi/2)$ and the substitutions

$$\sin \gamma = \frac{\sin(\psi/2)}{\sin(\psi_0/2)} \qquad \text{and} \qquad \sin\left(\frac{\psi_0}{2}\right) = k$$

Differentiating,

$$\frac{1}{2}\cos\left(\frac{\psi}{2}\right) d\psi = \sin\left(\frac{\psi_0}{2}\right) \cos \gamma \, d\gamma \tag{6.136}$$

and Eq. (6.135) is now written,

$$t = \int_0^{\gamma} \frac{2 \cos \gamma' \, d\gamma'/\cos(\psi/2) \sin(\psi_0/2)}{\omega_0\left[2\left(2\sin^2\frac{\psi_0}{2} - \sin\frac{\psi}{2}\right)\right]^{\frac{1}{2}}}$$

$$= \int_0^{\gamma} \frac{\cos \gamma' \, d\gamma'}{\omega_0 \cos\left(\frac{\psi}{2}\right)\left[1 - \frac{\sin^2(\psi/2)}{\sin^2(\psi_0/2)}\right]^{\frac{1}{2}}} \tag{6.137}$$

---

[3] Mathews, J., and Walker, R. L., *Mathematical Methods of Physics*, W. A. Benjamin, Inc., New York, 1961.

or

$$t = \int_0^\gamma \frac{d\gamma'}{\omega_0[1 - \sin^2(\psi_0/2)\sin^2\gamma]^{1/2}} = \int_0^\gamma \frac{d\gamma'}{\omega_0[1 - k^2\sin^2\gamma]^{1/2}} \tag{6.138}$$

where the limits have been changed to adhere to standard form. Since k and $\sin\gamma$ are each less than unity, the integrand may be expanded in a binomial series:

$$t = \frac{1}{\omega_0}\int_0^\gamma \left(1 + \frac{k^2}{2}\sin^2\gamma' + \frac{3}{8}k^4\sin^4\gamma' + \cdots\right)d\gamma' \tag{6.139}$$

Equation (6.139) may be integrated term by term and the motion readily represented by a plot of t as a function of γ.

The period is simply the time corresponding to $\gamma = 2\pi$. Since pendulous motion may be divided into four equal intervals per period, Eq. (6.139) may be rewritten

$$T = \frac{4}{\omega_0}\int_0^{\pi/2}\left(1 + \frac{k^2}{2}\sin^2\gamma' + \frac{3}{8}k^4\sin^4\gamma' + \cdots\right)d\gamma' \tag{6.140}$$

Term by term integration yields

$$\begin{aligned}
T &= \frac{4}{\omega_0}\left(\frac{\pi}{2} + \frac{\pi k^2}{8} + \frac{9\pi k^4}{144} + \cdots\right) \\
&= \frac{2\pi}{\omega_0}\left(1 + \frac{k^2}{4} + \frac{9k^4}{72} + \cdots\right)
\end{aligned} \tag{6.141}$$

Since $k = \sin(\psi_0/2)$, where ψ_0 represents the amplitude of oscillation,

$$T = \frac{2\pi}{\omega_0}\left[1 + \frac{\sin^2(\psi_0/2)}{4} + \frac{9}{72}\sin^4\left(\frac{\psi_0}{2}\right) + \cdots\right] \tag{6.142}$$

For small ψ_0 we expand Eq. (6.142) to compare the exact results with those of the previous approximate analysis:

$$T = \frac{2\pi}{\omega_0}\left(1 + \frac{\psi_0^2}{16} + \cdots\right)$$

This result agrees with the approximate analysis, Eq. (6.131).

Values of the complete elliptic integral of the first kind, Eq. (6.138), for $\gamma = \pi/2$ are tabulated in many standard references.[4]

6.14 THE SPHERICAL PENDULUM

We now investigate the nonplanar pendulum, usually referred to as the *spherical pendulum* because all possible positions of the pendulous mass lie on the surface of a sphere. The spherical polar coordinate system is most convenient for the

---

[4] Janke, E., and Emde, F., *Tables of Functions*, Dover Publications, Inc., New York, 1943.

purposes of analysis because of the condition of constraint, r = constant; that is, the distance from the support to the mass is fixed. As a result, only two coordinates are required to specify the mass position. There are only two forces acting on the mass: the weight mg and the tension in the massless supporting rod τ, as shown in Fig. 6.25.

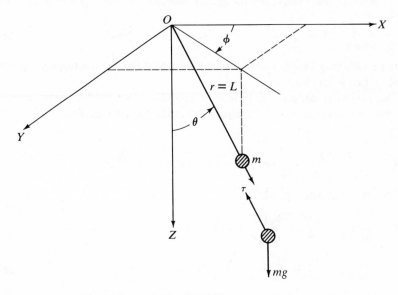

Fig. 6.25

The equations of motion in spherical polar coordinates are therefore

$$F_r = mg \cos \theta - \tau = m(-L\dot{\phi}^2 \sin^2 \theta - L\dot{\theta}^2) \tag{6.143a}$$

$$F_\theta = -mg \sin \theta = m(L\ddot{\theta} - L\dot{\phi}^2 \sin \theta \cos \theta) \tag{6.143b}$$

$$F_\phi = 0 = m(L\ddot{\phi} \sin \theta + 2L\dot{\phi}\dot{\theta} \cos \theta) \tag{6.143c}$$

where the angles θ and ϕ and the tension τ are unknown functions of time. Equations (6.143b) and (6.143c) represent two coupled differential equations in the variables θ and ϕ which are now solved subject to the appropriate initial conditions. Multiplying Eq. (6.143c) by $\sin \theta / m$,

$$0 = L\ddot{\phi} \sin^2 \theta + 2L\dot{\phi}\dot{\theta} \cos \theta \sin \theta = \frac{d}{dt} L\dot{\phi} \sin^2 \theta \tag{6.144}$$

Thus $L\dot{\phi} \sin^2 \theta$ is a constant of the motion which we call A_ϕ,

$$A_\phi = L\dot{\phi} \sin^2 \theta = L\dot{\phi}_0 \sin^2 \theta_0 \tag{6.145a}$$

or

$$\dot{\phi} = \frac{A_\phi}{L \sin^2 \theta} \tag{6.145b}$$

where $\dot{\phi}_0$ and θ_0 are the initial values of the functions.

Substitution of $\dot{\phi}$ [Eq. (6.145b)] into Eq. (6.143b) serves to uncouple ϕ and θ,

$$\ddot{\theta} + \frac{g}{L} \sin \theta - \frac{A_\phi^2}{L^2} \frac{\cos \theta}{\sin^3 \theta} = 0 \tag{6.146}$$

where we have divided through by mL.

Note that when $\dot{\phi}$ and hence A_ϕ are zero, Eq. (6.146) reduces to that of the simple pendulum; there is no out of plane motion. Another special case occurs when θ remains constant. From Eq. (6.143c) we obtain $\ddot{\phi} = 0$, the solution of which is

$$\phi = \omega_\phi t + \phi_0 \tag{6.147}$$

where $\omega_\phi = \dot{\phi} = $ constant and $\phi_0 = $ initial value of ϕ.

From Eq. (6.143b), therefore,

$$-mg \sin \theta_0 = -mL\dot{\phi}^2 \sin \theta_0 \cos \theta_0 \tag{6.148}$$

and hence

$$\cos \theta_0 = \frac{g}{L\omega_\phi^2} \tag{6.149}$$

This interesting result demonstrates that the constant angle θ_0 depends upon the ratio of g to the centripetal acceleration. This is, of course, the result obtained for the *conical pendulum*.

To arrive at a general solution of Eq. (6.146), a first integral of the motion will be required. Multiplying by $\dot{\theta} \, dt$,

$$\ddot{\theta}\dot{\theta} \, dt + \frac{g}{L} \sin \theta (\dot{\theta} \, dt) - \frac{A_\phi^2 \cos \theta (\dot{\theta} \, dt)}{L^2 \sin^3 \theta} = d\left(\frac{\dot{\theta}^2}{2} - \frac{g}{L} \cos \theta + \frac{A_\phi^2}{2L^2 \sin^2 \theta} \right) = 0 \tag{6.150}$$

Thus

$$\frac{\dot{\theta}^2}{2} - \frac{g}{L} \cos \theta + \frac{A_\phi^2}{2L^2 \sin^2 \theta} = \text{constant} = B_\theta \tag{6.151}$$

Solving for dt and integrating,

$$t = \int dt = \frac{1}{\sqrt{2}} \int \frac{d\theta}{[B_\theta + (g/L) \cos \theta - (A_\phi^2/2L^2 \sin^2 \theta)]^{1/2}} + C \tag{6.152}$$

Although Eq. (6.152) cannot be integrated in closed form, it can be treated by numerical techniques. A special case, solvable in closed form, occurs for small angular oscillations of a nearly conical pendulum. From Eq. (6.149),

$$\omega_\phi^2 = \frac{g}{L \cos \theta_0} \tag{6.153}$$

Combining Eqs. (6.145a) and (6.149),

$$A_\phi^2 = \frac{gL \sin^4 \theta_0}{\cos \theta_0} \tag{6.154}$$

Employing the above substitutions, Eq. (6.146), the equation of motion in θ, now becomes

$$\ddot{\theta} + \frac{g}{L} \sin \theta - \frac{g}{L} \frac{\sin^4 \theta_0}{\cos \theta_0} \frac{\cos \theta}{\sin^3 \theta} = 0 \tag{6.155}$$

From Eq. (6.155) it is found that when $\theta = \theta_0$, $\ddot{\theta} = 0$. Expanding the term,

$$\sin \theta - \frac{\sin^4 \theta_0}{\cos \theta_0} \frac{\cos \theta}{\sin^3 \theta}$$

about $\theta = \theta_0$,

$$\ddot{\theta} + \frac{g}{L} \left[(\theta - \theta_0) \left(\cos \theta_0 + \frac{\sin^2 \theta_0}{\cos \theta_0} + 3 \cos \theta_0 \right) + \frac{(\theta - \theta_0)^2}{2} (\ ) + \cdots \right] = 0 \tag{6.156}$$

To first order Eq. (6.156) is written

$$\ddot{\theta} + \frac{g}{L} \frac{1 + 3 \cos^2 \theta_0}{\cos \theta_0} (\theta - \theta_0) = 0 \tag{6.157}$$

the solution of which is

$$\theta - \theta_0 = a \cos \left[\left(\frac{g}{L} \frac{1 + 3 \cos^2 \theta_0}{\cos \theta_0} \right)^{\frac{1}{2}} t + \alpha \right] \tag{6.158}$$

where a is small relative to θ_0.

The period of the angle θ follows from the above:

$$T_\theta = 2\pi \left(\frac{L}{g} \frac{\cos \theta_0}{1 + 3 \cos^2 \theta_0} \right)^{\frac{1}{2}} \tag{6.159}$$

The period of the angle ϕ is determined from Eq. (6.149).

$$T_\phi = 2\pi \left(\frac{L}{g} \cos \theta_0 \right)^{\frac{1}{2}} \tag{6.160}$$

The ratio

$$\frac{T_\theta}{T_\phi} = \left(\frac{1}{1 + 3 \cos^3 \theta_0} \right)^{\frac{1}{2}}$$

is not in general a rational number.

An interesting approach to the description of motion obtains when the coordinates of the mass are written in the form

$$x = L \sin \theta \cos \phi$$

$$= L \sin \left\{ \theta_0 + a \cos \left[\left(\frac{g}{L} \frac{1 + 3 \cos^2 \theta_0}{\cos \theta_0} \right)^{1/2} t + \alpha \right] \right\} \cos(\omega_\phi t + \phi_0) \quad \textbf{(6.161)}$$

$$y = L \sin \theta \sin \phi$$

$$= L \sin \left\{ \theta_0 + a \cos \left[\left(\frac{g}{L} \frac{1 + 3 \cos^2 \theta_0}{\cos \theta_0} \right)^{1/2} t + \alpha \right] \right\} \sin(\omega_\phi t + \phi_0) \quad \textbf{(6.162)}$$

$$z = L \cos \theta = L \cos \left\{ \theta_0 + a \cos \left[\left(\frac{g}{L} \frac{1 + 3 \cos^2 \theta_0}{\cos \theta_0} \right)^{1/2} t + \alpha \right] \right\} \quad \textbf{(6.163)}$$

The bob is periodic in z with period T_θ. The path in the XY-plane, $R = (x^2 + y^2)^{1/2}$, describes a precessing ellipse, with the limits of R corresponding to θ_{max} and θ_{min}, $\theta_0 + a$ and $\theta_0 - a$, respectively. Because the ratio of the periods of θ and ϕ motion is not a rational number, when the angle ϕ has repeated itself ($\omega_0 t + \phi_0 = 2\pi$), x and y have not, so that the path of R is not closed. The extent of the precession may be determined by permitting θ to vary by 2π and ascertaining the corresponding change in ϕ. The path of R is a Lissajous figure, displaying the motion of a two-dimensional oscillator.

References

Fowles, G. R., *Analytical Mechanics*, Holt, Rinehart and Winston, Inc., New York, 1962.

Haag, J., *Oscillatory Motions*, Wadsworth Publishing Co., Inc., Belmont, Calif., 1962.

Housner, G. W., and Hudson, D. E., *Applied Mechanics*, D. Van Nostrand Company, Inc., Princeton, N.J., 1954.

Joos, G., *Theoretical Physics*, Third Edition, Hafner Publishing Company, New York, 1956.

Lindsay, R. B., *Mechanical Radiation*, McGraw-Hill, Inc., New York, 1960.

Stephenson, R. J., *Mechanics and Properties of Matter*, Second Edition, John Wiley & Sons, Inc., New York, 1960.

Tse, F. S., Morse, I. E., and Hinkle, R. T., *Mechanical Vibrations*, Allyn and Bacon, Inc., Boston, 1963.

EXERCISES

6.1 Determine the damping coefficient required in a spring–mass system so that the velocity decreases by 50 percent each time the mass passes its equilibrium position.

6.2 A mass m moves on a plane tangent to the earth. Find the natural frequency of oscillation of the mass for small displacements along the plane. Refer to Fig. P6.2.

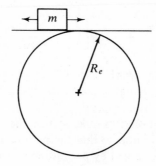

Fig. P6.2

6.3 Determine the time for an overdamped oscillator to return to within (a) 10, (b) 5, and (c) 1 percent of its equilibrium position when given an initial displacement x_0.

6.4 A mass of 25 g is attached to a spring of stiffness 400 dyn/cm. What should the damping coefficient be for critical damping?

6.5 Determine the natural frequencies for the arrangement shown in Fig. P6.5.

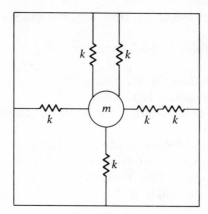

Fig. P6.5

6.6 The logarithmic decrement δ is defined for a damped unforced oscillator as the logarithm of the ratio of two successive maxima on the same side of equilibrium. Prove that

$$\delta = \frac{2\pi(b/b_{cr})}{[1 - (b/b_{cr})^2]^{1/2}}$$

Demonstrate that δ is independent of initial conditions.

Design a system of mass 5 kg so that $\delta = 0.1$ when the time between two successive maxima is 0.03 sec.

6.7 A ball of mass m rests inside a hoop of radius R. Find the natural frequency of oscillation of the mass for small displacements about the bottom of the hoop. Refer to Fig. P6.7.

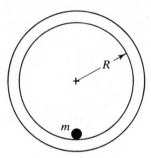

Fig. P6.7

6.8 A mass m slides without friction on two inclined planes, as shown in Fig. P6.8. Find the frequency of oscillation for small displacement about the equilibrium position.

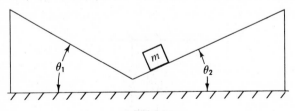

Fig. P6.8

6.9 A mass m attached to a rigid wall by means of a linear spring k is acted upon by a viscous damping force $-b\dot{x}$ and Coulomb friction μN. Determine the position of the mass as a function of time if the initial displacement is x_0 and the mass starts from rest.

6.10 A mass of 400 g is acted upon by a spring of stiffness 16,000 dyn/cm. (a) What is the natural frequency of oscillation? (b) What initial displacement, starting from rest, is required so that the displacement is 2 cm when one fourth of the system energy is kinetic?

6.11 Suppose the damping force acting on a linear oscillator is in the direction of the velocity, $+b\dot{x}$. What would the position be as a function of time for "underdamping," "overdamping," and "critical damping"?

6.12 A mass of 900 g is connected to a spring of stiffness 18,000 dyn/cm. If the damping constant is 3600 dyn-sec/cm, how long does it take for the mass to undergo 10 oscillations? What is its amplitude after this many oscillations?

6.13 A mass m is connected to two linear springs as shown in Fig. P6.13. Find the natural frequency of vibration of the mass if it only moves in the vertical direction.

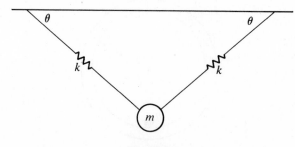

Fig. P6.13

6.14 Derive an expression for the mechanical energy loss per cycle for a damped free oscillator.

6.15 Determine the natural frequency of oscillation for the system shown in Fig. P6.15.

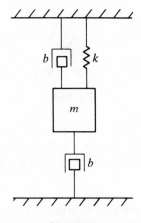

Fig. P6.15

6.16 Determine the natural frequency of oscillation for the system shown in Fig. P6.16.

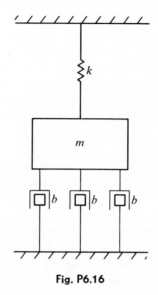

Fig. P6.16

6.17 Determine the natural frequency of oscillation for the system shown in Fig. P6.17.

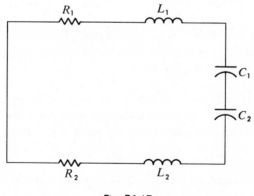

Fig. P6.17

6.18 Find the frequency of oscillation for the mass of Exercise 6.8 subject to large displacements.

6.19 A mass of 15 kg drops 15 cm onto a massless platform, suffering an inelastic collision. Design a spring damper system so that the platform moves to its new equilibrium position without overshoot in a minimum time. The new equilibrium position is 5 cm below the original position.

6.20 Redesign the spring in Exercise 6.19 for the mass dropping a distance of 7 m.

6.21 A particle of mass m is acted upon by a force

$$F = F_{max} \left(\frac{a^2}{x^2} - \frac{x^2}{a^2} \right)$$

Find the frequency of small oscillations about all equilibrium positions.

6.22 A mass m moves subject to the following forces: a linear restoring force $-kx$, a damping force $-b\dot{x}$, and a driving force $F_{max}e^{-\alpha t} \cos \omega t$. Find the position as a function of time.

6.23 Find the frequencies at which the maximum velocity is $1/\sqrt{2}$ times the velocity at resonance for (a) an underdamped and (b) an overdamped forced linear oscillator. These frequencies are called the half-power points.

6.24 The "Q" or quality of a series RLC electric circuit to which a sinusoidal voltage $V_{max} \sin \omega t$ is impressed can be defined as the maximum energy stored per cycle divided by the average power dissipated in the steady state: $Q = \omega L/R$. Find the Q of a damped mechanical oscillator.

6.25 Show that $x = A \sinh \omega t + B \cosh \omega t$ is a solution of the differential equation of an unforced damped oscillator and find ω in terms of k, m, and b. What are A and B in terms of x_0 and v_0?

6.26 Show that for a damped, forced oscillator, the velocity is in phase with a sinusoidally applied force when the natural frequency of undamped oscillation equals the frequency of the applied force.

6.27 An undamped oscillator is driven at its resonant frequency by a force $F = F_{max} \sin \omega_0 t$. What should the initial condition be so that the time for the amplitude of the oscillator to reach the breaking point is a maximum? The breaking force $F = 10F_{max}$.

6.28 Prove that $e^{i\theta} = \cos \theta + i \sin \theta$ by expanding both sides in a Taylor series about $\theta = 0$ and equating the real and complex parts.

6.29 What is $(i)^i$? Prove that if $A_1 + A_2$ is real and $A_1 - A_2$ is pure complex, A_1 and A_2 are complex conjugates.

6.30 Find the frequency of a sinusoidal driving force applied to an underdamped linear oscillator so that the amplitude of the steady-state motion is a maximum. What is the phase relationship between the applied force and the velocity?

6.31 Determine the phase angle α of an applied force

$$F = F_{max} \cos (\omega t + \alpha)$$

such that an undamped oscillator starting from rest experiences no transient motion.

6.32 What initial conditions are required so that an overdamped linear oscillator experiences no transient motion when subject to a sinusoidal driving force?

6.33 Prove that the time average of the kinetic and potential energies of an undamped unforced linear oscillator are the same. Find the ratio of the two energies if the undamped oscillator is subject to a driving force

$$F = F_{max} \cos \omega t$$

where $\omega \neq (k/m)^{1/2}$. The oscillator starts from rest at its equilibrium position.

6.34 A force $F_{max}(t/\tau) \cos (\omega_0 t + \alpha)$ is applied to a one-dimensional linear oscillator of mass m and stiffness k; τ, α, ω_0 are constants, and $\omega_0 = (k/m)^{1/2}$. If the mass starts from rest at the origin, find its position as a function of time.

6.35 A linear underdamped one-dimensional oscillator is subject to a force

$$F = F_{max} \cos \omega_1 t + \frac{F_{max}}{3} \cos 3\omega_1 t$$

Find the velocity and position of the mass for all t if it starts from its equilibrium position at rest.

6.36 An underdamped oscillator is excited at $\omega = (k/m)^{1/2}$. What change in the amplitude of oscillation will occur if k is reduced by 10 percent while ω remains constant?

6.37 Prove that for an undamped, unforced oscillator with spring constant $k = k(x)$, the amplitude is a function of the period of oscillation unless $k(x)$ is a constant.

6.38 An undamped oscillator is acted upon by

$$F = F_{max}(e^{-2\omega_0 t} - e^{-\omega_0 t}) \qquad \omega_0 = \left(\frac{k}{m}\right)^{1/2}$$

Find the position of the oscillator if the initial conditions are x_0 and $-v_0$.

6.39 A critically damped oscillator is acted upon by a force

$$F = F_{max}\left[e^{-(b/2m)t} - \cos \left(\frac{b}{2m}\right)^{1/2} t \right]$$

Determine the motion for initial conditions $x_0 = 0$, $v_0 = 0$.

6.40 How much energy per cycle is required to cause an underdamped forced linear oscillator to vibrate at $\omega = 0.1\omega_0$? Is the power input constant? At what applied frequency is the power input a maximum?

6.41 One cycle of the force

$$F = F_{max} \cos \omega_0 t$$

is applied to an undamped oscillator (mass m, spring constant k). Find the position of the mass as a function of time if it starts from rest at its equilibrium position.

6.42 An undamped oscillator is acted upon by the force shown in Fig. P6.42.

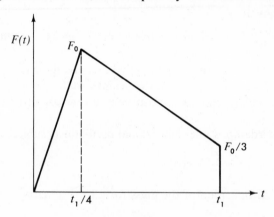

Fig. P6.42

What is the position of the mass as a function of time if the displacement is zero and the initial velocity

$$v_0 = \frac{F_0 t_1}{m}$$

6.43 Determine the period of a simple pendulum for amplitudes ranging from 2 to 52 degrees in increments of 10 degrees.

CHAPTER 7

GENERAL
MOTION
OF A PARTICLE

7.1 INTRODUCTION

In this chapter we have grouped a number of important applications of Newton's second law and the energy approach. These include the motion of projectiles acted upon by gravitational and frictional forces and the motion of charged particles subjected to electrostatic and magnetic influences. The cases cited are neither instances of rectilinear motion nor central force motion, in which the net force is always directed toward a fixed point. Thus the term "general motion" is often applied.

7.2 MOTION OF PROJECTILES

In studying the motion of projectiles, we begin with a much simplified case. As the original assumptions are changed to improve the approximation of the "real case," the equations become increasingly complex. In practice, a point is eventually reached in which numerical techniques suitable for computer solution are employed. The reader must be aware that approximations and simplifications limit the applicability of the results.

A shell is fired from a cannon as shown in Fig. 7.1. The factors governing the impact point include the wind velocity, frictional resistance to motion offered by the air, earth's rotation, as well as the gravitational field, initial velocity, and firing angle. In the first of three approximate analyses, the following assumptions and idealizations are made to simplify the equations of motion:

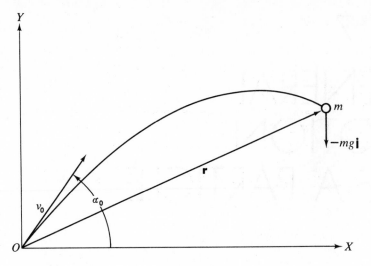

Fig. 7.1

a. *The projectile is a point mass or particle.* In a more accurate analysis, it would be considered a body possessing finite volume and a definite surface configuration. Our concern would then be with the motion of the mass center. The *attitude* of the projectile, described by the angles between reference axes in the projectile and a convenient external coordinate reference, is related to the air drag and would therefore enter into the formulation of the problem.

b. *The earth is nonrotating.* If greater accuracy is required, the accelerated or noninertial motion of the earth beneath the projectile must be taken into account. In this chapter the earth is used as a reference for which Newton's laws are assumed valid. This is a very good approximation for short ranges.

c. *The gravitational field is constant and acts perpendicular to the surface of a flat earth.* For distances small in comparison with the earth's radius, the flat-earth assumption yields good results. The nature of the variation of g has been discussed in Chapter 3.

d. *The air offers no resistance to motion; that is, motion occurs as it would in a vacuum.* Actually, air friction is important. It depends upon projectile attitude, wind velocity, air density, air viscosity, projectile configuration, and projectile speed.

e. *Motion occurs in a plane.*

Although the limitations of the analysis discussed above appear overwhelming, the results obtained are surprisingly accurate for short ranges.

7.3 PROJECTILE ANALYSIS I

Consider the projectile somewhere along its trajectory as shown in Fig. 7.1. Because there are no propulsive forces and air friction is neglected, only the gravitational force acts on the projectile. Thus

$$\mathbf{F} = -mg\hat{\mathbf{j}} = m\mathbf{a} \tag{7.1}$$

or

$$\mathbf{a} = \ddot{\mathbf{r}} = \frac{d^2}{dt^2}(x\hat{\mathbf{i}} + y\hat{\mathbf{j}}) = -g\hat{\mathbf{j}} \tag{7.2}$$

The acceleration is therefore $-g$ throughout the trajectory. This is, of course, because all forces except gravity have been neglected. Since g is independent of mass, for this special case, the motion is likewise independent of the projectile mass.

Each side of Eq. (7.2) is a vector. For the equality to hold, the respective X- and Y-components must be equal:

$$\ddot{x} = 0 \tag{7.3}$$

$$\ddot{y} = -g \tag{7.4}$$

The X- and Y-accelerations are uncoupled, permitting independent determination of the X- and Y-velocities and positions. Vector position and velocity are then found through vector addition of the respective components. Integrating Eq. (7.3),

$$\dot{x} = C_1 \tag{7.5}$$

$$x = C_1 t + C_2 \tag{7.6}$$

Integrating Eq. (7.4),

$$\dot{y} = -gt + C_3 \tag{7.7}$$

$$y = -\tfrac{1}{2}gt^2 + C_3 t + C_4 \tag{7.8}$$

The constants of integration are conveniently determined from the initial conditions. Referring to Fig. 7.1, the following initial conditions apply:

At $t = 0$: $x = 0, v_x = v_{0x} = v_0 \cos \alpha_0$ \hfill (7.9a)

At $t = 0$: $y = 0, v_y = v_{0y} = v_0 \sin \alpha_0$ \hfill (7.9b)

Note that at $t = 0$, x and y need not be zero. Their initial values depend upon the location of the origin.

Applying the conditions given by Eq. (7.9a) to Eqs. (7.5) and (7.6):

$$C_1 = v_{0x} \qquad\qquad C_2 = x_{t=0} = 0$$

from which the X-equations of motion are

$$x = v_{0x}t = (v_0 \cos \alpha_0)t \tag{7.10}$$

$$\dot{x} = v_x = v_0 \cos \alpha_0 \tag{7.11}$$

Equation (7.11) states that the X-velocity is constant. This follows directly from the assumption that no forces act in that direction.

Applying Eqs. (7.9b) to Eqs. (7.7) and (7.8),

$$C_3 = v_{0y} \quad \text{and} \quad C_4 = y_{t=0} = 0$$

from which the Y-equations of motion are

$$y = -\tfrac{1}{2}gt^2 + (v_0 \sin \alpha_0)t \tag{7.12}$$

$$\dot{y} = v_y = -gt + v_0 \sin \alpha_0 \tag{7.13}$$

Unlike the motion in the X-direction, the Y-component of velocity is not constant because the gravitational force acts at all times.

Illustrative Example 7.1
A shell is fired at an angle of 60 degrees from the horizontal as shown in Fig. 7.2. If v_0, the initial velocity, is 1000 ft/sec, determine the time of flight, the X-distance traveled (range), the maximum altitude, and the orientation of the velocity vector just before impact. Assume the shell to land at the same elevation at which it is fired.

SOLUTION
The time of flight can be determined by noting that in this case impact occurs at $y = 0$. Substituting $y = 0$ into Eq. (7.12) yields two values of t,

$$0 = -\frac{32.2}{2}t^2 + (1000)(0.866)t$$

from which $t = 0$ (corresponding to $y = 0$ and $x = 0$) and $t = 53.8$ sec (corresponding to $y = 0$ and $x = x_{max}$).

The X-distance traveled may now be calculated from Eq. (7.10):

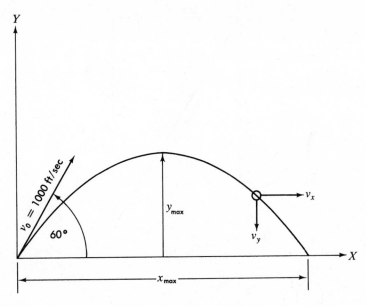

Fig. 7.2

$$x = (500)(53.8) = 2.69 \times 10^4 \text{ ft}$$

The maximum altitude may be determined by noting that at y_{max} the Y-velocity is zero. The time at which this occurs is found from Eq. (7.13):

$$0 = -32.2t + (1000)(0.866) \qquad t = 26.9 \text{ sec}$$

Note that when the impact and initial Y-coordinates are the same, the maximum altitude corresponds to the midpoint of travel, so the same result as above is obtained by halving the total time of flight. The maximum altitude from Eq. (7.12) is

$$y_{max} = -\frac{32.2}{2}(26.9)^2 + (866)(26.9) = 1.165 \times 10^4 \text{ ft}$$

Just before impact, $v_x = v_{0x} = 500$ ft/sec and v_y is given by Eq. (7.13):
$v_y = -(32.2)(53.8) + 866 = -866$ ft/sec $= -v_{0y}$.

From the velocity diagram, Fig. 7.3, $\tan \alpha = 866/500 = 1.732$; $\alpha = 60$ degrees.

To specify the *trajectory* which is defined in terms of position variables only, time must be eliminated between Eqs. (7.10) and (7.12). Substituting the expression for the time from Eq. (7.10), $t = x/v_{0x}$, into Eq. (7.12):

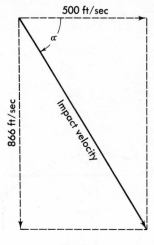

500 ft/sec

866 ft/sec

α

Impact velocity

Fig. 7.3

$$y = -\frac{g}{2}\left(\frac{x}{v_{0x}}\right)^2 + v_{0y}\frac{x}{v_{0x}} \tag{7.14}$$

Since $v_{0y}/v_{0x} = v_0 \sin \alpha_0/v_0 \cos \alpha_0 = \tan \alpha_0$, Eq. (7.14) becomes

$$y = \left[-\frac{g}{2}\left(\frac{1}{v_0 \cos \alpha_0}\right)^2\right]x^2 + (\tan \alpha_0)x \tag{7.15}$$

which describes a parabola. Thus the resultant motion depends only upon the initial conditions, which in Eq. (7.15) have been specified in terms of the initial velocity and angle. (In Section 8.4 it is demonstrated that a *radial* gravitational force results in hyperbolic and elliptic, as well as parabolic, trajectories.)

For a projectile beginning and ending its travel at $y = 0$, Eq. (7.15) yields

$$0 = \left[-\frac{g}{2}\left(\frac{1}{v_0 \cos \alpha_0}\right)^2\right]x^2 + (\tan \alpha_0)x \tag{7.16}$$

The roots of Eq. (7.16) are $x = 0$ and

$$x_{\text{max}} = \frac{2v_0^2 \tan \alpha_0 \cos^2 \alpha_0}{g} = \frac{2v_0^2 \sin \alpha_0 \cos \alpha_0}{g} = \frac{v_0^2 \sin 2\alpha_0}{g} \tag{7.17}$$

Equation (7.17) is a useful expression for the range.

By differentiating x in Eq. (7.17) with respect to α_0, the initial firing angle, and equating the result to zero, the firing angle corresponding to maximum range is determined,

$$\frac{dx}{d\alpha_0} = \frac{2v_0^2 \cos 2\alpha_0}{g} = 0$$

or

$$\cos 2\alpha_0 = 0$$

which is satisfied by $\alpha_0 = \pi/4$ rad.

At the maximum altitude, the slope of the trajectory is zero. Differentiating Eq. (7.15) with respect to x and equating to zero,

$$\frac{dy}{dx} = 0 = -g\left(\frac{1}{v_0 \cos \alpha_0}\right)^2 x + \tan \alpha_0$$

from which

$$x_{dy/dx=0} = \frac{v_0^2 \sin \alpha_0 \cos \alpha_0}{g} = \frac{v_0^2 \sin 2\alpha_0}{2g} \tag{7.18}$$

This corresponds to half the range for the case of firing and impact at $y = 0$. Substituting this value of x into Eq. (7.15),

$$y_{\max} = \frac{v_0^2 \sin^2 \alpha_0}{2g} \tag{7.19}$$

Illustrative Example 7.2

A projectile, fired with velocity v_0 and angle α_0 from the horizontal, is to impact at point P as shown in Fig. 7.4. What is the range measured along the straight line connecting the firing point and the target?

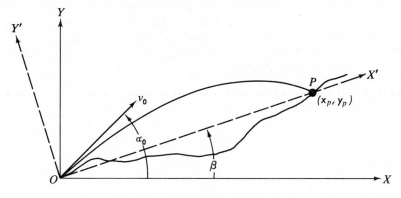

Fig. 7.4

SOLUTION

The point $P(x_p, y_p)$ lies on the trajectory, and therefore x_p and y_p are related by Eq. (7.15). If this equation is divided by x_p, the result is

$$\frac{y_p}{x_p} = \tan \alpha_0 - \frac{g}{2}\left(\frac{1}{v_0 \cos \alpha_0}\right)^2 x_p = \tan \beta$$

where $\beta = \tan^{-1}(y_p/x_p)$. On the basis of the above expression, the coordinate x_p may be expressed in terms of β and the initial conditions α_0 and v_0 as follows:

$$x_p = \frac{2}{g}(v_0 \cos \alpha_0)^2(\tan \alpha_0 - \tan \beta)$$

Finally, the range along OP is given by

$$R_\beta = \frac{x_p}{\cos \beta} = \frac{2}{g \cos \beta}(v_0 \cos \alpha_0)^2(\tan \alpha_0 - \tan \beta)$$

If β is zero, this equation reduces to Eq. (7.17), the expression for horizontal range for motion beginning and ending at $y = 0$.

We have left, as a more interesting solution for the reader, the trajectory determination in terms of variables in the $X'Y'$-coordinate system shown in Fig. 7.4.

7.4 PROJECTILE DRAG FORCE

The weakest assumption of analysis I is the neglect of fluid resistance. Consider now the motion through a fluid of a totally immersed object. Two types of forces act because of this relative motion, one related to frictional effects, tangent to the surface, the other normal to the surface, attributable to fluid pressure. These forces may be resolved into components parallel and perpendicular to the relative velocity of approach of the fluid and submerged object. The perpendicular fluid force component is termed the *lift* and the parallel component the *drag*. An enormous body of literature exists describing the theoretical and experimental work related to determining these forces as a function of geometry, fluid properties, and velocity. Much effort is devoted to determining the dimensionless coefficients of lift (C_L) and drag (C_D) defined below:

$$C_D = \frac{F_D}{\frac{1}{2}\rho u_\infty^2 A} \tag{7.20a}$$

and

$$C_L = \frac{F_L}{\frac{1}{2}\rho u_\infty^2 A} \tag{7.20b}$$

where F_L and F_D are the lift and drag forces of the fluid on the object, ρ the mass density of the fluid, u_∞ the velocity of the undisturbed fluid relative to the object, and A the area of the object projected on a plane perpendicular to the undisturbed fluid velocity.

For purposes of illustration, we shall confine our discussion to spheres, for which the lift is zero, because the geometry is always symmetrical about a plane parallel to the flow.

It is usual to relate the coefficients to another dimensionless variable termed the Reynolds number, defined

$$\mathrm{Re} = \frac{\rho u_\infty L}{\mu}$$

where L represents a characteristic length of the object, such as the diameter of a sphere, and μ is the viscosity of the fluid. In Fig. 7.5, C_D is presented as a function of Re for a sphere. It is noteworthy that for very low Reynolds numbers, fluid mechanical theory yields the result known as Stokes' law,

$$F_D = 6\pi R_0 \mu u_\infty \tag{7.21}$$

corresponding to a drag coefficient,

$$C_D = \frac{6\pi R_0 \mu u_\infty}{\frac{1}{2}\rho u_\infty^2 \pi R_0^2} = \frac{12\mu}{\rho u_\infty R_0} \tag{7.22a}$$

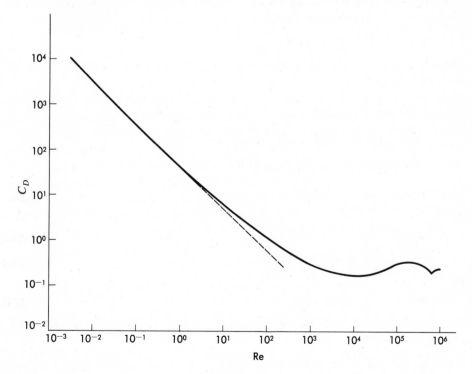

Fig. 7.5

or in terms of Re (based upon diameter),

$$C_D = \frac{24}{\text{Re}}$$

(7.22b)

This result corresponds to the range of Re from 5×10^{-2} to 1, the Stokes' law region shown in the figure. In this region the force is proportional to the first power of velocity. Beyond this the drag coefficient continues to decrease with increasing Re until Re = 1000, marking the approximate beginning of a nearly constant Reynolds number range ending at Re $\cong 2.5 \times 10^5$. In this regime, the drag force is proportional to the square of the velocity.

It is clear that it is the relative magnitudes of the variables forming Re which determine C_D. Consequently, what would ordinarily be considered a low velocity might be considerable with regard to C_D if, for example, the object were large and the viscosity quite low.

7.5 PROJECTILE ANALYSIS II

For very low Reynolds numbers, the absence of air currents, and uniform fluid properties, the drag force is approximately proportional to the projectile velocity. The equation of motion now becomes

$$\mathbf{F} = -mg\hat{\mathbf{j}} - b\mathbf{v} = m(\ddot{x}\hat{\mathbf{i}} + \ddot{y}\hat{\mathbf{j}})$$

(7.23)

where

$$\mathbf{v} = \dot{x}\hat{\mathbf{i}} + \dot{y}\hat{\mathbf{j}}$$

Rewriting Eq. (7.23),

$$m(\ddot{x}\hat{\mathbf{i}} + \ddot{y}\hat{\mathbf{j}}) = -b(\dot{x}\hat{\mathbf{i}} + \dot{y}\hat{\mathbf{j}}) - mg\hat{\mathbf{j}}$$

Equating coefficients of $\hat{\mathbf{i}}$ and $\hat{\mathbf{j}}$,

$$m\ddot{x} = -b\dot{x}$$

(7.24a)

or

$$m\dot{v}_x = -bv_x$$

(7.24b)

$$m\ddot{y} = -b\dot{y} - mg$$

(7.25a)

or

$$m\dot{v}_y = -bv_y - mg$$

(7.25b)

Again, the X- and Y-motions are uncoupled. Separating variables in Eq. (7.24b),

$$\frac{dv_x}{v_x} = -\frac{b}{m} dt$$

(7.26)

Integration yields

$$\ln v_x = -\frac{b}{m}t + C_1 \tag{7.27}$$

Substituting the initial condition: at $t = 0$, $v_x = v_{0x}$, into Eq. (7.27), $C_1 = \ln v_{0x}$, and, therefore,

$$\ln\left(\frac{v_x}{v_{0x}}\right) = -\frac{b}{m}t \tag{7.28}$$

In exponential form,

$$v_x = v_{0x}e^{-(b/m)t} \tag{7.29a}$$

or

$$\frac{v_x}{v_{0x}} = e^{-(b/m)t} \tag{7.29b}$$

Unlike the frictionless case, the X-velocity diminishes as $e^{-(b/m)t}$, approaching zero only for very large time. In Fig. 7.6, v_x/v_{0x} is shown as a function of time for $b = 0$ and $b > 0$.

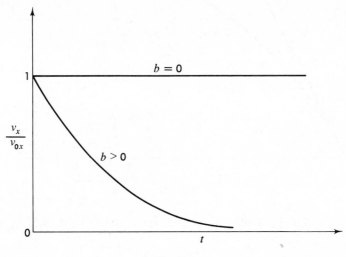

Fig. 7.6

Integrating Eq. (7.29),

$$x = -\frac{m}{b}v_{0x}e^{-(b/m)t} + C_2 \tag{7.30}$$

Employing the initial condition that at $t = 0$, $x = 0$: $C_2 = (m/b)v_{0x}$, which when substituted into Eq. (7.30) yields

$$x = \frac{mv_{0x}}{b}(1 - e^{-(b/m)t}) = x_{\lim}(1 - e^{-(b/m)t}) \tag{7.31}$$

Once again it is interesting to compare this case with the frictionless case. For $b > 0$, the X-displacement is observed to increase less and less rapidly with time (as the velocity diminishes). Since the X-velocity approaches zero for large time, the X-displacement has a limiting value corresponding to t approaching infinity:

$$x_{\text{lim}} = \frac{mv_{0x}}{b} \tag{7.32}$$

For $t = 0$, the slope of the X-displacement–time curve (v_x), Fig. 7.7, does not depend upon b and is v_{0x}. As shown, the slope remains constant for $b = 0$ but diminishes for $b > 0$.

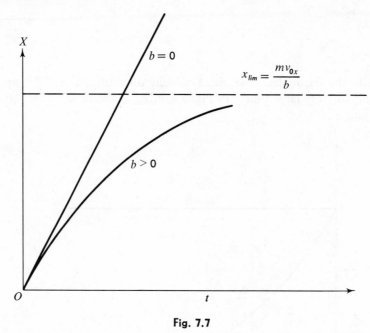

Fig. 7.7

A solution for the Y-motion requires integration of Eq. (7.25b), which may be written

$$m \, dv_y = -bv_y \, dt - mg \, dt \tag{7.33}$$

In separated form, Eq. (7.33) becomes

$$dt = \frac{-m \, dv_y}{bv_y + mg} \tag{7.34}$$

so that

$$t = -m \int \frac{dv_y}{bv_y + mg} + C_3 \tag{7.35}$$

Since

$$\int \frac{dv_y}{bv_y + mg} = \frac{1}{b} \ln (bv_y + mg)$$

Eq. (7.35) becomes

$$t = -\frac{m}{b} \ln (bv_y + mg) + C_3 \tag{7.36}$$

Introducing into Eq. (7.36) the initial condition, at $t = 0$, $v_y = v_{0y}$, $C_3 = (m/b) \ln (bv_{0y} + mg)$, which when substituted into Eq. (7.36) gives

$$t = -\frac{m}{b} \ln \left(\frac{bv_y + mg}{bv_{0y} + mg} \right) \tag{7.37}$$

Writing Eq. (7.37) in exponential form,

$$\frac{bv_y + mg}{bv_{0y} + mg} = e^{-(b/m)t}$$

or

$$\dot{y} = v_y = \left(v_{0y} + \frac{mg}{b} \right) e^{-(b/m)t} - \frac{mg}{b} \tag{7.38}$$

For very large values of t, $e^{-(b/m)t}$ approaches zero, and v_y, its asymptotic or terminal value,

$$v_{y,\text{term}} = -\frac{mg}{b} \tag{7.39}$$

When the particle "achieves" its terminal velocity, the acceleration is zero and Y-equilibrium exists. The result given by Eq. (7.39) may also be obtained, therefore, by equating the Y-drag force to the gravitational force [$\dot{v}_y = 0$ in Eq. (7.25b)].

The Y-position is obtained by integrating Eq. (7.38):

$$y = -\frac{m}{b} \left(v_{0y} + \frac{m}{b} g \right) e^{-(b/m)t} - \frac{mgt}{b} + C_4 \tag{7.40}$$

Employing the initial condition, at $t = 0$, $y = 0$,

$$C_4 = \frac{m}{b} \left(v_{0y} + \frac{m}{b} g \right)$$

Therefore,

$$y = \left(\frac{m}{b} v_{0y} + \frac{m^2}{b^2} g \right) (1 - e^{-(b/m)t}) - \frac{mgt}{b} \tag{7.41}$$

or

$$y = \frac{m}{b} [(v_{0y} - v_{y,\text{term}})(1 - e^{-(b/m)t}) - gt] \tag{7.42}$$

The trajectory is determined by eliminating t between Eqs. (7.31) and (7.41). From Eq. (7.31),

$$t = -\tau \ln\left(1 - \frac{x}{x_{\text{lim}}}\right) \tag{7.43}$$

where $\tau = m/b$. Substituting this value into Eq. (7.41),

$$y = \tau\left[(v_{0y} + g\tau)\frac{x}{x_{\text{lim}}} + g\tau \ln\left(1 - \frac{x}{x_{\text{lim}}}\right)\right] \tag{7.44}$$

The logarithm may be expanded in the following power series (since $x/x_{\text{lim}} < 1$):

$$\ln(1 - \zeta) = \zeta - \tfrac{1}{2}\zeta^2 - \tfrac{1}{3}\zeta^3 - \cdots - \qquad (\zeta^2 < 1) \tag{7.45}$$

Therefore, Eq. (7.44) is written

$$y = (v_{0y}\tau + g\tau^2)\left(\frac{x}{x_{\text{lim}}}\right) + g\tau^2\left[-\frac{x}{x_{\text{lim}}} - \frac{1}{2}\left(\frac{x}{x_{\text{lim}}}\right)^2 - \frac{1}{3}\left(\frac{x}{x_{\text{lim}}}\right)^3 - \cdots\right] \tag{7.46}$$

or

$$y = v_{0y}\tau\left(\frac{x}{x_{\text{lim}}}\right) - \tfrac{1}{2}g\tau^2\left(\frac{x}{x_{\text{lim}}}\right)^2 - \tfrac{1}{3}g\tau^2\left(\frac{x}{x_{\text{lim}}}\right)^3 - \cdots \tag{7.47}$$

The above clearly indicates that as x increases, the trajectory departs more and more from a parabolic path. For large x the corresponding value of y is smaller than it would be for a parabola. This is depicted in Fig. 7.8.

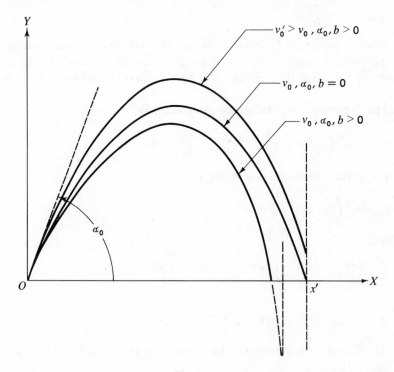

Fig. 7.8

By substituting the expansion

$$e^{-(b/m)t} = 1 - \frac{bt}{m} + \frac{(bt/m)^2}{2!} - \frac{(bt/m)^3}{3!} + \cdots \tag{7.48}$$

into Eqs. (7.31) and (7.41),

$$x = v_{0x}t - \frac{bv_{0x}t^2}{2m} + \cdots \tag{7.49}$$

and

$$y = v_{0y}t - \frac{gt^2}{2} - \frac{b}{m}\left(\frac{v_{0y}t^2}{2} - \frac{gt^3}{6}\right) + \cdots \tag{7.50}$$

It is clear that for sufficiently small values of b and t in Eqs. (7.49) and (7.50), the omission of air drag is justified. The magnitude of "sufficiently small" depends upon the desired accuracy.

Comparison of Eq. (7.49) with Eqs. (7.10) and of Eq. (7.50) with Eq. (7.12) indicates that the effect of air resistance is to diminish the X- and Y-coordinates for all values of t.

In order for a projectile subject to drag, to have the same range x' as a frictionless projectile fired at the same initial angle, its initial velocity must be increased. The trajectory for this case is also shown in Fig. 7.8.

Since the X- and Y-motions are independent of one another, the analysis of Y-motion is essentially identical with that pertaining to the motion of a particle projected vertically in a viscous medium.

It is of interest to explore the analogy between the X-projectile motion and the electric circuit shown in Fig. 7.9. For $E(t) = 0$, Kirchhoff's law gives,

$$-L\frac{dI}{dt} = RI \tag{7.51a}$$

or

$$\frac{dI}{dt} = -\frac{R}{L}I \tag{7.51b}$$

The solution of Eq. (7.51) is

$$I = I_0 e^{-(R/L)t} = I_0 e^{-(t/\tau)} \tag{7.52}$$

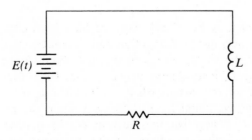

Fig. 7.9

where I_0 represents the current flowing at $t = 0$. This is analogous to the solution obtained for the X-velocity:

$$v_x = v_{0x}e^{-(b/m)t} = v_{0x}e^{-(t/\tau)} \tag{7.53}$$

Thus L/R and m/b play similar roles. Each is termed the *time constant* τ, a quantity of great importance in first-order systems, because it is a measure of the rapidity of system response. Other first-order systems are those characterizing radioactive decay, the attenuation of radiation by matter, and population growth, the latter implying a positive exponent.

Substituting $I = \dot{Q}$ permits Eq. (7.52) to be integrated, and Q, the charge, determined as a function of time. Thus charge and displacement are analogous. It is left as an exercise for the reader to demonstrate that when $E(t) = 0$ for $t < 0$, and $E(t) = $ constant for $t \geq 0$, the current and charge equations are analogous to the Y-velocity and Y-displacement equations, for $b > 0$.

7.6 ENERGY CONSIDERATIONS

If Eqs. (7.3) and (7.4) are multiplied by $m\dot{x}$ and $m\dot{y}$, respectively, we obtain the following:

$$m\dot{x}\ddot{x} = 0 \tag{7.54}$$

and

$$m\dot{y}\ddot{y} = -mg\dot{y} \tag{7.55}$$

The left side of Eq. (7.54) has been shown to represent the time rate of change of the kinetic energy associated with the X-motion,

$$m\dot{x}\ddot{x} = \frac{d}{dt}\tfrac{1}{2}m(\dot{x})(\dot{x}) = \frac{d}{dt}\tfrac{1}{2}mv_x^2 = \dot{T}_x \tag{7.56}$$

where the subscript on $\dot{T}_x$ does not imply a vector component; energy is, of course, a scalar. Thus Eq. (7.54) is now written

$$\dot{T}_x = 0 \qquad T_x = \text{constant} \tag{7.57}$$

Similarly,

$$\dot{T}_y = -mg\dot{y} \tag{7.58}$$

The result given by Eq. (7.57) may also have been obtained by noting that since zero net force acts in the X-direction and the motion is uncoupled, no work is done to influence the kinetic energy in that direction. Since $\dot{y}$ may be positive, negative, or zero, the Y-kinetic energy may decrease (if $\dot{y} > 0$), increase (for $\dot{y} < 0$), or remain constant for the instant when $\dot{y} = 0$. The term $-mg\dot{y}$ represents the rate at which the gravitational force does work upon the projectile. When the projectile is rising, $\dot{T}_y$ is negative, inasmuch as gravity opposes the velocity. The contrary is true for a falling projectile. In the foregoing instances, no dissipative forces act. To evaluate the influence of friction, multiply each side of Eqs. (7.24a) and (7.25a) by $\dot{x}$ and $\dot{y}$, respectively, to obtain

$$m\dot{x}\ddot{x} = -b(\dot{x})^2 \qquad (7.59)$$

and

$$m\dot{y}\ddot{y} = -b(\dot{y})^2 - mg\dot{y} \qquad (7.60)$$

Interpreting the left sides as before,

$$\dot{T}_x = -b(\dot{x})^2 \qquad (7.61)$$

and

$$\dot{T}_y = -b(\dot{y})^2 - mg\dot{y} \qquad (7.62)$$

The X-equation above states, for $b > 0$, that the X-kinetic energy diminishes with time. This is because $-b(\dot{x})^2$, the rate at which mechanical energy is dissipated, is always negative.

The Y-equation is more interesting. For $\dot{y} > 0$, $-mg\dot{y} < 0$, and both the drag and gravity forces serve to decrease the system kinetic energy. At the maximum altitude, $\dot{y} = 0$, and as we encountered in frictionless motion, the rate of change of kinetic energy is instantaneously zero. For the downward motion, $\dot{y} < 0$, and $-mg\dot{y} > 0$, thereby causing $-mg\dot{y}$ and $-b(\dot{y})^2$ to represent opposing influences. We may, therefore, anticipate that a downward velocity may be attained (if the projectile does not first strike the ground) for which both effects cancel ($\dot{T}_y = 0$):

$$b(\dot{y})^2 = -mg\dot{y} \qquad \dot{y} = -\frac{mg}{b}$$

This is, of course, the Y-terminal velocity previously discussed. When this velocity is achieved, the decrease in gravitational potential energy manifests itself as an increase in the internal energy of the projectile and its surroundings.

7.7 PROJECTILE ANALYSIS III

At higher projectile Reynolds numbers, there is considerable theoretical and experimental justification for assuming air drag proportional to the square of projectile speed. To find closed-form solutions of the resulting differential equations, the introduction of a rather restrictive simplifying condition will be necessary. The remaining assumptions are identical with those of the previous analyses.

For the assumed air drag,

$$\mathbf{F}_D = -cv^2\hat{\mathbf{v}} \qquad (7.63)$$

where $\hat{\mathbf{v}}$ is a unit vector along $\mathbf{v}$, and c is a factor dependent upon air properties, projectile geometry, and projectile attitude, all assumed constant. Since

$$\hat{\mathbf{v}} = \frac{\mathbf{v}}{v} = \frac{v_x\hat{\mathbf{i}} + v_y\hat{\mathbf{j}}}{v}$$

Eq. (7.63) may be written

$$\mathbf{F}_D = -cv^2\frac{v_x}{v}\hat{\mathbf{i}} - cv^2\frac{v_y}{v}\hat{\mathbf{j}} = -cvv_x\hat{\mathbf{i}} - cvv_y\hat{\mathbf{j}} \tag{7.64}$$

The differential equation becomes

$$-mg\hat{\mathbf{j}} - cv(v_x\hat{\mathbf{i}} + v_y\hat{\mathbf{j}}) = m(\dot{v}_x\hat{\mathbf{i}} + \dot{v}_y\hat{\mathbf{j}}) \tag{7.65}$$

The X- and Y-equations of motion are

$$-cvv_x = m\dot{v}_x \tag{7.66}$$

$$-mg - cvv_y = m\dot{v}_y \tag{7.67}$$

Since $v = (v_x^2 + v_y^2)^{1/2}$, the X- and Y-motions are now coupled in the above equations. There is no uncomplicated way of uncoupling the motion.

A partial uncoupling may be accomplished by assuming that the trajectory is quite flat, so that $v \cong v_x$. Simplified versions of Eqs. (7.66) and (7.67) are then, respectively,

$$-cv_x^2 = m\dot{v}_x \tag{7.68}$$

and

$$-mg - cv_xv_y = m\dot{v}_y \tag{7.69}$$

Equation (7.68) may be solved by separating variables and integrating,

$$\int \frac{dv_x}{v_x^2} = -\int \frac{c}{m}\,dt \tag{7.70}$$

resulting in

$$-\frac{1}{v_x} = -\frac{c}{m}t + C_1 \tag{7.71}$$

Employing the initial condition, at $t = 0$, $v_x = v_{0x}$, $C_1 = -1/v_{0x}$. The equation for X-velocity is, therefore,

$$v_x = \frac{v_{0x}}{1 + v_{0x}(c/m)t} \tag{7.72}$$

Integrating Eq. (7.72) and noting that the constant of integration is zero (for the initial condition $t = 0$, $x = 0$),

$$x = \frac{m}{c}\ln\left(1 + v_{0x}\frac{c}{m}t\right) \tag{7.73}$$

An expression for v_x [Eq. (7.72)] is now available for substitution into Eq. (7.69) (to eliminate v_x):

$$-mg - c\left[\frac{v_{0x}}{1 + v_{0x}(c/m)t}\right]v_y = m\dot{v}_y \tag{7.74}$$

This may be written

$$\dot{v}_y + \left[\frac{1}{t + (m/cv_{0x})} \right] v_y = -g \tag{7.75}$$

Replacing $-g$ with zero in Eq. (7.75) (a first-order differential equation with a non-constant coefficient) yields the homogeneous equation

$$\dot{v}_y + \left[\frac{1}{t + (m/cv_{0x})} \right] v_y = 0 \tag{7.76}$$

Separating variables and simplifying,

$$\frac{dv_y}{v_y} = - \left[\frac{1}{t + (m/v_{0x}c)} \right] dt \tag{7.77}$$

Integrating Eq. (7.77),

$$\ln v_y = -\ln \left(t + \frac{m}{cv_{0x}} \right) + \ln C_3 \tag{7.78}$$

or

$$v_y = \frac{C_3}{t + (m/cv_{0x})} \tag{7.79}$$

An additional solution is required to satisfy the inhomogeneous equation. The reader may verify that Eq. (7.80) satisfies the inhomogeneous equation when $A = -g/2$:

$$v_y = A \left(t + \frac{m}{cv_{0x}} \right) \tag{7.80}$$

The complete solution for v_y is, therefore,[1]

$$v_y = \frac{C_3}{t + (m/cv_{0x})} - \frac{g}{2} \left(t + \frac{m}{cv_{0x}} \right) \tag{7.81}$$

For the initial condition $t = 0$, $v_y = v_{0y}$,

$$C_3 = \frac{m}{cv_{0x}} \left(v_{0y} + \frac{gm}{2cv_{0x}} \right)$$

and the general expression for v_y is

$$v_y = \frac{m/cv_{0x}[v_{0y} + (g/2)(m/cv_{0x})]}{t + (m/cv_{0x})} - \frac{g}{2} \left(t + \frac{m}{cv_{0x}} \right) \tag{7.82}$$

Finally, integrating Eq. (7.82),

---

[1] Equation (7.75) may be solved by the use of an integrating factor. See, for example, Mathews, J., and Walker, R. L., *Mathematical Methods of Physics*, W. A. Benjamin, Inc., New York, 1964.

$$y = \frac{m}{cv_{0x}}\left(v_{0y} + \frac{gm}{2cv_{0x}}\right)\ln\left(t + \frac{m}{cv_{0x}}\right) - \frac{gt^2}{4} - \frac{gmt}{2cv_{0x}} + C_4 \qquad (7.83)$$

where

$$C_4 = -\frac{m}{cv_{0x}}\left(v_{0y} + \frac{gm}{2cv_{0x}}\right)\ln\left(\frac{m}{cv_{0x}}\right)$$

is obtained from the initial condition $t = 0$, $y = 0$.

7.8 FURTHER TRAJECTORY CONSIDERATIONS

The planar trajectories thus far discussed have been based upon several more or less restrictive conditions. The equations of motion may be improved, of course, by taking into account other factors, such as the altitude variation of air density (making c a function of height) and the effect of wind velocity. The first consideration usually takes the form

$$c = c_0 e^{-(y/h)}$$

where c_0 represents the value at sea level, y the distance above sea level, and h a scale height. This model, although far from perfect, offers the advantage of simple mathematical form.

The second consideration, wind velocity, may be treated by a term

$$c|\mathbf{v} - \mathbf{v}_w|^2$$

where $\mathbf{v} - \mathbf{v}_w$ represents the velocity of the projectile relative to the wind. Including both density variation and wind velocity, the drag force becomes

$$-c_0 e^{-(y/h)}|\mathbf{v} - \mathbf{v}_w|^2$$

adding significantly to the complexity of the equations of motion.

Since the wind is not limited to in-plane velocity, a cross-plane motion may result, and the projectile may thus undergo a three-dimensional trajectory.

Before deciding to solve the equations of motion with the above factors taken into account, consideration should be given the error associated with assuming the earth a secondary inertial frame of reference. In so doing the motion of the earth beneath the projectile is neglected. This effect is explored in Chapter 9, where it is shown that projectile motion viewed from an earth-fixed reference appears nonplanar, even in the absence of air currents.

7.9 AN ELECTRON IN AN ELECTROSTATIC FIELD

Consider the motion of an electron (assumed to be a particle) in a cathode-ray tube, a cross-sectional view of which is shown in Fig. 7.10.

The operation of the cathode-ray tube may be briefly described as follows:

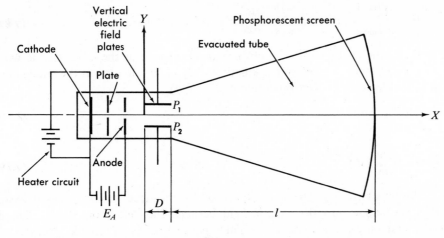

Fig. 7.10

a. The specially coated cathode is heated (by passing a current through a thin resistance wire) and the electrons liberated are accelerated toward the anode because of a potential difference between anode and cathode.

b. The plate serves to produce a collimated electron beam.

c. After passing through a small hole in the anode, the electrons are subjected to forces imposed upon them by the electric field between deflection plates P_1 and P_2 and another pair at right angles to those shown.

d. The beam of electrons hits the phosphorescent screen at the end of the tube, causing a bright spot to appear.

The cathode and anode form the basis of an electron gun in which the d-c source, designated E_a in Fig. 7.10, maintains a constant difference of potential ΔV. The electric field E at any point between the cathode and anode is defined as the force which acts on a unit positive charge at that point. This force is conservative, and as such is describable as the negative gradient of a potential V. Thus

$$\mathbf{E} = -\operatorname{grad} V \tag{7.84}$$

Since the above expression gives the force acting on a unit positive charge, it is necessary to multiply by $-e$, the electronic charge, to obtain the force acting on an electron:

$$\mathbf{F} = -e\mathbf{E} = e \operatorname{grad} V \tag{7.85}$$

The Cartesian components of $\mathbf{F}$ are, therefore,

$$F_x = e\frac{\partial V}{\partial x} \qquad F_y = e\frac{\partial V}{\partial y} \qquad F_z = e\frac{\partial V}{\partial z} \tag{7.86}$$

The work done on a single electron passing from cathode to anode (in the X-direction) is

$$W = \int_{\text{cathode}}^{\text{anode}} F_x \, dx = \int_{\text{cathode}}^{\text{anode}} e \frac{\partial V}{\partial x} \, dx = e(V_a - V_k) = e \, \Delta V \qquad \text{(7.87)}$$

Since the kinetic energy of the electron is *assumed* zero at the cathode, the work done on the electron is equal to the kinetic energy of the electron at the anode (also equal to the decrease in electronic potential energy),

$$e \, \Delta V = \tfrac{1}{2} m v^2 \qquad \text{(7.88)}$$

and

$$v = \left(\frac{2e}{m} \Delta V \right)^{\frac{1}{2}} \qquad \text{(7.89)}$$

We thus have a quantitative expression for the electron velocity as it passes through the anode into the electric field between the deflection plates. Let us consider first only the action of the plates parallel to the XZ-plane. Neglecting "fringing," we assume the field uniform between the plates and zero elsewhere. Figure 7.11 shows a free-body diagram of an electron between the plates.

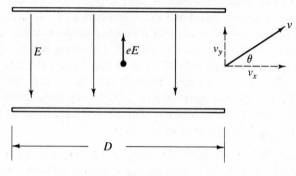

Fig. 7.11

After the electron passes through the anode, but before entering the field of the plates, no forces act, so that its velocity remains constant in this small region. When the electron moves between the plates, a force eE acts in the Y-direction. The X- and Y-equations of motion are, therefore,

$$F_x = m\ddot{x} = 0 \qquad \ddot{x} = \frac{F_x}{m} = 0 \qquad \text{(7.90)}$$

and

$$F_y = m\ddot{y} = eE_y \qquad \ddot{y} = \frac{eE_y}{m} \qquad \text{(7.91)}$$

Integrating,

$$\dot{x} = v_x = C_1 \qquad \text{(7.92)}$$

$$x = C_1 t + C_2 \qquad \text{(7.93)}$$

$$\dot{y} = v_y = \frac{eE_y}{m} t + C_3 \tag{7.94}$$

$$y = \frac{eE_y}{2m} t^2 + C_3 t + C_4 \tag{7.95}$$

Applying the initial conditions: at $t = 0$ (corresponding to entering between plates), $x = y = 0$, $v_y = 0$, $v_x = v_{0x}$, $C_1 = v_{0x}$, $C_2 = 0$, $C_3 = 0$, $C_4 = 0$, and therefore

$$\dot{x} = v_x = v_{0x} \tag{7.96}$$

$$x = v_{0x} t \tag{7.97}$$

and

$$\dot{y} = v_y = \frac{eE_y}{m} t \tag{7.98}$$

$$y = \frac{eE_y}{2m} t^2 \tag{7.99}$$

Since the X-velocity is constant, the time of passage through the plates

$$T = \frac{D}{v_{0x}} \tag{7.100}$$

The value of T thus obtained is also applicable to the Y-equations [Eqs. (7.98) and (7.99)]:

$$\dot{y}_e = \frac{eE_y}{m} \frac{D}{v_{0x}} \qquad \text{(at exit from the plates)} \tag{7.101}$$

and

$$y_e = \frac{eE_y}{2m} \left(\frac{D}{v_{0x}}\right)^2 \tag{7.102}$$

The angle θ the resultant electron velocity makes at exit with the X-axis is given by

$$\tan \theta = \left(\frac{v_y}{v_x}\right)_{\text{exit}} = \frac{(eE_y/m)(D/v_{0x})}{v_{0x}} = \frac{eE_y}{m} \frac{D}{v_{0x}^2} \tag{7.103}$$

The electron, having emerged from the influence of the field, now proceeds in a straight line to the phosphorescent screen. The distance above the X-axis, y_S, at which the beam strikes the screen is obtained by adding y_e to the Y-distance traversed at constant Y-velocity after leaving the plates, during time l/v_{0x}. Therefore,

$$y_S = \frac{eE_y}{2m} \left(\frac{D}{v_{0x}}\right) + \frac{eE_y}{m} \frac{D}{v_{0x}} \frac{l}{v_{0x}} \tag{7.104}$$

or

$$y_S = \left[\frac{e}{2m} \left(\frac{D}{v_{0x}}\right)^2 + \frac{e}{m} \frac{lD}{v_{0x}^2}\right] E_y$$

$$= \text{(constant)} \ E_y = K_y E_y \tag{7.105}$$

Similarly, a pair of plates producing a uniform electric field E_z will produce a Z-deflection at the screen:

$$z_S = K_z E_z \tag{7.106}$$

The linearity between the plate voltage and the deflection observed on the screen [Eqs. (7.105) and (7.106)] is, of course, dependent upon careful design.

The cathode-ray tube is used to display physical quantities, often time-dependent, which through suitable transducers result in voltages applied to the plates.

Consider an extension of the foregoing development in which E_y and E_z are not constant, as, for example,

$$E_y = E_A \sin \omega t \tag{7.107a}$$

$$E_z = E_B \cos \omega t \tag{7.107b}$$

The derivation of Eqs. (7.105) and (7.106) is predicated upon constant plate voltages, and one is therefore obliged to justify their application, as below, under circumstances in which E_y and E_z change quite rapidly. The rationale lies with the high velocity of the electron as it passes through the influence of the plates. Over the very short time of passage, the plate voltage may be regarded as essentially constant. It must be appreciated, however, that a limit on the rate of change of E_y and E_z is eventually reached for which the foregoing explanation cannot suffice, and a new analysis is then in order.

Subject to the limitations discussed above, we may combine Eqs. (7.105), (7.106), and (7.107),

$$y_S = K_y E_A \sin \omega t \quad \text{or} \quad \sin^2 \omega t = \frac{y_S^2}{K_y^2 E_A^2} \tag{7.108a}$$

$$z_S = K_z E_B \cos \omega t \quad \text{or} \quad \cos^2 \omega t = \frac{z_S^2}{K_z^2 E_B^2} \tag{7.108b}$$

Since $\sin^2 \omega t + \cos^2 \omega t = 1$,

$$\frac{y_S^2}{K_y^2 E_A^2} + \frac{z_S^2}{K_z^2 E_B^2} = 1 \tag{7.109}$$

and the trace is therefore an ellipse. The reader will recall the Lissajous figures discussed in Chapter 6, which depend upon the relationship of the Y- and Z-frequencies.

7.10 A CHARGED PARTICLE IN A MAGNETIC FIELD

The force acting on a particle of charge q resulting from a field of magnetic induction $\mathbf{B}$ is related to the particle velocity $\mathbf{v}$ by

$$\mathbf{F} = q(\mathbf{v} \times \mathbf{B}) = m\dot{\mathbf{v}} \tag{3.82}$$

Before solving the above equation, consider the following scalar product:

$$\mathbf{F} \cdot \mathbf{v} = q(\mathbf{v} \times \mathbf{B}) \cdot \mathbf{v} = 0 \tag{7.110}$$

This expression represents the power supplied by the magnetic force, that is, the rate at which work is done on the charged particle. Since the power is zero, the kinetic energy and hence particle speed v are constant. One draws the same conclusion by noting that as a consequence of the perpendicularity of $\mathbf{v}$ and $\dot{\mathbf{v}}$, there can be no acceleration along the path of the particle. Therefore, the only acceleration is that associated with change in direction, and it is reasonable to anticipate part of the motion to be circular.

Returning to Eq. (3.82), let us examine the motion of a charged particle in a uniform magnetic field acting in the positive Z-direction (magnetic induction $\mathbf{B} = B\hat{\mathbf{k}}$):

$$\mathbf{F} = m\ddot{\mathbf{r}} = q(\mathbf{v} \times B\hat{\mathbf{k}}) = (\dot{x}\hat{\mathbf{i}} + \dot{y}\hat{\mathbf{j}} + \dot{z}\hat{\mathbf{k}}) \times (qB\hat{\mathbf{k}}) \tag{7.111}$$

or

$$m\ddot{\mathbf{r}} = m(\ddot{x}\hat{\mathbf{i}} + \ddot{y}\hat{\mathbf{j}} + \ddot{z}\hat{\mathbf{k}}) = qB\dot{y}\hat{\mathbf{i}} - qB\dot{x}\hat{\mathbf{j}} \tag{7.112}$$

Since two vectors are equal only if their components are equal,

$$m\ddot{x} = qB\dot{y} \tag{7.113a}$$

$$m\ddot{y} = -qB\dot{x} \tag{7.113b}$$

$$\ddot{z} = 0 \tag{7.113c}$$

Integrating,

$$\dot{x} = \frac{qB}{m}y + C_1 \tag{7.114a}$$

$$\dot{y} = -\frac{qB}{m}x + C_2 \tag{7.114b}$$

$$\dot{z} = C_3 = \dot{z}_0 \tag{7.114c}$$

Substituting Eq. (7.114b) into Eq. (7.113a) to eliminate coupling,

$$\ddot{x} = \frac{qB}{m} - \frac{qB}{m}x + \frac{qB}{m}C_2 \tag{7.115}$$

Setting $qB/m = \omega$, the *cyclotron frequency*, and $k_1 = C_2/\omega$, Eq. (7.115) may be rearranged:

$$\ddot{x} + \omega^2 x = \omega^2 k_1 \tag{7.116}$$

This expression has the solution

$$x = k_1 + A\cos(\omega t + \beta_0) \tag{7.117}$$

where A and β_0 represent constants of integration. Equation (7.117) may be verified by substitution in Eq. (7.116) and has been explored in greater detail in Chapter 6.

Equation (7.117) when substituted into Eq. (7.114b) results in

$$\dot{y} = -\frac{qB}{m}[k_1 + A\cos(\omega t + \beta_0)] + C_2 = -A\omega\cos(\omega t + \beta_0) \qquad \textbf{(7.118)}$$

The solution of Eq. (7.118) is

$$y = k_2 - A\sin(\omega t + \beta_0) \qquad \textbf{(7.119)}$$

where k_2 is another constant of integration.

If t is eliminated between Eqs. (7.117) and (7.119), the projection of the trajectory on the XY-plane is found to be a circle:

$$(x - k_1)^2 + (y - k_2)^2 = A^2 \qquad \textbf{(7.120)}$$

Recall, however, that the particle also has a constant velocity component of motion in the Z-direction. The resultant motion is therefore a helix in the direction of **B** as shown in Fig. 7.12.

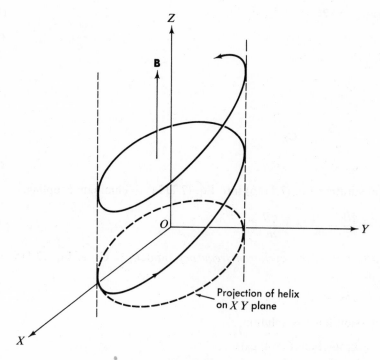

Projection of helix
on XY plane

Fig. 7.12

Illustrative Example 7.3

a. A particle of mass m and charge q is injected into a uniform magnetic field, **B**. If the particle velocity is initially perpendicular to the field, determine its trajectory.

b. Determine the cyclotron frequency of an electron in the atmosphere.

SOLUTION

From the analysis of Section 7.10 (with $\dot{z} = 0$), it is clear that the motion is circular and the acceleration is therefore v^2/R. For **v** perpendicular to **B**, the magnitude of the force is qvB. Thus $qvB = mv^2/R$, and the radius of the trajectory $R = mv/qB$. The time to travel a single revolution at constant speed,

$$T = \frac{2\pi R}{v} = \frac{2\pi mv}{qBv} = \frac{2\pi m}{qB} \text{ sec}$$

and the frequency is the reciprocal of the period determined above,

$$f = \frac{qB}{2\pi m}$$

The circular frequency

$$\omega = 2\pi f = \frac{qB}{m}$$

It is of interest to note that the period and frequency are independent of the speed of the particle as well as the radius of the orbit. The above frequency, as we have noted, is called the cyclotron frequency. The cyclotron, a device for accelerating charged particles, employs the principles here discussed to contain the particles.

The geomagnetic field at the earth's surface has an average strength of 4×10^{-5} Wb/m^2, or 0.4 G. The electronic charge and mass are, respectively, 1.6×10^{-19} C and 9.1×10^{-31} kg. The cyclotron frequency is, therefore,

$$\omega = \frac{(1.6 \times 10^{-19})(4 \times 10^{-5})}{9.1 \times 10^{-31}} = 7.03 \times 10^6 \text{ rad/sec}$$

When the electron attains a high-enough speed to be "relativistic," its mass will increase. Nevertheless, all results previously obtained remain valid provided the mass m is replaced by

$$m = \frac{m_0}{[1 - (v/c)^2]^{\frac{1}{2}}}$$

when m_0 represents the mass of the particle at rest and c is the speed of light.

Illustrative Example 7.4

An electron is released from rest at the cathode of a vacuum tube containing two parallel plates. The potential between cathode and anode is V volts. Determine the minimum uniform magnetic induction field required to prevent the electron from reaching the anode. Also determine the trajectory of the electron. Consider the electric field between cathode and anode to be uniform.

SOLUTION

Referring to Fig. 7.13, the electric field produces a force on the electron in the Y-direction. The magnetic induction field is perpendicular to the electric field, in the Z-direction. Applying Newton's second law,

$$-e\mathbf{E} - e(\mathbf{v} \times \mathbf{B}) = m\mathbf{a}$$

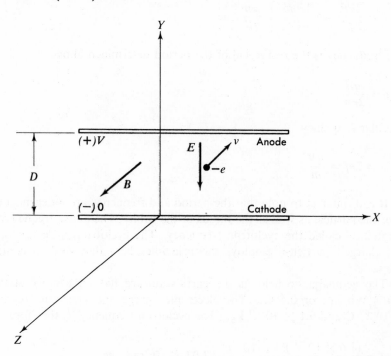

Fig. 7.13

The magnetic induction field $\mathbf{B} = B\hat{\mathbf{k}}$ and the electric field $\mathbf{E} = -(V/D)\hat{\mathbf{j}}$. The equations of motion are, therefore,

$$-e\dot{y}B = m\ddot{x}$$

$$\frac{eV}{D} + e\dot{x}B = m\ddot{y}$$

$$0 = m\ddot{z}$$

Because the electron starts from rest, $\dot{x}(0) = \dot{y}(0) = \dot{z}(0) = 0$. Selecting the initial position as the origin of the coordinate system: $x(0) = y(0) = z(0) = 0$. The solution of the Z-motion is $z = 0$. Integrating the X-equation of motion once,

$$-eyB = m\dot{x} + C_1$$

On the basis of the initial conditions, $C_1 = 0$, and

$$\dot{x} = -\frac{eB}{m}y = -\omega y$$

Substituting this value into the equation of motion for the Y-direction,

$$\frac{eV}{D} - \omega e By = m\ddot{y}$$

Dividing by m and rearranging,

$$\ddot{y} + \omega^2 y = \frac{eV}{Dm}$$

The solution of the homogeneous part of this equation is

$$y = C_2 \sin \omega t + C_3 \cos \omega t$$

and the solution to the inhomogeneous part is

$$y = \frac{eV}{Dm\omega^2}$$

The complete solution is therefore

$$y = \frac{eV}{Dm\omega^2} + C_2 \sin \omega t + C_3 \cos \omega t$$

From the initial conditions,

$$y(0) = \frac{eV}{Dm\omega^2} + C_3 = 0$$

$$C_3 = -\frac{eV}{Dm\omega^2}$$

At $t = 0$,

$$\dot{y}(0) = C_2\omega = 0$$

and therefore $C_2 = 0$. The solution for y is, therefore,

$$y = \frac{eV}{Dm\omega^2}(1 - \cos \omega t)$$

Recall that

$$\dot{x} = -\omega y = -\omega \left[\frac{eV}{Dm\omega^2} (1 - \cos \omega t) \right]$$

Integrating,

$$x = -\frac{eV}{Dm\omega} \left(t - \frac{\sin \omega t}{\omega} \right) + C_4$$

Applying the initial condition, at $t = 0$, $x = 0$: $C_4 = 0$.   Thus

$$x = -\frac{eV}{Dm\omega^2} (\omega t - \sin \omega t)$$

To determine the minimum magnetic induction field to cause the electron to just miss the anode, the condition $\dot{y} = 0$ is employed,

$$\dot{y} = \frac{eV}{Dm\omega} \sin \omega t = 0$$

which is satisfied when $\omega t = 0, \pi, 2\pi, \cdots, n\pi$. The values of $0, 2\pi, \cdots, 2n\pi$ correspond to the return of the electron to the cathode, while $\omega t = \pi, 3\pi, \cdots, (2n - 1)\pi$ correspond to the electron at the anode.   When $\omega t = \pi$, $y = D$, and

$$D = \frac{2eV}{Dm\omega^2}$$

or

$$\omega = \left(\frac{2eV}{mD^2} \right)^{\frac{1}{2}}$$

Since $\omega = eB/m$,

$$B = \left(\frac{2mV}{eD^2} \right)^{\frac{1}{2}}$$

the minimum field so that the electron just misses the anode.

As an alternative method of solution, let $x + iy = \eta$.   Multiplying the equation for Y-motion by $i = \sqrt{-1}$ and adding the equation of X-motion,

$$m(\ddot{x} + i\ddot{y}) = \frac{ieV}{D} + eB(i\dot{x} - \dot{y})$$

or

$$\ddot{\eta} = \frac{ieV}{Dm} + i\omega\dot{\eta}$$

At $t = 0$, $\eta = 0$, and $\dot{\eta} = 0$. Integrating directly,

$$\dot{\eta} - i\omega\eta = \frac{ieV}{Dm} t$$

and the solution for η is

$$\eta = \alpha t + \beta(1 - e^{i\omega t})$$

where α and β are determined from the differential equation for η:

$$\alpha = -\frac{eV}{D\omega m} \qquad\qquad \beta = \frac{ieV}{D\omega^2 m}$$

Therefore,

$$\eta = -\frac{eV}{D\omega m} t + \frac{ieV}{Dm\omega^2}(1 - e^{i\omega t}) = x + iy$$

where

$$x = \text{real part } \eta = -\frac{eV}{D\omega m}\left(t - \frac{\sin \omega t}{\omega}\right)$$

and

$$y = \text{imag part } \eta = \frac{eV}{Dm\omega^2}(1 - \cos \omega t)$$

as previously obtained.

References

Becker, R. A., *Introduction to Theoretical Mechanics*, McGraw-Hill, Inc., New York, 1954.

Fowles, G. R., *Analytical Mechanics*, Holt, Rinehart and Winston, Inc., New York, 1962.

Housner, G. W., and Hudson, D. E., *Applied Mechanics*, Second Edition, D. Van Nostrand Company, Inc., Princeton, N.J., 1959.

EXERCISES

7.1 A shell is fired with an initial velocity of 300 m/sec at an angle of 50 degrees with respect to the horizontal. If the cannon is located atop a 100-m mountain, determine the maximum altitude of the shell relative to the plain below, the range of the trajectory, and the time of flight. Assume that no frictional resistance is offered by the air.

7.2 Develop a general expression for the firing angle corresponding to maximum range as well as an expression for the maximum range given the situation described in Exercise 7.1.

7.3 An aircraft is in horizontal flight at a velocity of 500 mph and an altitude of 4000 ft. If an antiaircraft gun is fired the instant the aircraft is directly overhead, what angle must the gun make with the horizontal to hit the target? The muzzle velocity is 1400 ft/sec and air resistance is neglected.

7.4 At what horizontal distance from the antiaircraft gun should the aircraft in Exercise 7.3 drop a bomb in order to hit the gun? What is the time of travel of the bomb?

7.5 A projectile is fired at an angle of 45 degrees with respect to the horizontal and an initial velocity of 500 m/sec. Determine the range of the projectile measured along a slope making an angle of 30 degrees with the horizontal.

7.6 Two projectiles are fired from the same point on a horizontal plain but separated in time by interval T. The initial speed of each projectile is the same. Determine the relationship connecting the initial speed, the firing angle of each projectile (θ_1 and θ_2), and the time interval for a collision to occur. Assume zero air resistance.

7.7 A particle is thrown vertically upward. What must the initial velocity be if a maximum height of 300 ft is to be achieved?

7.8 Two particles are thrown vertically with the same initial velocity. The second particle is released T sec after the first. Determine the distance above the ground at which the particles meet.

7.9 A projectile is fired at an angle of 50 degrees above the horizontal with an initial velocity of 400 m/sec. Determine its position, velocity, and acceleration after 8 sec.

7.10 A particle weighing 1.5 lb is dropped from a 500-ft-high building. Determine the velocity with which the particle strikes the ground, assuming:

    a. zero air resistance.
    b. air resistance varying linearly with velocity, $b = 0.01$ lb/ft/sec.

7.11 A 5-kg projectile is thrown at an angle of 45 degrees above the horizontal with a velocity of 10 m/sec. Determine the position and velocity at impact, assuming:

    a. zero air resistance.
    b. air resistance varying linearly with velocity, $b = 0.001$ N/m/sec.

7.12 A 50-lb object is dropped vertically through a viscous medium with an initial velocity of 0.5 ft/sec. Asssuming that the resisting force varies linearly with velocity ($b = 0.1$ lb/ft/sec), determine the terminal velocity. How much time elapses before the velocity is within 5 percent of its terminal value?

7.13 A missile is fired with a velocity of 500 ft/sec. The firing angle is 10 degrees above a horizontal plane. If the air resistance varies as the square of the magnitude of the velocity with $c = 0.003$ lb/(ft/sec)^2, determine the range, the maximum height, the velocity at impact, and the time of flight.

7.14 Referring to Illustrative Example 7.2, determine the range R_β by employing the $X'Y'$-coordinate system shown in Fig. 7.4.

7.15 Demonstrate that a charged particle in a crossed electromagnetic field, where, **E** and **B** are constant, cannot be contained because of a drift velocity perpendicular to both **E** and **B**. Find the magnitude of the drift velocity.

7.16 An electron is subject to an electromagnetic wave

$$\mathbf{E} = E_0 \cos (\omega t - kz)\hat{\mathbf{i}}$$

and

$$\mathbf{B} = B_0 \sin (\omega t - kz)\hat{\mathbf{j}}$$

Determine the position of the electron as a function of time if it starts from rest.

7.17 Demonstrate that for the circuit shown in Fig. 7.9, for $E(t) = 0$, $t < 0$, and $E(t) =$ constant, $t \geq 0$, the charge and current are analogous to Y-displacement and Y-velocity for $b > 0$.

7.18 A ball is thrown vertically upward with velocity v_0. If the drag force is proportional to the second power of velocity, determine the maximum height, the time of ascent, and the time of descent. What is the velocity upon striking the ground?

7.19 A particle is projected upward at a velocity

$$v_0 = \alpha v_{term}$$

where $\alpha > 1$. If the drag force is proportional to the first power of velocity, determine the maximum height, the time of ascent, and the time of descent. What is the velocity upon striking the ground?

7.20 A particle is projected downward at twice the terminal velocity from h height. How long does it take to strike the ground and what is the velocity upon impact for:

    a. drag proportional to velocity?
    b. drag proportional to the square of velocity?

CHAPTER 8

CENTRAL FORCE MOTION

8.1 INTRODUCTION

A central force is one that is directed along the line running from a point (fixed or moving with constant velocity) to the particle on which it acts. Important instances of motion arising from central forces include planetary and satellite motion, and electrons in hydrogen-like atoms. These exemplify gravitational and electrostatic central forces, respectively.

8.2 THE TWO-BODY PROBLEM

Consider the motion of two bodies, subject to the condition that the only forces acting are equal and opposite, directed along the line connecting the masses. We shall show that the motion divides into two distinct parts, one of which is central force motion. For two particles, m_1 and m_2, at positions $\mathbf{r}_1$ and $\mathbf{r}_2$ with respect to the origin of an inertial coordinate frame as shown in Fig. 8.1, the equations of motion are

$$-F\frac{\mathbf{r}_2 - \mathbf{r}_1}{|\mathbf{r}_2 - \mathbf{r}_1|} = m_1\ddot{\mathbf{r}}_1 \tag{8.1}$$

$$F\frac{\mathbf{r}_2 - \mathbf{r}_1}{|\mathbf{r}_2 - \mathbf{r}_1|} = m_2\ddot{\mathbf{r}}_2 \tag{8.2}$$

Let us introduce two new coordinates, $\mathbf{r}_{CM}$ and $\mathbf{r}$, defined by

$$\mathbf{r}_{CM} = \frac{m_1\mathbf{r}_1 + m_2\mathbf{r}_2}{m_1 + m_2} \tag{8.3}$$

$$\mathbf{r} = \mathbf{r}_2 - \mathbf{r}_1 \tag{8.4}$$

Solving for $\mathbf{r}_1$ and $\mathbf{r}_2$ in terms of $\mathbf{r}_{CM}$ and $\mathbf{r}$,

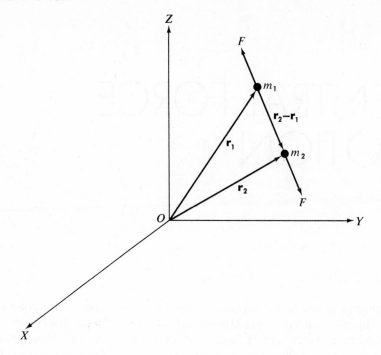

Fig. 8.1

$$\mathbf{r}_1 = \mathbf{r}_{CM} - \mathbf{r}\,\frac{m_2}{m_1 + m_2} \tag{8.5}$$

$$\mathbf{r}_2 = \mathbf{r}_{CM} + \mathbf{r}\,\frac{m_1}{m_1 + m_2} \tag{8.6}$$

Adding Eqs. (8.1) and (8.2),

$$m_1\ddot{\mathbf{r}}_1 + m_2\ddot{\mathbf{r}}_2 = \mathbf{0} = (m_1 + m_2)\ddot{\mathbf{r}}_{CM} \tag{8.7a}$$

or

$$\ddot{\mathbf{r}}_{CM} = \mathbf{0} \tag{8.7b}$$

The vector $\mathbf{r}_{CM}$ locates the center of mass of the two particles. Since the acceleration of the center of mass is zero [Eq. (8.7b)], it experiences constant velocity (possibly zero), regardless of the individual motions of m_1 and m_2.

Dividing Eq. (8.1) by m_1 and Eq. (8.2) by m_2 and subtracting,

$$\ddot{\mathbf{r}}_2 - \ddot{\mathbf{r}}_1 = \left(\frac{F}{m_1} + \frac{F}{m_2}\right)\frac{\mathbf{r}_2 - \mathbf{r}_1}{|\mathbf{r}_2 - \mathbf{r}_1|} \tag{8.8}$$

Substituting $\ddot{\mathbf{r}} = \ddot{\mathbf{r}}_2 - \ddot{\mathbf{r}}_1$, Eq. (8.8) becomes

$$\ddot{\mathbf{r}} = F\left(\frac{1}{m_1} + \frac{1}{m_2}\right)\frac{\mathbf{r}}{|\mathbf{r}|} = F\left(\frac{1}{m_1} + \frac{1}{m_2}\right)\hat{\mathbf{e}}_r \tag{8.9}$$

where $\hat{\mathbf{e}}_r$ is a unit vector directed from m_1 to m_2. Setting

$$\frac{1}{m_1} + \frac{1}{m_2} = \frac{1}{\mu} \qquad (8.10)$$

Eq. (8.9) is written

$$\mu\ddot{\mathbf{r}} = F\hat{\mathbf{e}}_r \qquad (8.11)$$

This equation describes the motion of a fictitious particle of mass μ, termed the *reduced mass*, subject to a central force $\mathbf{F}$, which, in general, may be time- as well as position- and velocity-dependent. Examining the form of Eq. (8.11) as well as the definition of $\mathbf{r}$, it is apparent that m_1 serves as a point fixed with respect to an inertial frame of reference, even though it is not such a point.

Thus, although we have somewhat anticipated Chapter 11, which treats multiparticle systems, the foregoing two-body analysis has produced two one-body equations. The first, Eq. (8.7b), describes the motion of the mass center. The second, Eq. (8.11), pertains to the motion of the reduced mass. The solution of these equations readily leads to the motions of the individual masses by substitution of $\mathbf{r}(t)$ and $\mathbf{r}_{CM}(t)$ into Eqs. (8.5) and (8.6). The motion of a particle subject to zero net force has been solved in Chapter 3, and we therefore concentrate on the solution of Eq. (8.11).

Central force motion manifests special properties which are independent of the form of $\mathbf{F}$. Taking the cross-product of $\mathbf{r}$ with Eq. (8.11),

$$\mathbf{r} \times (\mu\ddot{\mathbf{r}}) = \mathbf{r} \times F\hat{\mathbf{e}}_r = \mathbf{0} \qquad (8.12)$$

since $\mathbf{r}$ and $\hat{\mathbf{e}}_r$ are directed along the same line. Recall from Chapter 4 that as a consequence of $\dot{\mathbf{r}} \times \dot{\mathbf{r}} = \mathbf{0}$,

$$\mathbf{r} \times \mu\ddot{\mathbf{r}} = \frac{d}{dt}(\mu\mathbf{r} \times \dot{\mathbf{r}}) = \mathbf{0} \qquad (8.13)$$

Therefore, $\mu(\mathbf{r} \times \dot{\mathbf{r}}) = \mathbf{r} \times \mu\dot{\mathbf{r}}$, the moment of linear momentum or angular momentum of mass μ about the origin of $\mathbf{r}$ as shown in Fig. 8.2, is a constant vector. For *all* central force motion, the angular momentum $\mathbf{H}$ is a constant vector, implying that its orientation in space is constant. Since $\mathbf{H}$ is perpendicular to the plane containing $\mathbf{r}$ and $\dot{\mathbf{r}}$, the orientation of this plane is likewise constant.

The area $d\mathbf{A}$ swept out by $\mathbf{r}$ during time dt is given by

$$d\mathbf{A} = \frac{1}{2}\mathbf{r} \times d\mathbf{r} \qquad (8.14)$$

as shown in Fig. 8.2. The rate at which area is swept out by the radius vector $\mathbf{r}$ is therefore

$$\frac{d\mathbf{A}}{dt} = \frac{1}{2}\mathbf{r} \times \frac{d\mathbf{r}}{dt} = \frac{1}{2}\frac{\mathbf{H}}{\mu} = \text{constant vector} \qquad (8.15)$$

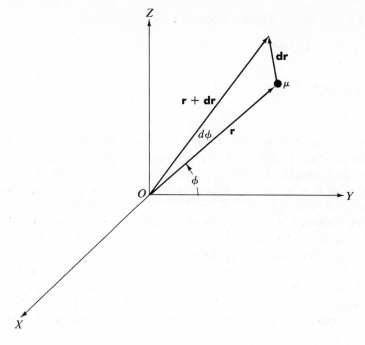

Fig. 8.2

Thus the *areal velocity* remains constant throughout the trajectory. When planetary motion is discussed, this result is demonstrated as equivalent to one of Kepler's three laws.

Since the motion of the reduced mass is planar and the force acting on μ is directed along the line from a point to the mass, polar coordinates are logically employed to describe the motion. Newton's second law, Eq. (8.11), in this representation is

$$F_R = \mu a_R = \mu(\ddot{R} - R\dot{\phi}^2) \tag{8.16}$$

$$F_\phi = \mu a_\phi = \mu(R\ddot{\phi} + 2\dot{R}\dot{\phi}) \qquad F_\phi = 0 \tag{8.17}$$

Multiplying Eq. (8.17) by R,

$$\mu(R^2\ddot{\phi} + 2R\dot{R}\dot{\phi}) = 0 = \frac{d}{dt}\mu R^2\dot{\phi} \tag{8.18}$$

which is a statement of conservation of angular momentum. Referring to Eq. (8.13),

$$\frac{d}{dt}\mu\mathbf{r} \times \dot{\mathbf{r}} = 0 = \frac{d}{dt}\mu[R\hat{\mathbf{e}}_R \times (\dot{R}\hat{\mathbf{e}}_R + R\dot{\phi}\hat{\mathbf{e}}_\phi)]$$

$$= \frac{d}{dt}\mu R^2\dot{\phi}\hat{\mathbf{k}} = \dot{\mathbf{H}}$$

Solving for $\dot{\phi}$,

$$\dot{\phi} = \frac{H}{\mu R^2} \tag{8.19}$$

Substituting Eq. (8.19) in Eq. (8.16), we have

$$F = \mu \ddot{R} - \frac{H^2}{\mu R^3} \tag{8.20}$$

Multiplying by $\dot{R}$, Eq. (8.20) becomes

$$F\dot{R} = \mu \ddot{R} \dot{R} - \frac{H^2 \dot{R}}{\mu R^3} = \frac{d}{dt}\frac{1}{2}\mu\dot{R}^2 + \frac{d}{dt}\frac{H^2}{2\mu R^2} \tag{8.21}$$

Substituting for H,

$$F\dot{R} = \frac{d}{dt}\frac{1}{2}\mu\dot{R}^2 + \frac{d}{dt}\frac{1}{2}\mu R^2 \dot{\phi}^2 = \frac{dT}{dt} \tag{8.22a}$$

where the kinetic energy

$$T = \tfrac{1}{2}\mu\dot{R}^2 + \tfrac{1}{2}\mu R^2 \dot{\phi}^2 = \tfrac{1}{2}\mu v^2 \tag{8.22b}$$

Equation (8.22a) states that the time rate of change of the kinetic energy is equal to the rate at which energy is supplied by the central force. Multiplying Eq. (8.22a) by dt and integrating,

$$\int_1^2 F\,dR = \tfrac{1}{2}\mu v_2^2 - \tfrac{1}{2}\mu v_1^2 \tag{8.23}$$

which is the familiar principle of work and energy. The analytic determination of the motion depends upon the integration of Eq. (8.23). If the force is derivable from a potential $V(R)$,

$$F = F(R) = -\frac{dV}{dR} \tag{8.24}$$

and

$$-\frac{dV}{dR}\frac{dR}{dt} = -\frac{dV}{dt} \tag{8.25}$$

Equation (8.22a) is therefore rewritten

$$-\frac{dV}{dt} = \frac{dT}{dt} \tag{8.26a}$$

or

$$\dot{V} + \dot{T} = 0 \tag{8.26b}$$

Integrating,

$$T + V = \text{constant} = E \tag{8.27}$$

where E is the total energy.

It must be emphasized that while E is constant, its value is not uniquely determined by Eq. (8.27). As has been noted, it is not energy but differences in energy which are of significance. The arbitrariness of E is eliminated by making a specific choice of coordinate reference and datum of potential energy. Depending upon these choices, $T + V$ may equal any constant — positive, negative, or zero. The kinetic energy cannot, of course, assume negative values. Consider the case in which V varies inversely as R. The zero of energy may then be selected to correspond to $T = 0$ and $R = \infty$. Should V vary directly as R, the zero of energy would correspond to $T = 0$ and $R = 0$. For other situations, such as V proportional to $\ln R$ or $(1/R) + R^2$, other choices of zero E must be elected. Rewriting Eq. (8.27) in terms of H, R, and $\dot{R}$,

$$E = \frac{1}{2}\mu\dot{R}^2 + \frac{1}{2}\frac{H^2}{\mu R^2} + V \tag{8.28}$$

A new function V', the *apparent* or *effective potential*, is defined by

$$V' = V + \frac{1}{2}\frac{H^2}{\mu R^2} \tag{8.29}$$

In terms of this new quantity, Eq. (8.28), the energy equation, involving R and $\dot{R}$, is

$$E = \frac{1}{2}\mu\dot{R}^2 + V'(R) \tag{8.30}$$

By introducing the effective potential, central force motion is seen to be analogous to one-dimensional motion (Chapter 5). Solving Eq. (8.30) for $\dot{R}$,

$$\dot{R} = \frac{dR}{dt} = \left\{\frac{2}{\mu}[E - V'(R)]\right\}^{1/2} \tag{8.31}$$

The solution for R in terms of t depends upon the form of V'. Once $R(t)$ is found, Eq. (8.19) is used to determine $\phi(t)$. Solving Eq. (8.31) for dt and integrating,

$$\int_{R_0}^{R} \frac{dR'}{\{(2/\mu)[E - V'(R')]\}^{1/2}} = \int_0^t dt' \tag{8.32}$$

Since the term inside the square root may never be negative, $E - V'(R') \geq 0$. The energy E may, however, be negative.

The form of V' and the value of E determine whether the motion is bound or unbound. For example, examine Fig. 8.3, where V' is plotted as a function of R for values of total energy E. If $E > E_1$, all values of R are allowed, and the motion is unbound. E cannot be less than E_2, because this would lead to negative kinetic energy. If E lies between E_2 and E_3, we have bound motion. The reader will recall that for a representative energy E_0 in this range, R_1 and R_2 are the maximum and minimum positions, called the turning points. For E between E_1 and E_3 (E_4), R can have any value greater than R_4. The motion is unbound, with R_4 the distance of closest approach.

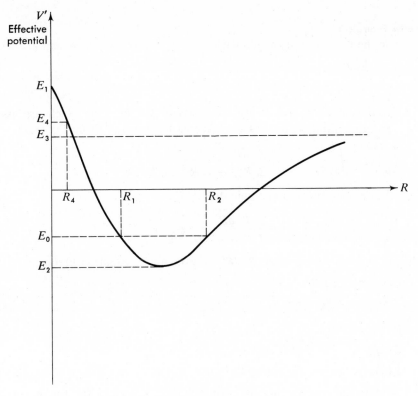

Fig. 8.3

The solution of R depends upon the form of $V(R)$ or $V'(R)$. Often, the potential can be expressed

$$V = V_0 R^\alpha \qquad (\alpha \neq 0) \tag{8.33}$$

where V_0 represents a constant. Then

$$F = -\frac{dV}{dR} = -F_0 R^{\alpha-1} \tag{8.34}$$

where $F_0 = \alpha V_0$. In this instance Eq. (8.32) takes the form

$$\left(\frac{2}{\mu}\right)^{\frac{1}{2}} t = \int_{R_0}^{R} \frac{dR'}{[E - (H^2/2\mu R'^2) - V_0 R'^\alpha]^{\frac{1}{2}}} \tag{8.35}$$

Not all values of α lead to closed-form solutions of the integral, and certain values have special significance. For instance, if $\alpha = 2$, the potential is characteristic of a two-dimensional linear oscillator. For $\alpha = -1$ we have a gravitational-type potential.

Illustrative Example 8.1

Determine the motion of a two-dimensional linear oscillator of potential energy

$$V = \tfrac{1}{2}kR^2$$

SOLUTION

For a given system, E and H are constants of the motion, which are assumed known, and the initial conditions are taken as R_0 and ϕ_0. From Eq. (8.35),

$$\left(\frac{2}{\mu}\right)^{\frac{1}{2}} t = \int_{R_0}^{R} \frac{dR'}{[E - (H^2/2\mu R'^2) - (kR'^2/2)]^{\frac{1}{2}}}$$

Multiplying numerator and denominator by R',

$$\left(\frac{2}{\mu}\right)^{\frac{1}{2}} t = \int_{R_0}^{R} \frac{R'\, dR'}{[ER'^2 - (H^2/2\mu) - (kR'^4/2)]^{\frac{1}{2}}}$$

Setting $x^2 = \tfrac{1}{2}kR'^4$, $dx = (2k)^{\frac{1}{2}}R'\, dR'$, and the above equation becomes

$$2\left(\frac{k}{\mu}\right)^{\frac{1}{2}} t = \int_{R_0}^{R} \frac{dx}{[E(2/k)^{\frac{1}{2}}x - (H^2/2\mu) - x^2]^{\frac{1}{2}}}$$

Consider the identity

$$\left[\frac{E}{(2k)^{\frac{1}{2}}} - x\right]^2 = \frac{E^2}{2k} - \frac{2Ex}{(2k)^{\frac{1}{2}}} + x^2$$

Adding and subtracting the term $E^2/2k$ to complete the square inside the radical, we have

$$2\left(\frac{k}{\mu}\right)^{\frac{1}{2}} t = \int_{R_0}^{R} \frac{dx}{\left\{\left(\dfrac{E^2}{2k} - \dfrac{H^2}{2\mu}\right) - \left[\dfrac{E}{(2k)^{\frac{1}{2}}} - x\right]^2\right\}^{\frac{1}{2}}}$$

This integral is of the form

$$-\int \frac{dy}{(A^2 - y^2)^{\frac{1}{2}}} = \arccos\left(\frac{y}{A}\right)$$

where

$$A = \left(\frac{E^2}{2k} - \frac{H^2}{2\mu}\right)^{\frac{1}{2}} \qquad y = \left[\frac{E}{(2k)^{\frac{1}{2}}} - x\right] \qquad dy = -dx$$

Thus

$$2\left(\frac{k}{\mu}\right)^{\frac{1}{2}} t = \arccos \frac{\left[\dfrac{E}{(2k)^{\frac{1}{2}}} - x\right]}{\left(\dfrac{E^2}{2k} - \dfrac{H^2}{2\mu}\right)^{\frac{1}{2}}} \Bigg|_{R_0}^{R}$$

or with x replaced with $(k/2)^{\frac{1}{2}}R'^2$,

$$2 \left(\frac{k}{\mu}\right)^{\frac{1}{2}} t = \arccos \frac{\left[\frac{E}{(2k)^{\frac{1}{2}}} - \left(\frac{k}{2}\right)^{\frac{1}{2}} R^2\right]}{\left(\frac{E^2}{2k} - \frac{H^2}{2\mu}\right)} - \beta$$

where

$$\beta = \arccos \frac{\left[\frac{E}{(2k)^{\frac{1}{2}}} - \left(\frac{k}{2}\right)^{\frac{1}{2}} R_0^2\right]}{\left(\frac{E^2}{2k} - \frac{H^2}{2\mu}\right)^{\frac{1}{2}}}$$

Solving for R^2, we obtain, after some manipulation,

$$R^2 = \frac{E}{k} - \left(\frac{E^2}{k^2} - \frac{H^2}{k\mu}\right)^{\frac{1}{2}} \cos\left[2\left(\frac{k}{\mu}\right)^{\frac{1}{2}} t + \beta\right]$$

To gain some insight into this rather cumbersome-appearing equation, let us examine the consequences of $H = 0$:

$$R^2 = \frac{E}{k} \left\{1 - \cos 2\left[\left(\frac{k}{\mu}\right)^{\frac{1}{2}} t + \frac{\beta}{2}\right]\right\} = \frac{E}{k} \left\{2 \sin^2\left[\left(\frac{k}{\mu}\right)^{\frac{1}{2}} t + \frac{\beta}{2}\right]\right\}$$

or

$$R = \left(\frac{2E}{k}\right)^{\frac{1}{2}} \sin\left[\left(\frac{k}{\mu}\right)^{\frac{1}{2}} t + \frac{\beta}{2}\right]$$

This is the familiar equation of one-dimensional simple-harmonic motion. Why should we have anticipated this result? Since $\mathbf{H} = \mathbf{0}$, $\mathbf{R} \times \dot{\mathbf{R}} = \mathbf{0}$, and we conclude that $\mathbf{R}$ and $\dot{\mathbf{R}}$ lie along the same line for all values of t. This information, coupled with the form of the potential, leads to harmonic motion.

Returning to the *general* solution, complete determination of the motion requires that we find $\phi(t)$, which is related to R^2 by Eq. (8.19):

$$\dot{\phi} = \frac{H}{\mu R^2} = \frac{H}{\mu \left\{\frac{E}{k} - \left(\frac{E^2}{k^2} - \frac{H^2}{k\mu}\right)^{\frac{1}{2}} \cos\left[2\left(\frac{k}{\mu}\right)^{\frac{1}{2}} t + \beta\right]\right\}}$$

Separating variables and integrating between appropriate limits,

$$\phi - \phi_0 = \frac{H}{\mu} \int_0^t \frac{dt'}{\frac{E}{k} - \left(\frac{E^2}{k^2} - \frac{H^2}{k\mu}\right)^{\frac{1}{2}} \cos\left[2\left(\frac{k}{\mu}\right) t' + \beta\right]}$$

Consider the integral

$$\int \frac{dz}{b + c \cos az} = \frac{1}{a(b^2 - c^2)^{\frac{1}{2}}} \tan^{-1}\left[\frac{(b^2 - c^2)^{\frac{1}{2}} \sin az}{c + b \cos az}\right]$$

where $b^2 > c^2$, $az = 2(k/\mu)^{1/2}t + \beta$, $b = E/k$, and $c = -[(E^2/k^2) - (H^2/k\mu)]^{1/2}$. Therefore,

$$\phi - \phi_0 = \tfrac{1}{2} \arctan \left\{ \frac{(H^2/k\mu)^{1/2} \sin [2(k/\mu)^{1/2}t + \beta]}{\dfrac{E}{k} \cos \left[2 \left(\dfrac{k}{\mu} \right)^{1/2} t + \beta \right] - \left(\dfrac{E^2}{k^2} - \dfrac{H^2}{k\mu} \right)^{1/2}} \right\} - \gamma_0$$

where

$$\gamma_0 = \arctan \left\{ \frac{(H^2/k\mu)^{1/2} \sin \beta}{\dfrac{E}{k} \cos \beta - \left(\dfrac{E^2}{k^2} - \dfrac{H^2}{k\mu} \right)^{1/2}} \right\}$$

If $H = 0$, $\phi = \phi_0 = $ constant. Thus we further demonstrate that the particle moves along a straight line for $\mathbf{H} = \mathbf{0}$. For the case in which $R = $ constant (circular motion), reference to the equation for R^2 gives

$$R^2 = \frac{E}{k}$$

since the coefficient of the cosine term must be zero for R^2 to remain constant for all values of t. Thus

$$\frac{E^2}{k^2} = \frac{H^2}{k\mu}$$

or

$$H = E \left(\frac{\mu}{k} \right)^{1/2} = \mu R^2 \dot{\phi}$$

Substituting $R^2 = E/k$ and solving for $\dot{\phi}$,

$$\dot{\phi} = \left(\frac{k}{\mu} \right)^{1/2}$$

and the angular velocity of the radius is found to equal the circular frequency of the motion.

It is often quite difficult or inconvenient to invert $t(R)$ as was done in the illustrative example. In this event, we may nevertheless determine $\phi(R)$ having first obtained $t(R)$, as is now demonstrated.

Solving Eqs. (8.19) and (8.31) for dt and equating,

$$\frac{\mu R^2 \, d\phi}{H} = \frac{dR}{\{(2/\mu)[E - V'(R)]\}^{1/2}} \tag{8.36}$$

Separating variables and integrating,

$$\phi - \phi_0 = \frac{H}{\mu} \int_{R_0}^{R} \frac{dR'}{R'^2 \{(2/\mu)[E - V'(R')]\}^{1/2}} \tag{8.37}$$

We now explore a standard technique for determining the trajectory $R(\phi)$ directly for a conservative force field. Making the substitution,

$$R = \frac{1}{u} \tag{8.38}$$

$$\dot{R} = \frac{d}{dt}\left(\frac{1}{u}\right) = \frac{d}{d\phi}\left(\frac{1}{u}\right)\frac{d\phi}{dt} = -\frac{1}{u^2}\frac{du}{d\phi}\frac{d\phi}{dt} \tag{8.39}$$

Employing $\dot{\phi}$ from Eq. (8.19),

$$\dot{R} = -\frac{1}{u^2}\frac{du}{d\phi}\frac{H}{\mu R^2} = -\frac{H}{\mu}\frac{du}{d\phi} \tag{8.40}$$

Substituting Eq. (8.40) into Eq. (8.28),

$$E = \frac{1}{2}\frac{H^2}{\mu}\left(\frac{du}{d\phi}\right)^2 + \frac{1}{2}\frac{H^2}{\mu}u^2 + V \tag{8.41}$$

Solving for $du/d\phi$,

$$\frac{du}{d\phi} = \left(\frac{2\mu E}{H^2} - \frac{2\mu V}{H^2} - u^2\right)^{\frac{1}{2}} \tag{8.42}$$

Separating variables and integrating,

$$\phi - \phi_0 = \int_{R_0}^{R} \frac{du}{\left(\dfrac{2\mu E}{H^2} - \dfrac{2\mu V}{H^2} - u^2\right)^{\frac{1}{2}}} \tag{8.43}$$

Again, we require $V(R)$ to integrate Eq. (8.43). If V is proportional to R or u raised to a power, Eq. (8.43) is integrable in closed form for many values of the exponent. In any event, a numerical integration can be performed.

Illustrative Example 8.2
Find the potential energy and the force, given the equation of the trajectory

$$R = R_0 e^{\phi}$$

SOLUTION
Since $u = 1/R = (1/R_0)e^{-\phi}$,

$$\frac{du}{d\phi} = -\frac{1}{R_0}e^{-\phi} = -u$$

Substituting for u and $du/d\phi$ in Eq. (8.41), the energy becomes

$$E = \frac{1}{2}\frac{H^2 u^2}{\mu} + \frac{1}{2}\frac{H^2 u^2}{\mu} + V$$

Solving for the potential,

$$V = E - \frac{H^2 u^2}{\mu} = E - \frac{H^2}{\mu R^2}$$

and the force is, therefore,

$$F = -\frac{dV}{dR} = -\frac{2H^2}{\mu R^3}$$

8.3 INVERSE-SQUARE FORCE

Let us now consider the solution of Eq. (8.43) when the potential is expressed

$$V = \frac{V_0}{R} = V_0 u \tag{8.44}$$

This potential energy corresponds to a force, V_0/R^2, clearly an inverse-square force. Planets and projectiles are examples of bodies with motions governed by inverse-square gravitational forces. Such forces are always attractive. The scattering of a positively charged ion by a nucleus is an instance of a repulsive electrostatic inverse-square force.

We employ Eq. (8.43) to determine the trajectory subject to the potential of Eq. (8.44):

$$\phi = \int \frac{du}{\left(\dfrac{2\mu E}{H^2} - \dfrac{2\mu V_0 u}{H^2} - u^2\right)^{1/2}} + C_1 \tag{8.45}$$

This integral is similar in form to the one encountered in Illustrative Example 8.1 for the linear oscillator. Applying the same techniques as before, Eq. (8.45) becomes

$$u = \frac{1}{R} = \left[\frac{2\mu E}{H^2} + \left(\frac{\mu V_0}{H^2}\right)^2\right]^{1/2} \cos(\phi + \pi - C_1) - \frac{\mu V_0}{H^2} \tag{8.46}$$

where $\pi - C_1 = -\phi_0$ is an integration constant to be evaluated from corresponding known values of R and ϕ along the path of motion. Equation (8.46) may be rewritten

$$u = \frac{1}{R} = A \cos(\phi - \phi_0) + B \tag{8.47}$$

where ϕ_0 determines the orientation of the trajectory in the XY-plane, as shown in Fig. 8.4. Since the angle the X-axis makes with the line of symmetry of the curve is arbitrary, we are free to orient the XY-axes so as to make $\phi_0 = 0$.

Equation (8.47) describes a conic section, that is, ellipse, parabola, or hyperbola, as we now demonstrate. An *ellipse* is the locus of points in a plane such that

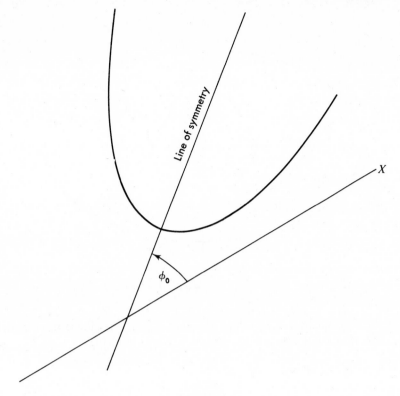

Fig. 8.4

the sum of the distances from two fixed points is a constant. Referring to Fig. 8.5, $\pm a$, the distance from the origin to the curve, is called the *semimajor axis* and $\pm \epsilon a$, the distances from the origin to the two fixed points, locate the *foci*, where ϵ is the *eccentricity* of the orbit. Evaluating $R + R_1$ for $\phi = 0$,

$$R + R_1 = (a - \epsilon a) + (\epsilon a + a) = 2a \tag{8.48}$$

Applying the law of cosines,

$$R_1^2 = R^2 + (2\epsilon a)^2 - 2R(2\epsilon a) \cos (\pi - \phi) \tag{8.49}$$

Solving Eq. (8.48) for R_1, squaring, and substituting Eq. (8.49) for R_1^2,

$$R^2 - 4aR + 4a^2 = R^2 + 4\epsilon^2 a^2 - 4\epsilon aR \cos (\pi - \phi) \tag{8.50}$$

Dividing by $4a$ and solving for $1/R$, we have

$$\frac{1}{R} = \frac{1 + \epsilon \cos \phi}{a(1 - \epsilon^2)} \qquad (\epsilon \geq 0) \tag{8.51}$$

where $-\cos \phi$ has been substituted for $\cos (\pi - \phi)$.

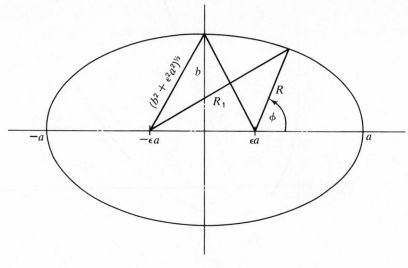

Fig. 8.5

In the special case $\epsilon = 0$, Eq. (8.51) gives $R = a$; the ellipse degenerates into a circle. The maximum value of R occurs when $\phi = \pi$, giving

$$R_{\max} = a(1 + \epsilon) = a + a\epsilon \tag{8.52}$$

Similarly,

$$R_{\min} = a(1 - \epsilon) = a - a\epsilon \tag{8.53}$$

at $\phi = 0$.

Elliptic motion is therefore bound, with R varying between the limits $R_{\max}$ and $R_{\min}$. The semiminor axis b is the maximum distance from the semimajor axis to the curve, as shown in Fig. 8.5. From the figure,

$$a^2 = b^2 + \epsilon^2 a^2$$

or

$$b = (a^2 - \epsilon^2 a^2)^{1/2} = a(1 - \epsilon^2)^{1/2} \tag{8.54}$$

The area of the ellipse, A_e, is given by the integral

$$A_e = \int_0^{2\pi} \frac{R^2 \, d\phi}{2} = \int_0^{2\pi} \frac{1}{2} \left[\frac{a(1 - \epsilon^2)}{1 + \epsilon \cos \phi} \right]^2 d\phi \tag{8.55}$$

or

$$A_e = \pi a^2 (1 - \epsilon^2)^{1/2} = \pi ab \tag{8.56}$$

The *parabola* is defined as the locus of all points in a plane such that the distances from a fixed point and a fixed line are equal. The fixed point and line are termed the *focus* and *directrix*, respectively. Referring to Fig. 8.6,

$$a = R - R \cos \phi \qquad\qquad (8.57)$$

or

$$R = \frac{a}{1 - \cos \phi} \qquad\qquad (8.58)$$

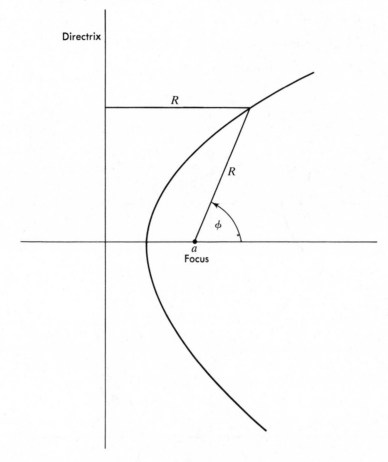

Fig. 8.6

There is no maximum value of R in this trajectory, owing to the nature of the denominator of Eq. (8.58). As ϕ approaches zero, R is observed to grow without limit. The minimum magnitude of R occurs when $\phi = \pi$; $R_{min} = a/2$. The reader may demonstrate that as the eccentricity of the elliptic orbit approaches unity, the trajectory becomes parabolic.

The *hyperbola* is the locus of all points in a plane such that the difference between the distances from two fixed points, each a focus, is a constant. There are two possibilities for the algebraic sign of this difference in distances. Referring to Fig. 8.7a, the foci are at $\pm \epsilon a$ and the curve intersects the horizontal axis at $-a$. Therefore,

$$R_1 - R = (\epsilon a - a) - (\epsilon a + a) = -2a \qquad \text{(8.59a)}$$

For the situation depicted in Fig. 8.7b,

$$R_1 - R = (\epsilon a + a) - (\epsilon a - a) = 2a \qquad \text{(8.59b)}$$

The law of cosines applied to triangle R_1, R, FF' gives

$$R_1^2 = R^2 + (2\epsilon a)^2 - 4\epsilon aR \cos(\pi - \phi) \qquad \text{(8.60)}$$

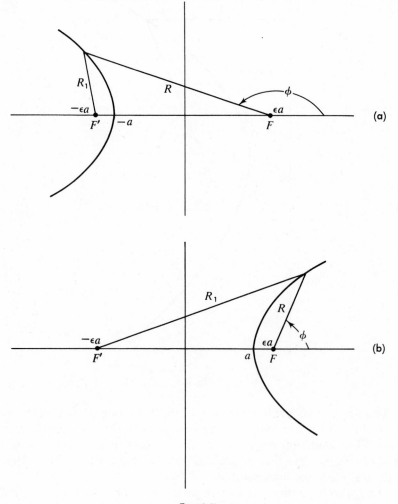

Fig. 8.7

Following the procedure for the ellipse, Eqs. (8.59) are solved for R_1, which is squared and substituted into Eq. (8.60),

$$(R - 2a)^2 = R^2 + 4\epsilon^2 a^2 + 4\epsilon\, aR \cos \phi \qquad\qquad \textbf{(8.61a)}$$

$$(R + 2a)^2 = R^2 + 4\epsilon^2 a^2 + 4\epsilon\, aR \cos \phi \qquad\qquad \textbf{(8.61b)}$$

yielding

$$\frac{1}{R} = \frac{-(\epsilon \cos \phi + 1)}{a(\epsilon^2 - 1)} \qquad\qquad \textbf{(8.62a)}$$

and

$$\frac{1}{R} = \frac{-(\epsilon \cos \phi - 1)}{a(\epsilon^2 - 1)} \qquad\qquad \textbf{(8.62b)}$$

For the hyperbola, $\epsilon > 1$, and we conclude that the numerators of Eqs. (8.62) must be positive (for positive R). The range of ϕ is, as a consequence, restricted to values given by

$$\phi = \pm \arccos \left(\frac{1}{\epsilon}\right) \qquad\qquad \textbf{(8.63)}$$

The maximum values of ϕ identify the angles made by the asymptotes of the hyperbolas with the axis of symmetry, as shown in Fig. 8.8.

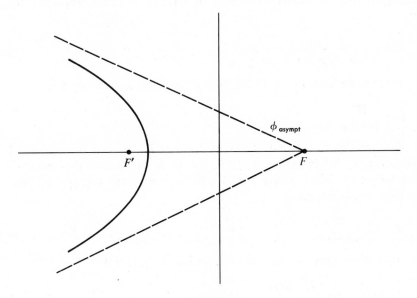

Fig. 8.8

8.4 GRAVITATIONAL CENTRAL FORCE MOTION

Having reviewed the properties of conics, we return now to interpretation of the equation of motion for inverse-square forces. In the gravitational field of the sun, the force between a mass m and the sun is $-Gmm_s/R^2$, where G represents the universal gravitational constant, m_s is the mass of the sun, and the minus sign indicates that the force is attractive. For planetary motion, we assume that:

a. The sun and planets are point masses. This may be justified either on the basis of the very large astronomical distances, relative to the solar diameter, or a spherically symmetric distribution of matter. Recall Section 3.9.

b. The sun and planets are rigid, so there are no tidal effects.

c. The density of matter in interplanetary space is so dilute that no frictional forces act.

d. Only two masses are present, and consequently the force system may be reduced to that of a single *central* force. Thus influences outside the two-mass system are excluded.

Returning to Eq. (8.46), $V_0 = -Gmm_s$, and the equation for the orbit becomes

$$R = \frac{1}{[(2\mu E/H^2) + (-Gmm_s\mu/H^2)^2]^{\frac{1}{2}}\cos\phi + (Gmm_s\mu/H^2)} \tag{8.64}$$

The energy E may be positive, negative, or zero. For zero energy, Eq. (8.64) is written

$$R = \frac{1}{(Gmm_s\mu/H^2)(1 - \cos\phi)} \tag{8.65a}$$

and is of the same form as Eq. (8.58), which relates to a parabola with

$$a = \frac{H^2}{Gm_sm\mu}$$

Zero energy means that the kinetic energy is equal to the negative of the potential energy. When the particle is an infinite distance from the center of force, its potential energy is zero (since $V = V_0/R$) as is its kinetic energy. For the zero-energy case, Eq. (8.64) gives

$$R_{\min} = \frac{H^2}{2Gmm_s\mu} \tag{8.65b}$$

For positive E we write Eq. (8.64) in the form of Eqs. (8.62),

$$R = \frac{H^2/Gmm_s\mu}{1 + [1 + (2EH^2/G^2m^2m_s^2\mu)]^{\frac{1}{2}}\cos\phi} \tag{8.66}$$

The coefficient of $\cos \phi$, the eccentricity ϵ, is greater than unity, because all terms in the bracket are positive. The motion is therefore hyperbolic, and the distance of closest approach is

$$\frac{H^2/Gmm_s\mu}{1 + [1 + (2EH^2/G^2m^2m_s^2\mu)]^{1/2}}$$

Some comets and meteors in the solar system execute parabolic and hyperbolic motion when they have sufficient energy $(E \geq 0)$ to escape the solar system. A particle within the solar system, possessing negative energy, is trapped and cannot escape unless additional energy is imparted to it. Referring to Eqs. (8.51) and (8.64) for negative energy,

$$\epsilon = \left(1 - \frac{2|E|H^2}{G^2m^2m_s^2\mu}\right)^{1/2} \tag{8.67}$$

giving $\epsilon < 1$. The values of R_{min} and R_{max}, called the perihelion and aphelion distances, respectively, are given by

$$R_{min} = \frac{H^2/Gmm_s\mu}{1 + [1 - (2H^2|E|/G^2m^2m_s^2\mu)]^{1/2}} \tag{8.68a}$$

$$R_{max} = \frac{H^2/Gmm_s\mu}{1 - [1 - (2H^2|E|/G^2m^2m_s^2\mu)]^{1/2}} \tag{8.68b}$$

The sum of R_{min} and R_{max} equals twice the semimajor axis, $2a$,

$$R_{min} + R_{max} = \frac{H^2}{Gmm_s\mu}\left[\frac{1}{1 + \left(1 - \frac{2H^2|E|}{G^2m^2m_s^2\mu}\right)^{1/2}} + \frac{1}{1 - \left(1 - \frac{2H^2|E|}{G^2m^2m_s^2\mu}\right)^{1/2}}\right] \tag{8.69}$$

or

$$R_{min} + R_{max} = \frac{Gmm_s}{|E|} = 2a \tag{8.70}$$

or, in terms of the energy,

$$|E| = \frac{Gmm_s}{2a} \tag{8.71}$$

The rate at which area is swept out by the radius vector is given by Eq. (8.15). The elliptic area is swept out once each period T:

$$\frac{1}{2}\frac{H}{\mu}T = \pi a^2(1 - \epsilon^2)^{1/2} \tag{8.72}$$

Substituting Eq. (8.67) for ϵ into Eq. (8.72) and solving for T,

$$T = \frac{2\pi \mu a^2}{H} \left(\frac{2|E| H^2}{G^2 m^2 m_s^2 \mu} \right)^{\frac{1}{2}} \tag{8.73}$$

or

$$T = 2\pi \mu^{\frac{1}{2}} a^2 \left(\frac{2|E|}{G^2 m^2 m_s^2} \right)^{\frac{1}{2}} \tag{8.74}$$

Since $a = Gmm_s/2|E|$,

$$T = 2\pi \left(\frac{\mu}{Gmm_s} \right)^{\frac{1}{2}} a^{3/2} \tag{8.75a}$$

or

$$T \cong \frac{2\pi}{(Gm_s)^{\frac{1}{2}}} a^{3/2} \tag{8.75b}$$

when the mass of the sun is very much larger than the planetary mass, so $\mu \cong m$.

Equations (8.15), (8.64), and (8.75b) are the three laws of planetary motion postulated by Kepler[1] (circa 1600) on the basis of careful astronomical measurements of Tycho Brahe. The laws were later derived by Newton,[2] assuming an inverse-square law of forces. They represent a brilliant confirmation of Newton's laws of motion.

Kepler's laws are:

a. For any planet, the radius vector from the sun to the planet sweeps out equal areas in equal times.

b. The planetary orbits about the sun are ellipses with the sun at a focus.

c. The square of the period is proportional to the cube of the semimajor axis.

Not only do Kepler's laws predict the motion of the planets around the sun, but that of the moon and artificial satellites about the earth as well. Actually, the earth is not quite spherical nor is its mass symmetrically distributed. The orbits are, nevertheless, nearly elliptical.

8.5 MOTION UNDER REPULSIVE INVERSE-SQUARE FORCE

In 1908 Geiger and Marsden,[3] under the direction of E. Rutherford, experimentally studied the scattering of alpha particles (doubly charged helium ions) by thin metallic foils. They employed an apparatus that may be schematically represented as in

[1] Kepler, J., *Astronomia Nova*, Prague, 1609; *Harmonice Mundi*, Linz, 1619.
[2] *Principia Mathematica Philosophiae Naturalis* of Sir I. Newton, Translated by A. Motte, Revised by F. Cajori, originally published 1687, University of California Press, Berkeley, Calif., 1947.
[3] Geiger, H., and Marsden, E., *Proc. Roy. Soc. London* **A82,** 495 (1909); *Phil. Mag.* **25,** 604 (1913).

Fig. 8.9. Their object was to verify the Thomson model[4] of the atom, in which positively charged matter filling the atomic volume contains a sufficient number of electrons to assure overall charge neutrality. The data differed to such an extent from the predictions based upon the Thomson atom that Rutherford[5] was led to consider alternative possibilities. He eventually proposed a solar-system model in which a highly concentrated positively charged nucleus is surrounded by electrons.

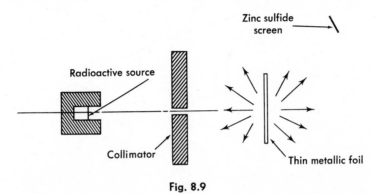

Fig. 8.9

To demonstrate the validity of his model, Rutherford calculated the scattering effect of an individual nucleus on alpha particles and compared his results with the data of Geiger and Marsden. The startling success of the Rutherford calculations represents what is probably the most significant example of a repulsive inverse-square force and gave great credence to the Rutherford model.

In accordance with the Rutherford approach, the only force acting on the alpha particle (shown in Fig. 8.10) is given by

$$F = \frac{2Ze^2}{R^2} \quad \text{(CGS units)} \tag{8.76}$$

where Z represents the atomic number and e is the electronic charge. Thus each alpha particle experiences the electrostatic influence of a single nucleus only in its passage through the foil; that is, no multiple scattering occurs, and the effects of electrons are negligible.

Alpha particles are monoenergetically emitted from radioactive nuclei, and for a particular species the kinetic energy is well known. Since we consider the potential energy zero at "infinity," this energy also represents the total energy E. Thus

$$E = \tfrac{1}{2}\mu\, v_0^2 \tag{8.77}$$

where v_0 is the alpha-particle velocity at infinity.

[4] Thomson, J. J., *Phil. Mag.* **7**, 237 (1904).
[5] Rutherford, E., *Phil. Mag* **21**, 669 (1911).

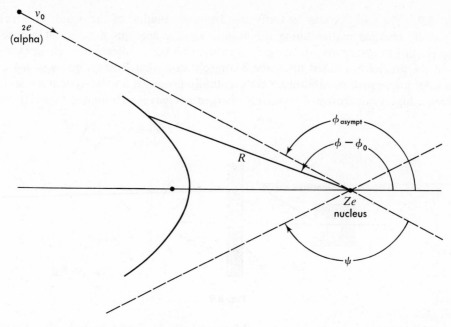

Fig. 8.10

To describe the orbit, another constant of the motion, the angular momentum H, is required:

$$H = \mu v_0 s \qquad (8.78)$$

or

$$H^2 = 2\mu E s^2 \qquad (8.79)$$

where s, the *impact parameter*, represents the distance of closest approach of the alpha particle in the *absence* of electrostatic forces, Fig. 8.11. The angular momen-

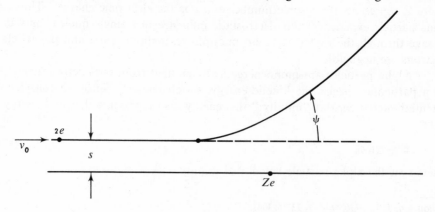

Fig. 8.11

tum thus depends upon the impact parameter, which in systems of atomic dimension cannot be measured. The scattering must, therefore, be determined as a function of s, treated as a parameter of the motion.

The potential energy associated with a force described in Eq. (8.76) is $V = 2Ze^2/R$. Substituting this expression together with Eqs. (8.77) and (8.79) into Eq. (8.46), we obtain

$$\frac{1}{R} = \left[\frac{1}{s^2} + \left(\frac{Ze^2}{Es^2}\right)^2\right]^{\frac{1}{2}} \cos\phi - \frac{Ze^2}{Es^2} \tag{8.80a}$$

This may be rewritten

$$\frac{1}{R} = \frac{[(E^2s^2/Z^2e^4) + 1]^{\frac{1}{2}} \cos\phi - 1}{Ze^2/Es^2} \tag{8.80b}$$

which is of the same general form as Eq. (8.62a), a hyperbola, with eccentricity

$$\epsilon = \left(\frac{E^2s^2}{Z^2e^4} + 1\right)^{\frac{1}{2}} \tag{8.81}$$

which clearly exceeds unity.

The asymptotes of the trajectory are given by Eq. (8.63):

$$\phi_{asympt} = \pm\cos^{-1}\left(\frac{1}{\epsilon}\right)$$

$$= \pm\cos^{-1}\left\{\frac{1}{[(E^2s^2/Z^2e^4) + 1]^{\frac{1}{2}}}\right\} \tag{8.82}$$

The asymptotic values of ϕ correspond to very large R, and Eq. (8.82) may therefore be verified by substituting $R = \infty$ in Eq. (8.80b). Hence the particle enters the field of the nucleus along one asymptote and leaves its influence along another (Fig. 8.10).

The angle of scatter ψ may be related to ϕ_{asympt} by examination of Fig. 8.10:

$$\psi = \pi - 2\phi_{asympt} \tag{8.83}$$

What we have derived in the foregoing development concerns a single alpha particle scattered by a single nucleus. In an actual experiment it is clear that there are many alpha particles present in the incident beam and many nuclei in the foil, although we retain the assumption that each alpha particle interacts with a single nucleus only. Since we cannot measure the impact parameter and can readily measure the angle of scatter, it serves us well to relate these two quantities.

Combining Eqs. (8.82) and (8.83),

$$\cos\left(\frac{\pi}{2} - \frac{\psi}{2}\right) = \frac{1}{[(E^2s^2/Z^2e^4) + 1]^{\frac{1}{2}}} \tag{8.84}$$

Referring now to Fig. 8.12, constructed on the basis of Eq. (8.84), we may write

$$\cot\left(\frac{\psi}{2}\right) = \frac{Es}{Ze^2} \tag{8.85}$$

from which

$$s = \frac{Ze^2}{E} \cot\left(\frac{\psi}{2}\right) \tag{8.86}$$

and

$$ds = -\frac{Ze^2}{E} \frac{d(\psi/2)}{\sin^2(\psi/2)} \tag{8.87}$$

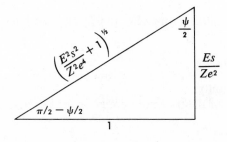

Fig. 8.12

Consider a foil of surface area A (perpendicular to the incident beam) and thickness T. The probability p of an alpha particle passing within a distance s of a nucleus, as shown in Fig. 8.13, is given by

$$p = \frac{\pi s^2}{A}$$

Therefore, the probability P that an alpha particle passes within a distance s of any of the nuclei is

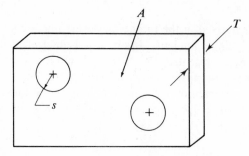

Fig. 8.13

$$P = pnTA = \pi s^2 nT \tag{8.88}$$

where n is the number of nuclei per unit volume of the foil and nTA therefore represents the total number of nuclei present.

An alpha particle with impact parameter less than s will be scattered through an angle greater than ψ. The probability of scattering through an angle exceeding ψ can, therefore, be determined by substituting Eq. (8.86) for s into Eq. (8.88). Thus

$$P = \pi \left(\frac{Ze^2}{E}\right)^2 \cot^2 \left(\frac{\psi}{2}\right) nT \tag{8.89}$$

The probability that an alpha particle will be scattered through an angle between ψ and $\psi + d\psi$ is equal to the probability of an alpha particle having an impact parameter between s and $s + ds$. The latter is the differential probability derived from Eq. (8.88),

$$dP = -\pi nT2s\ ds \tag{8.90}$$

where the negative sign denotes a decreased probability with increasing impact parameter.

Substituting for s and ds, Eqs. (8.86) and (8.87), we have

$$dP = 2\pi \left(\frac{Ze^2}{E}\right)^2 nT \frac{\cot (\psi/2)\, d(\psi/2)}{\sin^2 (\psi/2)}$$

$$= \pi \left(\frac{Ze^2}{E}\right)^2 nT \frac{\cos (\psi/2)\, d\psi}{\sin^3(\psi/2)} \tag{8.91}$$

Using the trigonometric identity

$$\cos \left(\frac{\psi}{2}\right) = \frac{1}{2} \frac{\sin \psi}{\sin (\psi/2)}$$

Eq. (8.91) becomes

$$dP = \frac{\pi}{2} \left(\frac{Ze^2}{E}\right)^2 nT \frac{\sin \psi\, d\psi}{\sin^4 (\psi/2)} \tag{8.92}$$

The probability of the alpha particle scattering through $d\psi$ and passing through a differential solid angle $d\Omega$ is, referring to Fig. 8.14,

$$\frac{dP}{d\Omega} = \frac{\dfrac{\pi}{2} \left(\dfrac{Ze^2}{E}\right)^2 \dfrac{nT \sin \psi\, d\psi}{\sin^4 (\psi/2)}}{2\pi \sin \psi\, d\psi}$$

or

$$\frac{dP}{d\Omega} = \frac{1}{4} \left(\frac{Ze^2}{E}\right)^2 \frac{nT}{\sin^4 (\psi/2)} \tag{8.93}$$

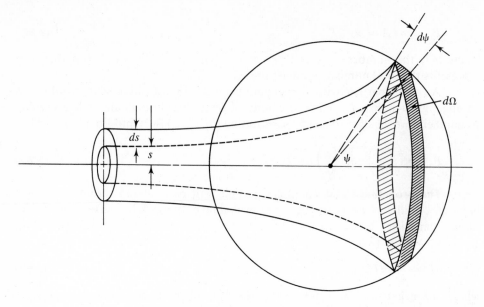

Fig. 8.14

If N_0 represents the number of alpha particles per second in the incident beam, the number of scintillations observed in the differential solid angle per unit time is

$$N_0 \frac{dP}{d\Omega} = \frac{N_0}{4} \left(\frac{Ze^2}{E} \right)^2 \frac{nT}{\sin^4(\psi/2)} \qquad (8.94)$$

which represents a form of Rutherford's law of scattering. Rutherford thus related a single alpha-particle–single nucleus interaction to the parameters of the laboratory experiment, conducted by Geiger and Marsden, involving many foils and alpha-particle energies. The same result as above is obtained when quantum mechanical techniques are employed.

According to Eq. (8.94), the number of particles scattered into $d\Omega$ per unit time should increase without limit as ψ approaches zero. This is, of course, impossible on physical grounds. The foregoing analysis pertained to an alpha particle which passes reasonably close to a nucleus, permitting interactions with other nuclei to be ignored. Small-angle scattering implies, however, a large impact parameter for a given interaction, meaning that the relative influence of a neighboring nucleus is significant and the results more characteristic of multiple scattering. Equation (8.94) simply does not take such considerations into account.

An abundance of additional experimental work supports the conclusions of Eq. (8.94):

a. The direct dependence upon the number of nuclei and the foil thickness.
b. The inverse dependence upon the square of the alpha particle energy.
c. The square dependence upon the atomic number of the foil nuclei.

The foregoing analysis of Rutherford scattering is predicated upon a coordinate system fixed to the nucleus, consistent with the reduced-mass concept already discussed. Obviously, the experimental data agree with Eq. (8.94) only to the extent that the nuclear mass is sufficiently great that the reduced mass is essentially equal to the mass of the alpha particle.

8.6 SCATTERING

Many processes, macroscopic as well as microscopic, involve the deflection of particles, from initial straight-line motion, by centers of force. The previous section analyzed an important instance of microscopic particles so deflected. The trajectory of a comet under the influence of the sun exemplifies scattering on a large scale. The concept of an *effective cross-sectional area* for scatter is quite useful in the interpretation and application of scattering data.

The *differential scatter cross section* $d\sigma$ is defined by the expression

$$\frac{d\sigma}{d\Omega} = \frac{N}{\eta I} \tag{8.95}$$

where N is the number of particles detected per second which are scattered into a solid angle between Ω and $\Omega + d\Omega$, η the number of scatterers, and I the number of particles per second per unit area in the incident beam. We shall assume all the particles in the incident beam identical in mass and energy. In Eq. (8.95), $d\sigma$ possesses dimensions of area and plays the role of an effective area for scattering. By this is meant that a particle passing through this area around the center of force will be scattered into the solid angle between Ω and $\Omega + d\Omega$.

The number of particles scattered into an angle between Ω and $\Omega + d\Omega$ is equal to the number of particles passing through the annular space between s and $s + ds$ where η is taken as unity. Therefore,

$$(2\pi s \, ds)I = -N \, d\Omega = -I \, d\sigma \tag{8.96}$$

where the minus sign indicates that as s is increased, the angle ψ decreases; hence $ds \sim -d\psi$ and $ds \sim -d\Omega$. From Eq. (8.96),

$$d\sigma = -2\pi s \, ds \tag{8.97}$$

Illustrative Example 8.3

Derive an expression for the differential scatter cross section of a mass m subject to the force

$$\mathbf{F} = \frac{K}{R^3} \hat{\mathbf{e}}_R$$

SOLUTION

The force given is conservative and the associated potential energy,

$$V = \frac{K}{2R^2} = \frac{Ku^2}{2}$$

where the zero of potential energy is "at $R = \infty$."

The second path equation or energy equation, Eq. (8.41), is

$$E = \frac{1}{2}\frac{H^2}{m}\left(\frac{du}{d\phi}\right)^2 + \frac{1}{2}\frac{H^2 u^2}{m} + \frac{Ku^2}{2}$$

where m and the reduced mass are identical, based upon an assumed infinite mass associated with the force center.

The energy equation may be written

$$E = \frac{1}{2}\frac{H^2}{m}\left(\frac{du}{d\phi}\right)^2 + \frac{1}{2}\left(\frac{H^2}{m} + K\right)u^2$$

which corresponds to the form

$$E = \tfrac{1}{2}m\left(\frac{dx}{dt}\right)^2 + \tfrac{1}{2}kx^2$$

for a one-dimensional linear oscillator. Thus the analogous quantities are mass and H^2/m, stiffness and $H^2/m + K$, displacement and u, and time and ϕ.

By analogy, the solution of the second path equation is

$$u = A \cos\left[\left(\frac{H^2 + mK}{H^2}\right)^{1/2}\phi + \lambda\right]$$

where substitution for u into the original energy equation gives $A = (2mE/H^2 + mK)^{1/2}$ and λ describes the orientation of the trajectory in the plane of the motion. The particle position is thus

$$\frac{1}{R} = \left(\frac{2mE}{H^2 + mK}\right)^{1/2}\cos\left[\left(\frac{H^2 + mK}{H^2}\right)^{1/2}\phi + \lambda\right]$$

Since the particle is originally a great distance from the scatterer ($1/R = 0$),

$$\cos\left[\left(\frac{H^2 + mK}{H^2}\right)^{1/2}\phi + \lambda\right] = 0$$

or

$$\left(\frac{H^2 + mK}{H^2}\right)^{1/2}\phi + \lambda = \pm\frac{\pi}{2}$$

When the particle leaves the influence of the scatterer, again $1/R = 0$. The angle of scatter

$$\psi = \pi - (\text{change of } \phi)$$

$$= \pi - \frac{\pi}{[H^2/(H^2 + mK)]^{1/2}} = \pi\left[1 - \left(\frac{H^2 + mK}{H^2}\right)^{1/2}\right]$$

Since the differential cross section is expressed in terms of the impact parameter s, we express ψ in terms of the same variable. The energy and angular momentum when the particle is very far removed from the force center are $E = \frac{1}{2}mv_0^2$ and $H = mv_0s$. Thus

$$\psi = \pi\left[1 - \left(\frac{2mEs^2 + mK}{2mEs^2}\right)^{\frac{1}{2}}\right] = \pi\left[1 - \left(1 + \frac{K}{2Es^2}\right)^{\frac{1}{2}}\right]$$

Solving for s^2,

$$s^2 = \frac{\pi^2 K}{2E[(\pi - \psi)^2 - \pi^2]}$$

Returning to Eq. (8.97),

$$d\sigma = -2\pi s\, ds = -\pi\, d(s^2)$$

and substitution of $d(s^2)$ gives

$$d\sigma = \frac{\pi^3 K(\pi - \psi)}{E(2\pi - \psi)^2\psi^2}\, d\psi$$

8.7 THE BOHR ATOM

The Rutherford model of the atom places the electrons a relatively great distance from the positively charged nucleus. The work of Geiger and Marsden substantiates this view of the atom. The electrons, subjected to an inverse-square electrostatic force of attraction, must, therefore, suffer acceleration. According to the predictions of classical electromagnetic theory, an accelerated charge radiates energy at all frequencies, except in the special case where the charge undergoes harmonic motion. Atomic spectra do not, however, display a continuum of frequencies or lines. Not only was this to be explained but the very existence of stable atoms as well, for if they radiate continuously, they cannot long exist. To explain the radiated spectra of atoms, Bohr[6] proposed that:

a. The motion of electrons is limited to particular orbits around the nucleus prescribed by

$$H = \frac{nh}{2\pi} = n\hbar$$

where H represents the angular momentum of the electron about the nucleus, h is Planck's constant[7] ($h = 6.62 \times 10^{-27}$ erg-sec), and n may be any positive integer.

b. The electron does not radiate as long as it remains in a permitted orbit. In moving from one orbit to another, the electron radiates or absorbs energy corresponding to the difference in the orbital energies.

---

[6] Bohr, N., *Phil. Mag.* **26**, 1 (1913).
[7] Planck, M., *Verhandl. Deut., Physikal.* **2**, 237 (1900).

A simple manner of evaluating the electron energy is to consider it in a circular orbit of radius a. The system energy consists of the electrostatic potential energy and the kinetic energy,

$$E = -\frac{Ze^2}{a} + \tfrac{1}{2}\mu v^2 = -\frac{Ze^2}{a} + \tfrac{1}{2}\mu a^2 \dot{\phi}^2 \tag{8.98}$$

where $v = a\dot{\phi}$.

Equating the force to the product of reduced mass and radial acceleration, and recalling that $\ddot{R} = 0$ for the circular orbit,

$$F = -\frac{Ze^2}{a^2} = \mu a_R = \mu(-a\dot{\phi}^2) \tag{8.99}$$

The angular momentum, according to the Bohr hypothesis, is

$$H = n\hbar (= \mu v a = \mu a^2 \dot{\phi}) \tag{8.100}$$

Simultaneous solution of Eqs. (8.99) and (8.100) for a and $\dot{\phi}$, and substitution into Eq. (8.98), yields

$$E_n = -\frac{\mu Z^2 e^4}{2\hbar^2 n^2} \tag{8.101}$$

Thus the electron according to the Bohr hypothesis exists in quantized or discrete energy states, E_n. If an analysis is performed for elliptic electronic orbits, the energy states are found not to depend upon eccentricity, and the result is the same as Eq. (8.101).

A more sophisticated technique developed subsequently by Bohr and Sommerfeld[8] quantizes, in addition, a function of the radial momentum $H_R = \mu \dot{R}$, but the result given by Eq. (8.101) is not appreciably altered. When relativistic considerations are introduced, however, the energy is found to depend upon the eccentricity to a small extent. Sommerfeld attempted to explain the so-called "fine structure" of the hydrogen spectrum by these added effects. His lack of success led to new approaches, including the Schrödinger–Heisenberg quantum-wave mechanics.

Returning to the Bohr atom (for hydrogen, $Z = 1$), in proceeding from state n to state m the energy of an electron changes by

$$E_m - E_n = -\frac{\mu e^4}{2\hbar^2}\left(\frac{1}{m^2} - \frac{1}{n^2}\right) \tag{8.102}$$

Assuming, as Bohr did, that there is radiated one photon of energy,

$$E_{nm} = \hbar\omega_{nm} \tag{8.103}$$

where ω_{nm} represents the angular frequency of the radiation,

---

[8] Sommerfeld, A., *Ann. Physik* **51,** 1 (1916).

$$\omega_{nm} = \frac{\mu e^4}{2\hbar^3} \left(\frac{1}{n^2} - \frac{1}{m^2} \right) \quad (m = 1, 2, \cdots; n = m + 1, m + 2, \cdots) \quad \textbf{(8.104)}$$

When the values of ω_{nm} are calculated for atomic hydrogen, they correspond to the empirical laws of spectroscopy formulated about 30 years before Bohr. The Bohr atom only explains the radiation from atomic hydrogen-like atoms. It also predicts hydrogen-like atoms to be two-dimensional, although experiment shows that they are actually three-dimensional. Bohr theory makes no predictions with regard to the intensity of the radiated spectrum. Nevertheless, the remarkable predictions of the Bohr theory have given rise to a more comprehensive theory of quantum mechanics, and the Bohr atom remains a useful concept today, providing a simple mechanical model of atomic systems.

References

Blass, G. A., *Theoretical Physics*, Appleton-Century Crofts, New York, 1962.

Corben, H. C., and Stehle, P., *Classical Mechanics*, Second Edition, John Wiley & Sons, Inc., New York, 1960.

Goldstein, H., *Classical Mechanics*, Addison-Wesley Publishing Company, Inc., Reading, Mass., 1950.

Landau, L. D., and Lifshitz, E. M., *Mechanics*, Addison-Wesley Publishing Company, Inc., Reading, Mass., 1960.

Sommerfeld, A., *Mechanics*, Translated by M. O. Stern, Academic Press, Inc., New York, 1957.

EXERCISES

8.1 A conic may be defined as the locus of points such that the distances from a fixed point (focus) and a fixed line (directrix) have a constant ratio. Demonstrate that this definition is consistent with the conics described in this chapter.

8.2 Show that the substitution $R = 1/u$ in the radial force equation [Eq. (8.16)] leads to the differential equation

$$\frac{d^2u}{d\phi^2} + u = - \frac{\mu}{H^2 u^2} f\left(\frac{1}{u}\right)$$

This expression, called the *first path equation*, is useful in solving for the perturbed motion about stable orbits.

8.3 Demonstrate that $\epsilon = 1$ represents the limiting case for escape of an earth satellite and prove that the minimum tangential velocity for escape from the earth's influence is

$$v_{escape} = \left(\frac{2Gm_e}{R_0}\right)^{\frac{1}{2}}$$

where R_0 represents the distance from the center of the earth to the satellite.

8.4 Prove that the velocity required to place a satellite into a circular orbit about the earth is

$$v_{circ} = \left(\frac{Gm_e}{R_0}\right)^{\frac{1}{2}}$$

8.5 An artificial satellite is placed in orbit by imparting to it a velocity of 8000 m/sec parallel to the earth's surface at an altitude of 700 km. Determine:

    a. the eccentricity of the orbit.

    b. the maximum and minimum altitudes of the satellite above the earth's surface (apogee and perigee).

    c. the radial and tangential velocities of the satellite as a function of ϕ.

8.6 Determine $R(t)$ and $\phi(t)$ for a mass subject to the force

$$\mathbf{F} = -\frac{k}{R^2}\,\hat{\mathbf{e}}_R$$

8.7 A mass is acted upon by a linear restoring force

$$\mathbf{F} = -kR\hat{\mathbf{e}}_R$$

Determine the orbit $R(\phi)$ and demonstrate that it is an ellipse. Discuss two degenerate cases, the circle and straight line.

8.8 What is the orbit of a mass subject to the force

$$\mathbf{F} = -\left(kR + \frac{K}{R^2}\right)\hat{\mathbf{e}}_R$$

8.9 A particle of mass m moves in the field of a Yukawa potential:

$$V = -\frac{V_0 e^{-R/\beta}}{R/\beta}$$

    a. Discuss the possible motions in terms of the apparent potential.

    b. Determine the radius of a circular orbit.

    c. Find the frequency of small oscillations about the circular orbit.

8.10 A particle moves in a potential field described by

$$V(R) = -\frac{k}{R} - \frac{a}{R^2}$$

and possesses energy E and angular momentum H.

  a. Describe the motion in terms of the apparent potential.
  b. Determine the orbit for negative values of energy.
  c. Determine the frequency of circular motion.
  d. Find the frequency of small oscillations about the circular orbit.

8.11 Prove that stable circular orbits are possible for a potential $V = V_0 R^\alpha$ only if $\alpha > -2$.

8.12 Determine the force acting on a particle if its orbit is described by

$$R = R_0 \left(\frac{\phi}{\phi_0}\right)^3$$

8.13 Determine the force acting on a particle if the orbit is

$$\frac{1}{R} = \frac{A}{\phi^4} - \frac{B}{\phi^2}$$

8.14 Prove that a parabola is generated from an ellipse when the eccentricity approaches unity from a value less than 1. Prove that a parabola is generated from a hyperbola when the eccentricity approaches unity from a value greater than 1.

8.15 What maximum impact parameter must an alpha particle have to just penetrate the nucleus if its energy is such that in a head-on encounter it penetrates 10 percent of the distance into the nucleus?

8.16 A projectile is fired from a point on the earth to another, diametrically opposite, point. Determine the required initial velocity and firing angle if the maximum altitude above the surface of the earth is 80 km.

8.17 It is required to place a satellite into an overhead equatorial orbit which has a period of exactly 1 day. Determine the minimum energy required to launch such a satellite. What speed will it have when in orbit?

8.18 Derive an expression for the differential scattering cross section of a negatively charged ion scattered by a positively charged nucleus.

8.19 Calculate the energy levels of positronium, an atom comprised of two electrons of equal mass and opposite charge. Compare the energy levels with those of hydrogen.

8.20 Derive an expression for the differential scattering cross section for a single nucleus and a single alpha particle.

8.21 Determine the distance of closest approach of 1.5- and 10-MeV alpha particles with respect to the force center of a gold nucleus ($Z = 79$), for angles of scatter of 10, 90, and 180 degrees. At what energy is the distance of closest approach (for $\psi = 180$ degrees) equal to 10^{-14} m (the approximate nuclear radius)? At this energy and greater, deviations from the predictions of Rutherford scattering may be anticipated because the nucleus is penetrated and the assumed force model no longer applies. (*Note:* 1 MeV $= 1.6 \times 10^{-13}$ J.)

8.22 For an accelerated electron, the rate of energy loss in CGS units is given by

$$-\frac{dE}{dt} = \frac{2}{3}\frac{a^2 e^2}{c^3}$$

where a represents the acceleration, e the electronic charge, and c the speed of light. Estimate the time for an electron to fall into the nucleus if it is initially in a large circular orbit.

CHAPTER 9

PARTICLE DYNAMICS IN NONINERTIAL FRAMES OF REFERENCE

9.1 INTRODUCTION

We have already pointed out that Newton's laws require that motion be referred to primary or secondary inertial frames of reference. Results obtained as a consequence of referring motion to noninertial frames, however, such as earth-fixed references, may be in error by an acceptable amount. In these instances, noninertial frames are employed as though they are inertial references. This is certainly the basis of solution of typical problems in elementary mechanics, where the matter of coordinate frame is never broached at all, the earth serving somehow as an intuitively satisfactory reference for all measurements.

In this chapter our primary concern is particle dynamics, written in terms of noninertial frames of reference, but including the effects of coordinate-frame motion, so that the final expressions are completely valid. Special emphasis is given earth-based coordinate frames, which are "natural" in many situations. The dynamic equations derived readily reveal the errors introduced by considering the earth as fixed in space.

9.2 GENERAL EQUATIONS OF MOTION

Consider $X'Y'Z'$ to be an inertial frame of reference. Then, as we have discussed, Newton's second law for a particle is written,

$$\mathbf{F} = m\mathbf{a}_{X'Y'Z'} = m\ddot{\mathbf{r}}' \tag{9.1}$$

where $\mathbf{F}$ represents the sum of all forces acting on the particle and $\mathbf{a}_{X'Y'Z'}$ is the acceleration produced by these forces, measured with respect to $X'Y'Z'$. In nonrelativistic mechanics, force is a unique vector in space, independent of the coordinate reference in which it is observed. Hence, in different coordinate frames, the individual components of force may vary, but the magnitude and orientation in space of the force vector are the same, regardless of where measured. This is not true of the acceleration, however, because this term represents the second time derivative of the position vector **r**. As demonstrated in Chapter 2, the time derivatives of the position vector are very much dependent upon the motion of the coordinate reference.

In Chapter 2 an expression was developed relating acceleration measured in coordinate frames of reference experiencing general relative motion. Let us consider $X'Y'Z'$ the frame to which we wish to refer the acceleration. Recall,

$$\mathbf{a}_{X'Y'Z'} = \mathbf{a}_{XYZ} + \ddot{\mathbf{R}} + \boldsymbol{\omega} \times (\boldsymbol{\omega} \times \boldsymbol{\rho}) + \dot{\boldsymbol{\omega}} \times \boldsymbol{\rho} + 2\boldsymbol{\omega} \times \mathbf{v}_{XYZ} \tag{9.2}$$

where $\mathbf{a}_{XYZ}$ is the acceleration of the particle measured relative to the XYZ reference, which itself experiences general motion with respect to $X'Y'Z'$, $\ddot{\mathbf{R}}$ is the acceleration of the origin of XYZ relative to that of $X'Y'Z'$, $\boldsymbol{\omega}$ represents the angular velocity of XYZ relative to $X'Y'Z'$, $\dot{\boldsymbol{\omega}}$ is the angular acceleration of XYZ with respect to $X'Y'Z'$, $\boldsymbol{\rho}$ is a vector running from the origin of XYZ to the particle, and $\mathbf{v}_{XYZ}$ is the velocity of the particle as measured in the XYZ-coordinate system.

Substituting for $\mathbf{a}_{X'Y'Z'}$ in Eq. (9.1), we have

$$\mathbf{F} = m[\mathbf{a}_{XYZ} + \ddot{\mathbf{R}} + \boldsymbol{\omega} \times (\boldsymbol{\omega} \times \boldsymbol{\rho}) + \dot{\boldsymbol{\omega}} \times \boldsymbol{\rho} + 2\boldsymbol{\omega} \times \mathbf{v}_{XYZ}] \tag{9.3}$$

In Eq. (9.3), $\mathbf{v}_{XYZ}$, $\mathbf{a}_{XYZ}$, and $\boldsymbol{\rho}$ represent the only kinematic quantities measured with respect to the noninertial reference; the remaining terms are calculated on the basis of separate determinations of $\ddot{\mathbf{R}}$, $\boldsymbol{\omega}$, and $\dot{\boldsymbol{\omega}}$.

Equation (9.3) is a valid statement of Newton's second law as written. It is often rewritten as below, to preserve the form of Newton's second law with particle acceleration referred *not* to the inertial frame of reference but to the unprimed reference. Thus

$$\mathbf{F} - m\ddot{\mathbf{R}} - m\boldsymbol{\omega} \times (\boldsymbol{\omega} \times \boldsymbol{\rho}) - m\dot{\boldsymbol{\omega}} \times \boldsymbol{\rho} - 2m\boldsymbol{\omega} \times \mathbf{v}_{XYZ} = m\mathbf{a}_{XYZ} \tag{9.4}$$

For a noninertial reference experiencing translation only, the terms containing angular velocity and angular acceleration of XYZ with respect to $X'Y'Z'$ vanish and we have

$$\mathbf{F} - m\ddot{\mathbf{R}} = m\mathbf{a}_{XYZ} \tag{9.5}$$

The left side of Eq. (9.4) is observed to contain real forces and fictitious forces. The real forces are attributable to interactions such as contact, gravitational attraction, and magnetic and electrostatic fields. The fictitious forces represent an adjustment, terms to be added to the left side of Eq. (9.4) to retain the form of Newton's second law, albeit somewhat artificially. These nonreal forces exist only because of the particular manner in which the dynamical equation is written. Note that if the XYZ-frame experiences constant linear velocity and no rotation ($\boldsymbol{\omega} = 0$, $\dot{\boldsymbol{\omega}} = 0$, $\ddot{\mathbf{R}} = 0$),

$$\mathbf{F} = m\mathbf{a}_{XYZ} \tag{9.6}$$

is valid, and XYZ is thus a secondary inertial frame of reference as discussed in Chapter 3.

The fictitious forces $-m\ddot{\mathbf{R}}$ and $-m\dot{\boldsymbol{\omega}} \times \boldsymbol{\rho}$ exist because of the linear and angular accelerations of XYZ with respect to $X'Y'Z'$ and are generally given no special names.

Of particular interest for reasons soon to be explored are $-2m\boldsymbol{\omega} \times \mathbf{v}_{XYZ}$, known as the *Coriolis* force, and $-m\boldsymbol{\omega} \times (\boldsymbol{\omega} \times \boldsymbol{\rho})$, the *centrifugal* force.

The Coriolis force is normal to the plane containing the angular velocity vector ($\boldsymbol{\omega}$ of XYZ) and the relative velocity $\mathbf{v}_{XYZ}$ in the direction of advance of a right-hand screw as $\mathbf{v}_{XYZ}$ is rotated in the direction of $\boldsymbol{\omega}$. This force is perpendicular to $\mathbf{v}_{XYZ}$, altering the motion of the particle as viewed from XYZ. The Coriolis force is not real to the extent that it does not exist in $X'Y'Z'$. When viewed from XYZ, however, its effects are readily observable. In the noninertial frame there must be a relative velocity $\mathbf{v}_{XYZ}$ for there to be a nonzero Coriolis force.

The centrifugal force (so called because of its literal meaning, "away from the center") is directed opposite the centripetal force. When viewed from the noninertial frame, therefore, this force appears to impel the particle outward from the axis of rotation.

9.3 THE EARTH-BASED NONINERTIAL FRAME

One of the simplest (and most important) formulations in noninertial coordinate frames involves a mass suspended above the earth's surface. Consider now such a case in which the XYZ-frame is attached to mass m as shown in Fig. 9.1. In this figure OX is east-directed, OY points to the north, and OZ represents an upward-directed vertical.

As for the inertial frame of reference to which motion of the earth-fixed XYZ-axes is referred, we begin by selecting $X''Y''Z''$, with origin at the center of the sun, as such a coordinate system, shown in Fig. 9.2. If the gravitational forces of the sun, moon, and the remainder of the universe are assumed negligible, the acceleration of the earth's center is zero ($\ddot{\mathbf{R}}'' = 0$). This is equivalent to stating that the nonrotating $X'Y'Z'$-system, with origin coincident with the center of the earth, is a secondary inertial frame of reference. The XYZ-system is noninertial because it experiences what we shall regard as significant rotation with respect to $X'Y'Z'$.

To justify our assumption regarding the acceleration of the origin of $X'Y'Z'$, consider the following. Since the angular speed of the earth about the sun is one revolution each 365 days, this contribution to the total angular speed of the earth,

$$\omega'' = 2.0 \times 10^{-7} \text{ rad/sec}$$

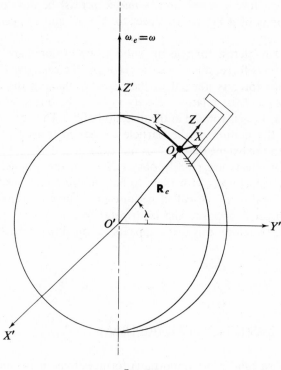

Fig. 9.1

The acceleration of O' relative to O'' is centripetal:

$$(\omega'')^2 R'' = 1.95 \times 10^{-2} \text{ ft/sec}^2$$

This represents less than one tenth of 1 percent of the gravitational acceleration.

The angular speed of the earth about an axis passing through the poles is one revolution per 24 hr:

$$\omega_e \cong 7.29 \times 10^{-5} \text{ rad/sec}$$

The acceleration of the origin of XYZ with respect to that of $X'Y'Z'$ has a maximum magnitude given by

$$\omega_e^2 R_e = 0.112 \text{ ft/sec}^2$$

which is approximately one third of 1 percent of the gravitational acceleration.

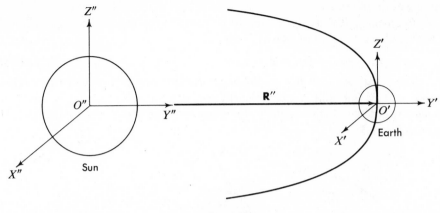

Fig. 9.2

The angular velocity of the earth is identical with the angular velocity of the XYZ-axes with respect to $X'Y'Z'$, since the noninertial system is attached to the mass, and the mass has zero motion relative to the earth. The angular acceleration of the earth and hence that of XYZ is approximately zero; this corresponds to $\dot{\boldsymbol{\omega}} = \mathbf{0}$ in Eq. (9.4). Thus the term $-m\dot{\boldsymbol{\omega}} \times \boldsymbol{\rho}$ in Eq. (9.4) is eliminated. The latter is also zero on the basis that $\boldsymbol{\rho}$, the vector from the origin of XYZ to the particle, is zero. For the same reason, the centrifugal force $-m\boldsymbol{\omega} \times (\boldsymbol{\omega} \times \boldsymbol{\rho})$ is zero. Since the mass experiences no velocity relative to XYZ, the Coriolis force is likewise zero.

What remains of Eq. (9.4) is

$$m\mathbf{a}_{XYZ} = \mathbf{F} - m\ddot{\mathbf{R}} = \boldsymbol{\tau} + \mathbf{F}_e - m\ddot{\mathbf{R}} \tag{9.7}$$

where all the applied forces have been included in $\mathbf{F}$. These are

$\boldsymbol{\tau}$ = supporting force of the string

$\mathbf{F}_e$ = gravitational pull of the earth

Other forces, such as the gravity of the sun and moon, have been neglected.

The term $\ddot{\mathbf{R}}$ is simply the acceleration of the origin of XYZ with respect to the inertial frame of reference $X'Y'Z'$, for which we have already computed a maximum value of 0.112 ft/sec^2. Since XYZ rotates at angular velocity $\boldsymbol{\omega}_e$ with respect to $X'Y'Z'$, $\ddot{\mathbf{R}}$ is the centripetal acceleration given by

$$\ddot{\mathbf{R}} = \boldsymbol{\omega}_e \times (\boldsymbol{\omega}_e \times \mathbf{R}_e) \tag{9.8}$$

Returning now to Eq. (9.7), recognizing that $\mathbf{a}_{XYZ} = \mathbf{0}$ since the mass is at rest with respect to earth,

$$\mathbf{0} = \boldsymbol{\tau} + \mathbf{F}_e - m\boldsymbol{\omega}_e \times (\boldsymbol{\omega}_e \times \mathbf{R}_e) \tag{9.9}$$

The real and fictitious forces in Eq. (9.9) add to zero, making it *appear* to be an equation of statics.

Recognizing the string tension to be equal to the negative of $m\mathbf{g}$, where $\mathbf{g}$ is the familiar "acceleration of gravity," we have

$$m\mathbf{g} = \mathbf{F}_e - m\boldsymbol{\omega}_e \times (\boldsymbol{\omega}_e \times \mathbf{R}_e) \tag{9.10}$$

Since $-\boldsymbol{\omega}_e \times (\boldsymbol{\omega}_e \times \mathbf{R}_e)$ is directed outward, normal to the earth's axis of rotation, and $\mathbf{F}_e$ is radial, $m\mathbf{g}$ can clearly not be a radial force; that is, $m\mathbf{g}$ is not directed toward the earth's center. The relationship of the forces in Eq. (9.10) is shown in Fig. 9.3.

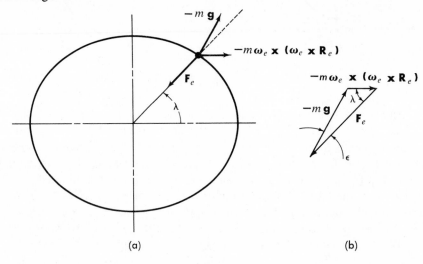

(a)   (b)

Fig. 9.3

It is of interest to recall that the earth is not a true sphere but an ellipsoid, flattened at the poles and bulging at the equator. As shown in Fig. 9.3(a), the "plumb-bob" vertical indicating the direction of $-m\mathbf{g}$ is actually normal to the earth's surface because of this departure from sphericity. It does not indicate, however, the *true* vertical directed toward the earth's center.

We shall now briefly explore the deviation of the "plumb-bob" vertical from the true vertical. Referring to Fig. 9.3(b) and applying the law of sines,

$$\frac{-m\omega_e^2 R_e \cos \lambda}{\sin \epsilon} = \frac{-mg}{\sin \lambda} \tag{9.11}$$

where $-m\omega_e^2 R_e \cos \lambda$ has been substituted for the cross-product.

Since we presume ϵ to be a small angle, $\sin \epsilon \cong \epsilon$:

$$\epsilon = \frac{\omega_e^2 R_e \cos \lambda \sin \lambda}{g} = \frac{\omega_e^2 R_e \sin 2\lambda}{2g} \tag{9.12}$$

The maximum deviation occurs at a latitude of 45 degrees, 17×10^{-4} rad.

9.4 PROJECTILE MOTION

The reader will recall that one of the important approximations employed in formulating the equations governing the general motion of a particle (Chapter 7) was that of a nonrotating earth. This approximation was tantamount to employing a coordinate system fixed at the earth's surface as an inertial frame of reference. We are now prepared to study the dynamics of a projectile over a rotating earth.

For an earth-fixed coordinate system with axes as directed in Fig. 9.4, the particle dynamics are described by

$$m(\ddot{x}\hat{i} + \ddot{y}\hat{j} + \ddot{z}\hat{k}) = \mathbf{F} - m\ddot{\mathbf{R}} - m\boldsymbol{\omega} \times (\boldsymbol{\omega} \times \boldsymbol{\rho})$$
$$- m\dot{\boldsymbol{\omega}} \times \boldsymbol{\rho} - 2m\boldsymbol{\omega} \times \mathbf{v}_{XYZ} \qquad (9.13)$$

where, unlike the case of the suspended mass, $\boldsymbol{\rho}$ is not zero. Note that the components of acceleration are parallel to the respective axes of the earth-fixed coordinate frame.

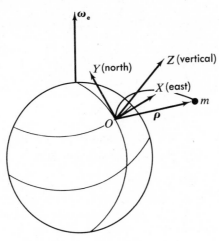

Fig. 9.4

The term $-m\ddot{\mathbf{R}}$, as before, equals $-m\boldsymbol{\omega} \times (\boldsymbol{\omega} \times \mathbf{R}_e)$, which, for distances relatively close to the earth, is very much larger than $-m\boldsymbol{\omega} \times (\boldsymbol{\omega} \times \boldsymbol{\rho})$. The latter is therefore neglected, as is $-\dot{\boldsymbol{\omega}} \times \boldsymbol{\rho}$. $\mathbf{F}$ consists of aerodynamic, propulsive, and gravitational forces. Of these three forces, if only the earth's gravitational pull is significant, we have

$$m\mathbf{a}_{XYZ} = (\mathbf{F}_e - m\ddot{\mathbf{R}}) - 2m\boldsymbol{\omega} \times \mathbf{v}_{XYZ} \qquad (9.14)$$

Again substituting $m\mathbf{g}$ for the term in parentheses above,

$$m\mathbf{a}_{XYZ} = m\mathbf{g} - 2m\boldsymbol{\omega} \times \mathbf{v}_{XYZ} \qquad (9.15)$$

and so the primary difference between Eq. (7.2) which ignores the earth's rotation, and Eq. (9.15) is the inclusion of the Coriolis force. This force must be considered when the time of flight is appreciable.

To evaluate the Coriolis force, we require $\omega = \omega_e$ and $\mathbf{v}_{XYZ}$. Writing in terms of components parallel to OX, OY, and OZ and unit vectors in XYZ,

$$\boldsymbol{\omega} = \boldsymbol{\omega}_e = (0)\hat{\mathbf{i}} + (\omega \cos \lambda)\hat{\mathbf{j}} + (\omega \sin \lambda)\hat{\mathbf{k}} \tag{9.16}$$

For $\mathbf{v}_{XYZ}$ we simply use

$$\mathbf{v}_{XYZ} = v_x\hat{\mathbf{i}} + v_y\hat{\mathbf{j}} + v_z\hat{\mathbf{k}} = \dot{x}\hat{\mathbf{i}} + \dot{y}\hat{\mathbf{j}} + \dot{z}\hat{\mathbf{k}} \tag{9.17}$$

The Coriolis force is therefore

$$(-2m)(\omega \cos \lambda \hat{\mathbf{j}} + \omega \sin \lambda \hat{\mathbf{k}}) \times (v_x\hat{\mathbf{i}} + v_y\hat{\mathbf{j}} + v_z\hat{\mathbf{k}}) \tag{9.18}$$

which when expanded is written

$$(2mv_y\omega \sin \lambda - 2mv_z\omega \cos \lambda)\hat{\mathbf{i}} - (2mv_x\omega \sin \lambda)\hat{\mathbf{j}} + (2mv_x\omega \cos \lambda)\hat{\mathbf{k}} \tag{9.19}$$

Substituting for the Coriolis force in Eq. (9.15) leads to the following scalar equations:

$$\ddot{x} = 2\omega(\dot{y} \sin \lambda - \dot{z} \cos \lambda) \tag{9.20}$$

$$\ddot{y} = -2\omega(\dot{x} \sin \lambda) \tag{9.21}$$

$$\ddot{z} = -g + 2\omega(\dot{x} \cos \lambda) \tag{9.22}$$

Integration of Eqs. (9.20), (9.21), and (9.22), employing the conditions that initially the velocity components are $\dot{x}_0$, $\dot{y}_0$, $\dot{z}_0$ and $x(0) = y(0) = z(0) = 0$,

$$\dot{x} = 2\omega(y \sin \lambda - z \cos \lambda) + \dot{x}_0 \tag{9.23}$$

$$\dot{y} = -2\omega(x \sin \lambda) + \dot{y}_0 \tag{9.24}$$

$$\dot{z} = -gt + 2\omega(x \cos \lambda) + \dot{z}_0 \tag{9.25}$$

It is implicit in the analysis that λ does not appreciably change during the flight.

Noting that the equation for $\ddot{x}$ contains $\dot{y}$ and $\dot{z}$ and that these are functions of the spatial variable x only in Eqs. (9.24) and (9.25), the X-acceleration may be obtained as a function of x alone by making the appropriate substitutions:

$$\ddot{x} = 2\omega[-2\omega(x \sin \lambda) + \dot{y}_0] \sin \lambda - 2\omega[-gt + 2\omega(x \cos \lambda) + \dot{z}_0] \cos \lambda$$

$$= -4\omega^2 x \sin^2 \lambda + 2\omega gt \cos \lambda + 2\omega \dot{y}_0 \sin \lambda - 2\omega \dot{z}_0 \cos \lambda - 4\omega^2 x \cos^2 \lambda \tag{9.26}$$

Dropping the terms in ω^2 as small in comparison with the others, Eq. (9.26), when integrated, becomes

$$\dot{x} = \omega gt^2 \cos \lambda + 2\omega t(\dot{y}_0 \sin \lambda - \dot{z}_0 \cos \lambda) + \dot{x}_0 \tag{9.27}$$

where $\dot{x}_0$ is the initial X-velocity, a constant of integration. Another integration yields the X-displacement as a function of time,

$$x = \tfrac{1}{3}\omega g t^3 \cos \lambda + \omega t^2 (\dot{y}_0 \sin \lambda - \dot{z}_0 \cos \lambda) + \dot{x}_0 t \qquad (9.28)$$

where the constant of integration is zero since it represents the initial X-position.

Again noting that Eqs. (9.24) and (9.25) for $\dot{y}$ and $\dot{z}$ depend upon only the space variable x, Eq. (9.28) may be substituted for x to obtain

$$\dot{y} = -2\omega \sin \lambda (\tfrac{1}{3}\omega g t^3 \cos \lambda + \omega t^2 \dot{y}_0 \sin \lambda - \omega t^2 \dot{z}_0 \cos \lambda + \dot{x}_0 t) + \dot{y}_0 \qquad (9.29)$$

$$\dot{z} = -gt + 2\omega \cos \lambda (\tfrac{1}{3}\omega g t^3 \cos \lambda + \omega t^2 \dot{y}_0 \sin \lambda - \omega t^2 \dot{z}_0 \cos \lambda + \dot{x}_0 t) + \dot{z}_0 \qquad (9.30)$$

Finally, again dropping terms in ω^2, Eqs. (9.29) and (9.30) are integrated, yielding

$$y = \dot{y}_0 t - \omega \dot{x}_0 t^2 \sin \lambda \qquad (9.31)$$

$$z = -\tfrac{1}{2}g t^2 + \omega \dot{x}_0 t^2 \cos \lambda + \dot{z}_0 t \qquad (9.32)$$

If a projectile is fired in an easterly direction (with an elevation), the initial components of velocity are $\dot{x}_0$, 0, $\dot{z}_0$ and the displacements are

$$x = \tfrac{1}{3}\omega g t^3 \cos \lambda - \dot{z}_0 \omega t^2 \cos \lambda + \dot{x}_0 t \qquad (9.33)$$

$$y = -\omega \dot{x}_0 t^2 \sin \lambda \qquad (9.34)$$

$$z = -\tfrac{1}{2}g t^2 + \omega \dot{x}_0 t^2 \cos \lambda + \dot{z}_0 t \qquad (9.35)$$

It is apparent from the above equations that the trajectory is not planar, as was the case in Chapter 7. We have only to drop the terms in ω, however, to reobtain the expressions derived for the nonrotating earth. Although there is no initial Y-velocity, a Y-displacement nevertheless occurs, owing to the rotation of the earth beneath the projectile. An observer at XYZ witnesses a drift of the projectile in the negative Y or southerly direction. Looking east, the mass thus drifts to the right. Similarly, a projectile with initial velocity components 0, $\dot{y}_0$, $\dot{z}_0$ will have an X-displacement:

$$x = \tfrac{1}{3}\omega g t^3 \cos \lambda + \omega t^2 (\dot{y}_0 \sin \lambda - \dot{z}_0 \cos \lambda) \qquad (9.36)$$

For large t or small $\dot{z}_0$, x is positive; the projectile drifts east (to the right) when fired in a northerly direction.

An object dropped with zero initial velocity, $\dot{x}_0 = \dot{y}_0 = \dot{z}_0 = 0$, will experience Z-displacement in accordance with the same relationship already developed for free fall, $z = \tfrac{1}{2}g t^2$. At the same time, there is again an easterly drift given by

$$x = \tfrac{1}{3}\omega g t^3 \cos \lambda \qquad (9.37)$$

From Eq. (3.35) it is clear that the time t required for an object dropped from rest to fall a distance $-h$ is

$$t_h = \left(\frac{2h}{g}\right)^{\frac{1}{2}} \tag{9.38}$$

Substituting t_h for t in Eq. (9.37) the easterly drift is now written

$$x = \tfrac{1}{3}\omega g \left(\frac{2h}{g}\right)^{3/2} \cos \lambda \tag{9.39}$$

From Eqs. (9.37) and (9.39) we observe the drift to be proportional to the cube of t_h and the distance of fall raised to the three-halves power. Note that the drift is zero at the poles.

Illustrative Example 9.1
A particle at rest is dropped from an altitude of 1000 ft. Determine the drift at the north pole, $\lambda = 45$ degrees, and the equator.

SOLUTION
The X- or easterly drift is given by Eq. (9.39). For $\omega = 7.29 \times 10^{-5}$ rad/sec, $g = 32.2$ ft/sec^2, and $h = 1000$ ft, we have: At the north pole ($\lambda = 90$ degrees), $x = 0$; at $\lambda = 45$ degrees, $x = 0.27$ ft or 3.2 in.; at the equator ($\lambda = 0$ degrees), $x = 0.38$ ft or 4.6 in.

In view of the relatively small drift, it is apparent that the influences of air currents and viscosity over the 7.9 sec of fall make experimental verification of the results obtained quite difficult.

9.5 CORIOLIS EFFECT UPON MOVING FLUID MASSES

The concepts discussed with reference to a projectile apply equally to a mass of fluid moving over the earth's surface, as, for example, air or water. In meteorological phenomena the relative importance of the Coriolis force is magnified by the long time periods over which it acts.

Consider an air mass as shown in Fig. 9.5, in which the pressure decreases in the direction of decreasing distance from the center. Since, in the absence of other effects fluids normally flow from regions of high to low pressure, the flow is initially radial in the direction of the dashed lines.

The Coriolis influence may readily be seen by substituting into the expression for the Coriolis force velocities of $\pm v_x \hat{\mathbf{i}}$ and $\pm v_y \hat{\mathbf{j}}$, and the earth's angular velocity $\boldsymbol{\omega} = \omega_y \hat{\mathbf{j}} + \omega_z \hat{\mathbf{k}}$. For an initial flow from the west (v_x positive), the Coriolis force causes a deflection toward the south. Similarly, the remaining flow paths as observed from a reference moving with the earth are curved as shown in Fig. 9.5. Note that the resultant flow is always to the right of the initial flow, regardless of the original direction.

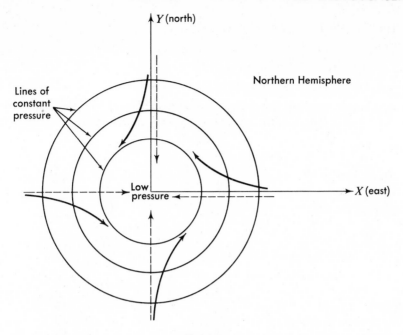

Fig. 9.5

The deflection of the fluid stream by the Coriolis force explains the counter-clockwise rotation of cyclones and whirlpools in the northern hemisphere. In the southern hemisphere the rotation is reversed. Another point of interest lies in the fact that the fluid tends to move tangent to lines of constant pressure, and not normal to such lines, as would be the case were the Coriolis force absent.

We have ignored the influence of fluid viscosity, which causes the actual motion to cross the isobars at a small angle when a steady state is achieved.

9.6 THE FOUCAULT PENDULUM [1]

The motion of a simple pendulum viewed from an earth-fixed frame of reference exhibits interesting features. Such a system is called a Foucault pendulum. Consider a mass suspended by a string, as in Section 9.3, undergoing relative velocity and acceleration with respect to the earth-fixed XYZ-axes. The only restriction on the motion of the pendulous mass is that the angle between the string and the vertical (Z-axis) remains small.

Equation (9.15) for the projectile applies here as well, provided a string tension τ is added:

$$m\mathbf{a}_{XYZ} = m\mathbf{g} + \tau - 2m\boldsymbol{\omega} \times \mathbf{v}_{XYZ} \tag{9.40}$$

---

[1] First proposed and built by J. L. Foucault in 1851. Analytic solution first performed by Haward, *Trans. Cambridge Phil. Soc.* **10** (1856).

The Coriolis force components are identical with those given by Eq. (9.19). The tension

$$\tau = \tau_x\hat{\mathbf{i}} + \tau_y\hat{\mathbf{j}} + \tau_z\hat{\mathbf{k}}$$
$$= \tau \cos(\tau, \hat{\mathbf{i}})\hat{\mathbf{i}} + \tau \cos(\tau, \hat{\mathbf{j}})\hat{\mathbf{j}} + \tau \cos(\tau, \hat{\mathbf{k}})\hat{\mathbf{k}} \qquad (9.41)$$

where the cosines represent the direction cosines between τ and the respective axes. Referring to Fig. 9.6, the cosines are readily determined:

$$\cos(\tau, \hat{\mathbf{i}}) = -\frac{x}{L}$$

$$\cos(\tau, \hat{\mathbf{j}}) = -\frac{y}{L}$$

$$\cos(\tau, \hat{\mathbf{k}}) = \frac{L-z}{L}$$

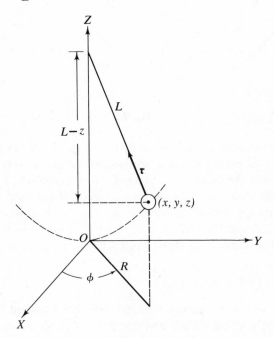

Fig. 9.6

Substituting for the direction cosines, Eq. (9.40) leads to three scalar equations in x, y, z, their derivatives, and τ:

$$m\ddot{x} = -\tau\frac{x}{L} + 2m\omega(\dot{y}\sin\lambda - \dot{z}\cos\lambda) \qquad (9.42)$$

$$m\ddot{y} = -\tau\frac{y}{L} - 2m\omega\dot{x}\sin\lambda \qquad (9.43)$$

$$m\ddot{z} = \tau\frac{L-z}{L} - mg + 2m\omega x\cos\lambda \qquad (9.44)$$

A fourth equation, relating x, y, and z, is the constraint equation stating that L is a constant.

Since it is the motion of the mass, rather than the variation of string tension which interests us most, it is desirable to eliminate τ between Eqs. (9.42) and (9.43). Assuming the Z-velocity to be quite small relative to $\dot{x}$ and $\dot{y}$, in view of the small-angle restriction, multiplication of Eq. (9.42) by y and Eq. (9.43) by $-x$ yields

$$my\ddot{x} = -\frac{\tau xy}{L} + 2my\dot{y}\omega \sin \lambda \tag{9.45}$$

and

$$-mx\ddot{y} = \frac{\tau xy}{L} + 2mx\dot{x}\omega \sin \lambda \tag{9.46}$$

which when added and rearranged gives (with m canceled)

$$(y\ddot{x} - x\ddot{y}) - \omega \sin \lambda(2x\dot{x} + 2y\dot{y}) = 0 \tag{9.47}$$

Equation (9.47) suggests the form

$$\frac{d}{dt}(y\dot{x} - x\dot{y}) - \omega \sin \lambda \frac{d}{dt}(x^2 + y^2) = 0 \tag{9.48}$$

the integral of which is

$$y\dot{x} - x\dot{y} - \omega \sin \lambda(x^2 + y^2) = A \tag{9.49}$$

where A is a constant of integration.

The tension may also be eliminated in Eqs. (9.42), (9.43), and (9.44) by employing a familiar technique; multiply the above equations by $\dot{x}$, $\dot{y}$, and $\dot{z}$, respectively:

$$m\ddot{x}\dot{x} = -\tau \frac{x}{L}\dot{x} + 2m\omega(\dot{y}\dot{x} \sin \lambda - \dot{z}\dot{x} \cos \lambda) \tag{9.50}$$

$$m\ddot{y}\dot{y} = -\tau \frac{y}{L}\dot{y} - 2m\omega(\dot{x}\dot{y} \sin \lambda) \tag{9.51}$$

$$m\ddot{z}\dot{z} = \tau \frac{L - z}{L}\dot{z} - mg\dot{z} + 2m\omega(\dot{x}\dot{z} \cos \lambda) \tag{9.52}$$

Adding we have

$$m(\ddot{x}\dot{x} + \ddot{y}\dot{y} + \ddot{z}\dot{z}) = -\frac{\tau}{L}[x\dot{x} + y\dot{y} - (L - z)\dot{z}] - mg\dot{z} \tag{9.53}$$

The bracketed term equals zero on the basis of the constraint equation

$$x^2 + y^2 + (L - z)^2 = L^2 \tag{9.54}$$

which when differentiated with respect to time yields

$$2x\dot{x} + 2y\dot{y} + 2(L - z)(-\dot{z}) = 0 \tag{9.55}$$

Again, a derivative form is suggested for Eq. (9.53), leading immediately to direct integration,

$$\tfrac{1}{2}m\frac{d}{dt}(\dot{x}^2 + \dot{y}^2 + \dot{z}^2) = -mg\dot{z} \tag{9.56}$$

and

$$\tfrac{1}{2}m(\dot{x}^2 + \dot{y}^2 + \dot{z}^2) + mgz = B' \tag{9.57}$$

where B' is the integration constant. Equation (9.57) should appear familiar to the reader. Subject to the approximations made, it represents the equation of conservation of mechanical energy. Expanding the constraint equation, Eq. (9.54),

$$x^2 + y^2 - 2Lz + z^2 = 0 \tag{9.58}$$

Dropping the z^2-term as a consequence of the small-angle approximation, and solving for z, we have

$$z = \frac{x^2 + y^2}{2L} \tag{9.59}$$

Again regarding $\dot{z}$ as negligible, substituting Eq. (9.59) for z, and canceling m, Eq. (9.57) becomes

$$\dot{x}^2 + \dot{y}^2 + \frac{g}{L}(x^2 + y^2) = B \tag{9.60}$$

where B is another constant.

Equation (9.60) may be easily solved by employing the polar form,

$$x = R\cos\phi \quad \text{and} \quad y = R\sin\phi$$

where R and ϕ are shown in Fig. 9.6.

Making the appropriate substitutions for the variables and their derivatives, we obtain

$$\dot{R}^2 + R^2\left(\dot{\phi}^2 + \frac{g}{L}\right) = B \tag{9.61}$$

Thus, if it can be demonstrated that

$$\frac{g}{L} \gg \dot{\phi}^2$$

we have

$$\dot{R}^2 + R^2\frac{g}{L} = B \tag{9.62}$$

which upon differentiation yields the familiar form describing simple harmonic motion of a pendulum:

$$\ddot{R} + \frac{g}{L}R = 0 \tag{9.63}$$

Expressing Eq. (9.49) in polar variables,

$$-\dot{\phi}R^2 - \omega R^2 \sin \lambda = A \qquad (9.64)$$

The constant of integration A may be evaluated from the initial condition: at $t = 0$, $x = y = 0$ $(R = 0)$, and therefore $A = 0$. This leaves

$$\dot{\phi} = -\omega \sin \lambda \qquad (9.65)$$

and therefore $\dot{\phi}^2$ is of the order of ω^2, negligible relative to g/L.

How are the results of the foregoing analysis to be interpreted? To be sure, Eq. (9.63) describes a pendulum, and R, the projection of L upon the XY-plane, varies sinusoidally in time. What can be said of the vertical plane of the motion? This motion is described by Eq. (9.65) or its integral,

$$\phi = -\omega t \sin \lambda \qquad (9.66)$$

where the constant of integration has been taken as zero.

Thus, if the mass is caused to swing in a plane at a pole of the earth $(\sin \lambda = 1)$, the plane would appear to rotate at $-\omega$ with respect to XYZ fixed to the pole. At a pole, therefore, the plane which contains the oscillation of the mass rotates approximately once each day. At locations other than the poles, the time for the plane of the oscillation to make one complete revolution, that is, the period, is obtained from Eq. (9.66) for $\phi = 2\pi$:

$$T = \frac{24 \text{ hr}}{\sin \lambda} \qquad (9.67)$$

At the equator, $\sin \lambda = 0$, and the period is infinite; that is, there is no apparent rotation of the plane of oscillation.

Once again, the phenomenon is but a creature of the reference. We of course know that the plane of the pendulum's motion at a pole must remain constant in space, because the pendulous mass is imparted no momentum perpendicular to the plane to alter its motion in inertial space.

Before leaving the Foucault pendulum, it is instructive to examine a different approach and in so doing gain a new insight into the Coriolis force. If we wish to apply the principle of torque and angular momentum to the pendulous mass, it is clear that unless both the real and the fictitious forces are employed, the origin of XYZ cannot be used as a point about which moments are taken. This is because XYZ is not an inertial frame of reference.

Considering only the vertical OZ-axis, neither the force of gravity nor the tension in the supporting cord contribute to moments of force about this line. The only "force" remaining, on the basis of the foregoing development, is the Coriolis force. The moment of this force about the origin of the accelerating XYZ-coordinate system is

$$\mathbf{M} = \mathbf{r} \times \mathbf{F} = (x\hat{\mathbf{i}} + y\hat{\mathbf{j}} + z\hat{\mathbf{k}}) \times (-2\boldsymbol{\omega} \times \mathbf{v}_{XYZ})$$

$$= (x\hat{\mathbf{i}} + y\hat{\mathbf{j}} + z\hat{\mathbf{k}}) \times [(2mv_y\omega \sin \lambda - 2mv_z\omega \cos \lambda)\hat{\mathbf{i}}$$

$$- (2mv_x\omega \sin \lambda)\hat{\mathbf{j}} + (2mv_x\omega \cos \lambda)\hat{\mathbf{k}}] \tag{9.68}$$

As before, assuming $v_z = \dot{z} = 0$, and carrying out the indicated operation for the Z-component of $\mathbf{M}$,

$$M_z = -2m\omega \sin \lambda(xv_x + yv_y) = -m\omega \sin\lambda\,(2x\dot{x} + 2y\dot{y}) \tag{9.69}$$

Since

$$\frac{d}{dt} R^2 = \frac{d}{dt}(x^2 + y^2) = 2x\dot{x} + 2y\dot{y} \tag{9.70}$$

one obtains

$$M_z = -m\omega \sin \lambda \frac{d}{dt} R^2 \tag{9.71}$$

Recall, however, that the moment of force is equal to the time rate of change of angular momentum. Employing polar coordinates,

$$M_z = \frac{d}{dt} mR^2\dot{\phi} \tag{9.72}$$

and therefore

$$-m\omega \sin \lambda \frac{d}{dt} R^2 = \frac{d}{dt} mR^2\dot{\phi} \tag{9.73}$$

The integral of Eq. (9.73) is

$$-m\omega \sin \lambda R^2 - mR^2\dot{\phi} = C' \tag{9.74}$$

or

$$R^2(-\dot{\phi} - \omega \sin \lambda) = C \tag{9.75}$$

which is identical with Eq. (9.64).

Now consider a new X-axis, say OX^*, located in accordance with Eq. (9.75); that is,

$$\dot{\phi}^* = -\dot{\phi} - \omega \sin \lambda \tag{9.76}$$

or

$$\phi^* = -\phi - \omega t \sin \lambda \tag{9.77}$$

Equation (9.74) may, in accordance with the above, be rewritten

$$mR^2\dot{\phi}^* = C' \tag{9.78}$$

and it is observed that the angular momentum is conserved in this new coordinate system, in which the vertical axis is as before but the X^*-axis rotates relative to X. The motion of the pendulum is thus entirely within the plane containing OZ and OX^*.

References

Becker, R. A., *Introduction to Theoretical Mechanics*, McGraw-Hill, Inc., New York, 1954.

Fowles, G. R., *Analytical Mechanics*, Holt, Rinehart and Winston, Inc., New York, 1962.

Goldstein, H., *Classical Mechanics*, Addison-Wesley Publishing Company, Inc., Reading, Mass., 1950.

Marion, J. B., *Classical Dynamics of Particles and Systems*, Academic Press, Inc., New York, 1965.

Osgood, W. F., *Mechanics*, The Macmillan Company, New York, 1956.

Symon, K. R., *Mechanics*, Second Edition, Addison-Wesley Publishing Company, Inc., Reading, Mass., 1960.

EXERCISES

9.1 Calculate the Coriolis force on a 500-lb shell traveling 1000 ft/sec in a north-easterly direction if it is fired at 40 degrees north latitude.

9.2 Estimate the Coriolis force on, and the transverse displacement of, a baseball in the midwestern United States.

9.3 A projectile is fired straight down from a height of 1000 ft at the equator with an initial velocity of 1500 ft/sec. Determine the landing point.

9.4 Calculate the Coriolis and centrifugal forces acting on a particle of mass m experiencing constant-speed motion u with respect to the rim of a wheel as shown in Fig. P9.4. Take the angular velocity ω of the wheel to be constant and refer motion to a frame of reference fixed to the wheel.

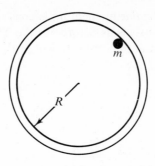

Fig. P9.4

9.5 The motion of a charged particle subject to either an electric or magnetic field has been discussed in connection with the cathode-ray tube and the cyclotron. Consider now a particle of charge q, mass m, and velocity v passing through an electric field of intensity $\mathbf{E}$ and at the same time a magnetic field of intensity $\mathbf{B}$. The equation of motion is

$$m\mathbf{a}_{X'Y'Z'} = q(\mathbf{E} + \mathbf{v} \times \mathbf{B})$$

where the term in parentheses is called the Lorentz force on a charged particle and where particle motion is referred to an inertial frame of reference. For the system described:

  a. Rewrite the equation of motion in a coordinate system rotating with constant angular velocity

$$\boldsymbol{\omega} = -\frac{q}{2m}\,\mathbf{B}$$

Thus the angular velocity of the rotating coordinate system is antiparallel to the direction of the magnetic field.

  b. Demonstrate that if the magnitude of the magnetic field intensity is small enough to justify dropping terms on the order of B^2, the equation of motion becomes

$$m\mathbf{a}_{XYZ} = q\mathbf{E}$$

where the motion is referred to the moving system. This expression represents Larmor's theorem, and the angular velocity referred to above is termed the Larmor precession.

9.6 A projectile is fired with an initial velocity $v_0 = 1000$ ft/sec relative to the earth and an angle $\alpha_0 = 45$ degrees. If the latitude at the firing point is 30 degrees south and the initial velocity vector is directed due north, determine the impact point.

9.7 Consider the motion of a 300-lb "horse" on a merry-go-round measured with respect to a frame of reference fixed to and rotating with the merry-go-round as shown in Fig. P9.7. The Z-position of the saddle is given by

$$z = 4 + \sin t \text{ ft}$$

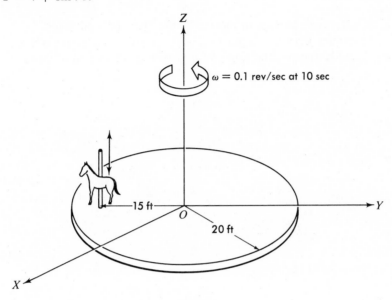

Fig. P9.7

Consider the earth as an inertial frame of reference.

    a. Determine the Coriolis force.

    b. What total force acts on the horse at the instant the angular velocity of the merry-go-round with respect to the earth is 0.05 rev/sec? Assume that the angular acceleration is uniform and that it takes 10 sec for the final angular velocity of 0.1 rev/sec to be achieved.

9.8 Discuss the forces acting on a stereophonic needle in terms of the frequency and amplitude of sound produced and the angular velocity of the turntable.

9.9 At what locations on the earth will the period of a Foucault pendulum be 2 days and 4 days?

9.10 If the period of a Foucault pendulum is 55 hr and the rotation of its plane of oscillation is counterclockwise when viewed from above, what is the latitude (north or south) of the pendulum?

9.11 A pendulous mass of 1000 g is suspended by means of a 5-m cord at a location where $\lambda = 40$ degrees north latitude. The maximum angle which the cord makes with the vertical is 10 degrees.

a. Determine the motion of the projection of the mass on the horizontal XY-plane.

b. What is the equation of motion of the plane of oscillation?

c. Determine the vertical position of the mass as a function of time.

d. Calculate the cord tension and direction as a function of time.

CHAPTER 10

SPECIAL RELATIVITY

10.1 INTRODUCTION

The measurements of Galileo and Brahe form much of the experimental basis for Newton's laws of motion. These postulates were thoroughly tested during the eighteenth and nineteenth centuries and were shown to apply to the mechanical systems then known. The theory thus far presented in this text develops logically from Newton's laws applied to a single particle.

Near the end of the nineteenth century, there were those who believed that the theoretical foundations of physics were well formulated and well understood. All that presumably remained was to obtain more precise values of the constants of physics. There were, of course, a few "minor" unsolved problems at that time, the solutions of which, as a matter of fact, were fundamental to the evolution of twentieth-century physics, and indeed have led to the reformulation of classical physics.

In 1864 Maxwell,[1] connecting the works of Gauss, Faraday, Ampère, and others, formulated what are now known as Maxwell's equations of electromagnetism. These classical laws remain in use today in much the same way as the laws of Newton. One of Maxwell's conclusions was that light is an electromagnetic wave, represented by orthogonal, periodically varying electric and magnetic fields. This was subsequently verified. Because all waves then known required a medium through which to travel (for example, water waves, sound waves in air and other media, waves on a vibrating string), Maxwell proposed that the *ether* satisfied similar requirements for light waves.

Soon after the ether was proposed, physicists began to plan experiments, some rather ingenious, to measure its properties. Since light was observed to travel at the same speed regardless of the source or its motion, the natural impli-

---

[1] Maxwell, J. C., *Phil. Trans. Roy. Soc. London* **155**, 459 (1865).

cation was that the ether which should pervade all space is at rest. In 1887 Michelson and Morley[2] performed an experiment, considered a landmark of physics, with the objective of measuring the velocity of the earth relative to the ether. The conclusion they reached, after a series of careful measurements, was that there is no velocity difference between the earth and the ether, or at least none which they could detect. To explain this *null* result, Fitzgerald[3] and Lorentz[4] proposed an ad hoc theory according to which the motion of the earth relative to the ether produces a compensatory effect, canceling the anticipated results. Their theory was judged unsatisfactory, and the world of physics was very unsettled.

In 1905 Einstein[5] proposed the theory of relativity, containing postulates that did not fit within the framework of classical physics. These are:

a. The laws of physics are the same in all inertial frames of reference.
b. The speed of light (in vacuo, that is, space devoid of matter, forces, or fields) is an invariant in all inertial frames of reference and independent of the motion of the source.

As corollaries of the postulates, we have:

a. No information or energy can be transmitted faster than the speed of light.
b. There is no way of measuring the motion of the ether.

Corollary b does not state that ether is nonexistent but implies rather that we are incapable of detecting motion relative to it. An extension of this idea has led to operational physics: Only those quantities which lend themselves to possible measurement are in the realm of physics. All others belong in metaphysics. Thus a discussion of the transmittance of signals faster than the speed of light is not properly within physics, according to corollary a. Should it be someday demonstrated that any of these postulates or corollaries has been violated, special relativity will then require reformulation. Until that time we shall continue to adhere to those principles which are successful in predicting natural phenomena.

10.2 CONSTANT-VELOCITY COORDINATE TRANSFORMATION

If two coordinate frames of reference experience constant velocity relative to one another, and one is an inertial system, the other is also an inertial system. As we have discussed at length, Newton's laws are valid only in such systems. Consider now the two coordinate frames shown in Fig. 10.1, in which the corresponding X, Y, Z and X', Y', Z' axes are parallel. The constant velocity of XYZ relative to $X'Y'Z'$ is directed parallel to Y. Visualization is facilitated if one regards the $X'Y'Z'$ system as fixed; this convention will be employed throughout this chapter.

[2] Michelson, A. A., and Morley, E. W., *Am. J. Sci.* **34,** 337 (1887).
[3] *Scientific Writings of G. F. Fitzgerald*, Dublin, 1902.
[4] Lorentz, H. A., *Versingen Zittingen Acad. Wetenschap. Amsterdam* **1,** 74 (1893).
[5] Einstein, A., *Ann. Physik* **17,** 891 (1905).

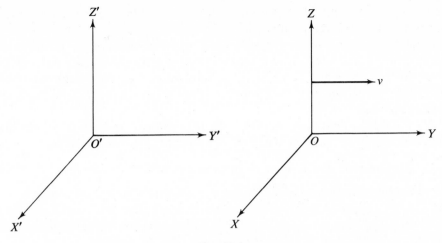

Fig. 10.1

Obviously, neither XYZ nor $X'Y'Z'$ need be fixed. The only limitation to which we must strictly adhere relates to the inertial nature of both frames of reference.

The Galilean transformation[6] equations connecting the position of a point in each system are

$$x = x' \tag{10.1}$$

$$y = y' - vt \tag{10.2}$$

$$z = z' \tag{10.3}$$

where the systems are coincident at time $t = 0$.

The inverse equations are

$$x' = x \tag{10.4}$$

$$y' = y + vt \tag{10.5}$$

$$z' = z \tag{10.6}$$

Since v is constant, the accelerations in either frame are identical:

$$\frac{d^2x}{dt^2} = \frac{d^2x'}{dt^2} \qquad \frac{d^2y}{dt^2} = \frac{d^2y'}{dt^2} \qquad \frac{d^2z}{dt^2} = \frac{d^2z'}{dt^2} \tag{10.7}$$

Thus the dynamical laws of motion are the same in either system provided velocity-dependent forces are not encountered. (For example, Maxwell's equations do not yield equal forces in both systems when the Galilean transformation is employed because the magnetic force acting on a charged particle is velocity-dependent.)

---

[6] Galileo, Galilei, *Dialogues Concerning Two New Sciences*, Translated by H. Crew and A. de Salvio, Dover Publications, Inc., New York, 1952.

Suppose a beam of light traveling at speed c is initiated in the positive Y-direction. An observer fixed in the XYZ-system measures c. An observer in $X'Y'Z'$ will, according to the Galilean transformation, measure $c + v$. According to Einstein, however, the primed observer must measure c, nothing more or less. Since the first prediction is incorrect, as substantiated by the Michelson–Morley experiment, we discard *for these purposes* the Galilean transformation and seek another approach.

Consider a light pulse starting from O at time $t = 0$ when O' is coincident with O as in Fig. 10.2(a). At time t the light will have traveled a distance

$$R = (x^2 + y^2 + z^2)^{\frac{1}{2}} = ct \tag{10.8a}$$

in the unprimed system.

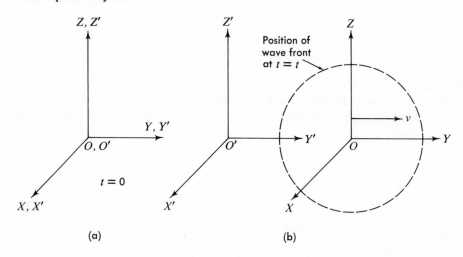

Fig. 10.2

Hence

$$x^2 + y^2 + z^2 - c^2 t^2 = 0 \tag{10.8b}$$

where ct represents the distance traveled by the light in the unprimed system.

In the primed system

$$R' = (x'^2 + y'^2 + z'^2)^{\frac{1}{2}} = ct \tag{10.9a}$$

or $\quad x'^2 + y'^2 + z'^2 - c^2 t^2 = 0 \tag{10.9b}$

Equating Eqs. (10.8b) and (10.9b),

$$x^2 + y^2 + z^2 - c^2 t^2 = x'^2 + y'^2 + z'^2 - c^2 t^2 \tag{10.10}$$

Introducing Eqs. (10.1) and (10.3), we are left with

$$y^2 = y'^2 \tag{10.11}$$

which is, of course, a contradiction of Eq. (10.2). We thus observe that the Galilean transformation is inconsistent with postulate b of special relativity. Einstein concluded that the times measured in XYZ and $X'Y'Z'$ are unequal, and Eq. (10.9) should read

$$x'^2 + y'^2 + z'^2 = c^2 t'^2 \tag{10.12}$$

where t' represents the time recorded in the primed system.

Our intuition tells us that time is independent of coordinate frame. It may thus seem difficult to conceive of two identical clocks, moving with constant relative velocity and recording different times. Since this astounding hypothesis leads to accurate predictions, we accept its validity. It is not too surprising that our intuition misleads, for such time effects are of significance only when the relative speed of the two clocks approaches that of light—rare indeed.

Subject to the foregoing discussion, Eq. (10.10) is now written

$$y^2 - c^2 t^2 = y'^2 - c^2 t'^2 \tag{10.13}$$

Any transformation from y to y' and from t to t' must embrace the concept that the speed of light is an invariant. There are conceivably many transformations which may succeed. Lorentz chose a linear transformation, because among other things it satisfies the principle of superposition for addition of velocities. More important, no contradictions of its results have been found.

To derive a set of transformation equations, we begin with the following expressions, which retain the form of Eq. (10.5):

$$y' = a_{11}y + a_{12}t \tag{10.14}$$

and

$$t' = a_{21}y + a_{22}t \tag{10.15}$$

Here t and t' are treated as coordinates, which in view of Eq. (10.8) appears quite reasonable. There are four constants requiring evaluation in Eqs. (10.14) and (10.15). Substitution of y' and t' into Eq. (10.13) yields

$$y^2 - c^2 t^2 = (a_{11}y + a_{12}t)^2 - c^2(a_{21}y + a_{22}t)^2 \tag{10.16}$$

Expanding the above expression, we have

$$y^2 - c^2 t^2 = a_{11}^2 y^2 + a_{12}^2 t^2 + 2a_{11}a_{12}yt - c^2 a_{21}^2 y^2 - c^2 a_{12}^2 t^2 - 2c^2 a_{21}a_{22}yt \tag{10.17}$$

Equating coefficients of y and t on both sides, since the above expression must hold for all y and t,

$$\text{For } y^2: \quad 1 = a_{11}^2 - c^2 a_{21}^2 \tag{10.18}$$

$$\text{For } yt: \quad 0 = 2a_{11}a_{12} - 2c^2 a_{21}a_{22} \tag{10.19}$$

$$\text{For } t^2: \quad -c^2 = a_{12}^2 - c^2 a_{22}^2 \tag{10.20}$$

There are four unknowns and as yet only three equations. To determine a fourth equation, consider the origin of the primed system as viewed from the unprimed system, Fig. 10.2(b):

$$a_{11}y + a_{12}t = 0 \qquad \text{for } y' = 0$$

where y represents the coordinate in the unprimed frame of the origin of the primed frame. The velocity of the origin of $X'Y'Z'$ as viewed from XYZ is, by differentiation of the above,

$$\frac{dy}{dt} = -\frac{a_{12}}{a_{11}} \tag{10.21}$$

but since $dy/dt = -v$, the velocity of $X'Y'Z'$ relative to XYZ, we have

$$v = \frac{a_{12}}{a_{11}} \tag{10.22}$$

Simultaneous solution of Eqs. (10.18), (10.19), (10.20), and (10.22) yields

$$a_{11} = \frac{1}{(1 - v^2/c^2)^{\frac{1}{2}}} \tag{10.23}$$

$$a_{12} = \frac{v}{(1 - v^2/c^2)^{\frac{1}{2}}} \tag{10.24}$$

$$a_{21} = \frac{v}{c^2} \frac{1}{(1 - v^2/c^2)^{\frac{1}{2}}} \tag{10.25}$$

$$a_{22} = \frac{1}{(1 - v^2/c^2)^{\frac{1}{2}}} \tag{10.26}$$

The transformation equations are therefore

$$y' = \frac{1}{(1 - v^2/c^2)^{\frac{1}{2}}} (y + vt) \tag{10.27}$$

$$t' = \frac{1}{(1 - v^2/c^2)^{\frac{1}{2}}} \left(\frac{yv}{c^2} + t\right) \tag{10.28}$$

Standard abbreviations employed are

$$\gamma = \left(1 - \frac{v^2}{c^2}\right)^{-\frac{1}{2}} = (1 - \beta^2)^{-\frac{1}{2}}$$

where $\beta = v/c$. In terms of these quantities,

$$y' = \gamma(y + vt) = \gamma(y + \beta ct) \tag{10.29}$$

$$t' = \gamma\left(\frac{vy}{c^2} + t\right) = \gamma\left(\frac{y\beta}{c} + t\right) \tag{10.30}$$

The inverse equations are

$$y = \gamma(y' - vt') = \gamma(y' - \beta ct') \tag{10.31}$$

$$t = \gamma\left(t' - \frac{y'v}{c^2}\right) = \gamma\left(t' - \frac{y'\beta}{c}\right) \tag{10.32}$$

The coordinates x and z are equal in each frame of reference, because the relative motion is along Y only. Therefore, $x = x'$, $z = z'$. Equations (10.29), (10.30), (10.31), and (10.32) are the famous Lorentz transformation equations. In the limit as v approaches zero, γ approaches unity and β, zero, and Eqs. (10.29) and (10.30) become

$$y' = y + vt$$

$$t' = t$$

It is thus not surprising to find that in the limiting case of zero velocity, the Lorentz and Galilean transformations yield identical results.

The speed of light, 3×10^{10} cm/sec, is extremely large when compared with the velocities of most macroscopic systems, so that a "small" velocity may be on the order of 10^8 cm/sec or 1 million mph. This is at least an order of magnitude greater than the fastest rockets of today. For the example cited, γ is roughly 1.000005 and β, 0.003.

It is possible to demonstrate that the transformation equations thus generated are unique, so that all other transformations should fail. Had, for example, a transcendental transformation been selected, the linear terms of a power-series expansion would have given the same results, and the coefficients of all the nonlinear terms would be zero.

10.3 LENGTH CONTRACTION

Applying the Lorentz transformation, one can predict the lengths of objects measured in coordinate systems which experience constant relative velocity. An observer at rest with respect to a meter stick measures its length to be L_0. What length will he measure if he moves with velocity v relative to the meter stick? Assume that the meter stick is at rest in the unprimed system and lies parallel to the Y-direction, one end at y_2 the other at y_1. When in XYZ he measures $L_0 = y_2 - y_1$. When the observer is in the primed system, he measures $L = y_2' - y_1'$, the observed length when the measurement is made at a single time t'. In order to compare L and L_0, the inverse transformations are applied:

$$y_2 = \gamma(y_2' - vt') \tag{10.33}$$

$$y_1 = \gamma(y_1' - vt') \tag{10.34}$$

Thus L and L_0 are related by

$$L_0 = \gamma(y_2' - y_1') = \gamma L \tag{10.35}$$

The measured length L therefore appears to be the original length divided by γ or

$$L = L_0 \left(1 - \frac{v^2}{c^2}\right)^{\frac{1}{2}}$$

(10.36)

Equation (10.36) states that the meter stick appears shortened in the direction of motion when the observer experiences motion relative to it. This contraction is precisely the amount necessary to explain the null result of the Michelson–Morley experiment.

The reader should demonstrate that placing the meter stick in either coordinate system results in identical contractions and that the sign of the relative velocity does not alter this result.

10.4 TIME DILATION

An observer measures a time interval T_0 by means of a clock with respect to which he experiences zero relative motion. What time interval T will he measure when he moves with velocity v relative to the clock? The time interval recorded by the clock fixed at point y_0' is $t_2' - t_1' = T_0$. The observer in the unprimed system measures $T = t_2 - t_1$. The times are related by the Lorentz transformation as follows:

$$t_2 = \gamma \left(t_2' - \frac{y_0'\beta}{c}\right)$$

(10.37)

$$t_1 = \gamma \left(t_1' - \frac{y_0'\beta}{c}\right)$$

(10.38)

Hence

$$T = t_2 - t_1 = \gamma(t_2' - t_1') = \gamma T_0 = \frac{T_0}{(1 - \beta^2)^{\frac{1}{2}}}$$

(10.39)

The observer moving relative to the clock records a longer time than one at rest relative to the clock, and therefore one speaks of *time dilation*.

10.5 ADDITION OF VELOCITY

Inherent in the Galilean transformation is the principle which permits the vector addition of velocity. When velocities approach that of light, it is apparent that the Lorentz transformation is appropriate. Consider a rocket ship traveling with velocity v relative to the earth, transmitting a beam of light (with velocity c) along its direction of motion as shown in Fig. 10.3. What velocity is recorded by a fixed observer on earth (assuming the earth is fixed)? According to Galileo, $c + v$; according to Einstein, c. Since $c + v$ is not measured whereas c is, we discard ordinary velocity addition and seek an approach consistent with special relativity.

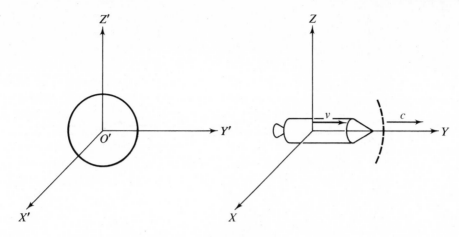

Fig. 10.3

To verify the velocity of the light beam as measured on earth, consider the differentials of y' and t', the position and time coordinates of the light as seen from the earth, the primed system.

Taking the differentials of Eqs. (10.29) and (10.30),

$$dy' = \gamma(dy + \beta c\, dt) \tag{10.40}$$

and

$$dt' = \gamma\left(\frac{\beta}{c}\, dy + dt\right) \tag{10.41}$$

Dividing,

$$\frac{dy'}{dt'} = \frac{dy + \beta c\, dt}{dt + (\beta\, dy/c)} = \frac{(dy/dt) + \beta c}{1 + (\beta/c)(dy/dt)} \tag{10.42}$$

Since $dy/dt = c$ and $\beta = v/c$,

$$\frac{dy'}{dt'} = \frac{c + v}{1 + (v/c)} = c \tag{10.43}$$

Thus $dy'/dt' = c$ is measured by an observer on earth whether v is positive or negative; the Lorentz transformation equations are indeed consistent with the invariance of the speed of light.

If instead of a ray of light, the rocket ship launches a projectile in the Y-direction at velocity u with respect to the rocket ship as in Fig. 10.4, what is the projectile velocity measured on earth? Following a procedure similar to that taken in the preceding development, Eq. (10.42)

$$\frac{dy'}{dt'} = \frac{(dy/dt) + \beta c}{1 + (\beta/c)(dy/dt)} = \frac{u + v}{1 + (uv/c^2)} \tag{10.44}$$

where dy/dt is now u rather than c.

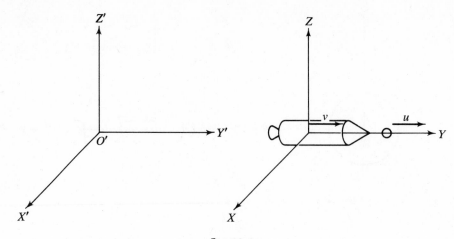

Fig. 10.4

Let us represent u and v as $k_1 c$ and $k_2 c$, respectively, where, according to the Einstein postulate, k_1 and k_2 must be less than unity. With this substitution, Eq. (10.44) becomes

$$\frac{dy'}{dt'} = \frac{k_1 c + k_2 c}{1 + k_1 k_2} = \frac{(k_1 + k_2)c}{1 + k_1 k_2} \tag{10.45}$$

To demonstrate that dy'/dt' is less than c (despite very large values of u and v) we must prove that

$$f(k_1, k_2) = \frac{k_1 + k_2}{1 + k_1 k_2} < 1 \tag{10.46}$$

This is equivalent to stating that $f(k_1, k_2)$ is less than unity for all values of k_1 and k_2 less than unity. Under what circumstances will $f(k_1, k_2)$ be a maximum (for if it is a maximum, it comes closest to unity)? For a function of more than one independent variable, it is a necessary test for maxima that the partial derivative with respect to each variable vanish. Thus

$$\frac{\partial f}{\partial k_1} = 0 \quad \text{and} \quad \frac{\partial f}{\partial k_2} = 0$$

or

$$\frac{\partial f}{\partial k_1} = \frac{(1 + k_1 k_2) - (k_1 + k_2)k_2}{(1 + k_1 k_2)^2} = 0 \tag{10.47}$$

and

$$\frac{\partial f}{\partial k_2} = \frac{(1 + k_1 k_2) - (k_1 + k_2)k_1}{(1 + k_1 k_2)^2} = 0 \tag{10.48}$$

Multiplying Eqs. (10.47) and (10.48) by $(1 + k_1 k_2)^2$ and expanding what remains,

$$1 + k_1 k_2 - k_1 k_2 - k_2^2 = 0$$

and

$$1 + k_1 k_2 - k_1 k_2 - k_1^2 = 0$$

or

$$1 - k_2^2 = 0$$

and

$$1 - k_1^2 = 0$$

Thus a maximum value of the function $f(k_1, k_2)$ may occur when $k_1 = k_2 = 1$, for which $f(1, 1) = 1$. If k_1 and k_2 are less than 1, $f(k_1, k_2) < 1$. We may conclude, therefore, that regardless of how fast the rocketship and its projectile are traveling, the relative velocity of projectile to earth is less than c, because both u and v are less than c. It is interesting to add that if either k_1 or k_2 equals unity, $f(k_1, k_2)$ is likewise unity.

At the other end of the velocity spectrum, for small u and v, Eq. (10.44) gives

$$\frac{dy'}{dt'} \cong v + u \tag{10.49}$$

which is the result of nonrelativistic addition of velocity.

Now consider the launching of a projectile at right angles to the rocketship as in Fig. 10.5. The velocity dy'/dt' now represents the rocketship motion only, inasmuch as in Eq. (10.44), u_y, the component of projectile velocity parallel to v, is now zero.

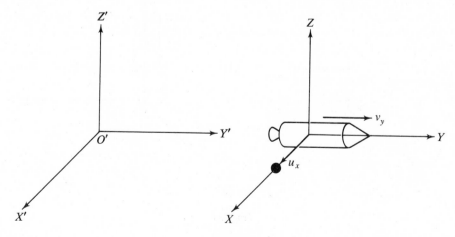

Fig. 10.5

The X-velocity of the projectile measured on the earth is

$$\frac{dx'}{dt'} = \frac{dx}{\gamma\left(\beta\frac{dy}{c} + dt\right)} = \frac{dx/dt}{\gamma\left(1 + \frac{\beta}{c}\frac{dy}{dt}\right)} \tag{10.50}$$

where $dx = dx'$ (since no relative motion occurs in the X-direction).

The above equation may be written

$$\frac{dx'}{dt'} = u_x\left(1 - \frac{v_y^2}{c^2}\right)^{\frac{1}{2}} \tag{10.51}$$

where β equals zero.

Here we have the interesting result that despite the fact that no magnitude change accompanies the transformation of dx', dt' changes as a result of the Lorentz transformation, and hence dx'/dt', the X-component of projectile velocity as measured from the earth, differs from the X-component as measured from the rocket.

10.6 THE VARIATION OF MASS WITH VELOCITY

Consider two observers, one fixed in system XYZ and the other fixed in $X'Y'Z'$, Fig. 10.6. The unprimed system experiences velocity v (in the Y-direction) relative to the primed system as shown. The observer in XYZ throws a mass m in the Z-direction with velocity $-V$ and the observer in $X'Y'Z'$ throws a mass m in the Z'-direction with velocity V. The objects are released so as to collide[7] along Z

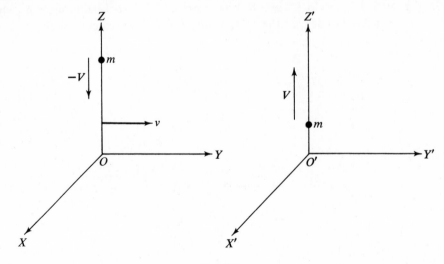

Fig. 10.6

---

[7] We are anticipating material of Chapter 11.

and Z' just as the origins coincide, as in Fig. 10.7. Before collision the observer in $X'Y'Z'$ sees his mass with components of velocity equal to zero along Y' and V along Z'. As seen from XYZ, the components of this velocity are $-v$ and $V(1 - v^2/c^2)^{1/2}$, respectively, the latter derived similarly to Eq. (10.51). The observer in XYZ sees his mass with components 0 and $-V$, whereas in $X'Y'Z'$ the components are v and $-V(1 - v^2/c^2)^{1/2}$. Each observer measures identical mass m when the objects possess velocity V. Because we have found that displacement, time, and velocity are different when measured in systems undergoing relative motion, it would be presumptuous to assume mass identical in two such systems. Thus we denote the mass measured and thrown in a given system by m, and that measured in one system but thrown in the other by m'. A collision between smooth spheres is assumed to occur along the line of centers. Neither mass suffers a change of momentum perpendicular to this line. Before the collision, the momentum of

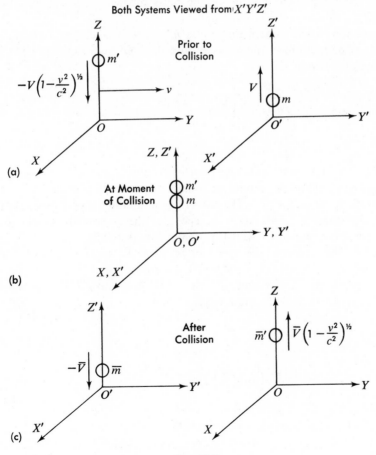

Fig. 10.7

the system of two particles along Z' measured in $X'Y'Z'$ is $mV - m'V(1 - v^2/c^2)^{1/2}$ and along Y', $m'v$.

After the collision, the observer in $X'Y'Z'$ must see his mass undergo the same change in momentum as the observer in XYZ observes for his, because for each observer, each mass undergoes an essentially identical experiment.

Conserving momentum in the Y'-direction,

$$m'v = \overline{m}'v \tag{10.52}$$

where $\overline{m}'$ represents the mass thrown in the unprimed system and viewed in $X'Y'Z'$ after the collision. Thus $m' = \overline{m}'$, and in a like manner $m = \overline{m}$.

Similarly, for the Z'-direction (as seen from $X'Y'Z'$),

$$mV - m'V\left(1 - \frac{v^2}{c^2}\right)^{1/2} = \overline{m}'\overline{V}\left(1 - \frac{v^2}{c^2}\right)^{1/2} - \overline{m}\overline{V} \tag{10.53}$$

where $-\overline{V}$ represents the velocity of the mass thrown in $X'Y'Z'$ as observed in $X'Y'Z'$ after collision.

Equation (10.52) states that mass m' remains unchanged throughout the collision. If one assumes that $m = m'$ and $\overline{m} = \overline{m}'$, as in classical physics, Eq. (10.53) gives the result

$$V\left[1 - \left(1 - \frac{v^2}{c^2}\right)^{1/2}\right] = \overline{V}\left[\left(1 - \frac{v^2}{c^2}\right)^{1/2} - 1\right] \tag{10.54}$$

or

$$\overline{V} = -V \tag{10.55}$$

But $\overline{V}$ and V represent *magnitudes* only and are both positive quantities. Thus the masses, m and m', cannot be equal if the concept of conservation of linear momentum is to be preserved. The reader has undoubtedly concluded, or recalled, however, that mass depends upon the speed of the mass with respect to the observer. It is this dependence which we shall soon ascertain.

Before the collision, the speed of m, $s_m = V$, as viewed in $X'Y'Z'$. The speed of m', $s_{m'} = [v^2 + V^2(1 - v^2/c^2)]^{1/2}$, as viewed in $X'Y'Z'$. Afterward, the respective speeds are $s_{\overline{m}} = \overline{V}$ and $s_{\overline{m}'} = [v^2 + \overline{V}^2(1 - v^2/c^2)]^{1/2}$.

Conservation of momentum requires that

$$\overline{m}'s_{\overline{m}'} = \overline{m}'\left[v^2 + \overline{V}^2\left(1 - \frac{v^2}{c^2}\right)\right]^{1/2} = m's_{m'}$$

$$= m'\left[v^2 + V^2\left(1 - \frac{v^2}{c^2}\right)\right]^{1/2} \tag{10.56}$$

We therefore conclude that

$$V = \overline{V} \tag{10.57}$$

having already ruled out the possibility that $V = -\overline{V}$. That is, the velocity mag-

nitudes along the direction of the collision are equal. Substituting Eq. (10.52) and
(10.57) in Eq. (10.53), we obtain

$$m' = \frac{m}{(1 - v^2/c^2)^{1/2}} \tag{10.58}$$

Thus the mass appears to increase as its speed increases. If V is small relative to
v, the mass which one measures in a given reference is essentially the rest mass and
it is clear that the principal velocity effect is attributable not to the velocity with
which the mass is thrown, but rather to that of the moving frame of reference.

Ample verification of this result exists. The first confirming experiment was
performed by Bucherer[8] in 1909. He measured the charge to mass ratio of the
electron and found it essentially constant at low velocities but decreasing more and
more rapidly as the velocity of the electron approached that of light. It is implicit,
of course, that the electronic charge remains constant.

10.7 RELATIVISTIC ENERGY

In terms of the linear momentum,

$$p = mv = \frac{m_0 v}{(1 - v^2/c^2)^{1/2}} \tag{10.59}$$

where m_0 represents the mass measured at zero velocity, the rest mass. Newton's
second law, in one dimension, in terms of the relativistic momentum, is thus

$$F = \dot{p} = \frac{d}{dt}\left[\frac{m_0 v}{(1 - v^2/c^2)^{1/2}}\right] \tag{10.60}$$

The work done by this force, on a particle initially at rest, is

$$W = \int_0^x F\, dx = \int_0^x \frac{d}{dt}\left[\frac{m_0 v}{(1 - v^2/c^2)^{1/2}}\right] dx \tag{10.61}$$

But $dx/dt = v$ and thus

$$W = \int_0^v v\, d\left[\frac{m_0 v}{(1 - v^2/c^2)^{1/2}}\right] \tag{10.62}$$

Integrating by parts,

$$W = \frac{m_0 v^2}{(1 - v^2/c^2)^{1/2}} - \int_0^v \frac{m_0 v\, dv}{(1 - v^2/c^2)^{1/2}}$$

$$= \frac{m_0 v^2}{(1 - v^2/c^2)^{1/2}} + m_0 c^2\left(1 - \frac{v^2}{c^2}\right)^{1/2} - m_0 c^2$$

$$= \Delta T = m_0 c^2\left[\frac{1}{(1 - v^2/c^2)^{1/2}} - 1\right] = (m - m_0)c^2 \tag{10.63}$$

---

[8] Bucherer, A. H., *Ann. Physik* **28**, 513 (1909).

Retaining the principle of work and energy, the work done is equated to the change in kinetic energy T of the mass. Attributing a *zero-speed* or *rest energy* to the particle, m_0c^2, the total energy of the mass is

$$E = mc^2 \tag{10.64}$$

Einstein predicted that rest energy could be converted into other forms, the most spectacular manifestation occurring in nuclear detonations.

Let us consider the kinetic energy for low velocities for which we anticipate unimportant relativistic effects. Expanding T in powers of v^2,

$$T = m_0c^2 \left[\frac{1}{(1 - v^2/c^2)^{1/2}} - 1 \right]$$

$$= m_0c^2 \left(1 + \frac{v^2}{2c^2} + \frac{3v^4}{8c^4} + \cdots - 1 \right) \tag{10.65}$$

or

$$T = \tfrac{1}{2} m_0 v^2 \qquad \text{for } v \ll c \tag{10.66}$$

This is, of course, the classical value of the kinetic energy, which is the low-velocity approximation to the relativistic kinetic energy.

10.8 APPLICATIONS OF RELATIVITY

When particles within a system interact in the absence of external forces, both the total energy and linear momentum are conserved. This result has been used to predict the fission of uranium by neutrons as well as the energy of decay of radioactive elements.

The laws of conservation of energy and linear momentum, more fully explored in Chapter 11, are now applied to demonstrate that when a single nucleus decays into three fragments, the energy of each is not uniquely determined, whereas the opposite is true in a two-fragment decay.

Consider the mass m at rest and not acted upon by any external forces, Fig. 10.8. When a decay occurs, resulting in particles of rest mass m_1, m_2, m_3, where $m_1 + m_2 + m_3 < m$, the vector sum of the momenta of m_1, m_2, m_3 must add to zero. Their kinetic energy, however, obtains from the conversion of rest mass to energy, and is not conserved.

Equating the rest energy before decay to the rest energy plus kinetic energy afterward,

$$mc^2 = \frac{m_1c^2}{(1 - v_1^2/c^2)^{1/2}} + \frac{m_2c^2}{(1 - v_2^2/c^2)^{1/2}} + \frac{m_3c^2}{(1 - v_3^2/c^2)^{1/2}} \tag{10.67}$$

where v_1, v_2, v_3 are the speeds of masses m_1, m_2, m_3. The momenta of m_1, m_2, m_3, in order to add to zero, must lie in a plane. The X- and Y-components of momentum from Fig. 10.8 are

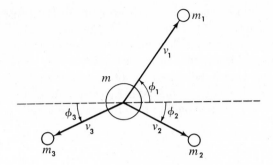

Fig. 10.8

$$0 = \frac{m_1 v_1 \cos \phi_1}{(1 - v_1^2/c^2)^{\frac{1}{2}}} + \frac{m_2 v_2 \cos \phi_2}{(1 - v_2^2/c^2)^{\frac{1}{2}}} + \frac{m_3 v_3 \cos \phi_3}{(1 - v_3^2/c^2)^{\frac{1}{2}}} \qquad (10.68)$$

$$0 = \frac{m_1 v_1 \sin \phi_1}{(1 - v_1^2/c^2)^{\frac{1}{2}}} + \frac{m_2 v_2 \sin \phi_2}{(1 - v_2^2/c^2)^{\frac{1}{2}}} + \frac{m_3 v_3 \sin \phi_3}{(1 - v_3^2/c^2)^{\frac{1}{2}}} \qquad (10.69)$$

From Eqs. (10.67), (10.68), and (10.69) there appear to be six unknowns, v_1, v_2, v_3, ϕ_1, ϕ_2, ϕ_3, but only three equations. There are really only five unknowns, however, because one angle is arbitrary. Nevertheless, the equations of relativity are insufficient to determine all the unknowns.

For the decay of m' into two masses, Fig. 10.9, the situation is different. Again, we apply conservation of energy,

$$m'c^2 = \frac{m_1' c^2}{(1 - v_1'^2/c^2)^{\frac{1}{2}}} + \frac{m_2' c^2}{(1 - v_2'^2/c^2)^{\frac{1}{2}}} \qquad (10.70)$$

The momentum of m_1' and m_2' must lie along the same line for these two vectors to add to zero. Thus

$$0 = \frac{m_1' v_1'}{(1 - v_1'^2/c^2)^{\frac{1}{2}}} - \frac{m_2' v_2'}{(1 - v_2'^2/c^2)^{\frac{1}{2}}}$$

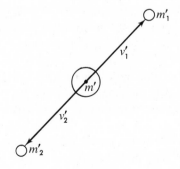

Fig. 10.9

There are only two unknowns in Eqs. (10.70) and (10.71); consequently, v'_1 and v'_2 may be uniquely determined. Thus a single particle, say U^{238}, decays into two parts, an alpha particle and Th^{234}, which always have the same energy. A beta emitter, say C^{14}, which decays into three parts, N^{14}, a beta particle, and a neutrino, does not always provide a beta particle of identical energy. Instead they are emitted in a range of energies from zero (corresponding to maximum neutrino energy) to a maximum (corresponding to zero neutrino kinetic energy).

References

Born, M., *Einstein's Theory of Relativity*, Revised Edition, Dover Publications, Inc., New York, 1962.

Eddington, A. S., *The Mathematical Theory of Relativity*, University Press, Cambridge, Mass., 1924.

Einstein, A., Lorentz, H. A., Minkowski, H., and Weyl, H., Translated by Perrett and Jeffery, *The Principle of Relativity*, Dover Publications, Inc., New York, 1952.

Goldstein, H., *Classical Mechanics*, Addison-Wesley Publishing Company, Inc., Reading, Mass., 1950.

EXERCISES

10.1 Referring to Sections 10.3 and 10.4, demonstrate that if a meter stick is placed in the primed system, an observer in the unprimed system measures L_0/γ, and that if a clock is placed in the unprimed system, an observer in the primed system measures γT_0.

10.2 The half-life of the pion is 2.5×10^{-8} sec. What is its half-life when its velocity is $0.9c$, $0.99c$, $0.999c$, $0.9999c$, and $0.99999c$?

10.3 A rocket ship travels from the earth to Alpha Centuri, 2.3 light-years away, with constant speed. What is the elapsed time for the trip, as measured from the earth and from the rocket ship, if the speed is $0.01c$, $0.1c$, $0.5c$, $0.9c$?

10.4 Two galaxies are observed to move away from the earth, one at $0.3c$, the other at $0.7c$. What is their relative velocity as measured from one galaxy if they are approaching or receding from one another along a straight line?

10.5 A particle has velocity components U_x, U_y, U_z. What are its velocity components in a frame of reference having a velocity V_y relative to the original frame of reference?

10.6 Light consists of photons, which, in vacuo, travel at c. Light sources emit photons of wavelength λ (and frequency ν) and their corresponding energy is $E = h\nu$, where h is Planck's constant (6.62×10^{-33} J-sec). Calculate the Doppler shift by considering the frequencies and wavelengths measured by an observer moving at constant velocity with respect to the source. Can you explain why, for light, only the relative motion between the source and observer is of consequence, whereas for sound the results depend upon whether it is the source or receiver which moves?

10.7 For a 200-BeV (2×10^{11} eV) proton accelerator, determine the momentum and speed of the protons.

10.8 Demonstrate that for a free particle, the total energy

$$E = (p^2c^2 + m_0^2c^4)^{\frac{1}{2}} = mc^2 = \frac{m_0c^2}{(1 - v^2/c^2)^{\frac{1}{2}}}$$

where

$$p = \frac{m_0v}{(1 - v^2/c^2)^{\frac{1}{2}}}$$

and that for $v \ll c$,

$$E = m_0c^2 + \frac{1}{2}\frac{p^2}{m_0}$$

10.9 A O^{16} nucleus spontaneously emits four alpha particles, two each along the X- and Y-axes. How much total energy is required of the oxygen nucleus if each alpha is to have 100 MeV of kinetic energy? What is the relative speed of an alpha particle with respect to each of the others? The masses are

$$\alpha = 6.636565 \times 10^{-27} \text{ kg}$$
$$O^{16} = 26.467141 \times 10^{-27} \text{ kg}$$

10.10 Charged particles move in circular motion, as a result of a magnetic field, according to the equation

$$F = qvB = \frac{mv^2}{R}$$

where $m = m_0/(1 - v^2/c^2)^{\frac{1}{2}}$. Determine the maximum energy a cosmic-ray proton may have if it is to be trapped within our galaxy, assumed a sphere of 300 light-year diameter, and possessing an average magnetic induction, $B = 10^{-9}\,\text{Wb/m}^2$. It is assumed that any cosmic rays found with higher energy come from outside our galaxy.

10.11 Referring to Exercise 10.10, what are the maximum electron and maximum alpha-particle energies?

10.12 The sun radiates energy at the rate of 2 ergs/g/sec, and the mass of the sun is approximately 2×10^{33} g. What is the mass loss of the sun per year? How long is required for the mass of the sun to decrease by 1 percent?

10.13 Show that the rest energy of an electron is equivalent to 0.511 MeV. One electron volt = 1.60×10^{-19} J and the electronic mass is 9.11×10^{-31} kg.

10.14 Compare the kinetic energies of an electron applying the following expressions: $\frac{1}{2}m_0v^2$, $\frac{1}{2}mv^2$, and $m_0c^2\{[1/(1 - v^2/c^2)^{1/2}] - 1\}$ when the speed of the electron is $0.001c$, $0.01c$, $0.1c$, $0.5c$, $0.9c$, $0.99c$, $0.9999c$, $0.999999c$.

10.15 Determine the speed of an electron having energies 1, 10, 10^2, 10^3, 10^4, 10^5, 10^6, 10^7, 10^8, 10^9 eV applying $\frac{1}{2}m_0v^2$, $\frac{1}{2}mv^2$, and $m_0c^2\{[1/(1 - v^2/c^2)^{1/2}] - 1\}$ and plot the results using a logarithmic scale.

CHAPTER 11

DYNAMICS OF A SYSTEM OF PARTICLES

11.1 INTRODUCTION

A system of particles is a collection of matter which exists within a prescribed boundary. The earlier chapters discussed various approaches and concepts for treating a special case, a single particle. These developments, extended in this chapter to multiparticle systems, provide the basis for the mechanics of rigid and deformable bodies.

11.2 THE CONCEPT OF A SYSTEM

A system of particles composed of four masses is shown in Fig. 11.1(a). Each external force represents the sum of forces acting because of influences located outside the boundary. The internal forces are of the gravitational type. They exist because of the mutual forces of attraction of the masses contained within the boundary. In Fig. 11.1(b) we have redrawn the boundary to exclude mass 4. As a consequence, the gravitational forces acting on masses 1, 2, and 3 due to mass 4 are no longer internal, and are added to the previously existing external forces to produce new external force resultants as shown.

Selecting a system boundary is similar to selecting a coordinate system. A judicious choice often facilitates analysis and solution but is not always simple or obvious.

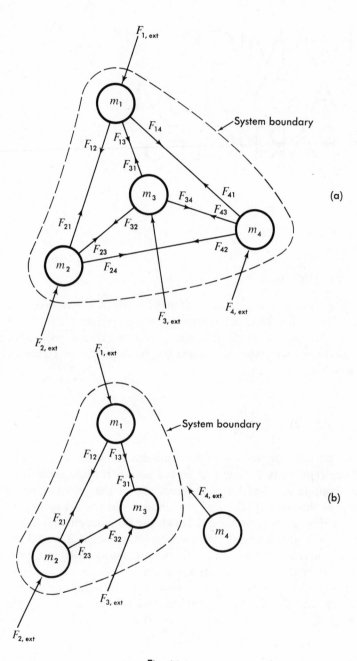

Fig. 11.1

11.3 NEWTON'S SECOND LAW

A system of particles comprised of masses m_1, $m_2 \cdots$, m_n is shown in Fig. 11.2. Newton's second law for particle 1 is

$$\mathbf{F}_1 = m_1\ddot{\mathbf{r}}_1 = \mathbf{F}_{1,\text{ext}} + \mathbf{F}_{1,\text{int}} \qquad (11.1)$$

where $\mathbf{F}_1$ = sum of all forces acting on particle 1

$\mathbf{F}_{1,\text{ext}}$ = sum of all forces acting on particle 1 due to influences external to the system

$\mathbf{F}_{1,\text{int}}$ = sum of all forces acting on particle 1 due to influences present within the system

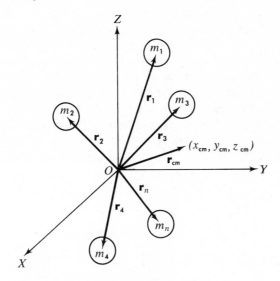

Fig. 11.2

The sum of the internal forces acting on mass 1 is

$$\mathbf{F}_{1,\text{int}} = \mathbf{F}_{12} + \mathbf{F}_{13} + \cdots + \mathbf{F}_{1n} \qquad (11.2)$$

where $\mathbf{F}_{12}$ is the force on particle 1 because of the presence of particle 2, and so on. For the ith particle,

$$\mathbf{F}_{i,\text{int}} = \sum_{j=1}^{n} \mathbf{F}_{ij} \qquad (i \neq j) \qquad (11.3)$$

For the entire system we have only to add an equation of the form of Eq. (11.3) for each particle to obtain the total effect of the internal forces:

$$\sum_{i=1}^{n} \mathbf{F}_{i,\text{int}} = \sum_{i=1}^{n} \sum_{j=1}^{n} \mathbf{F}_{ij} \qquad (i \neq j) \qquad (11.4)$$

Therefore, including both internal and external forces,

$$\sum_{i=1}^{n} \mathbf{F}_{i,\text{ext}} + \sum_{i=1}^{n} \sum_{j=1}^{n} \mathbf{F}_{ij} = \sum_{i=1}^{n} m_i \ddot{\mathbf{r}}_i \tag{11.5}$$

Since the internal forces act in accordance with Newton's third law, they cancel in pairs; that is, $\mathbf{F}_{ij} = -\mathbf{F}_{ji}$. Consequently, for the entire system, only the external forces remain,

$$\mathbf{F} = \sum_{i=1}^{n} \mathbf{F}_{i,\text{ext}} = \sum_{i=1}^{n} m_i \ddot{\mathbf{r}}_i \tag{11.6}$$

where $\mathbf{F}$ represents the sum of all forces acting on the system. Equation (11.6) may also be written

$$\mathbf{F} = \sum_{i=1}^{n} \frac{d^2}{dt^2}(m_i \mathbf{r}_i) = \frac{d^2}{dt^2} \sum_{i=1}^{n} m_i \mathbf{r}_i \tag{11.7}$$

where the order of differentiation and summation has been interchanged as permitted by the distributive nature of differentiation, and where $m_i \mathbf{r}_i$ is the first moment of mass of m: about the origin of the coordinate system, an inertial frame of reference. It is often useful to replace a multiparticle system with a single equivalent particle of equal mass. The question immediately arises as to where to locate this "single particle" with respect to the origin of XYZ, Fig. 11.2. It follows from Eq. (11.7) that the position is obtained by equating the first moment of the total system mass m to the sum of the individual moments of mass, since

$$\frac{d^2}{dt^2}(m\mathbf{r}_{\text{CM}}) = \frac{d^2}{dt^2} \sum_{i=1}^{n} m_i \mathbf{r}_i$$

That is,

$$m\mathbf{r}_{\text{CM}} = \sum_{i=1}^{n} m_i \mathbf{r}_i \tag{11.8}$$

where $\mathbf{r}_{\text{CM}}$ locates the *center of mass* with respect to the origin of XYZ. From Eq. (11.8),

$$\mathbf{r}_{\text{CM}} = \frac{m_1 \mathbf{r}_1 + m_2 \mathbf{r}_2 + \cdots + m_n \mathbf{r}_n}{m_1 + m_2 + \cdots + m_n} = \frac{\sum_{i=1}^{n} m_i \mathbf{r}_i}{\sum_{i=1}^{n} m_i} \tag{11.9}$$

Since $\mathbf{r}_{\text{CM}} = x_{\text{CM}}\hat{\mathbf{i}} + y_{\text{CM}}\hat{\mathbf{j}} + z_{\text{CM}}\hat{\mathbf{k}}$ ($x_{\text{CM}}, y_{\text{CM}}, z_{\text{CM}}$ are the coordinates of the center of mass), Eq. (11.9) leads to three scalar equations:

$$x_{\text{CM}} = \frac{\sum_{i=1}^{n} m_i x_i}{m} \qquad y_{\text{CM}} = \frac{\sum_{i=1}^{n} m_i y_i}{m} \qquad z_{\text{CM}} = \frac{\sum_{i=1}^{n} m_i z_i}{m} \tag{11.10}$$

On the basis of Eqs. (11.7) and (11.8) Newton's second law may be written in the very useful form

$$\mathbf{F} = \frac{d^2}{dt^2}(m\mathbf{r}_{CM}) = \frac{d}{dt}(m\mathbf{v}_{CM}) \qquad (11.11)$$

or

$$\mathbf{F} = m\ddot{\mathbf{r}}_{CM} = m\mathbf{a}_{CM} \qquad (11.12)$$

Equation (11.11) states that the sum of forces acting on a system of particles is equal to the time rate of change of linear momentum of the system, evaluated on the basis of the system mass m and the velocity of the mass center, $\mathbf{v}_{CM}$. For a system of constant mass, Eq. (11.12) states that the net force acting on the system equals the product of system mass and the acceleration of the mass center, $\mathbf{a}_{CM}$. If $\mathbf{F}$ represents the *external* force resultant, the validity of the foregoing statements depends upon the total cancellation of internal forces.

Our discussion of the center of mass has applied to a system of discrete masses. In assuming the mass of each discrete particle concentrated at its individual mass center, and in writing Newton's second law for each particle, we have already implied the existence of Eqs. (11.11) and (11.12). In other words, each discrete mass actually represents a spatial distribution of mass which has itself been replaced by a single mass concentrated at its mass center. While Eqs. (11.11) and (11.12) are extremely useful, they describe the motion of the mass center only, and yield no information concerning the motion of the individual particles with respect to either the XYZ-coordinate system or the center of mass.

Newton's second law as stated in Eqs. (11.11) and (11.12) is valid only when the motion of the mass center is determined with respect to an inertial frame of reference. However, the location of the mass center, relative to the system of particles, is a function of the mass distribution and not of the coordinate system. Thus, whatever coordinate reference is selected to determine $\mathbf{r}_{CM}$, the same point in space will always be defined, although the coordinates of the point depend upon the choice of reference.

11.4 ATWOOD'S MACHINE [1]

Atwood's machine is a mechanical system consisting of two masses connected by a cord passing over a pulley as depicted in Fig. 11.3(a). For the purposes of performing simple yet revealing analyses, Atwood's machine is here treated as a two-particle system characterized by the following important simplifications:

    a. The connecting cord is massless and inextensible.
    b. The pulley possesses no inertia and rotates on frictionless bearings. As a variation of the above, consider the cord as passing over a stationary

[1] Atwood, G., *On the Rectilinear Motion and Rotation of Bodies*, Cambridge, 1784.

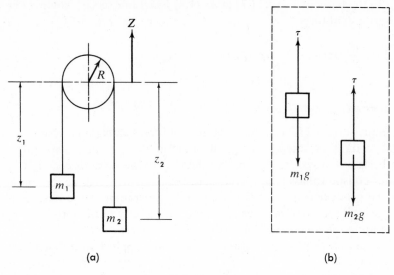

(a)   (b)

Fig. 11.3

peg, assuming a frictionless interface between cord and the surface over which it slides.

We first determine the acceleration of each mass by straightforward second-law analysis. From a free-body diagram [see Fig. 11.3(b)] we have

$$\tau - m_1 g = m_1 \ddot{z}_1 \tag{11.13a}$$

and

$$\tau - m_2 g = m_2 \ddot{z}_2 \tag{11.13b}$$

where τ is the cord tension, assumed constant throughout its entire length, and z_1, z_2 are the instantaneous distances from the center of the pulley to the respective masses.

As a direct consequence of cord inextensibility, only one coordinate is necessary to completely specify the system, the quantities z_1 and z_2 being related by the expression

$$z_1 = L - z_2 - \pi R \tag{11.14}$$

where L is the total cord length and R the pulley radius.

Differentiating the above constraint equation with respect to time, the following expressions relate the velocities and accelerations:

$$\dot{z}_1 = -\dot{z}_2 \quad \text{and} \quad \ddot{z}_1 = -\ddot{z}_2$$

If the masses were connected by a cord displaying spring-like characteristics, two independent variables, z_1 and z_2, would be required to describe the two-degree-of-freedom system.

Substitution of $\ddot{z}_2 = -\ddot{z}_1$ in Eq. (11.13b), subtracting Eq. (11.13b) from Eq. (11.13a), and solving for the acceleration of m_1 we have

$$\ddot{z}_1 = \frac{g(m_2 - m_1)}{m_1 + m_2} \tag{11.15}$$

For $m_2 > m_1$, $\ddot{z}_1 > 0$, that is, the acceleration of m_1 is upward; for $m_1 = m_2$, $\ddot{z}_1 = 0$; for $m_1 > m_2$, $\ddot{z}_1 < 0$. Note that the acceleration remains constant in all cases.

Having derived an expression for $\ddot{z}_1$, the tension τ is found by substitution of Eq. (11.15) into Eq. (11.13a):

$$\tau = m_1\ddot{z}_1 + m_1g \tag{11.16}$$

or

$$\tau = m_1g\frac{m_2 - m_1}{m_1 + m_2} + m_1g \tag{11.17}$$

Simplifying,

$$\tau = \frac{2m_1m_2g}{m_1 + m_2} \tag{11.18}$$

From the above, note that $m_1 = m_2$ leads to $\tau = m_1g = m_2g$, which is, of course, the equilibrium case. At the other extreme, if one mass is very much greater than the other, the effect is essentially a free-fall acceleration of the system of mass. If, for example, $m_1 \gg m_2$, $\ddot{z}_1 \cong -g$ and $\tau = 2m_2g$.

Despite the fact that one mass accelerates upward and one downward, as long as any mass unbalance exists, the net system acceleration must be downward. That this is so can be readily verified by considering the acceleration of the mass center. For the system of two masses, Fig. 11.3,

$$2\tau - m_1g - m_2g = (m_1 + m_2)\ddot{z}_{CM} \tag{11.19}$$

where $\ddot{z}_{CM}$ represents the acceleration of the mass center.

Substituting for the tension,

$$\ddot{z}_{CM} = g\left[\frac{4m_1m_2}{(m_1 + m_2)^2} - 1\right] = -g\frac{(m_1 - m_2)^2}{(m_1 + m_2)^2} \tag{11.20}$$

Thus all values of m_1 and m_2 (except $m_1 = m_2$) produce a negative acceleration of the mass center. For $m_1 = m_2$, $\ddot{z}_{CM} = 0$, since the net force acting upon the system of particles is zero.

Illustrative Example 11.1
Consider the system of particles m_1 and m_2 shown in Fig. 11.4(a). If the masses slide under the influence of gravity, derive an expression for the acceleration of the mass center along the incline (X-direction) in terms of m_1, m_2, θ, and the coefficients of sliding friction μ_1 and μ_2.

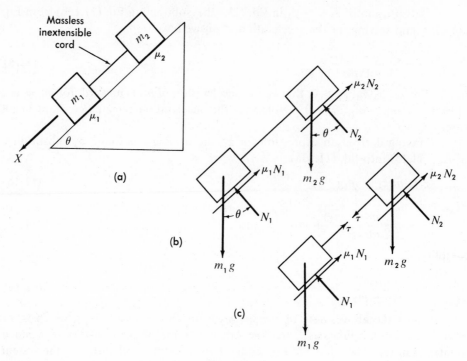

Fig. 11.4

SOLUTION

The sum of external forces along the incline, referring to the free-body diagram of Fig. 11.4(b), is equal to the product of system mass and the acceleration of the mass center (the internal forces cancel):

$$m_1g \sin \theta - \mu_1 m_1 g \cos \theta + m_2 g \sin \theta - \mu_2 m_2 g \cos \theta = (m_1 + m_2)a_{CM,x}$$

or

$$a_{CM,x} = \frac{(m_1 + m_2)g \sin \theta - (\mu_1 m_1 + \mu_2 m_2)g \cos \theta}{m_1 + m_2}$$

For the frictionless case, $\mu_1 = \mu_2 = 0$, and $a_{CM,x} = g \sin \theta$.

Let us now consider several numerical examples:

a. Calculate the acceleration of the mass center, the acceleration of each mass, and the cord tension if $m_1 = m_2 = 0.1$ kg, $\theta = 45$ degrees, $\mu_1 = 0.1$, and $\mu_2 = 0.2$.

b. Calculate the same quantities as in (a) with the data changed so that $\mu_1 = \mu_2 = 0.2$.

c. Calculate the same quantities as in (a) with $\mu_1 = 0.2$ and $\mu_2 = 0.1$.

SOLUTION (a)

A free-body diagram of each mass isolates the tensile force τ as shown in Fig. 11.4(c). For mass 1,

$$-\tau - \mu_1 m_1 g \cos \theta + m_1 g \sin \theta = m_1 a_{1x}$$

For mass 2,

$$\tau - \mu_2 m_2 g \cos \theta + m_2 g \sin \theta = m_2 a_{2x}$$

Since it is apparent that mass 2 tends to retard the motion of mass 1, the cord must therefore be in tension, and $a_{1x} = a_{2x}$. Simultaneous solution of the equations of motion gives $a_{1x} = a_{2x} = 5.9$ m/sec^2. The acceleration of the mass center should be the same as that of each mass, there being no motion relative to the mass center. Calculation based upon the expression for $a_{CM,x}$ yields $a_{CM,x} = 5.9$ m/sec^2.

SOLUTION (b)

If it is assumed that τ is tensile or zero, $a_{1x} = a_{2x}$, and solution of the above equations yields $\tau = 0$, $a_{1x} = a_{2x} = 5.5$ m/sec^2. Since the masses have no motion relative to the mass center, $a_{CM,x} = 5.5$ m/sec^2.

SOLUTION (c)

Clearly the cord cannot be in tension, because mass 2 moves faster than mass 1. Since the cord cannot sustain compression, we conclude that $\tau = 0$. Solution of the equations of motion yields $a_{1x} = 5.5$ m/sec^2 and $a_{2x} = 6.2$ m/sec^2. Thus mass 2 moves toward mass 1. The acceleration of the mass center, $a_{CM,x} = 5.9$ m/sec^2. It is observed, therefore, that both masses experience motion relative to the mass center and application of the expression for $a_{CM,x}$ alone can shed no light upon this relative motion.

11.5 ELECTRICAL CONDUCTIVITY

It is interesting to explore, by means of Newtonian mechanics, the a-c and d-c conductivities of a partially ionized gas. We begin with the defining equation for electrical conductivity σ (Ohm's law),[2]

$$\mathbf{J} = \sigma \mathbf{E} \tag{11.21}$$

where $\mathbf{E}$ is the electric field intensity and $\mathbf{J}$ the current density. The latter is the resultant of the various species of charges flowing, and may be expressed

$$\mathbf{J} = \sum_{i=1}^{n} n_i q_i \mathbf{v}_i \tag{11.22}$$

---

[2] Lorentz, H. A., *The Theory of Electrons and Its Applications to the Phenomena of Light and Radiant Heat*, Second Edition, Dover Publications, Inc., New York, 1909.

where n_i represents the number density of the ith species, and q_i and $\mathbf{v}_i$ are the charge and velocity associated with this species. We assume that because the electronic mass is so much smaller than that of the molecules, the impressed electric field results in only electron motion. Because of collisions between electrons and other particles, a "viscous" force $-m\nu_c\mathbf{v}$ exists wherein ν_c is the collision frequency and m is the electronic mass.

Newton's second law written for a single electron is

$$m\dot{\mathbf{v}} = (-e)\mathbf{E} - m\nu_c\mathbf{v} \qquad (11.23)$$

which for a constant electric field has the following steady-state solution for velocity:

$$\mathbf{v} = \left(\frac{-e}{m\nu_c}\right)\mathbf{E} \qquad (11.24)$$

The transient motion of the electron corresponds to the *source current*, that which is present without the electric field, resulting from initial motion. Returning to Eq. (11.22), and substituting for $\mathbf{v}$ (recalling that we are considering only one species, the electron),

$$\mathbf{J} = \frac{n(-e)(-e)\mathbf{E}}{m\nu_c} = \frac{ne^2\mathbf{E}}{m\nu_c} \qquad (11.25)$$

From Ohm's law,

$$\sigma_{\text{dc}} = \frac{ne^2}{m\nu_c} \qquad (11.26)$$

To obtain the a-c conductivity, let $\mathbf{E} = \mathbf{E}_0 e^{i\omega t}$ where ω is the impressed frequency and $\mathbf{E}_0$ the maximum field strength. The equation of motion is

$$m\dot{\mathbf{v}} = (-e)\mathbf{E}_0 e^{i\omega t} - m\nu_c\mathbf{v} \qquad (11.27)$$

for which the steady-state solution is

$$\mathbf{v} = \frac{-e\mathbf{E}_0 e^{i\omega t}}{m(\nu_c + i\omega)} = \frac{-e\mathbf{E}}{m(\nu_c + i\omega)} \qquad (11.28)$$

The a-c conductivity is, therefore,

$$\sigma_{\text{ac}} = \frac{ne^2}{m(\nu_c + i\omega)} \qquad (11.29)$$

To interpret the complex conductivity, we note that the real part is associated with energy dissipation, while the imaginary part plays the role of an inductor in an *RL* circuit.

11.6 OSCILLATIONS IN TWO-PARTICLE SYSTEMS

We now examine a frictionless system of two particles connected by linear springs as shown in Fig. 11.5. Since complete description of system position requires the location of each mass, $x_1(t)$ and $x_2(t)$, the system possesses two degrees of freedom. Implicit in limiting the total number of degrees of freedom is the existence of constraints, inasmuch as two unconstrained particles have six degrees of freedom $(x_1, y_1, z_1, x_2, y_2, z_2)$. The constraint equations state that y_1, z_1, y_2, z_2 are constants.

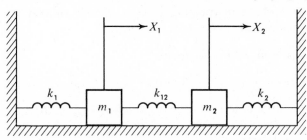

Fig. 11.5

Applying Newton's second law to each mass, we have

$$\sum F_{x1} = m_1\ddot{x}_1 \qquad -k_1x_1 - k_{12}(x_1 - x_2) = m_1\ddot{x}_1 \tag{11.30a}$$

$$\sum F_{x2} = m_2\ddot{x}_2 \qquad -k_2x_2 - k_{12}(x_2 - x_1) = m_2\ddot{x}_2 \tag{11.30b}$$

where $|x_1 - x_2|$ represents the deformation of the center spring.
Rearranging Eqs. (11.30),

$$m_1\ddot{x}_1 + (k_1 + k_{12})x_1 - k_{12}x_2 = 0 \tag{11.31a}$$

$$m_2\ddot{x}_2 + (k_2 + k_{12})x_2 - k_{12}x_1 = 0 \tag{11.31b}$$

We have already dealt at length with similar equations describing single-degree-of-freedom systems but which did not, of course, contain the terms associated with the elastic coupling of the masses.

The reader will recall that in Chapter 6 the solution $x = Be^{st}$ led eventually to solutions of the form

$$x = A \cos(\omega t + \theta) \tag{11.32}$$

Assuming solutions of this form for x_1 and x_2,

$$x_1 = A \cos(\omega t + \theta) \tag{11.33a}$$

$$x_2 = B \cos(\omega t + \theta) \tag{11.33b}$$

Examination of Eqs. (11.31) indicates that both ω and θ must be identical in x_1 and x_2 in order to satisfy these equations for all t.

Performing the indicated operations and substituting in Eqs. (11.31), we obtain

$$m_1[-A\omega^2 \cos(\omega t + \theta)] + (k_1 + k_{12})A \cos(\omega t + \theta) - k_{12}B \cos(\omega t + \theta) = 0$$

(11.34a)

$$m_2[-B\omega^2 \cos(\omega t + \theta)] + (k_2 + k_{12})B \cos(\omega t + \theta) - k_{12}A \cos(\omega t + \theta) = 0$$

(11.34b)

Canceling $\cos(\omega t + \theta)$ and rearranging leads to the following set of algebraic linear homogeneous equations in A and B,

$$(k_1 + k_{12} - m_1\omega^2)A - k_{12}B = 0$$

(11.35a)

$$-k_{12}A + (k_2 + k_{12} - m_2\omega^2)B = 0$$

(11.35b)

which may be rewritten in terms of the *amplitude ratio*, $\mu = B/A$,

$$\mu = \frac{B}{A} = \frac{k_1 + k_{12} - m_1\omega^2}{k_{12}}$$

(11.36a)

$$\mu = \frac{B}{A} = \frac{k_{12}}{k_2 + k_{12} - m_2\omega^2}$$

(11.36b)

Equating B/A above and multipying,l an expression in ω^4 is obtained. Alternatively, to obtain solutions for A and B, other than the trivial values $A = B = 0$, the determinant of the coefficients of A and B from Eqs. (11.35) must equal zero. In either case, the following *characteristic* or *frequency* equation is obtained:

$$\omega^4 - \left(\frac{k_1 + k_{12}}{m_1} + \frac{k_2 + k_{12}}{m_2}\right)\omega^2 + \frac{k_1k_2 + k_1k_{12} + k_2k_{12}}{m_1m_2} = 0$$

(11.37)

The above equation, being quadradic in ω^2, leads to

$$\omega^2 = \frac{1}{2}\left\{\left(\frac{k_1 + k_{12}}{m_1} + \frac{k_2 + k_{12}}{m_2}\right) \mp \left[\left(\frac{k_1 + k_{12}}{m_1} + \frac{k_2 + k_{12}}{m_2}\right)^2 \right.\right.$$
$$\left.\left. - 4\left(\frac{k_1k_2 + k_1k_{12} + k_2k_{12}}{m_1m_2}\right)\right]^{1/2}\right\}$$

(11.38)

The reader may verify by expansion that the above equation may be written

$$\omega^2 = \frac{1}{2}\left\{\left(\frac{k_1 + k_{12}}{m_1} + \frac{k_2 + k_{12}}{m_2}\right) \mp \left[\left(\frac{k_1 + k_{12}}{m_1} - \frac{k_2 + k_{12}}{m_2}\right)^2 + \frac{4k_{12}^2}{m_1m_2}\right]^{1/2}\right\}$$

(11.39)

In the form of Eq. (11.39), it is clear that the number under the square root is positive for all values of the constants, and consequently both values of ω^2 are real.

In Eq. (11.38),

$$\frac{k_1 + k_{12}}{m_1} + \frac{k_2 + k_{12}}{m_2} > \left[\left(\frac{k_1 + k_{12}}{m_1} + \frac{k_2 + k_{12}}{m_2} \right)^2 \right.$$
$$\left. - 4 \left(\frac{k_1 k_2 + k_1 k_{12} + k_2 k_{12}}{m_1 m_2} \right) \right]^{\frac{1}{2}}$$

and therefore ω^2 is positive as well. We are thus led to two values of ω^2, ω_1^2 and ω_2^2, and four real roots of ω^4: $\mp \omega_1$, $\mp \omega_2$. The negative frequencies do not provide solutions different from those given by the positive ω, and they may therefore be discarded. The lower of the two natural frequencies ω_1 is called the fundamental; ω_2 is the second natural frequency. Since ω_1 and ω_2 pertain to the motion of both masses, the complete solution may now be written

$$x_1 = A_1 \cos(\omega_1 t + \theta_1) + A_2 \cos(\omega_2 t + \theta_2) \tag{11.40a}$$

$$x_2 = B_1 \cos(\omega_1 t + \theta_1) + B_2 \cos(\omega_2 t + \theta_2) \tag{11.40b}$$

Although it appears that there are six constants, this number is reduced to four, since the amplitudes are related by Eqs. (11.36). If ω_1^2 and ω_2^2 are substituted for ω^2 in Eqs. (11.36), the following are obtained:

$$\mu_1 = \frac{B_1}{A_1} = \frac{k_1 + k_{12} - m_1 \omega_1^2}{k_{12}} = \frac{k_{12}}{k_2 + k_{12} - m_2 \omega_1^2} \tag{11.41a}$$

$$\mu_2 = \frac{B_2}{A_2} = \frac{k_1 + k_{12} - m_1 \omega_2^2}{k_{12}} = \frac{k_{12}}{k_2 + k_{12} - m_2 \omega_2^2} \tag{11.41b}$$

Finally, the equations of motion are, after substitution of Eqs. (11.41) into Eqs. (11.40):

$$x_1 = A_1 \cos(\omega_1 t + \theta_1) + A_2 \cos(\omega_2 t + \theta_2) \tag{11.42a}$$

$$x_2 = \mu_1 A_1 \cos(\omega_1 t + \theta_1) + \mu_2 A_2 \cos(\omega_2 t + \theta_2) \tag{11.42b}$$

where A_1, A_2, θ_1, θ_2 are the constants of integration to be evaluated by applying the initial conditions relating to the position and the velocity of m_1 and m_2. The constants are therefore functions of how the motion of the system is initiated rather than basic system parameters (the masses and the spring constants). In summary, a two-degree-of-freedom system is characterized by two natural frequencies and requires the knowledge of four initial conditions.

We have referred to a fundamental or lower natural frequency and a second natural frequency. If the motion is initiated in such a way that $A_2 = 0$, it is clear from Eqs. (11.42) that both masses oscillate at the same frequency, ω_1. It is left for the reader to verify that the amplitude ratio μ_1 is always positive; hence the motions of m_1 and m_2 are in phase as shown in Fig. 11.6. The foregoing describes the *fundamental mode* of vibration. Similarly, a second natural mode of vibration

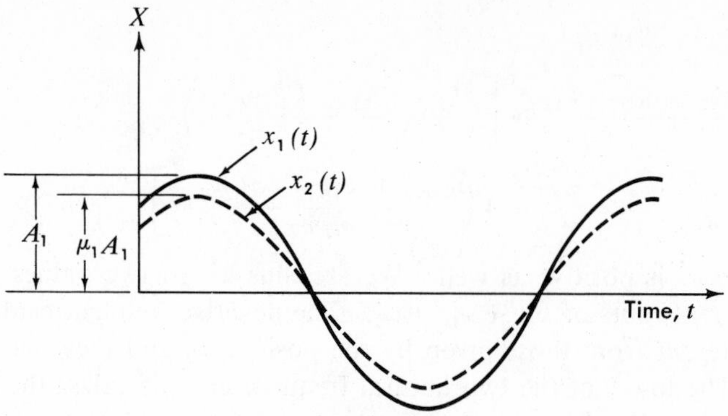

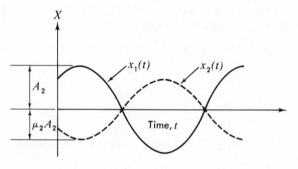

Fig. 11.6

exists in which the system oscillates at the higher frequency, ω_2. The reader may verify that μ_2 is always negative, and therefore second-mode vibration is distinguished by the motion of m_1 and m_2 out of phase by π rad, as shown in Fig. 11.7.

Fig. 11.7

In general, A_1 and A_2 are not zero, and the displacement of each mass is composed of the fundamental and second mode oscillations combined in such a way as to satisfy the initial conditions.

The foregoing analysis is now applied to two special and somewhat simpler two-mass systems to facilitate visualization of the motions. First consider a system in which the masses are equal and the end springs are of equal stiffness: $m_1 = m_2 = m$ and $k_1 = k_2 = k$. Substitution into Eq. (11.38) yields

$$\omega_1^2 = \frac{k}{m}$$

$$\omega_2^2 = \frac{k + 2k_{12}}{m}$$

These natural frequencies correspond to the amplitude ratios:

$$\mu_1 = \frac{k + k_{12} - m(k/m)}{k_{12}} = 1$$

$$\mu_2 = \frac{k + k_{12} - m[(k + 2k_{12})/m]}{k_{12}} = -1$$

The fundamental mode is thus described by

$$x_1 = A_1 \cos(\omega_1 t + \theta_1)$$

$$x_2 = A_1 \cos(\omega_1 t + \theta_1) \qquad \textbf{(11.43)}$$

Since the motion of each mass is identical, the masses moving in phase and having equal amplitudes, the deformation of the center spring, $x_1 - x_2$, is zero. This may also be deduced by performing a free-body analysis of the two-mass system. The sum of all *external* forces equals the product of system mass, $2m$, and the acceleration of the mass center $\ddot{x}_{CM}$. Since the center spring causes no external force to act on the system of two masses:

$$-kx_1 - kx_2 = 2m\ddot{x}_{CM}$$

or

$$-k[A_1 \cos(\omega_1 t + \theta_1) + A_1 \cos(\omega_1 t + \theta_1)] = 2m\ddot{x}_{CM}$$

Therefore,

$$\ddot{x}_{CM} = -\frac{k}{m} A_1 \cos(\omega_1 t + \theta_1) = -\omega_1^2 A_1 \cos(\omega_1 t + \theta_1)$$

which is identical with the first-mode acceleration of each mass. Consequently, each mass oscillates as though the other were absent, that is, at the individual natural frequencies $\omega_1 = (k/m)^{1/2}$.

In the second mode, the motion is

$$x_1 = A_2 \cos(\omega_2 t + \theta_2)$$

$$x_2 = -A_2 \cos(\omega_2 t + \theta_2) \qquad \textbf{(11.44)}$$

While the amplitudes are equal, the displacements are out of phase by π rad. The masses therefore display opposite motion with the deformation of the center spring given by

$$x_1 - x_2 = 2A_2 \cos(\omega_2 t + \theta_2)$$

The motion of the center of mass, which for the symmetrical second mode corresponds to the displacement of the midpoint of the center spring, may likewise be obtained from a free-body analysis,

$$-k_1 x_1 - k_2 x_2 = 2m\ddot{x}_{CM} = -k[A_2 \cos(\omega_2 t + \theta_2) - A_2 \cos(\omega_2 t + \theta_2)]$$

or

$$\ddot{x}_{CM} = 0$$

In view of the symmetry of the motion of m_1 and m_2, the midpoint of the center spring is stationary and is therefore a *nodal point*. Thus the second mode of oscillation may be regarded as the separate motion of two isolated systems, each of mass m, attached to rigid walls by springs of stiffness k and $2k_{12}$, as shown in Fig. 11.8. That halving the length of a spring is equivalent to doubling the spring constant is easily verified. Since the deformation of each half of the spring is exactly half the total spring distortion, and the spring force is the full value throughout the entire length of the spring,

$$k'_{12} = \frac{k_{12}(x_1 - x_2)}{\frac{1}{2}(x_1 - x_2)} = 2k_{12}$$

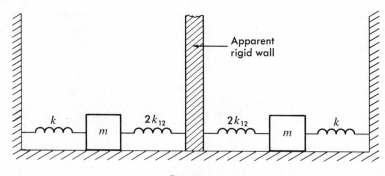

Fig. 11.8

Next we consider the case in which the masses are equal but the outer springs are not attached to the walls. Thus $m_1 = m_2 = m$ and $k_1 = k_2 = 0$. Applying Eq. (11.38), the natural frequencies are $\omega_1 = 0$ and $\omega_2 = (2k_{12}/m)^{1/2}$. Since the amplitude ratios are $\mu_1 = 1$ and $\mu_2 = -1$, the displacements are,

$$x_1 = A_1 \cos \theta_1 + A_2 \cos (\omega_2 t + \theta_2)$$

and

$$x_2 = A_1 \cos \theta_1 - A_2 \cos (\omega_2 t + \theta_2)$$

A zero natural frequency, although surprising at first, is to be expected on the basis of the lack of coupling between the masses and the rigid walls. The in-phase mode is thus characterized by zero frequency (infinite period), meaning that this motion is not oscillatory. A system exhibiting a zero natural frequency as above is termed *semidefinite*. It is clear that no external forces act on the system and, as a consequence, the acceleration of the mass center is zero. The system may therefore experience zero or constant velocity.

11.7 NORMAL COORDINATES

It is always possible, in dealing with a system of linear oscillators, to transform from the ordinary particle coordinates to another set which renders the equations of motion far less complicated in that each of these new coordinates possesses its distinctive frequency of oscillation. Such coordinates are the *normal coordinates* of a system of particles. In the example we next discuss, the normal coordinates are a simple linear function of $x_1(t)$ and $x_2(t)$. In general, the functional dependence of the normal coordinates upon the ordinary coordinates may require considerable algebraic manipulation to obtain.

Let us return to the system for which $m_1 = m_2 = m$ and $k_1 = k_2 = k$, and seek a set of coordinates ξ and η such that

$$\xi = C_1 \cos(\omega_1 t + \phi_1) \tag{11.45a}$$

$$\eta = C_2 \cos(\omega_2 t + \phi_2) \tag{11.45b}$$

These expressions are distinguished by a single frequency as well as a single mode of oscillation associated with each coordinate. The solution in terms of x_1 and x_2 for this special case has already been found. Adding the modes given by Eqs. (11.43) and (11.44), and recalling that $\mu_1 = 1$ and $\mu_2 = -1$:

$$x_1 = A_1 \cos(\omega_1 t + \theta_1) + A_2 \cos(\omega_2 t + \theta_2) \tag{11.46a}$$

$$x_2 = A_1 \cos(\omega_1 t + \theta_1) - A_2 \cos(\omega_2 t + \theta_2) \tag{11.46b}$$

Adding and subtracting Eqs. (11.46),

$$x_1 + x_2 = 2A_1 \cos(\omega_1 t + \theta_1) \tag{11.47a}$$

$$x_1 - x_2 = 2A_2 \cos(\omega_2 t + \theta_2) \tag{11.47b}$$

One is led to conclude that

$$\xi = \frac{x_1 + x_2}{2} \tag{11.48a}$$

$$\eta = \frac{x_1 - x_2}{2} \tag{11.48b}$$

or

$$x_1 = \xi + \eta \tag{11.49a}$$

$$x_2 = \xi - \eta \tag{11.49b}$$

What has been accomplished is to reduce the coupled physical system described in terms of x_1 and x_2 to two uncoupled systems described by ξ and η. This is readily observed by substituting Eqs. (11.49a) and (11.49b) into Eqs. (11.31) (for $m_1 = m_2 = m$ and $k_1 = k_2 = k$):

$$m(\ddot{\xi} + \ddot{\eta}) + (k + k_{12})(\xi + \eta) - k_{12}(\xi - \eta) = 0$$

$$m(\ddot{\xi} - \ddot{\eta}) + (k + k_{12})(\xi - \eta) - k_{12}(\xi + \eta) = 0$$

Adding and subtracting the above, we arrive at two differential equations, each expressed in terms of a single variable:

$$m\ddot{\xi} + k\xi = 0 \tag{11.50a}$$

$$m\ddot{\eta} + (k + 2k_{12})\eta = 0 \tag{11.50b}$$

The solutions of these separated equations (describing the equivalent of two single-degree-of-freedom systems of natural frequencies $(k/m)^{1/2}$ and $[(k + 2k_{12}/m)]^{1/2}$) are

$$\xi = C_1 \cos\left[\left(\frac{k}{m}\right)^{1/2} + \phi_1\right] \tag{11.51a}$$

$$\eta = C_2 \cos\left[\left(\frac{k + 2k_{12}}{m}\right)^{1/2} + \phi_2\right] \tag{11.51b}$$

These are, of course, identical with Eqs. (11.45). The use of ξ and η, the normal coordinates, thus leads to differential equations in which the variables are uncoupled; the equations are solved independently of one another. We should not be surprised at this result, for we have already seen that the general vibration may be constructed through the superposition of the principal modes, each of which represents an independent harmonic motion. The relationship between the normal coordinates and the principal modes in the foregoing example is clear.

Another advantage of the use of normal coordinates is related to considerations of energy. The effect is to simplify the form of the potential- and kinetic-energy functions, thereby reducing the complexity of the mathematical description of coupled multimass systems such as the atoms in a crystal lattice.

11.8 LINEAR MOMENTUM

The definition of linear momentum for a single particle may be extended to a system of particles by summation of the individual momenta,

$$\mathbf{p} = \sum_{i=1}^{n} \mathbf{p}_i = \sum_{i=1}^{n} m_i \dot{\mathbf{r}}_i = \sum_{i=1}^{n} m_i \mathbf{v}_i \tag{11.52}$$

where $\mathbf{p}$ is the total system linear momentum. To obtain the vector momentum, individual momenta within the system are added. During an explosion, for example, high-velocity particles are projected in all directions, and if the distribution favors no direction with regard to the number of particles or their speeds, one may expect the system momentum to total zero, despite the extremely high individual momenta. The foregoing description applies immediately after detonation. With the passage of time, the forces of gravity and air drag cause the net momentum to change from its zero value.

Equation (11.6) may be written in a more general form,

$$\mathbf{F} = \sum_{i=1}^{n} \frac{d}{dt}(m_i \mathbf{v}_i) = \sum_{i=1}^{n} \dot{\mathbf{p}}_i = \dot{\mathbf{p}} \qquad (11.53)$$

11.9 IMPULSE AND MOMENTUM

The relationship between impulse and the change of linear momentum for a system of particles is treated as an extension of the single-particle case. From Eq. (11.53),

$$\mathbf{F}\,dt = d\left[\sum_{i=1}^{n}(m_i \mathbf{v}_i)\right] \qquad (11.54)$$

and therefore

$$\underbrace{\int_{t_1}^{t_2} \mathbf{F}\,dt}_{\text{system impulse}} = \underbrace{\left(\sum_{i=1}^{n} m_i \mathbf{v}_i\right)_{t_2}}_{\substack{\text{system momentum}\\ \text{at } t_2}} - \underbrace{\left(\sum_{i=1}^{n} m_i \mathbf{v}_i\right)_{t_1}}_{\substack{\text{system momentum}\\ \text{at } t_1}} \qquad (11.55)$$

or

$$\int_{t_1}^{t_2} \mathbf{F}\,dt = (m\mathbf{v}_{\text{CM}})_{t_2} - (m\mathbf{v}_{\text{CM}})_{t_1} \qquad (11.56)$$

Equations (11.55) and (11.56) are vector expressions and may each be written as three independent scalar equations, as was done in the case of a single particle.

It is often useful to replace the general time-varying force $\mathbf{F}(t)$ with a constant average force $\overline{\mathbf{F}}$ acting throughout the entire time interval $\Delta t = t_2 - t_1$. This average force is given by

$$\overline{\mathbf{F}} = \frac{\sum_{i=1}^{n}(m_i \mathbf{v}_i)_2 - \sum_{i=1}^{n}(m_i \mathbf{v}_i)_1}{\Delta t} \qquad (11.57)$$

since

$$\overline{\mathbf{F}}\,\Delta t = \int_{t_1}^{t_2} \mathbf{F}\,dt$$

11.10 CONSERVATION OF LINEAR MOMENTUM

When the sum of all forces acting on a system of particles is zero, Eq. (11.53) shows that the system momentum must remain constant (since $\dot{\mathbf{p}} = \mathbf{0}$). This leads directly to the *principle of conservation of linear momentum: If the net force acting on a system of particles is zero, linear momentum is conserved.* The following example illustrates this principle for a two-particle system.

Illustrative Example 11.2

Two masses m_1 and m_2, initially at rest on a frictionless surface, are connected by a linear spring, as shown in Fig. 11.9(a). The spring is compressed an amount δ and released. Derive an expression for the velocity of each mass when the spring returns to its undeformed length L.

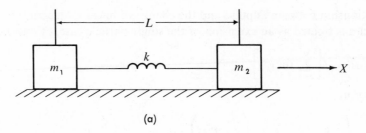

(a)

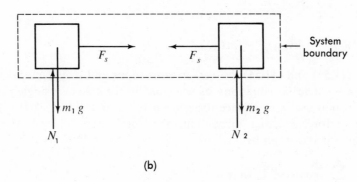

(b)

Fig. 11.9

SOLUTION

The system of particles after the spring is released is isolated in Fig. 11.9(b), in which the internal and external forces are shown. The internal forces (spring forces) cancel in pairs, and the sum of external forces (weights and normal reactions) is zero. The vector impulse is zero by virtue of the zero net force, and therefore linear momentum is conserved. Owing to the system constraints, only X-motion can occur. At any arbitrary time, the sum of the linear momenta must equal a constant, which in this case is zero, because the initial system momentum is zero,

$$m_1 \dot{x}_1 + m_2 \dot{x}_2 = 0 \qquad \dot{x}_1 = -\left(\frac{m_2}{m_1}\right)\dot{x}_2$$

Another equation is required since there are two unknowns, $\dot{x}_1$ and $\dot{x}_2$. In the absence of dissipative forces, mechanical energy is conserved and the system exchanges kinetic and elastic potential energy as discussed in Chapter 5. The total system energy at any time must equal the maximum elastic potential energy (obtained by initially compressing the spring, that is, doing work on the system),

$$E = \tfrac{1}{2}k\delta^2 = \tfrac{1}{2}m_1\dot{x}_1^2 + \tfrac{1}{2}m_2\dot{x}_2^2$$

where $\dot{x}_1$ and $\dot{x}_2$ represent the velocities corresponding to the undeformed state of the spring (maximum velocities). Simultaneous solution of the equations gives

$$\dot{x}_2 = \left[\frac{k\delta^2}{(m_2^2 + m_1m_2)/m_1}\right]^{\frac{1}{2}}$$

and

$$\dot{x}_1 = -\left(\frac{m_2}{m_1}\right)\left[\frac{k\delta^2}{(m_2^2 + m_1m_2)/m_1}\right]^{\frac{1}{2}} = -\left[\frac{k\delta^2}{(m_1^2 + m_1m_2)/m_2}\right]^{\frac{1}{2}}$$

11.11 VARIABLE-MASS SYSTEM

In treating the dynamics of a single particle and a system of particles, we have assumed individual particle mass to remain constant. This assumption was based upon the fact that in most applications, speeds are but a small fraction of the speed of light. Total system mass, however, need not remain constant. It may change through the process of mass transfer across a system boundary. This type of variable-mass system should be distinguished from one in which the mass of a single particle varies.

Consider the impulse–momentum relationship as it applies to a rocket propulsion system, Fig. 11.10. We examine the rocket and the expelled gases at times t and $t + \Delta t$, assuming all forces to act in the Y-direction and the gas force F_g and jet velocity v_j (velocity of jet relative to rocket) constant. Applying Eq. (11.55) to the gas:

$$\underbrace{-F_g\,\Delta t}_{} \quad = \quad \underbrace{\Delta m(v + \Delta v - v_j)}_{} \quad - \quad \underbrace{(\Delta m)v}_{} \qquad \text{(11.58)}$$

impulse due to gas force acting on expelled mass ($F_g \gg$ gravitational force on gas)	momentum at $t + \Delta t$ of expelled gases where $v + \Delta v - v_j$ is gas velocity relative to an inertial frame of reference	momentum at t of expelled gases (now part of fuel)

Thus

$$-F_g\,\Delta t = \Delta m\,\Delta v - v_j\,\Delta m \qquad \text{(11.59)}$$

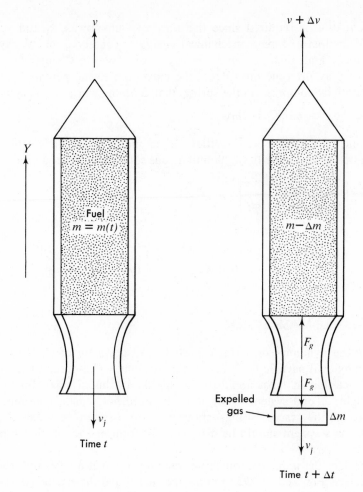

Fig. 11.10

where Δm represents the change in rocket mass and Δv the change in rocket velocity during the time interval Δt. The impulse–momentum equation for the rocket is

$$\underbrace{F_{\text{ext}}\,\Delta t + F_g\,\Delta t}_{\substack{\text{impulse due to gas} \\ \text{force acting on} \\ \text{rocket plus external} \\ Y\text{-force on rocket}}} = \underbrace{(m - \Delta m)(v + \Delta v)}_{\substack{\text{momentum of} \\ \text{rocket at } t + \Delta t}} - \underbrace{(m - \Delta m)v}_{\substack{\text{momentum of} \\ \text{rocket at } t}} \qquad \text{(11.60)}$$

or

$$F_{\text{ext}}\,\Delta t + F_g\,\Delta t = m\,\Delta v - \Delta m\,\Delta v \qquad \text{(11.61)}$$

Adding Eqs. (11.59) and (11.61),

$$F_{ext} \Delta t = m \Delta v - v_j \Delta m \qquad (11.62)$$

Dividing through by Δt and taking the limit as Δt approaches zero,

$$F_{ext} + v_j \frac{dm}{dt} = m \frac{dv}{dt} \qquad (11.63)$$

where $v_j(dm/dt)$ is often referred to as the rocket thrust. If the actual circumstances permit the assumption that $F_{ext} = 0$, as would be the case in outer space, Eq. (11.63) becomes

$$v_j \frac{dm}{dt} = m \frac{dv}{dt} \qquad (11.64)$$

The same result is obtained by observing that in the absence of *external* forces linear momentum is conserved:

$$\underbrace{mv}_{\substack{\text{momentum of rocket} \\ \text{and gas at } t}} = \underbrace{(m - \Delta m)(v + \Delta v)}_{\substack{\text{momentum of rocket} \\ \text{at } t + \Delta t}} + \underbrace{\Delta m(v + \Delta v - v_j)}_{\substack{\text{momentum of gas} \\ \text{at } t + \Delta t}} \quad (11.65)$$

11.12 IMPACT [3]

The principle of conservation of linear momentum applies when either no forces act or when the sum of forces acting is zero. The latter encompasses a large number of related situations involving impact or collision. The key to the solution of many of these problems is that during impact, the only significant forces acting on the system are due to impact and cancel, occurring as they do in equal and opposite pairs.

Consider the dynamics during impact of a system composed of two masses. As the masses contact, each deforms until maximum deformation occurs. The time interval from initial contact to maximum deformation is the *period of deformation*, and from maximum deformation to the point of just separating, the *period of restitution*. The total time of impact, deformation plus restitution, is quite short, usually measured in thousandths or millionths of a second. The description of the forces and motions during this period requires complex analytical concepts of solid-state physics, but fortunately the techniques described below relate only to states of the system immediately before and after material deformation. Since the formulation is still incomplete, much work in this area is experimental in nature.

---

[3] Goldsmith, W., *Impact*, Edward Arnold & Co., London, 1960.

11.13 DIRECT CENTRAL IMPACT

Examine the impact of the two masses shown in Fig. 11.11, limiting the collision to *central impact*, during which the mass centers lie on the line of impact. The classifications *direct* central impact and *oblique* central impact relate to whether or not the velocity vectors of the approaching masses lie along the line of impact as shown in Figs. 11.11 and 11.12.

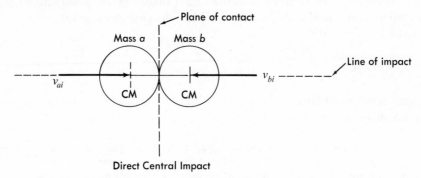

Direct Central Impact

Fig. 11.11

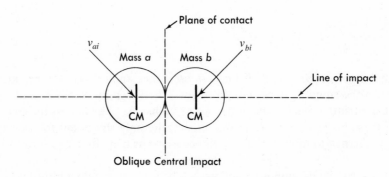

Oblique Central Impact

Fig. 11.12

Writing the impulse–momentum equation for masses a and b individually,

$$\int \mathbf{F}_a \, dt = m_a \mathbf{v}_{af} - m_a \mathbf{v}_{ai} \tag{11.66a}$$

$$\int \mathbf{F}_b \, dt = m_b \mathbf{v}_{bf} - m_b \mathbf{v}_{bi} \tag{11.66b}$$

where the subscripts i and f refer to the initial and final velocities.

If the masses are reasonably smooth, the final velocities also lie along the line of impact, and consequently the equations are one-dimensional.

Because $\int F_b\,dt = -\int F_a\,dt$, adding Eqs. (11.66a) and (11.66b) yields

$$m_a v_{af} - m_a v_{ai} + m_b v_{bf} - m_b v_{bi} = 0$$

or

$$\underbrace{m_a v_{ai} + m_b v_{bi}}_{\text{initial system momentum}} = \underbrace{m_a v_{af} + m_b v_{bf}}_{\text{final system momentum}} = \text{constant momentum} \qquad (11.67)$$

The impulse–momentum equations for masses a and b clearly demonstrate that momentum is not conserved for each mass but rather for the system of two masses.

Although we shall not delve into the complex internal processes that accompany a collision, it is of value to divide the interaction into the deformation and restitution periods. In so doing, use is made of the fact that at the time of maximum deformation, the masses possess no relative velocity. Stated differently, when each mass has suffered maximum deformation, each must possess the identical velocity: $v_a' = v_b' = v'$.

Applying the impulse–momentum equation to mass a for the deformation period, $t = 0$ to $t = t_d$, we have

$$\int_0^{t_d} F_a\,dt = m_a v' - m_a v_{ai} = I_{da} \qquad (11.68)$$

where I_{da} represents the linear impulse of mass a during deformation.

Similarly, for mass b,

$$\int_0^{t_d} F_b\,dt = m_b v' - m_b v_{bi} = I_{db} = -I_{da} \qquad (11.69)$$

Solving Eqs. (11.68) and (11.69) for v_{ai} and v_{bi}, respectively, we have

$$v_{ai} = \frac{-I_{da}}{m_a} + v' \qquad (11.70a)$$

$$v_{bi} = \frac{I_{da}}{m_b} + v' \qquad (11.70b)$$

The difference in the initial velocities of masses a and b may therefore be expressed in terms of the impulse during deformation and the masses, as follows:

$$v_{bi} - v_{ai} = I_{da}\left(\frac{1}{m_a} + \frac{1}{m_b}\right) = I_{da}\frac{m_a + m_b}{m_a m_b} \qquad (11.71)$$

The period of restitution begins at time t_d and ends at time t_r. Each mass possesses velocity v' at the beginning of this period and individual velocities v_{af} and v_{bf} at its conclusion. For mass a,

$$\int_{t_d}^{t_r} F_a\,dt = m_a v_{af} - m_a v' = I_{ra} \qquad (11.72)$$

where I_{ra} represents the impulse during restitution for mass a. For mass b during restitution,

$$\int_{t_d}^{t_r} F_b \, dt = m_b v_{bf} - m_b v' = I_{rb} = -I_{ra} \tag{11.73}$$

The final velocities, obtained from Eqs. (11.72) and (11.73), are thus

$$v_{af} = \frac{I_{ra}}{m_a} + v' \tag{11.74a}$$

$$v_{bf} = \frac{-I_{ra}}{m_b} + v' \tag{11.74b}$$

and their difference is

$$v_{bf} - v_{af} = -I_{ra}\left(\frac{1}{m_a} + \frac{1}{m_b}\right) = -I_{ra}\frac{m_a + m_b}{m_a m_b} \tag{11.75}$$

The relationship between I_{da}, the impulse associated with deformation, and I_{ra}, the restitution impulse, depends upon many factors such as geometry, material properties, and velocity, and one must usually be content with an experimental determination. These quantities are generally connected by the expression,

$$e = \frac{I_{ra}}{I_{da}} \tag{11.76}$$

where e, the *coefficient of restitution*, is experimentally determined and may be regarded as constant within a range of parameters reasonably close to those of the experiment. It embraces, in a deceptively simple manner, a multitude of complex phenomena and should be regarded with the same attitude as the coefficient of friction. Dividing Eq. (11.75) by Eq. (11.71) yields

$$-\frac{v_{bf} - v_{af}}{v_{bi} - v_{ai}} = \frac{I_{ra}}{I_{da}} \tag{11.77}$$

or, in view of Eq. (11.76),

$$e = -\frac{v_{bf} - v_{af}}{v_{bi} - v_{ai}} = \frac{-(\text{velocity of separation})}{\text{velocity of approach}} \tag{11.78}$$

It is clear from the above expression that $e \geq 0$.

For the special case in which there are no losses in mechanical energy associated with impact, the system kinetic energy before and after remains constant, and the collision is termed *perfectly elastic*. For this case,

$$\tfrac{1}{2}m_a v_{ai}^2 + \tfrac{1}{2}m_b v_{bi}^2 = \tfrac{1}{2}m_a v_{af}^2 + \tfrac{1}{2}m_b v_{bf}^2 \tag{11.79}$$

After algebraic manipulation, Eq. (11.79) is written

$$m_a(v_{ai} + v_{af})(v_{ai} - v_{af}) = m_b(v_{bf} + v_{bi})(v_{bf} - v_{bi}) \qquad (11.80)$$

Dividing Eq. (11.80) by Eq. (11.67),

$$v_{ai} + v_{af} = v_{bf} + v_{bi}$$

or

$$v_{bf} - v_{af} = -(v_{bi} - v_{ai}) \qquad (11.81)$$

The ratio, $-(v_{bf} - v_{af})/(v_{bi} - v_{ai})$, is, of course, the coefficient of restitution, equal to unity for a perfectly elastic collision.

For direct central impact other than perfectly elastic, $e < 1$; there is a loss of translational energy associated with impact, and therefore an energy-loss term must be added to the right side of Eq. (11.79). If the premise that system forces total zero is valid, conservation of momentum applies for all collisions, whether or not kinetic energy is conserved. In collision problems of the type discussed, the two final velocities are usually unknown and two independent equations are therefore required. One such equation is provided by momentum considerations. For a perfectly elastic collision, the second equation is provided by either the equation describing energy conservation, Eq. (11.79) or its equivalent, Eq. (11.78), with $e = 1$. For cases in which e lies between zero and unity, Eq. (11.78) is used.

When the impact is perfectly inelastic (perfectly plastic), the velocity of separation is zero; the two masses possess the same final velocity. The impact of two pieces of putty or soft clay, in which they stick together, is an example of perfectly inelastic impact. For this case $e = 0$, as can be confirmed by substituting $v_{bf} = v_{af}$ into Eq. (11.78).

11.14 OBLIQUE CENTRAL IMPACT

Although the approach velocities in general each make an angle with the line of impact, Fig. 11.12, the impact forces lie entirely along this line, and ideally the impact has no effect upon those components of velocity normal to the line of impact. Clearly some information must be available with regard to the orientation of the line of impact. If the masses are smooth, the same coefficient of restitution may be used for oblique central impact, as for direct central impact.

Illustrative Example 11.3
A mass m strikes a smooth flat surface as shown in Fig. 11.13. If the coefficient of restitution is e, determine the angle of rebound ϕ, the final velocity v_f, the energy loss per impact in terms of initial velocity v_i and θ.

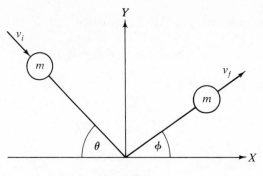

Fig. 11.13

SOLUTION

Only the vertical component of velocity is affected by the collision, because the impact forces are assumed entirely vertical, $v_{ix} = v_{fx}$. Applying Eq. (11.78) and assuming the velocity of the flat surface (with which is associated a very large mass) to be zero before and after impact,

$$e = -\frac{v_{fy} - 0}{v_{iy} - 0}$$

and consequently, the Y-component of the final velocity is given by

$$v_{fy} = -ev_{iy}$$

The angle of rebound is

$$\tan \phi = \left|\frac{v_{fy}}{v_{fx}}\right| = \left|\frac{ev_{iy}}{v_i \cos \theta}\right| = \frac{ev_i \sin \theta}{v_i \cos \theta} = e \tan \theta$$

The final velocity,

$$v_f = (v_{fx}^2 + v_{fy}^2)^{1/2} = [(v_i \cos \theta)^2 + (ev_i \sin \theta)^2]^{1/2}$$

and the loss of kinetic energy per collision is

$$\Delta T = \tfrac{1}{2}m(v_i^2 - v_f^2) = \tfrac{1}{2}m[v_i^2 - (v_i \cos \theta)^2 - (ev_i \sin \theta)^2]$$
$$= \tfrac{1}{2}mv_i^2[1 - (\cos^2 \theta + e^2 \sin^2 \theta)] = \tfrac{1}{2}mv_i^2 \sin^2 \theta(1 - e^2)$$

When $e = 0$, a maximum loss of kinetic energy occurs; when $e = 1$, the loss is zero.

Illustrative Example 11.4

The conditions prior to an oblique central impact are described in Fig. 11.14. If the line of impact is parallel to the X-direction, determine the final velocity of each mass and the loss of kinetic energy for (a) $e = 0$, (b) $e = 0.5$, (c) $e = 1$.

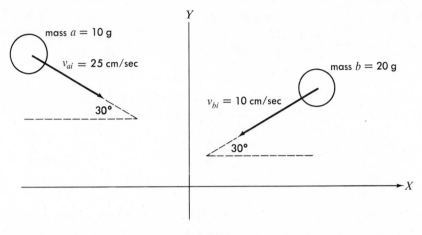

Fig. 11.14

SOLUTION

In all three cases, impact is assumed to occur only in the X-direction. Consequently, the final Y-components of velocity of the masses are equal to the respective initial Y-components:

$$v_{afy} = v_{aiy} = -12.5 \text{ cm/sec}$$

$$v_{bfy} = v_{biy} = -5 \text{ cm/sec}$$

(a) $e = 0$

The X-component of the velocity of separation is zero, and the masses therefore possess identical final X-velocities. The momentum equation is

$$m_a v_{aix} + m_b v_{bix} = m_a v_{afx} + m_b v_{bfx} = (m_a + m_b) v_{afx}$$

Substituting $v_{aix} = 21.7$ cm/sec and $v_{bix} = -8.66$ cm/sec, as well as the known masses, $v_{afx} = v_{bfx} = 1.43$ cm/sec.

The initial system X-momentum is 43 g-cm/sec and is the same in all three cases. The magnitude of the final velocity of mass a is

$$[(1.43)^2 + (12.5)^2]^{1/2} = 12.6 \text{ cm/sec}$$

Similarly for mass b, the final speed is 5.2 cm/sec. The total initial kinetic energy is $\frac{1}{2}m_a v_{ai}^2 + \frac{1}{2}m_b v_{bi}^2$ and is calculated to be 4125 ergs. The final kinetic energy for $e = 0$ is $\frac{1}{2}m_a v_{af}^2 + \frac{1}{2}m_b v_{bf}^2 = 1065$ ergs and the loss of kinetic energy is 74 percent.

(b) $e = 0.5$

The momentum equation has already been written for case (a). Applying Eq. (11.78),

$$e = 0.5 = -\frac{v_{bfx} - v_{afx}}{-8.66 - 21.7}$$

Solving the equations simultaneously, $v_{abx} = 6.5$ cm/sec and $v_{afx} = -8.7$ cm/sec, from which the final speeds are $v_{af} = 15.3$ cm/sec and $v_{bf} = 8.2$ cm/sec. The final kinetic energy is 1840 ergs and the kinetic energy loss is 56 percent.

(c) $e = 1$

Solving Eqs. (11.67) and (11.78) simultaneously, $v_{afx} = -18.8$ cm/sec and $v_{bfx} = 11.5$ cm/sec. The final speeds are therefore $v_{af} = 22.6$ cm/sec and $v_{bf} = 12.5$ cm/sec. The final kinetic energy is 4125 ergs and the loss of kinetic energy is zero, as anticipated.

11.15 KINETIC ENERGY

The total kinetic energy of a system of particles with respect to a prescribed coordinate reference is obtained by summing the kinetic energy of each particle comprising the system,

$$T = \sum_{i=1}^{n} T_i = \sum_{i=1}^{n} \tfrac{1}{2} m_i v_i^2 \tag{11.82}$$

where v_i is the velocity magnitude of the ith particle with respect to XYZ. The velocity $\mathbf{v}_i$ may be expressed as the velocity of the mass center of the system $\mathbf{v}_{CM}$ plus the velocity of the ith particle relative to the mass center $\dot{\boldsymbol{\rho}}_{CM,i}$, Fig. 11.15:

$$\mathbf{v}_i = \mathbf{v}_{CM} + \dot{\boldsymbol{\rho}}_{CM,i} \tag{11.83}$$

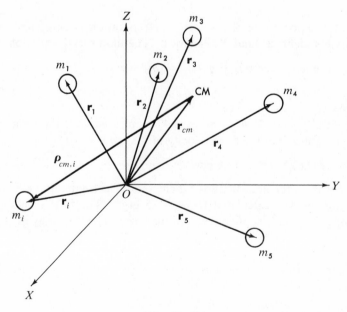

Fig. 11.15

For the entire system,

$$T = \tfrac{1}{2} \sum_{i=1}^{n} m_i |\, \mathbf{v}_{CM} + \dot{\boldsymbol{\rho}}_{CM,i}|^2 = \tfrac{1}{2} \sum_{i=1}^{n} m_i (\mathbf{v}_{CM} + \dot{\boldsymbol{\rho}}_{CM,i}) \cdot (\mathbf{v}_{CM} + \dot{\boldsymbol{\rho}}_{CM,i})$$

$$(11.84)$$

Expanding Eq. (11.84),

$$T = \tfrac{1}{2} \left(\sum_{i=1}^{n} m_i \right) v_{CM}^2 + \mathbf{v}_{CM} \cdot \left(\sum_{i=1}^{n} m_i \dot{\boldsymbol{\rho}}_{CM,i} \right) + \tfrac{1}{2} \sum_{i=1}^{n} m_i (\dot{\rho}_{CM,i})^2 \qquad (11.85)$$

The term $\mathbf{v}_{CM} \cdot (\sum_{i=1}^{n} m_i \dot{\boldsymbol{\rho}}_{CM,i})$ may be expressed $\mathbf{v}_{CM} \cdot (d/dt) \sum_{i=1}^{n} m_i \boldsymbol{\rho}_{CM,i}$. The summation is clearly zero, representing, as it does, the sum of the first moments of mass taken about the mass center.

The kinetic energy of the entire system is, therefore,

$$T = \tfrac{1}{2} m v_{CM}^2 + \tfrac{1}{2} \sum_{i=1}^{n} m_i \dot{\rho}_{CM,i}^2 \qquad (11.86)$$

where m is the total system mass.

According to Eq. (11.86), the system kinetic energy may be divided into the energy of the total mass m traveling at a speed v_{CM} plus the sum of the kinetic energies of the individual masses relative to the mass center. This concept has particular appeal in describing the kinetic energy of a rigid body.

Let us consider examples in which one of the terms of Eq. (11.86) is zero. The kinetic energy of a formation of aircraft flying with identical velocities may be calculated on the basis of the total system mass and the velocity of the mass center $\mathbf{v}_{CM}$, since $\dot{\boldsymbol{\rho}}_{CM,i} = \mathbf{0}$. On the other hand, consider four airplanes participating in stunt flying. If the aircraft fly away from one another at identical speeds, heading east, west, north, and south, respectively, the velocity and hence the kinetic energy associated with the mass center of the system of masses is zero. The system kinetic energy is in this case the kinetic energy of each aircraft based upon the velocities with respect to the stationary mass center.

Illustrative Example 11.5

Consider the system of particles described below:

$$m_1 = 5 \text{ g} \qquad \mathbf{v}_1 = 3\hat{\mathbf{j}} + 4\hat{\mathbf{k}} \text{ cm/sec}$$

$$m_2 = 6 \text{ g} \qquad \mathbf{v}_2 = 2\hat{\mathbf{j}} + 5\hat{\mathbf{k}} \text{ cm/sec}$$

$$m_3 = 4 \text{ g} \qquad \mathbf{v}_3 = \hat{\mathbf{j}} + 3\hat{\mathbf{k}} \text{ cm/sec}$$

$$m_4 = 4 \text{ g} \qquad \mathbf{v}_4 = -4\hat{\mathbf{j}} - 5\hat{\mathbf{k}} \text{ cm/sec}$$

$$m_5 = 1 \text{ g} \qquad \mathbf{v}_5 = -15\hat{\mathbf{j}} - 2\hat{\mathbf{k}} \text{ cm/sec}$$

Determine the kinetic energy of the system two ways: (a) applying Eq. (11.82), and (b) applying Eq. (11.86).

SOLUTION

(a) The kinetic energy of a system of particles is simply the sum of the particle energies based upon velocity with respect to the reference as described by Eq. (11.82).

$$T = \tfrac{1}{2}m_1 v_1^2 + \tfrac{1}{2}m_2 v_2^2 + \tfrac{1}{2}m_3 v_3^2 + \tfrac{1}{2}m_4 v_4^2 + \tfrac{1}{2}m_5 v_5^2$$

where $v_1^2 = 3^2 + 4^2 = 25$ cm^2/sec^2, $v_2^2 = 29$, $v_3^2 = 10$, $v_4^2 = 41$, $v_5^2 = 229$. Using these figures and the values of mass given, $T = 366$ ergs.

(b) To apply Eq. (11.86) it is necessary to determine the velocity of the mass center. Since its position is given by,

$$\mathbf{r}_{CM} = \frac{m_1 \mathbf{r}_1 + m_2 \mathbf{r}_2 + m_3 \mathbf{r}_3 + m_4 \mathbf{r}_4 + m_5 \mathbf{r}_5}{m_1 + m_2 + m_3 + m_4 + m_5}$$

the velocity of the mass center is obtained by differentiating,

$$\mathbf{v}_{CM} = \dot{\mathbf{r}}_{CM} = \frac{m_1 \mathbf{v}_1 + m_2 \mathbf{v}_2 + m_3 \mathbf{v}_3 + m_4 \mathbf{v}_4 + m_5 \mathbf{v}_5}{m_1 + m_2 + m_3 + m_4 + m_5}$$

Substituting the given information, $\mathbf{v}_{CM} = 2\hat{\mathbf{k}}$ cm/sec and the particle velocities relative to the mass center are therefore

$$\dot{\boldsymbol{\rho}}_{CM,1} = (3\hat{\mathbf{j}} + 4\hat{\mathbf{k}}) - 2\hat{\mathbf{k}} = 3\hat{\mathbf{j}} + 2\hat{\mathbf{k}} \qquad (\dot{\rho}_{CM,1}^2 = 13)$$

$$\dot{\boldsymbol{\rho}}_{CM,2} = (2\hat{\mathbf{j}} + 5\hat{\mathbf{k}}) - 2\hat{\mathbf{k}} = 2\hat{\mathbf{j}} + 3\hat{\mathbf{k}} \qquad (\dot{\rho}_{CM,2}^2 = 13)$$

Similarly,

$$\dot{\boldsymbol{\rho}}_{CM,3} = \hat{\mathbf{j}} - \hat{\mathbf{k}} \qquad (\dot{\rho}_{CM,3}^2 = 2)$$

$$\dot{\boldsymbol{\rho}}_{CM,4} = -4\hat{\mathbf{j}} - 7\hat{\mathbf{k}} \qquad (\dot{\rho}_{CM,4}^2 = 65)$$

$$\dot{\boldsymbol{\rho}}_{CM,5} = -15\hat{\mathbf{j}} - 4\hat{\mathbf{k}} \qquad (\dot{\rho}_{CM,5}^2 = 241)$$

The sum of the kinetic-energy terms associated with motion relative to the mass center is, therefore,

$$\tfrac{1}{2}(5)(13) + \tfrac{1}{2}(6)(13) + \tfrac{1}{2}(4)(2) + \tfrac{1}{2}(4)(65) + \tfrac{1}{2}(1)(241)$$

totaling 326 ergs. The kinetic-energy contribution of the total mass (20 g) traveling at the velocity of the mass center, $\mathbf{v}_{CM} = 2\hat{\mathbf{k}}$ cm/sec, is $\tfrac{1}{2}(20)(4)$, or 40 ergs. The total kinetic energy is therefore 366 ergs, as before.

11.16 THE WORK-ENERGY PRINCIPLE

The work done by internal and external forces acting on a system of particles is

$$W = \sum_{i=1}^{n} \int_1^2 \mathbf{F}_i \cdot d\mathbf{r}_i = \int_1^2 \sum_{i=1}^{n} (\mathbf{F}_{i,\text{int}} + \mathbf{F}_{i,\text{ext}}) \cdot d\mathbf{r}_i \qquad \textbf{(11.87)}$$

Note that the summation over a finite number of discrete particles has been placed inside the integral by virtue of the distributive nature of integration. Since $\mathbf{r}_i = \mathbf{r}_{CM} + \boldsymbol{\rho}_{CM, i}$, Fig. 11.15,

$$\mathbf{dr}_i = \mathbf{dr}_{CM} + \mathbf{d}\boldsymbol{\rho}_{CM,i}$$

and Eq. (11.87) may therefore be written

$$W = \int_1^2 \sum_{i=1}^n (\mathbf{F}_{i,\text{ext}} + \mathbf{F}_{i,\text{int}}) \cdot (\mathbf{dr}_{CM} + \mathbf{d}\boldsymbol{\rho}_{CM,i}) \tag{11.88}$$

Expanding Eq. (11.88) we have

$$W = \int_1^2 \left(\sum_{i=1}^n \mathbf{F}_{i,\text{ext}} \right) \cdot \mathbf{dr}_{CM} + \int_1^2 \left(\sum_{i=1}^n \mathbf{F}_{i,\text{int}} \right) \cdot \mathbf{dr}_{CM}$$

$$+ \int_1^2 \sum_{i=1}^n [(\mathbf{F}_{i,\text{ext}} + \mathbf{F}_{i,\text{int}}) \cdot \mathbf{d}\boldsymbol{\rho}_{CM,i}] \tag{11.89}$$

The first term on the right side of Eq. (11.89) may be expressed

$$\int_1^2 \mathbf{F} \cdot \mathbf{dr}_{CM}$$

where $\mathbf{F}$ represents the sum of external forces. Referring to Section 11.3, one can easily demonstrate that

$$\int_1^2 \mathbf{F} \cdot \mathbf{dr}_{CM} = \tfrac{1}{2} m (v_{CM,2}^2 - v_{CM,1}^2) \tag{11.90}$$

The second term of Eq. (11.89) is zero, in accordance with Newton's third law, as already shown. Equating the work, described by Eq. (11.89) to the change of kinetic energy as given by Eq. (11.86), we obtain the result

$$\int_1^2 \mathbf{F} \cdot \mathbf{dr}_{CM} + \int_1^2 \sum_{i=1}^n [(\mathbf{F}_{i,\text{int}} + \mathbf{F}_{i,\text{ext}}) \cdot \mathbf{d}\boldsymbol{\rho}_{CM,i}]$$

$$= \tfrac{1}{2} m (v_{CM,2}^2 - v_{CM,1}^2) + \tfrac{1}{2} \sum_{i=1}^n m_i (\dot{\rho}_{CM,i,2}^2 - \dot{\rho}_{CM,i,1}^2) \tag{11.91}$$

In view of Eq. (11.90), the second term on the left side of Eq. (11.91) must thus equal the kinetic energy associated with motion relative to the mass center:

$$\int_1^2 \sum_{i=1}^n [(\mathbf{F}_{i,\text{int}} + \mathbf{F}_{i,\text{ext}}) \cdot \mathbf{d}\boldsymbol{\rho}_{CM,i}] = \tfrac{1}{2} \sum_{i=1}^n m_i (\dot{\rho}_{CM,i,2}^2 - \dot{\rho}_{CM,i,1}^2) \tag{11.92}$$

The reader may recall that in Illustrative Example 11.2 the internal forces, while canceling in pairs, nevertheless do work on the system as the spring proceeds from one state of deformation to another.

11.17 CENTER-OF-MASS COORDINATE SYSTEM

Because of the peculiarities of geometry and constraint, a problem is often *solved* more easily in one coordinate system than another. Physical *measurements* are generally made, however, in the *laboratory coordinate system*. For example, in a scattering experiment, where projectiles are fired at a fixed target, the measurement of projectile scattering angles is made in a system at rest with respect to the position of the target before the interaction, the laboratory system. A more convenient system, the *center-of-mass system*, is available for *describing* (rather than measuring) particle motion in scattering.

To describe the motion of the colliding particles, we must know something of the interaction. In atomic and nuclear collisions, it is often warranted to assume conservation of translational kinetic energy. Inelastic processes, on the other hand, such as those associated with electron excitation or ionization, lend themselves to general treatment employing quantum mechanics.

In the absence of complete knowledge of the interaction, the projectile is considered as arriving from minus infinity along the X-axis, scattering far from the target in some other direction. Measurements are made far enough away from the target where there are no longer significant interactions.

An important example of the application of center-of-mass coordinates is the scattering of neutrons by various nuclei, as in nuclear reactors. Uranium-fueled reactors operate efficiently when neutrons are present at low energies (<1 eV) to cause the fission of uranium-235 and the subsequent release of enormous energy per fission. In the process of fission, an average of 2.5 neutrons are released at a mean kinetic energy of roughly 2 MeV. Unless the neutrons are drastically reduced in energy, it is highly probable that they will be captured by the uranium-238, thus precluding another fission. It is desirable to have a moderator containing nuclei efficient in slowing down fission neutrons without absorption. We now study the general properties of the neutron–nucleus collision to guide in the selection of a moderator.

A neutron of mass m and velocity $\mathbf{v}_i$ approaches a target nucleus of mass M, initially at rest, Fig. 11.16. In a frame of reference moving with the velocity of the

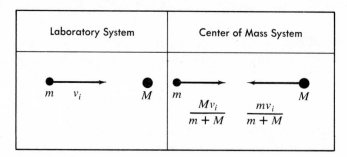

Fig. 11.16

center of mass, the velocity of the mass center is obviously zero. In the laboratory system (LS), the velocity of the mass center, obtained by differentiating Eq. (11.9), is

$$U_{CM} = \frac{m\mathbf{v}_i}{m + M} \tag{11.93}$$

The initial velocity of the neutron in the center of mass system (CMS) $\mathbf{v}_{i,CM}$ is $\mathbf{v}_i$ minus the velocity of the center of mass (Fig. 11.16):

$$\mathbf{v}_{i,CM} = \mathbf{v}_i - \frac{m\mathbf{v}_i}{m + M} = \frac{M\mathbf{v}_i}{m + M} \tag{11.94}$$

As to the nucleus, its initial velocity $\mathbf{V}_i$ in the LS is zero, and in the CMS is zero minus the velocity of the center of mass:

$$\mathbf{V}_{i,CM} = \mathbf{0} - \frac{m\mathbf{v}_i}{m + M} \tag{11.95}$$

Since the net force acting on the system of particles is zero (in any coordinate system), the acceleration of the mass center (origin of CMS) is likewise zero, and the velocity of the origin of the CMS is constant.

In addition, the CMS and LS experience no relative rotation, and, therefore, if the LS is an inertial frame of reference, the CMS is also such a system.

In terms of quantities measured in the LS, conservation of kinetic energy and momentum written in the CMS are

$$\tfrac{1}{2}m\left(\frac{M v_i}{m + M}\right)^2 + \tfrac{1}{2}M\left(\frac{-m v_i}{m + M}\right)^2 = \tfrac{1}{2}m v_{f,CM}^2 + \tfrac{1}{2}M V_{f,CM}^2 \tag{11.96}$$

$$m\left(\frac{M\mathbf{v}_i}{m + M}\right) + M\left(\frac{-m\mathbf{v}_i}{m + M}\right) = \mathbf{0} = m\mathbf{v}_{f,CM} + M\mathbf{V}_{f,CM} \tag{11.97}$$

where $\mathbf{v}_{f,CM}$ and $\mathbf{V}_{f,CM}$ are the final velocities of the neutron and nucleus in the CMS, as shown in Table 11.1.

Since the linear momentum in the CMS is always zero, the individual momenta of the neutron and nucleus (assumed point masses) must lie along the same straight line to the scale shown, and Eq. (11.97) may therefore be written as a scalar equation.

Table 11.1

BEFORE INTERACTION				AFTER INTERACTION			
NEUTRON		TARGET NUCLEUS		NEUTRON		TARGET NUCLEUS	
LS	CMS	LS	CMS	LS	CMS	LS	CMS
v_i	$v_{i,CM}$	V_i	$V_{i,CM}$	v_f	$v_{f,CM}$	V_f	$V_{f,CM}$

Solving Eqs. (11.96) and (11.97) for $v_{f,\mathrm{CM}}$ and $V_{f,\mathrm{CM}}$, we have

$$v_{f,\mathrm{CM}} = \frac{Mv}{m + M} \tag{11.98}$$

and

$$V_{f,\mathrm{CM}} = -\frac{m}{m + M} v_i \tag{11.99}$$

In Fig. 11.17 are shown the velocities of the neutron and nucleus in the CMS before and after the interaction. The reader will note that the angle φ_{CM}, between the initial and final directions in the CMS, is undetermined. To the scale shown there appears to be zero angular momentum (the impact parameter appears to be zero). Nuclear interactions do, of course, occur for nonzero impact parameters and are depicted in a manner similar to Fig. 11.17. Since motion occurs in a plane, there are four components of the velocities, but only three equations: energy and two scalar momentum equations, thus the undetermined angle φ_{CM}. Contrast this with the collision problems of the billiard-ball type, in which additional information was presumably available, describing the nature of the interaction in terms of the direction of the line of impact. Since φ_{CM} cannot be determined on the basis of classical mechanics, the velocities after the interaction remain a function of this angle.

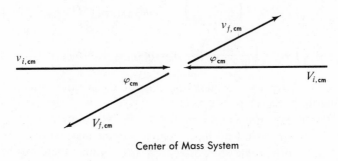

Center of Mass System

Fig. 11.17

How are the measurements in the laboratory system to be related to the foregoing analysis performed in the center-of-mass system? Examine Fig. 11.18, based upon the vector addition:

$$\mathbf{v}_f = \mathbf{v}_{f,\mathrm{CM}} + \mathbf{U}_{\mathrm{CM}} \tag{11.100}$$

In words, the final neutron velocity in the laboratory system is equal to the final neutron velocity in the center-of-mass system plus the velocity of the center-of-mass system ($\mathbf{U}_{\mathrm{CM}}$) with respect to the laboratory system.

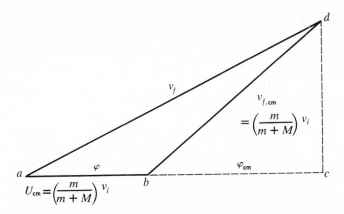

Fig. 11.18

According to Fig. 11.18 then, the relationship between φ, the angle of scatter in the LS, and φ_{CM}, the angle of scatter in the CMS, may be expressed by equating side cd in triangles bcd and acd,

$$\frac{M}{m + M} v_i \sin \varphi_{CM} = \left[\frac{m}{m + M} v_i + \frac{M}{m + M} v_i \cos \varphi_{CM} \right] \tan \varphi \qquad \textbf{(11.101a)}$$

or $\quad \tan \varphi = \dfrac{M \sin \varphi_{CM}}{m + M \cos \varphi_{CM}} = \dfrac{\sin \varphi_{CM}}{(m/M) + \cos \varphi_{CM}}$ $\qquad \textbf{(11.101b)}$

The final speed of the neutron in the LS from Fig. 11.18 may be found from the law of cosines for triangle abd:

$$v_f = \left[\frac{m}{m + M} v_i^2 + \frac{M}{m + M} v_i^2 + \frac{2mMv_i^2}{(m + M)^2} \cos \varphi_{CM} \right]^{1/2} \qquad \textbf{(11.102)}$$

The quantities V_f and ϕ (Fig. 11.19) are related to v_i, v_f, and φ by conservation of momentum equations, parallel and normal to the original direction of the neutron in the LS:

$$mv_i = mv_f \cos \varphi + MV_f \cos \phi \qquad MV_f \cos \phi = mv_i - mv_f \cos \varphi \qquad \textbf{(11.103a)}$$

$$0 = mv_f \sin \varphi - MV_f \sin \phi \qquad MV_f \sin \phi = mv_f \sin \varphi \qquad \textbf{(11.103b)}$$

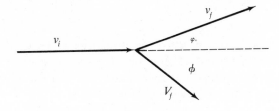

Laboratory System

Fig. 11.19

Squaring and adding Eqs. (11.103):

$$M^2 V_f^2 (\sin^2 \phi + \cos^2 \phi) = m^2 v_i^2 - 2m^2 v_i v_f \cos \varphi + m^2 v_f^2 (\sin^2 \varphi + \cos^2 \varphi)$$

(11.104a)

or

$$V_f = \frac{m}{M} (v_i^2 - 2 v_i v_f \cos \varphi + v_f^2)^{\frac{1}{2}}$$

(11.104b)

Division of Eq. (11.103b) by (11.103a) yields

$$\tan \phi = \frac{m v_f \sin \varphi}{m v_i - m v_f \cos \varphi} = \frac{\sin \varphi}{(v_i / v_f) - \cos \varphi}$$

(11.105)

Thus measurement of v_i, v_f, and φ makes possible the determination of the recoil velocity of the nucleus in the laboratory system (calculating V_f and ϕ).

The final neutron energy in terms of v_i, Eq. (11.102), is

$$\tfrac{1}{2} m v_f^2 = \tfrac{1}{2} m v_i^2 \left[\left(\frac{m}{m + M} \right)^2 + \left(\frac{M}{m + M} \right)^2 + \frac{2mM}{(m + M)^2} \cos \varphi_{CM} \right]$$

(11.106)

We are now in a position to evaluate the fractional energy loss for the neutron associated with each interaction with a nucleus:

$$\frac{\tfrac{1}{2} m v_i^2 - \tfrac{1}{2} m v_f^2}{\tfrac{1}{2} m v_i^2} = 1 - \left[\left(\frac{m}{m + M} \right)^2 + \left(\frac{M}{m + M} \right)^2 + \frac{2mM}{(m + M)} \cos \varphi_{CM} \right]$$

$$= \frac{2mM(1 - \cos \varphi_{CM})}{(m + M)^2}$$

(11.107)

The angle resulting in maximum energy loss is $\varphi_{CM} = \pi$. For minimum fractional loss, $\varphi_{CM} = 0$. Differentiating Eq. (11.107) with respect to M and equating to zero, the value of M resulting in a maximum loss per interaction at any φ_{CM} is found to be m. For our choice of moderator nucleus, therefore, we desire one of mass as close as possible to that of the neutron.

For $m = M$, Eq. (11.101b) gives

$$\tan \varphi = \frac{\sin \varphi_{CM}}{1 + \cos \varphi_{CM}} = \tan \left(\frac{\varphi_{CM}}{2} \right)$$

(11.108a)

or

$$\varphi = \frac{\varphi_{CM}}{2}$$

(11.108b)

Using the mass criterion, a hydrogen nucleus or proton is best. Protons, however, also *capture* neutrons. Therefore, we seek a moderator of less favorable mass but one that does not exhibit appreciable neutron capture. Deuterium or heavy hydrogen represents one such choice.[4]

---

[4] Glasstone, S., and Edlund, M. C., *The Elements of Nuclear Reactor Theory*, D. Van Nostrand Company, Inc., Princeton, N.J., 1952.

11.18 ANGULAR MOMENTUM (MOMENT OF MOMENTUM)

Consider the system of particles shown in Fig. 11.20. Taking moments about O, the origin of an inertial frame of reference, of each term in Eq. (11.5), written in momentum form, we have

$$\sum_{i=1}^{n} \mathbf{r}_i \times \mathbf{F}_{i,\text{ext}} + \sum_{i=1}^{n} \mathbf{r}_i \times \mathbf{F}_{i,\text{int}} = \sum_{i=1}^{n} \frac{d}{dt}(\mathbf{r}_i \times \mathbf{p}_i) \qquad (11.109)$$

where $\mathbf{r}_i \times \mathbf{p}_i$ is the moment of linear momentum of the ith particle about O.

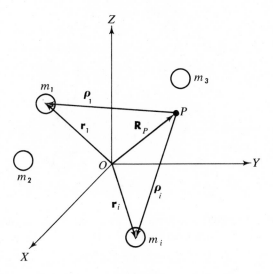

Fig. 11.20

Recall the notation $\mathbf{F}_{i,\text{int}} = \sum_{j=1}^{n} \mathbf{F}_{ij}$. Equation (11.109) is, therefore,

$$\sum_{i=1}^{n} \mathbf{r}_i \times \mathbf{F}_{i,\text{ext}} + \sum_{i=1}^{n} \sum_{j=1}^{n} \mathbf{r}_i \times \mathbf{F}_{ij} = \frac{d}{dt}\left(\sum_{i=1}^{n} \mathbf{r}_i \times \mathbf{p}_i\right) \qquad (11.110)$$

In Eq. (11.110) the order of the summation and differentiation has been interchanged, as permitted by the distributive property of differentiation. Referring to Fig. 11.21, let us examine the nature of the second term in Eq. (11.110) as it applies to particles i and j. The moment of the internal forces for the two particles taken about O is

$$\mathbf{M} = \mathbf{r}_i \times \mathbf{F}_{ij} + \mathbf{r}_j \times \mathbf{F}_{ji} \qquad (11.111)$$

Assuming that $\mathbf{F}_{ij} = -\mathbf{F}_{ji}$, Eq. (11.111) may be written

$$\mathbf{M} = (\mathbf{r}_i - \mathbf{r}_j) \times \mathbf{F}_{ij} \qquad (11.112)$$

When the internal forces obey Newton's third law in the *strong form*, that is, action

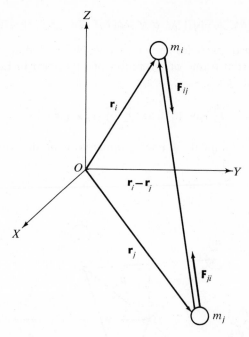

Fig. 11.21

and reaction forces are equal and opposite and act along the same line, $\mathbf{r}_i - \mathbf{r}_j$ is colinear with $\mathbf{F}_{ij}$; and the moment must be zero. In a like manner, the sum of moments of internal forces is zero, and the second term of Eq. (11.110) vanishes, leaving

$$\sum_{i=1}^{n} \mathbf{r}_i \times \mathbf{F}_i = \frac{d}{dt}\left(\sum_{i=1}^{n} \mathbf{r}_i \times \mathbf{p}_i\right) \tag{11.113}$$

where the subscript "ext" is no longer necessary.

Equation (11.113) states that the sum of all moments of external force taken about the origin of an inertial frame of reference is equal to the time derivative of the sum of the angular momenta. In more compact form, Eq. (11.113) is written

$$\mathbf{M}_0 = \dot{\mathbf{H}}_0 \tag{11.114}$$

When $\mathbf{M}_0 = 0$, $\mathbf{H}_0$ is constant; angular momentum is conserved. The reader will recall this result in central force motion.

In the foregoing discussion, moments of force and linear momentum were taken about the origin of an inertial frame of reference; $\mathbf{r}_i$ represented a vector running directly from the origin to the mass in question. A similar development will now be carried out in which, referring to Fig. 11.20, the vector $\boldsymbol{\rho}_i$ plus $\mathbf{R}_P$ will be substituted for $\mathbf{r}_i$, where the vector $\mathbf{R}_P$ runs from the origin to P, an arbitrary point, not necessarily fixed. Equation (11.113) may now be written

$$\sum_{i=1}^{n} (\mathbf{R}_P + \boldsymbol{\rho}_i) \times \mathbf{F}_i = \sum_{i=1}^{n} \frac{d}{dt} [(\mathbf{R}_P + \boldsymbol{\rho}_i) \times \mathbf{p}_i] \qquad (11.115)$$

Recall from Chapter 4 that

$$\frac{d}{dt} [(\mathbf{R}_P + \boldsymbol{\rho}_i) \times \mathbf{p}_i] = (\mathbf{R}_P + \boldsymbol{\rho}_i) \times \dot{\mathbf{p}}_i \qquad (11.116)$$

Carrying out the indicated operations, Eq. (11.115) therefore becomes

$$\sum_{i=1}^{n} \mathbf{R}_P \times \mathbf{F}_i + \sum_{i=1}^{n} \boldsymbol{\rho}_i \times \mathbf{F}_i = \sum_{i=1}^{n} \mathbf{R}_P \times \dot{\mathbf{p}}_i + \sum_{i=1}^{n} \mathbf{p}_i \times \dot{\mathbf{p}}_i \qquad (11.117)$$

Since $\mathbf{F}_i = \dot{\mathbf{p}}_i$, the first terms on either side of Eq. (11.117) cancel, leaving

$$\sum_{i=1}^{n} \boldsymbol{\rho}_i \times \mathbf{F}_i = \sum_{i=1}^{n} \boldsymbol{\rho}_i \times \dot{\mathbf{p}}_i \qquad (11.118)$$

Since

$$\frac{d}{dt} (m_i \mathbf{v}_i) = \frac{d}{dt} (m_i \dot{\mathbf{r}}_i) = m_i \ddot{\mathbf{r}}_i = m_i (\ddot{\mathbf{R}}_P + \ddot{\boldsymbol{\rho}}_i) \qquad (11.119)$$

Equation (11.118) may now be written

$$\sum_{i=1}^{n} \boldsymbol{\rho}_i \times \mathbf{F}_i = \sum_{i=1}^{n} \boldsymbol{\rho}_i \times [m_i (\ddot{\mathbf{R}}_P + \ddot{\boldsymbol{\rho}}_i)] \qquad (11.120)$$

or

$$\sum_{i=1}^{n} \boldsymbol{\rho}_i \times \mathbf{F}_i = \left[\sum_{i=1}^{n} m_i \boldsymbol{\rho}_i \right] \times \ddot{\mathbf{R}}_P + \sum_{i=1}^{n} \boldsymbol{\rho}_i \times m_i \ddot{\boldsymbol{\rho}}_i \qquad (11.121)$$

where $\ddot{\mathbf{R}}_P$, a common factor, has been removed from the brackets. The second term on the right side of Eq. (11.121) may be expressed

$$\sum_{i=1}^{n} \boldsymbol{\rho}_i \times m_i \dot{\mathbf{v}}_{Pi} = \sum_{i=1}^{n} \boldsymbol{\rho}_i \times \dot{\mathbf{p}}_{Pi} \qquad (11.122)$$

where $\mathbf{v}_{Pi}$ is the velocity of the ith particle relative to point P. Now consider the operation

$$\sum_{i=1}^{n} \frac{d}{dt} (\boldsymbol{\rho}_i \times \mathbf{p}_{Pi}) = \sum_{i=1}^{n} \boldsymbol{\rho}_i \times \dot{\mathbf{p}}_{Pi} + \sum_{i=1}^{n} \dot{\boldsymbol{\rho}}_i \times \mathbf{p}_{Pi} \qquad (11.123)$$

The second term on the right side must be zero, since $\mathbf{p}_{Pi} = m_i \mathbf{v}_{Pi}$, and $\dot{\boldsymbol{\rho}}_i = \mathbf{v}_{Pi}$ are colinear. Making the indicated substitution in Eq. (11.121) and interchanging the order of summation and differentiation on the left side of Eq. (11.123) as already discussed,

$$\sum_{i=1}^{n} \boldsymbol{\rho}_i \times \mathbf{F}_i = \left[\sum_{i=1}^{n} m_i \boldsymbol{\rho}_i \right] \times \ddot{\mathbf{R}}_P + \frac{d}{dt} \sum_{i=1}^{n} \boldsymbol{\rho}_i \times \mathbf{p}_{Pi} \qquad (11.124)$$

where $\sum_{i=1}^{n} \boldsymbol{\rho}_i \times \mathbf{F}_i$ is the sum of moments of force taken about P, designated $\mathbf{M}_P$, and $(d/dt) \sum_{i=1}^{n} \boldsymbol{\rho}_i \times \mathbf{p}_{Pi}$, designated $\dot{\mathbf{H}}_P$, is the time derivative of the sum of the moments of linear momentum taken about P as viewed from the origin of the inertial frame of reference. Bear in mind that $\mathbf{p}_{Pi} = m_i \dot{\boldsymbol{\rho}}_i$ is a linear momentum relative to point P. In more succinct form, Eq. (11.124) is

$$\mathbf{M}_P = \left[\sum_{i=1}^{n} m_i \boldsymbol{\rho}_i \right] \times \ddot{\mathbf{R}}_P + \dot{\mathbf{H}}_P \tag{11.125}$$

If the first term on the right side of Eq. (11.125) is zero, the resulting equation is simply

$$\mathbf{M}_P = \dot{\mathbf{H}}_P \tag{11.126}$$

This is, of course, identical in form with Eq. (11.114).

The circumstances under which $\sum_{i=1}^{n} m_i \boldsymbol{\rho}_i \times \ddot{\mathbf{R}}_P$ vanishes are:

a. If point P is either fixed or moves with constant velocity with respect to the reference, thus making $\ddot{\mathbf{R}}_P = \mathbf{0}$.

b. If point P is coincident with the center of mass of the system (for which $\ddot{\mathbf{R}}_P$ need not be zero), then by definition of the mass center, $\sum_{i=1}^{n} m_i \boldsymbol{\rho}_i = 0$. The subscript CM then replaces P in Eq. (11.126).

c. If $\sum_{i=1}^{n} m_i \boldsymbol{\rho}_i$ and $\ddot{\mathbf{R}}_P$ are parallel, the cross product will vanish. $(1/\sum m_i) \times \sum_{i=1}^{n} m_i \boldsymbol{\rho}_i$ is a vector extending from point P to the mass center, since it locates the mass center with respect to a coordinate reference located at P. If the acceleration of P is parallel to $\sum_{i=1}^{n} m_i \boldsymbol{\rho}_i$, that is, either directed toward or away from the mass center, the cross product must be zero.

Anticipating Chapter 12, let us examine briefly a collection of particles so constrained as to assure the constancy of all distances between particles. Assume this *rigid body* to experience rotation about a fixed line in space, which we shall take to be the Z-axis. Applying Eq. (4.24c) to the ith particle, located at position x_i, y_i with respect to a coordinate system with origin on the axis of rotation,

$$M_{zi} = \frac{d}{dt} (m_i x_i \dot{y}_i - m_i y_i \dot{x}_i) \tag{11.127}$$

where M_{zi} is the torque associated with particle m_i. If r_i is the distance from the axis to m_i ($r_i^2 = x_i^2 + y_i^2$): $x_i = r_i \cos \theta_i$; $y_i = r_i \sin \theta_i$; $\dot{x}_i = -r_i \sin \theta_i \dot{\theta}_i = -y_i \dot{\theta}_i$; $\dot{y}_i = \dot{\theta}_i \cos \theta_i \dot{\theta}_i = x_i \dot{\theta}_i$. Equation (11.127) may now be written

$$M_{zi} = \frac{d}{dt} (m_i x_i^2 \dot{\theta}_i + m_i y_i^2 \dot{\theta}_i) = \frac{d}{dt} (m_i r_i^2 \dot{\theta}_i) \tag{11.128}$$

For the entire collection of particles,

$$M_z = \sum_{i=1}^{n} M_{zi} = \frac{d}{dt} \left(\omega \sum_{i=1}^{n} m_i r_i^2 \right) \tag{11.129}$$

where $\omega = \dot{\theta}_i$, the angular velocity, is common to each r_i because of rigidity. In the above expression, $\sum_{i=1}^{n} m_i r_i^2$, the sum of each particle mass multiplied by the square of its distance from the axis of rotation, is termed the moment of inertia, I. Equation (11.129) is thus

$$M_z = \frac{d}{dt}(I\omega) \qquad (11.130)$$

where there is a clear parallel between $I\omega$, the angular momentum of a rigid body, and mv, the linear momentum.

Illustrative Example 11.6

Four equal masses m are connected by massless spokes to a massless pulley as shown in Fig. 11.22. Wrapped around the pulley is a cord subject to tension τ. Determine the angular velocity of the pulley at time t.

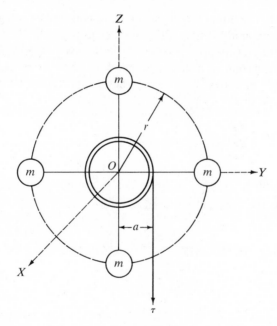

Fig. 11.22

SOLUTION

The sum of the external moments of force about 0 is τa, since the moments of the gravitational forces associated with masses m cancel. According to Eq. (11.114),

$$\mathbf{M} = a\hat{\mathbf{j}} \times -\tau\hat{\mathbf{k}} = \frac{d}{dt}\left(\sum_{i=1}^{n} \mathbf{r}_i \times m_i \mathbf{v}_i\right)$$

where $\mathbf{v}_i = \boldsymbol{\omega} \times \mathbf{r}_i$, or

$$-\tau a\hat{\mathbf{i}} = \frac{d}{dt} \{r\hat{\mathbf{j}} \times m(-\omega\hat{\mathbf{i}} \times r\hat{\mathbf{j}}) + r\mathbf{k} \times m(-\omega\hat{\mathbf{i}} \times r\hat{\mathbf{k}})$$

$$+ (-r\hat{\mathbf{j}}) \times m[(-\omega\hat{\mathbf{i}}) \times (-r\hat{\mathbf{j}})] + (-r\hat{\mathbf{k}}) \times m[(-\omega\hat{\mathbf{i}}) \times (-r\hat{\mathbf{k}})]\}$$

$$= \frac{d}{dt} (4m\omega r^2\hat{\mathbf{i}})$$

Thus

$$\frac{d\omega}{dt} = \frac{\tau a}{4mr^2}$$

and therefore

$$\omega - \omega_0 = \frac{\tau a}{4mr^2} t$$

where ω_0 represents the initial angular velocity, a constant of integration.

The foregoing analysis may also have begun with Eq. (11.130), in which the total moment of inertia about the axis of rotation is simply $4mr^2$. The angular momentum for the system of particles is therefore $H = -4mr^2\omega$.

The moment of force, $M = -\tau a$, when equated to time rate of change of angular momentum, gives

$$-\tau a = -4mr^2 \frac{d\omega}{dt}$$

or

$$M = I\dot{\omega} = I\alpha$$

where α represents the angular acceleration of the system.

11.19 ADDITIONAL APPLICATIONS TO ATOMIC AND NUCLEAR SYSTEMS

Beta Emission

The interpretation of beta emission is presented to further demonstrate the universality of the laws of mechanics. As early as 1910, alpha, beta, and gamma rays were known to be emitted from radioactive nuclei.[5] Alpha rays were identified as doubly charged helium ions, beta rays as negative electrons, and gamma rays as high-energy electromagnetic radiation. In the emission of alpha and gamma rays from single species, the radiation was found to occur at a few discrete, or what was later to be termed quantized, energies. On the other hand, beta particles were

---

[5] Rutherford, E., *Phil. Mag.* **5**, 177 (1903).

observed to be emitted at all energies up to a maximum value. The question facing physicists of that day was: Why should beta emission exhibit a continuous spectrum? On the basis of classical mechanics, it is natural to seek an answer based upon consideration of the conservation laws: energy, linear momentum, angular momentum, and charge.

In the decay of a parent nucleus into a daughter and an electron, as in the simple beta decay of a neutron, $n^0 \rightarrow p^+ + e^-$, the energy available to the proton and electron must equal the difference in total system rest mass before and after the reaction. Since the rest masses are quite well known, the available energy, equal for all neutron decays, is also known.

Conservation of linear momentum dictates that the recoil velocity of the proton be directed opposite that of the beta in order that the final momentum (beta plus proton) equal the initial momentum, that of the neutron.

When the laws of conservation of energy and linear momentum are considered together, the energy of the electron is prescribed for a given reaction. This, by itself, represents a direct contradiction of the experimental facts.

Angular momentum considerations add to the paradox. Since the electron, proton, and neutron each have equal intrinsic angular momenta, it is not possible, using the addition laws of quantum mechanics, to equate the angular momenta before and after the decay.

Some solace may be gained in that the charge is conserved since the neutron is uncharged, and the charges of the electron and proton are equal and opposite and therefore cancel.

When physicists sought to account for beta decay, they had the alternatives of either discarding the laws of conservation of energy, linear momentum, and angular momentum, or discovering something that was not apparent, which would explain the known facts and agree with the laws of physics.

In 1935 Fermi[6] developed a theory involving a new particle, which he called the neutrino, for "little neutral." The idea for this particle was first proposed in 1929 by Pauli.[7] The neutrino was to have zero charge but angular momentum, kinetic energy, and linear momentum. It was to be emitted at the same time as the beta particle. The existence of more than one additional particle could also have been postulated to explain beta decay, but the principle was adhered to that in the absence of rationale to the contrary, the simplest theory should be advanced. In Chapter 10 it was shown that when one mass decays into three with a certain total energy E, the one representing a beta can have a wide range of energy from zero to a maximum approaching E, with the particles representing the neutrino and nucleus carrying off the recoil momentum.

It was not until 1955 that Cowan and Reines[8] experimentally discovered

[6] Fermi, E., *Z. Physik* **88**, 161 (1934).
[7] Pauli, W., *Rapp. Septieme Conseil Phys.*, Solvay, Brussels (1933).
[8] Cowan, C. L., and Reines, F., *Phys. Rev.* **113**, 273 (1959).

the neutrino. Although the experimental proof did not come for 20 years, physicists nevertheless had such confidence in the laws of physics that many complex theories involving the neutrino were developed before its discovery.

Compton Scattering

To explain the scattering of gamma rays by free electrons, Compton[9] proposed that the momentum of a gamma ray, or short-wavelength electromagnetic radiation, is h/λ, where λ represents the wavelength of the radiation. Planck had already put forth his bold hypothesis that the energy of a photon is $h\nu$ or hc/λ, where ν is the frequency of the radiation and c the speed of light. To confirm the validity of the Compton momentum, a beam of monoenergetic gamma rays was scattered by free electrons, and the new wavelength of the gammas after scattering was measured as a function of the angle of scatter. Relativistic energy and momentum relationships are required to describe the phenomenon under discussion.

Before the collision, the photon possesses kinetic energy hc/λ, and the electron is at rest. After the collision, the photon has an energy hc/λ', and the electron possesses energy T. Conservation of energy gives

$$\frac{hc}{\lambda} = \frac{hc}{\lambda'} + T \tag{11.131}$$

Initially the momentum of the photon is h/λ and that of the electron is zero. After the collision the photon and electron are assumed to move at angles φ and φ_e, respectively, as shown in Fig. 11.23. Conserving momentum along the original direction of motion,

$$\frac{h}{\lambda} = \frac{h}{\lambda'} \cos\varphi + p_e \cos\varphi_e \tag{11.132}$$

where p_e is the relativistic momentum of the electron. Conserving momentum perpendicular to the original direction of motion:

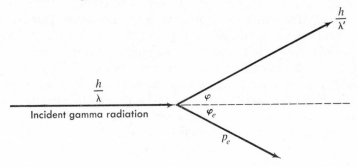

Fig. 11.23

[9] Compton, A. H., *Phys. Rev.* **21**, 715 (1923); **22**, 409 (1923).

$$0 = \frac{h}{\lambda'} \sin \varphi - p_e \sin \varphi_e \qquad \textbf{(11.133)}$$

since there is initially no momentum along this direction and no external forces act. The relativistic kinetic energy of an electron may be expressed

$$T = (p_e^2 c^2 + m_0^2 c^4)^{\frac{1}{2}} - m_0 c^2 \qquad \textbf{(11.134)}$$

where m_o is the electronic rest mass. We thus have three equations and four unknowns: φ, φ_e, p_e, λ'; m, c, h, λ are presumed known. Compton solved these equations for λ' as a function of φ and verified these results experimentally. To solve for λ' as a function of φ, it is necessary to solve Eqs. (11.132) and (11.133) for $\cos \varphi_e$ and $\sin \varphi_e$, respectively. Squaring these new equations and adding eliminates φ_e. If next the resulting equation is solved for p_e, combined with Eq. (11.132) and substituted into Eq. (11.131), we obtain the well-known relationship

$$\lambda' - \lambda = \frac{h}{mc}(1 - \cos \varphi) \qquad \textbf{(11.135)}$$

We thus have a manifestation of the particle nature of electromagnetic radiation.

References

Bohm, D., *Quantum Theory*, Prentice-Hall, Inc., Englewood Cliffs, N.J., 1951.

Born, M., *Atomic Physics*, Sixth Edition, Hafner Publishing Company, New York, 1956.

Bryant, P. M., Lavik, M., and Salomon, G., *Mechanisms of Solid Friction*, Elsevier Publishing Company, Amsterdam, 1964.

Greenwood, D. T., *Principles of Dynamics*, Prentice-Hall, Inc., Englewood Cliffs, N.J., 1965.

Landau, L. D., and Lifshitz, E. M., *Mechanics*, Addison-Wesley Publishing Company, Inc., Reading, Mass., 1960.

Shames, I. H., *Engineering Mechanics*, Second Edition, Prentice-Hall, Inc., Englewood Cliffs, N.J., 1967.

Symon, K. R., *Mechanics*, Second Edition, Addison-Wesley Publishing Company, Inc., Reading, Mass., 1960.

Von Kármán, T., and Biot, M. A., *Mathematical Methods in Engineering*, McGraw-Hill, Inc., New York, 1940.

EXERCISES

11.1 Determine the location and acceleration of the center of mass given the system of particles shown in Fig. P11.1.

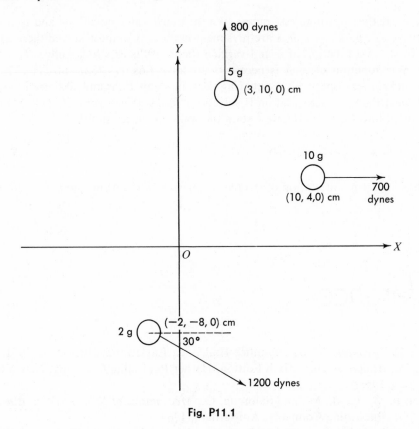

Fig. P11.1

11.2 Determine the Cartesian components of acceleration of the mass center of the following system of particles. Each particle is located by a radius vector **r** from the origin of an inertial frame of reference. Also locate the instantaneous mass center.

$$m_1 = 5 \text{ g, } F_1 = 3\hat{i} - 2\hat{j} + 6\hat{k} \text{ dyn, } r_1 = 3\hat{i} + 3\hat{j} \text{ cm}$$

$$m_2 = 10 \text{ g, } F_2 = 7\hat{i} \text{ dyn, } r_2 = 4\hat{i} - 3\hat{k} \text{ cm}$$

$$m_3 = 5 \text{ g, } F_3 = 2\hat{i} + 3\hat{j} \text{ dyn, } r_3 = 7\hat{k} \text{ cm}$$

11.3 A double Atwood's machine is depicted in Fig. P11.3. Derive expressions for the tension in each cord, the acceleration of each mass, and the acceleration of the mass center of the entire system. Neglect friction, pulley mass, and pulley rotation.

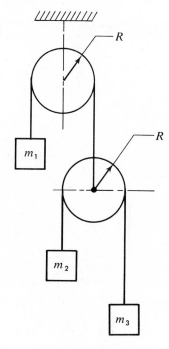

Fig. P11.3

11.4 A jet of liquid of mass density ρ, cross-sectional area A, and velocity v_1 strikes a vane as shown in Fig. P11.4. What are the magnitude and direction of the force required to prevent the vane from moving?

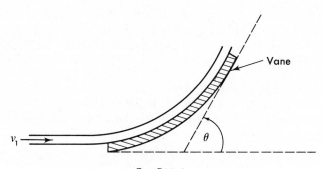

Fig. P11.4

11.5 A chain of length L and mass per unit length ρ_L is held as shown in Fig. P11.5 and then released. Determine the velocity of the chain as the final link leaves the table. Neglect friction. Also derive an expression for the time required for the chain to just reach the point of departure.

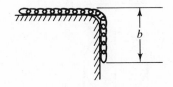

Fig. P11.5

11.6 For the system described in Fig. P11.6, determine the natural frequencies if $m = 10$ g and $k = 2000$ dyn/cm.

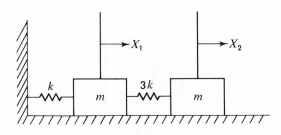

Fig. P11.6

11.7 Referring to Exercise 11.6, determine $x_1(t)$ and $x_2(t)$ for the initial conditions $x_1(0) = 2$ cm, $\dot{x}_1(0) = 0$, $x_2(0) = 0$, $\dot{x}_2(0) = 0$.

11.8 Repeat Exercise 11.7 for initial conditions $x_1(0) = 0$, $\dot{x}_1(0) = 10$ cm/sec, $x_2(0) = 0$, $\dot{x}_2(0) = 0$.

11.9 A 200-g block sliding along a table top with a velocity to the right of 200 cm/sec is struck by a 5-g bullet traveling toward the left which passes completely through. If the initial velocity of the bullet is 5000 cm/sec and the final velocity of the block is 150 cm/sec, determine:

    a. the final velocity of the bullet.
    b. the loss of kinetic energy of the bullet and the block.
    c. the average force of friction on the bullet and the block as a result of the interaction.
    d. the work done by the frictional forces.

Assume the contact between the block and the table top to be frictionless.

11.10 Using the same initial conditions as in Exercise 11.9, determine the final velocity of the bullet and block as well as the loss of kinetic energy for each of the following collisions:

    a. perfectly elastic.
    b. perfectly inelastic.
    c. for a coefficient of restitution of 0.5.

11.11 A simple pendulum of mass m is connected to a block of mass M, constrained to move on a horizontal plane as shown in Fig. P11.11. Neglecting friction, derive the equations of motion for the block and the pendulum. Motion is initiated by displacing the pendulum from $\theta = 0$.

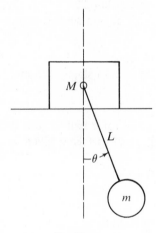

Fig. P11.11

11.12 What maximum velocity is attained by a vertical, single-stage, variable-mass rocket subject to constant gravitational acceleration and a drag force proportional to the first power of velocity?

11.13 Determine the velocity at the end of 5 sec and the maximum velocity of a rocket fired vertically upward. Aerodynamic forces may be neglected and the gravitational acceleration may be assumed constant. The following data apply:

    Total loaded mass = 80 kg
    Empty rocket mass = 40 kg
    Jet velocity v_j = 1200 m/sec
    Time to burnout = 16 sec

11.14 Referring to Exercise 11.13, calculate the altitude at the time of burnout.

11.15 Examination of the results of Exercise 11.13 leads one to conclude that if a high velocity at burnout is to be achieved, the final rocket mass must be as small as possible. Multistage rockets are used to reduce the unproductive rocket mass associated with storing and burning the spent fuel. In a two-stage arrangement, a second rocket is the payload of the first, so that after first-stage burnout, the second rocket is fired, leaving behind the empty first stage. This concept has been extended to several stages.

Derive an expression, ignoring gravity, for the velocity at second-stage burnout in terms of the following parameters:

v_j = velocity of exhaust gases with respect to rocket (assumed the same for each stage)

m_A = total mass of rocket at launch

m_B = total mass remaining at first-stage burnout (including mass of second stage)

m_C = mass of second stage

m_D = mass remaining at second-stage burnout

11.16 Show that the time required for a ball to stop bouncing after being dropped from a height h onto a smooth, level surface is given by

$$t = \left(\frac{2h}{g}\right)^{\frac{1}{2}} \frac{1 + e}{1 - e}$$

where e is the coefficient of restitution.

11.17 Referring to the situation described in Exercise 11.16, demonstrate that the sum of the energy loss associated with each bounce equals the initial potential energy of the ball. Also show that the total distance traveled is given by

$$H = h \frac{1 + e^2}{1 - e^2}$$

11.18 A hammer pivoted at A is released from a horizontal position at B, strikes the block as shown in Fig. P11.18, and rebounds to a 45-degree angle. Assume that the 10-lb head represents essentially all the hammer weight, and that there is no friction between the 100-lb block and the supporting surface. Determine:

a. the velocities of hammer and block after impact.
b. the coefficient of restitution.

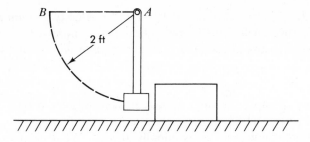

Fig. P11.18

11.19 At the same time that a mass m is fired directly upward with velocity v_0, an identical mass is dropped from a height H in line with the first. If the coefficient of restitution for the collision is e, what initial velocity v_i of the second mass is required to cause it to just return to its initial position?

11.20 Two particles of mass m_a and m_b and initial velocities v_{ai} and v_{bi} suffer direct central impact. Prove that their velocities after impact are

$$v_{af} = \frac{m_b v_{bi}(1 + e) + (m_a - em_b)v_{ai}}{m_a + m_b}$$

$$v_{bf} = \frac{m_a v_{ai}(1 + e) + (m_b - em_a)v_{bi}}{m_a + m_b}$$

11.21 Two objects of equal mass moving in the same plane experience a collision for which the coefficient of restitution may be taken as 0.7. The initial velocities are

$$\mathbf{v}_{ai} = 30\hat{\mathbf{i}} + 40\hat{\mathbf{j}} \text{ ft/sec}$$
$$\mathbf{v}_{bi} = -50\hat{\mathbf{i}} \text{ ft/sec}$$

Determine the velocity of each object after impact as well as the loss of energy experienced by each. Take the line of impact to be parallel to the X-axis.

11.22 Repeat Exercise 11.21 for impact along a line parallel to the Y-axis.

11.23 Repeat Exercise 11.21 for $e = 0$.

11.24 A chain collision of three automobiles of approximately equal mass m occurs, as in Fig. P11.24. Vehicles b and c are initially at rest when b is struck by

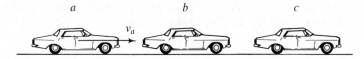

Fig. P11.24

a traveling at v_a. Ignoring friction, determine the velocity of *c* immediately after the collision for two coefficients of restitution $e = 0$ and $e = 0.5$. How much of the kinetic energy is lost in each case? What impulse is imparted to vehicle *c* in both cases?

11.25 Assuming an impact in which no loss of energy occurs, determine the velocity of each mass in Fig. P11.25 immediately after impact if mass m_2 is released from rest in the position shown and m_1 is initially at rest. What maximum heights will m_1 and m_2 attain after impact?

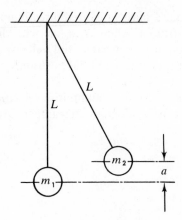

Fig. P11.25

11.26 Derive an expression for the initial velocity of a bullet as determined by a ballistic pendulum. A mass *M* supported by a string of length *L* is observed to be vertically displaced by an amount *h*, as a result of a bullet of mass *m* fired into it. Also determine the loss of kinetic energy for this system of particles. For what segments of the motion is mechanical energy conserved and why? Where is linear momentum conserved and why?

11.27 Repeat Exercise 11.5 taking into account Coulomb friction between the chain and the horizontal surface.

11.28 Analyze Atwood's machine by applying the work–energy equation to the two-particle system. Explain why the cord–pulley arrangement is said to provide a *workless constraint* upon the motion of the masses.

11.29 Determine the natural frequencies of the double pendulum shown in Fig. P11.29.

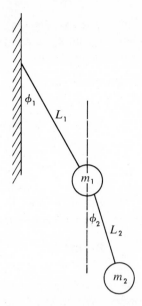

Fig. P11.29

11.30 Determine the natural frequencies of the elastically coupled simple pendulums of Fig. P11.30. Use the angular coordinates.

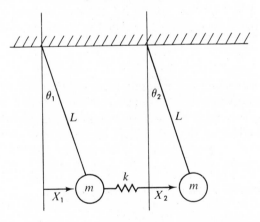

Fig. P11.30

11.31 Rederive the equations of motion for the coupled pendulum employing the coordinates x_1 and x_2.

11.32 Demonstrate that the normal coordinates for the coupled pendulum are

$$\xi = x_1 + x_2$$
$$\eta = x_1 - x_2$$

11.33 For the coupled pendulum, write expressions for the system kinetic energy and potential energy (a) in terms of x_1 and x_2, and (b) in terms of ξ and η.

11.34 What initial conditions lead to first-mode oscillation in Exercise 11.30?

11.35 Determine the natural frequencies of the three-degree-of-freedom system described in Fig. P11.35.

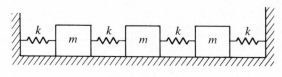

Fig. P11.35

11.36 Determine the energy loss of an alpha particle undergoing scattering by a nucleus of charge Z in the laboratory system as a function of the angle of scatter.

11.37 Rederive the Rutherford scattering cross-section formula in the laboratory system. What occurs if the mass of the nucleus is much greater than that of the alpha particle? If $Z = 80$, what should the ratio of the alpha particle mass to nucleus be so that the cross sections calculated in the CMS and LS differ by less than 1 percent?

11.38 Neutron–nucleus scattering is approximately isotropic in the center-of-mass system; therefore, the mean angle of scatter is $\pi/2$. Calculate the mean angle of scatter in the laboratory system and the energy loss for the following nuclei: proton, same mass as the neutron, m; deuterium, $2m$; beryllium, $9m$; carbon, $12m$; oxygen, $16m$.

11.39 A system of masses a and b is described as follows:

$m_a = 10$ g, located at $(5, -10, 3)$ cm
$m_b = 25$ g, located at $(-2, -3, 5)$ cm
$\mathbf{v}_a = 3\hat{\mathbf{i}} + 2\hat{\mathbf{j}} - \hat{\mathbf{k}}$ cm/sec
$\mathbf{v}_b = 2\hat{\mathbf{i}}$ cm/sec

Determine:

a. the location of the center of mass relative to the origin.
b. the moment of momentum relative to the origin and the point $(1, 1, 1)$ cm.
c. the kinetic energy relative to the origin.
d. the kinetic energy relative to the mass center.

11.40 An Atwood's machine, Fig. P11.40, is released from rest in the position shown. Determine the following quantities at time t:

    a. cord tension.
    b. acceleration of the mass center.
    c. acceleration of each mass.
    d. velocity of the mass center.
    e. velocity of each mass.
    f. position of the mass center.
    g. position of each mass.
    h. linear momentum of the system.
    i. system kinetic energy as sum of the terms associated with velocity of mass center and velocity relative to mass center.

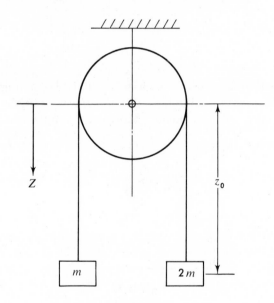

Fig. P11.40

11.41 Consider the following system of four point masses

$$m_1 = 10 \text{ g}, \ \mathbf{v}_1 = 3\hat{\mathbf{i}} + 2\hat{\mathbf{j}} \text{ cm/sec}, \ \mathbf{r}_1 = -2\hat{\mathbf{i}} - 2\hat{\mathbf{j}} \text{ cm}$$
$$m_2 = 3 \text{ g}, \ \mathbf{v}_2 = -7\hat{\mathbf{j}} \text{ cm/sec}, \ \mathbf{r}_2 = 7\hat{\mathbf{i}} + 2\hat{\mathbf{j}} \text{ cm}$$
$$m_3 = 2 \text{ g}, \ \mathbf{v}_3 = -2\hat{\mathbf{i}} - 2\hat{\mathbf{j}} \text{ cm/sec}, \ \mathbf{r}_3 = -6\hat{\mathbf{j}} \text{ cm}$$
$$m_4 = 7 \text{ g}, \ \mathbf{v}_4 = 8\hat{\mathbf{i}} + 7\hat{\mathbf{j}} \text{ cm/sec}, \ \mathbf{r}_4 = -2\hat{\mathbf{j}} \text{ cm}$$

where $\mathbf{r}_i$ represents a vector from the origin to the mass. Determine:

    a. the location of the center of mass.
    b. the velocity of the mass center.
    c. the angular momentum of the system taken about the origin.
    d. the kinetic energy of the system as the sum of terms associated with the motion of the mass center and motion relative to the mass center.

11.42 A system consists of two objects each weighing 12 lb connected by an elastic member with spring constant $k = 100$ lb/in. If at $t = 0$ an 800-lb horizontal force is applied as shown in Fig. P11.42, determine for a frictionless system:

    a. the acceleration of the mass center.
    b. the displacement of the mass center 10 sec after the force is applied assuming zero initial conditions on position and velocity.
    c. the displacement of each mass as a function of time.

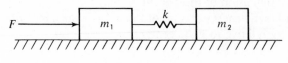

Fig. P11.42

11.43 For the system of particles shown in Fig. P11.43, determine the velocity of the masses when m_2 is permitted to fall a distance H if it is initially at rest. Assume a coefficient of friction μ for the contact between m_1 and the inclined plane.

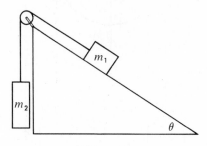

Fig. P11.43

11.44 Determine the moment of momentum vector about the origin for the system of particles described in Fig. P11.44.

11.45 A system of three particles, each of mass 10 g, is described in Fig. P11.45. Determine:

    a. the angular momentum of the system about the origin of XYZ.
    b. the angular momentum of the system about point P.

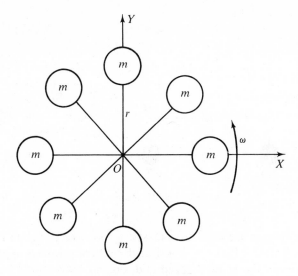

Fig. P11.44

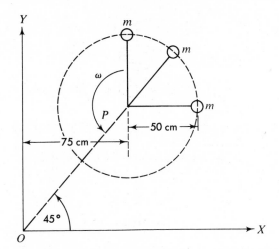

Fig. P11.45

11.46 A system of two equal particles in the YZ-plane is connected by a rigid massless rod as shown in Fig. P11.46. A constant force $\mathbf{F} = -F\hat{\mathbf{k}}$ begins to act at $t = 0$ when the system is at rest in the position shown.

    a. Determine the position, velocity, and acceleration of the mass center as functions of time considering the entire system of particles.

    b. Determine the position, velocity, and acceleration of each particle as functions of time.

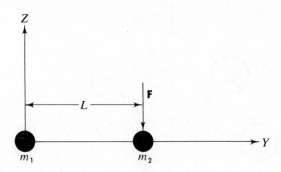

Fig. P11.46

11.47 Two 20-g masses are fixed to a massless rod, free to rotate about axis *A-A* as shown in Fig. P11.47. If at $t = 0$, a time-dependent torque is applied, $M = 5 + 3t$ dyn-cm, what will the angular speed and angular acceleration be at $t = 10$ sec if the initial angular velocity is zero?

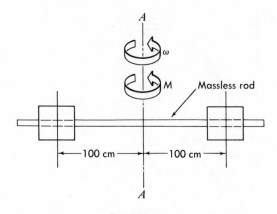

Fig. P11.47

11.48 If in Problem P11.47 no torque is applied, and the masses slide away from the axis of rotation at a speed of 125 cm/sec relative to the rod, what is the angular acceleration of the rod when $\omega = 1$ rad/sec and the masses are in the position shown?

11.49 If at the instant the masses are in the position shown in Fig. P11.47 the angular velocity is 1 rad/sec, and a torque given in Exercise 11.47 is applied, what angular velocity will the masses have when they have moved to a position 150 cm from the axis of rotation? The velocity of the masses relative to the rod is 125 cm/sec.

11.50 Determine the gamma-ray energy which results when a stationary electron absorbs all the energy in Compton scattering.

11.51 Gamma-ray-energy measurements are made in solid, optically clear, scintillating crystals. The gammas interact with essentially free electrons in the crystal with the result that the electrons have all their kinetic energy converted to light. Measurement of the light is then related to the gamma-ray energy. Assuming each gamma ray loses maximum energy in a single interaction and perfectly efficient light conversion and detection occur, construct a scale relating the energy corresponding to the light intensity detected to the incident gamma-ray energy.

11.52 Derive the expression for gamma ray–nucleus scattering and show that the nucleus serves to deflect the gamma ray without changing its energy appreciably.

CHAPTER 12

DYNAMICS
OF A RIGID BODY

12.1 INTRODUCTION

The preceding chapters have been almost exclusively devoted to systems approximated by one or more point masses. The point mass proved to be a simple vehicle for developing the momentum and energy approaches from the postulates of Newton. Many systems, ranging in component size from atoms to planets, could thus be satisfactorily treated as aggregates of point masses.

In this chapter we deal with a special case of the system of particles, the rigid body. As the term implies, this idealization requires that the distances between the individual particles comprising the whole remain fixed, despite the action of external forces. The rigid body is evidently subjected to internal forces supplying the requisite amount of internal constraint. In principle, all the approaches developed for the aggregate of particles apply to the rigid body. We can carry our idealization one step further if the assumption is added that particle spacing is so small as to permit the substitution of a continuous distribution of matter for the system of discrete masses, thus leading to the concept of a *rigid continuum*. This requires that physical properties such as mass density be averaged over volumes large enough to contain a statistically significant number of molecules. Continuous distributions permit the summations of Chapter 11 to be replaced by integrations over mass or volume.

12.2 EQUIVALENT SYSTEMS

In many applications, analysis is significantly simplified by substituting an equivalent system of forces and moments of force for the actual system. Equivalent or equipollent systems are characterized by identical force and moment resultants. These moments must, of course, be taken about the same point in comparing the actual system with the equipollent system or the comparison will be meaningless.

Equal-force resultants in the actual and equivalent systems assure the same acceleration of the mass center. The acceleration of the mass center *does not* depend upon the arrangement of forces but rather the resultant. It is, of course, the *actual* distribution of forces upon which deformation of a real body depends. Similarly, equal net moments of force indicate that each system experiences the same "twisting effect." Clearly the individual moments of force, as well as the resultant of these moments, depend upon the position of forces relative to the point about which moments are taken.

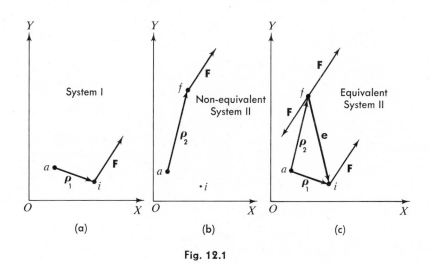

Fig. 12.1

One of the simplest situations encountered involves the relocation of a single force to some parallel new position. What will assure the equivalence of the two force systems shown in Figs. 12.1(a) and 12.1(b)? From the viewpoint of force, system I is clearly identical to system II. The moment of force in system I about any arbitrary point a is not equal to the moment in system II about the same point; $\rho_2 \times F \neq \rho_1 \times F$. To assure equivalence, therefore, a couple moment C, produced by a pair of equal and opposite forces separated by a distance, is added to system II:

$$\rho_2 \times F + C = \rho_1 \times F \qquad C = (\rho_1 - \rho_2) \times F = e \times F \qquad (12.1)$$

The principle is therefore established that accompanying the parallel displacement of a force, equivalence dictates that a couple moment be added to the system. No force is thereby added, since the net force associated with a couple is zero.

Viewed differently, the system of three forces shown in Fig. 12.1(c), achieved by adding equal and opposite forces F to system I, is equivalent to force F in its initial position. The reader can easily verify equivalence by taking moments of each system about any point, say point i.

More complex systems of forces and couple moments can be made equivalent to a single resultant force acting along a given line and a single couple moment. The principle involved in determining the resultant of a general force system can be derived by again referring to Fig. 12.1. In moving the force $\mathbf{F}$ from i to f, a couple moment equal to the moment of $\mathbf{F}$ about f was added to ensure equivalency. The system of forces and couple moments shown in Fig. 12.2 may be similarly reduced. For example, the force and moment resultants at the origin are

$$\mathbf{F}_R = \mathbf{F}_1 + \mathbf{F}_2 + \cdots + \mathbf{F}_n = \sum_{i=1}^{n} \mathbf{F}_i \tag{12.2}$$

$$\mathbf{M}_R = \mathbf{r}_1 \times \mathbf{F}_1 + \mathbf{r}_2 \times \mathbf{F}_2 + \cdots + \mathbf{r}_n \times \mathbf{F}_n + \mathbf{C}_1 + \mathbf{C}_2 + \cdots + \mathbf{C}_m$$

$$= \sum_{i=1}^{n} \mathbf{r}_i \times \mathbf{F}_i + \sum_{i=1}^{m} \mathbf{C}_i \tag{12.3}$$

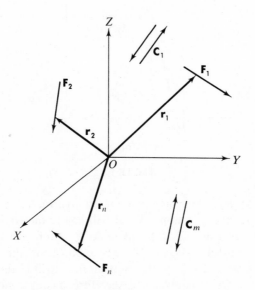

Fig. 12.2

Note that in Eq. (12.3), the moments associated with the force couples have merely been moved to the origin and added vectorially to the moments of force about that point. The basis for this rests in the fact that moments associated with force couples are free vectors. This may be verified by observing that the couple moment $\mathbf{C} = \mathbf{e} \times \mathbf{F}$ is independent of the location of point a in Fig. 12.3.

Equations (12.2) and (12.3) specify that the equivalent system must contain $\mathbf{F}_R$ and couple moment $\mathbf{M}_R$, but they do not direct us as to how $\mathbf{M}_R$ is to be achieved. Since no additional force is to be added, a pair of equal and opposite forces, properly separated, can provide the requisite $\mathbf{M}_R$ without influencing $\mathbf{F}_R$. We have thus used the term couple moment in referring to the moment resultant.

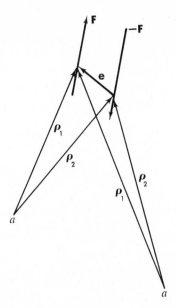

Fig. 12.3

Is a system composed of $\mathbf{F}_R$ and $\mathbf{M}_R$ (arbitrarily oriented relative to $\mathbf{F}_R$), the simplest equivalent system? Recall that the parallel displacement of a force requires the simultaneous addition of a couple moment to assure equipollence. In a system containing $\mathbf{F}_R$ and $\mathbf{M}_R$, appropriate parallel displacement of $\mathbf{F}_R$ can be effected to require the addition of a couple moment of such magnitude and direction as to cancel the component of $\mathbf{M}_R$ perpendicular to $\mathbf{F}_R$. That only the perpendicular component of $\mathbf{M}_R$ can be eliminated is clear from the fact that $\mathbf{e} \times \mathbf{F}_R$ is perpendicular to $\mathbf{F}_R$. What remains is thus a force resultant and a parallel couple moment, and this must suffice as the simplest equivalent of a general force system. Such a system is called a *wrench*.

12.3 CONCURRENT FORCE SYSTEMS

In a *concurrent force system* the lines of action of all forces intersect at a common point. The moments of force about this point are zero, and therefore a single force resultant placed there produces the same moment about any other point as the individual forces. This represents the simplest system equivalent to a set of concurrent forces.

12.4 COPLANAR FORCE SYSTEMS

Consider a system of forces lying in a single plane as shown in Fig. 12.4(a). Since any two nonparallel vectors in a plane are concurrent, $\mathbf{F}_1$ and $\mathbf{F}_2$ may be replaced by a resultant at point a, Fig. 12.4(b). Similarly, this resultant has a point of concurrency

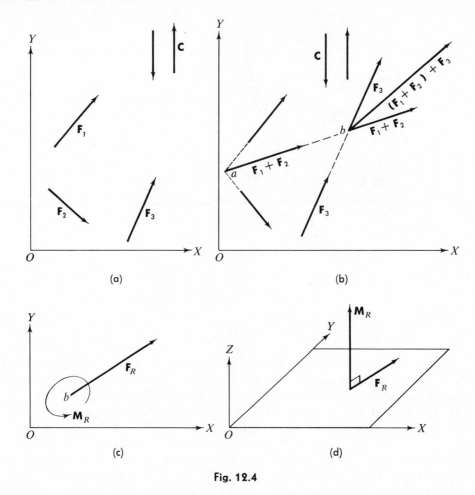

Fig. 12.4

with F_3 at point b, at which the resultant $(F_1 + F_2) + F_3$ is placed. Consideration must also be given any force couples which exist, and their moments added at point b. The equipollent system is thus a force resultant F_R at point b plus M_R, the sum of the couple moments of the original force system, Fig. 12.4(c).

An equivalent system may also be determined by applying Eqs. (12.2) and (12.3). Since the forces, including the couple forces, lie entirely in the XY-plane, it is clear that M_R is normal to the XY-plane, Fig. 12.4(d). Is this the simplest equivalent system? The answer is no, because F_R can be displaced parallel to itself in the plane in such a way as to eliminate M_R. We conclude therefore that if the force resultant is not zero, a coplanar system of forces may be replaced by a single force properly located. If, on the other hand, the force resultant equals zero, we have no way of eliminating M_R, and the equivalent system is simply a couple moment, which may itself be zero.

12.5 PARALLEL-FORCE SYSTEMS

In the case of a parallel system of forces, depicted in Fig. 12.5, an equivalent system may be found in which a single force resultant acts at some point, say O, together with a couple moment in the XY-plane. The couple moment is given by Eq. (12.3) with $\mathbf{r}_i = x_i\hat{\mathbf{i}} + y_i\hat{\mathbf{j}} + z_i\hat{\mathbf{k}}$:

$$\mathbf{M}_R = (x_1\hat{\mathbf{i}} + y_1\hat{\mathbf{j}} + z_1\hat{\mathbf{k}}) \times F_1\hat{\mathbf{k}} + \cdots + \mathbf{C}_1 + \mathbf{C}_2 + \cdots \tag{12.4}$$

Therefore,

$$\mathbf{M}_R = \sum_{i=1}^{n}(-F_ix_i\hat{\mathbf{j}} + F_iy_i\hat{\mathbf{i}}) + \sum_{i=1}^{m}C_{ix}\hat{\mathbf{i}} + \sum_{i=1}^{m}C_{iy}\hat{\mathbf{j}} \tag{12.5}$$

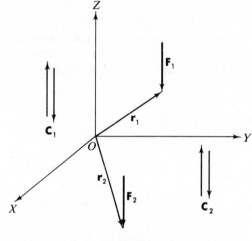

Fig. 12.5

By displacing $\mathbf{F}_R$ parallel to itself, the resultant moment is made zero, and $\mathbf{F}_R$ is all that remains of the original parallel force system. If the force resultant is zero, the moment cannot be eliminated, and it remains as the simplest system.

12.6 DISTRIBUTED SYSTEMS

Consider the one-dimensional distribution of mass shown on the beam in Fig. 12.6, where *mass per unit length* $m(x)$ is plotted as a function of x, the beam-length coordinate. Here we have a simple distributed-mass system. In a gravitational field, a corresponding distributed-force system exists, $w(x) = gm(x)$. By integrating $w(x)$ over the beam, the total force W can be obtained:

$$W = \int_0^L w(x)\,dx \tag{12.6}$$

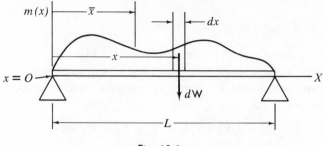

Fig. 12.6

Thus, from the viewpoint of equivalence, a concentrated force W can be used in place of the force distribution. Equivalence also requires that the distributed system and the concentrated force system have the same moment about any point. The force W can be thought of as acting at some position $\bar{x}$ which is the X-coordinate of the *center of gravity* of the distributed mass. The moment of force W about any point must be equal to the sum of the moments of all the elemental weights dW about that point. If $x = 0$ is selected as the point about which moments are summed,

$$W\bar{x} = \int_0^L x\, dW = \int_0^L xw(x)\, dx \tag{12.7}$$

or in view of Eq. (12.6),

$$\bar{x} = \frac{\int_0^L xw(x)\, dx}{\int_0^L w(x)\, dx} \tag{12.8}$$

The above expression is very similar to that defining the X-coordinate of the center of mass. In a uniform gravitational field, the center of mass and center of gravity coincide.

12.7 STATICS OF A RIGID BODY

The relations describing the statics or equilibrium of a rigid body stem directly from the dynamical equations for an aggregate of particles [Eqs. (11.11) and (11.126)]. If a rigid body is in equilibrium, linear and angular momentum are unchanging in time, and therefore Eqs. (11.11) and (11.126) become

$$\sum_{i=1}^{n} \mathbf{F}_i = 0 \tag{12.9}$$

$$\sum_{i=1}^{m} \mathbf{M}_i = 0 \tag{12.10}$$

with the usual restrictions as to frames of reference and points about which moments are taken.

The single vector-force equation of equilibrium may be written as three scalar equations:

$$\sum_{i=1}^{n} F_{xi} = 0 \qquad \sum_{i=1}^{n} F_{yi} = 0 \qquad \sum_{i=1}^{n} F_{zi} = 0 \qquad\qquad \textbf{(12.11)}$$

From the vector-moment equation of equilibrium we may write

$$\sum_{i=1}^{m} M_{xi} = 0 \qquad \sum_{i=1}^{m} M_{yi} = 0 \qquad \sum_{i=1}^{m} M_{zi} = 0 \qquad\qquad \textbf{(12.12)}$$

We thus possess six *independent* equations to describe equilibrium of the general spatial force system. It is important to emphasize the word independent, because it is possible to write an infinite number of equations of equilibrium simply by summing forces in arbitrary directions. However, these force equations are not all independent, and simple algebraic manipulation results in, at most, three independent force equations. Likewise, moments may be taken about an infinite number of lines or axes in space. Figure 12.7 shows, for various force systems, the number of independent equations that can be written, as well as the nature of these equations. Formal proof is left to the reader.

12.8 CONSTRAINT OF A RIGID BODY

Consider a rigid body supported by a number of connecting elements which are hinged and rigid. Assuming hinged elements is equivalent to specifying that each element can only exert forces that act along the line joining its hinges. The question arises as to whether the supporting elements completely constrain the rigid body in space or, more generally, what is the minimum number of elements required for complete constraint? From Fig. 12.8, which shows a rigid body undergoing increasingly more constraint, it is possible to make some generalizations. Although six elements or linkages are required to achieve complete constraint of a rigid body in space, it is important to note that this is a necessary but not a sufficient condition. For complete constraint, it is also required that the six elements have lines of action which cannot be intersected by a single straight line. Only partial constraint will be achieved if all six elements or linkages are parallel. For complete constraint of a rigid body in a plane, three nonparallel elements are required which do not intersect at a common point. Only partial constraint will be achieved with three parallel elements or with three elements with lines of action intersecting at a common point.

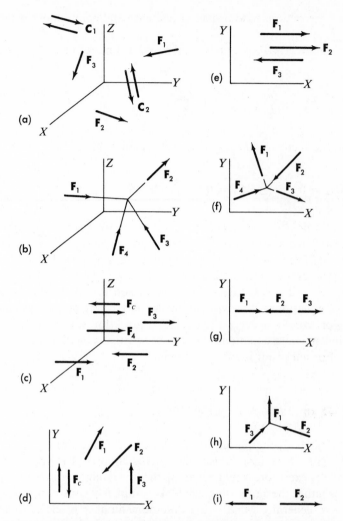

Figure 12.7 (a) General spatial system (three-dimensional force system), six independent equations. Maximum of three force equations may be used; all six equations may be moment equations. (b) Concurrent spatial force system, three independent equations. Lines of action of all forces intersect at a common point; equations of equilibrium may be only force equations, only moment equations, or a combination. (c) System of parallel forces, not necessarily coplanar, three independent equations. Generally one force equation and two moment equations are used; three independent moment equations may be written. (d) Coplanar force system, three independent equations. No more than two force equations may be written. (e) Coplanar parallel system, two independent equations. One force equation and one moment equation, or two moment equations, describe equilibrium. (f) Coplanar concurrent system, two independent equations. Equilibrium is described by one force equation and one moment equation, by two force equations, or by two moment equations. (g) Colinear system, one independent equation. One force equation or one moment equation may be written. (h) Coplanar three-force system, two independent equations. Same comment as for (f); note that for equilibrium this system must be concurrent. (i) Two-force system, one independent equation. Same comment as for (g); note that for equilibrium this system must be colinear.

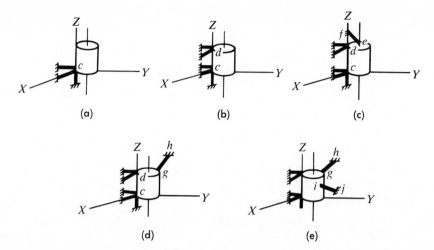

Figure 12.8 (a) Point c is constrained because constraining elements (linkages) are not all contained in a single plane. Rotation of the entire body about point c is still possible, however. Only partial constraint has been achieved. (b) In addition to point c, point d is now constrained so that the line c-d is completely constrained. Rotation about line c-d is still possible, however. Only partial constraint has been achieved. (c) Element e-f has been added. Since it intersects line c-d, the axis of rotation, no additional constraint has been achieved. (d) Element g-h does not intersect the axis of rotation. Complete constraint has been achieved. Constraining forces can be determined by employing equations of statics only; that is, a statically determinate structure exists. (e) Line i-j is a redundant constraint because the rigid body is already completely constrained before its addition. The constraining forces cannot be evaluated by the equations of statics alone; that is, a statically indeterminate structure exists.

12.9 BEAMS

The conditions of equilibrium are now applied in examining the internal forces and moments existing within the common structural element known as the beam. These members are characteristically loaded in a direction primarily transverse to the longitudinal axis, as shown in Fig. 12.9. Note that, in this case, the left support is capable of sustaining a vertical load only, whereas the right support can sustain both vertical and horizontal forces. The fact that a support is capable of sustaining a particular load or moment does not necessarily mean that a nonzero load or moment exists. The arbitrary loading shown in Fig. 12.9 consists of concentrated loads P_1 and P_2 which we assume ideally to act over zero area, and the distributed load $w(x)$. It is usual in simple beam calculations to locate the origin at the left side of the beam, although some unique loading arrangements may make another choice more suitable.

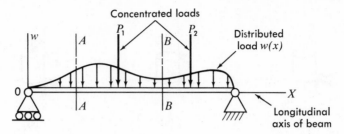

Fig. 12.9

The reactions, which are the forces and moments exerted by the supports, are determined by considering the entire beam as a free body and applying the equations of equilibrium. It is usually simplest to apply the conditions of equilibrium to vertical forces, horizontal forces, and the moments of force taken about a support (or moments about each support and summation of horizontal forces).

If the beam is "cut" at section A-A, the free-body diagram is as shown in Fig. 12.10(a). When the section of beam under consideration is isolated in a free-body diagram, the effect of the remainder of the beam must be replaced by suitable moments and forces. Usually a vertical force V and a moment M must be supplied by the beam to maintain equilibrium of the section under examination. The mechanics by which the beam is able to resist applied moments and maintain equilibrium is discussed in Chapter 13.

Applying the condition of equilibrium to vertical forces,

$$R_1 - V - \int_0^x w(x)\, dx = 0 \tag{12.13}$$

In addition, we may sum moments taken about point a to obtain

$$M - R_1 x + (x - \bar{x}) \int_0^x w(x)\, dx = 0 \tag{12.14}$$

where $\bar{x}$ is given by Eq. (12.8). The directions selected for the *vertical shearing force V* and the *bending moment M* are in accordance with convention for a left-hand beam section.[1]

The free-body diagram that applies to a left-hand beam section taken at B-B in Fig. 12.9 is shown in Fig. 12.10(b). For this section the equations of equilibrium are

---

[1] On such a section, downward shear and counterclockwise bending moment, as shown in Fig. 12.10, are positive. On a right-hand section, the opposite is the convention. These conventions are consistent with those described in terms of beam deformation and deflection associated with positive and negative vertical shear and bending moments shown in Fig. 12.11. If V or M is found to be of negative sign after substitution of numerical values, the shear or bending moment is actually opposite in direction to that assumed. In performing numerical work, it is important to observe that the algebraic signs must be in accordance with the coordinate system. Thus a positive shearing force on a left-hand beam section is a negative force in the coordinate system chosen in Fig. 12.10.

$$R_1 - P_1 - \int_0^x w(x)\,dx - V = 0 \qquad\qquad (12.15)$$

$$M - R_1 x + (x - \bar{x})\int_0^x w(x)\,dx = 0 \qquad\qquad (12.16)$$

where moments have been taken about point b.

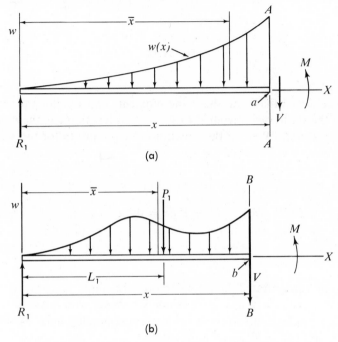

(a)

(b)

Fig. 12.10

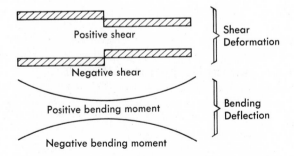

Fig. 12.11

An interesting relationship between the shear and bending moment can be derived on closer examination of the beam. Figure 12.12(b) shows a free body of a beam section of length dx which exists in a beam such as shown in Fig. 12.12(a).

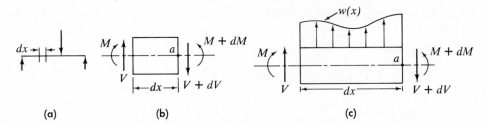

Fig. 12.12

It is initially assumed that both shear and moment can vary along the length of the beam; hence the shear varies from V to $V + dV$ in length dx as the moment varies from M to $M + dM$. Writing the equations of equilibrium for the section in Fig. 12.12(b), we have $V - (V + dV) = 0$, or

$$dV = 0 \tag{12.17}$$

From $\sum M_a = 0$, $(M + dM) - M - V\,dx = 0$, or

$$V = \frac{dM}{dx} \tag{12.18}$$

Between concentrated loads, therefore, the shear is constant and at any point is equal to the rate of change of bending moment with position variable.

For a beam section subjected to distributed loading only, as shown in Fig. 12.12(c), $V + w(x)\,dx - (V + dV) = 0$ or

$$w(x) = \frac{dV}{dx} \tag{12.19}$$

and

$$- M - V\,dx - w(x)\,dx \left(\frac{dx}{2}\right) + (M + dM) = 0 \tag{12.20a}$$

or

$$V = \frac{dM}{dx} \tag{12.20b}$$

where the third term of the left side of Eq. (12.20a) is negligibly small in comparison with the other terms.

For a section of a beam subject to distributed loading, the rate of change of shear is thus found to equal the loading rate $w(x)$ at any point, and, as before, the derivative of the bending moment with respect to x is equal to the shear at any point.

The foregoing can be restated from a different viewpoint. In a beam subjected to distributed loading, the change in shear between any two sections is equal to the area of the loading-rate diagram between the same two sections. If x_1 and x_2 are the respective X-coordinates of sections 1 and 2, then integrating Eq. (12.19) yields

$$\Delta V = V_2 - V_1 = \int_{x_1}^{x_2} w(x)\,dx \qquad (12.21)$$

From the expression $V = dM/dx$ it is clear that the bending moment attains a maximum absolute value when $V = 0$ when the shear is a continuous function of x (as, for example, in the case of a distributed load), or when V passes through zero (as in the case of a discontinuity in the shear).

From Eq. (12.20b) the change in bending moment between any two sections is equal to the area of the shear diagram between the same sections:

$$\Delta M = M_2 - M_1 = \int_{x_1}^{x_2} V\,dx \qquad (12.22)$$

An exception to this statement must be noted in the case of a couple moment acting between the given beam sections. In this event, the couple moment must be added to the change in moment obtained above.

Illustrative Example 12.1
Determine the variation of vertical shear and bending moment across the entire length of the simply supported beam loaded as shown in Fig. 12.13(a).

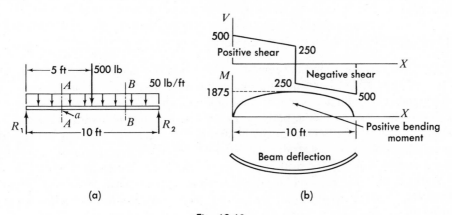

Fig. 12.13

SOLUTION
Applying equilibrium conditions to the entire beam as a free body, we find that $R_1 = R_2 = 500$ lb. For an arbitrary section A-A, a distance x from the left end of the beam, the shear equation is

$$R_1 - \int_0^x w(x)\, dx - V = 0 \qquad (x < 5 \text{ ft})$$

Since $w(x) = 50$ lb/ft (a constant), $\int_0^x w(x)\, dx = 50x$. Therefore, $V = 500 - 50x$. Taking moments about point a yields

$$M = 500x - 50x\frac{x}{2} = 500x - 25x^2$$

The above equation is applicable in the range $0 \le x < 5$ ft.

The equations written for a section cut at B-B are

$$V = R_1 - 50x - P_1 = 500 - 50x - 500 = -50x$$

$$M = R_1 x - (x - \bar{x})\int_0^x w(x)\, dx - P_1(x - 5)$$

$$= 500x - 25x^2 - 500(x - 5) = 2500 - 25x^2$$

The region of applicability of these equations is described by $5 \le x \le 10$ ft.

From the first equation for shear (section A-A), substitution of $x = 5$ ft yields $V = 250$ lb. The shear calculated at $x = 5$ ft from the second shear equation (section B-B) is $V = -250$ lb. Because of the concentrated load, a discontinuity in the shearing force exists at $x = 5$ ft and, since V passes through zero there, the bending moment is a maximum at this point. The bending moment at $x = 5$ ft is calculated to be 1875 lb-ft from both bending-moment equations. The shear and bending-moment diagrams are plots of the above equations and are shown in Fig. 12.13(b).

12.10 RIGID-BODY KINEMATICS — INDEPENDENT COORDINATES

Prior to embarking upon the geometric description of rigid-body motion, the number and type of independent coordinates required must be well understood. By independent coordinates is meant the minimum number necessary at any time to describe the position of the body in a complete and unambiguous manner. The number of such coordinates equals the number of degrees of freedom of the body.

If a body is so constrained as to permit only rectilinear translation, clearly the position along the direction of motion measured from an arbitrary origin represents the only required position variable. In the case of motion about a fixed line, here again a single coordinate suffices, for example, the angle which a given line on the body makes with a reference direction.

Relaxing the constraint somewhat, we turn now to general motion in a plane. Referring to Fig. 12.14(a), it is our task to select the minimum number of coordinates to uniquely specify the position at any time. If the location with respect to a given reference of two points P_1 and P_2 fixed on the body is known at all times, then, by definition of a rigid body, the location of all other points on the body will also be known. Thus we may be reasonably confident that knowledge of

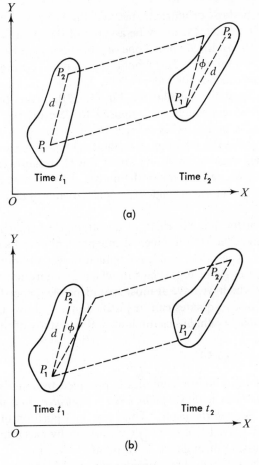

(a)

(b)

Fig. 12.14

the X- and Y-coordinates of P_1 and P_2, a total of four coordinates, can be used to describe the motion. Is this the minimum number of coordinates; that is, is this really a four-degree-of-freedom system? The answers may be found in the assumption of rigidity: the distance between points 1 and 2 equals a constant, d. Thus one additional equation is provided:

$$d = [(x_2 - x_1)^2 + (y_2 - y_1)^2]^{1/2} \qquad (12.23)$$

Since d is known, only three coordinates need be specified to describe the position at any time, say x_1, y_1, and x_2. As an alternative choice of a third coordinate, consider the angle ϕ between line d and the positive X-direction. Once P_1 is known, knowledge of d and ϕ lead directly to point P_2.

The problem of describing the position of a rigid body undergoing nonplanar motion is resolved in much the same way. Knowledge of the coordinates of only two points on the body is insufficient, because the body may rotate about an axis passing through the points. We may be assured of the exact specification of position if the coordinates of three points, one of which does not lie on the line connecting the other two, are known. This would require nine coordinates for description of position. These are not all independent, however, in view of the three constraint equations available [similar to Eq. (12.23)]. Thus the specification of a minimum of only six coordinates is required to describe the position of a rigid body in space; such a body has six degrees of freedom. It is usual to select three coordinates to describe the position of a point such as the mass center, and three angular coordinates such as rotations about the Cartesian axes to determine the orientation in space of a given line on the body. A more convenient set of angular coordinates known as the Euler angles is discussed later in describing the dynamics of a top.

It is noteworthy that the effect of constraints, whether internal or external to the system, is to reduce the number of independent coordinates required to describe the motion. The most significant reduction was of course accomplished in going from the aggregate of n particles in which $3n$ coordinates were required, to the rigid body, for which we have at most six independent coordinates.

The choice of coordinates can be justified in a slightly different manner, but first we must consider two theorems fundamental to a study of rigid-body motion.

12.11 EULER'S THEOREM [2]

The foundation of rigid-body kinematics is provided by *Euler's theorem*. *Any displacement of a rigid body with one point fixed is equivalent to a single rotation about some axis passing through the point.* Thus any final position of a rigid body fixed at a point may be derived from any prior position by executing a single rotation about some line passing through the fixed point.

To prove the theorem, first consider the rigid body shown in Fig. 12.15(a). We know that the location of three noncolinear points establishes its configuration. For convenience the points selected are the fixed point O and two others, A and B, equidistant from O, and therefore lying on the surface of a sphere as depicted. Our concern is now with the displacements, compatible with the fixity of a single point, of two points on the surface of a sphere.

Consider the displacement of the rigid body to be one which moves A to the position occupied by B, and B to some new position C; all points are, of course, equidistant from O and therefore lie on the surface of the sphere. It follows that arcs AB and BC of the great circle are equal in view of the fixed relationship between A and B.

The remainder of the proof depends upon the construction of two great circles centered at O bisecting the arcs AB and BC perpendicularly as shown. The great circles so constructed intersect at point O'.

---

[2] Euler, L., *Novi Comment. Petrop.* XX, 189 (1776).

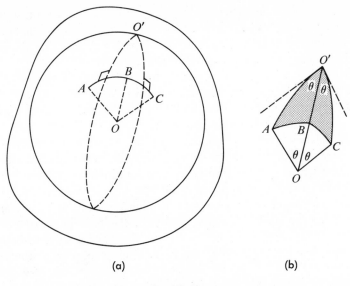

(a)   (b)

Fig. 12.15

Spherical triangles ABO' and BCO' are congruent since their sides are equal, and consequently the angles θ formed by tangents drawn to arcs AO', BO', and CO' at O' are equal. These angles are, of course, also equal to the angles θ formed by OA, OB, OC in Fig. 12.15(b). In effecting the displacement, all points lying on line OO' maintain exactly the same orientation in space. It is clear, then, that a rotation about axis OO' brings A to the position occupied by B, and B to that occupied by C, and Euler's theorem is thus proved.

The continuous motion of a rigid body about a point may be regarded then as rotation about a line passing through the point which constantly changes its orientation in space. As the time period in question approaches zero as a limit, the axis becomes an *instantaneous axis of rotation*.

12.12 CHASLES' THEOREM [3]

Euler's theorem, dealing with general motion about a fixed point, leads quite naturally to *Chasles' theorem*, fundamental to the analysis of general motion with no points fixed. The complexity inherent in a description of motion in which the position of each particle comprising the body is changed as a consequence of a general displacement is enormously reduced if one considers the displacement equivalent to a translation plus a rotation about some axis. This is Chasles' theorem, which, to be properly applied, requires that the translation of a particular point, the base point, through which the axis of rotation passes, be specified. Ap-

[3] Chasles, M., *Bull. Univ. Sci. (F. E. Russac)* **XIV**, 321 (1830).

plication of Chasles' theorem thus reduces any displacement to translation, which has no effect upon orientation, inasmuch as every straight line on the body remains parallel to its original direction, and rotation, which serves to alter the orientation.

The selection of a base point is usually a matter of convenience. We have demonstrated that the motion of a system of particles may be viewed as motion of the mass center plus motion relative to the mass center. In particular, the kinetic energy and angular momentum have been so treated. The center of mass is therefore a likely candidate for the base point.

Since the base point is unaffected by the rotation, being itself the point about which rotation occurs, and because the translation has no effect upon orientation, it is clear that the order of translation and rotation is immaterial in proceeding from an initial to a final configuration of the body.

Clearly, the rigid body is rotated through the same angular displacement, and in the same direction, regardless of where the base point is located. The translational displacement required to achieve the same final rigid-body position does, however, depend upon where the base point is located, as the reader may verify.

Chasles' theorem thus provides another basis for choosing coordinates to describe rigid-body motion. Referring again to Fig. 12.14(a), consider the position of the body at times t_1 and t_2. We may view the body as undergoing first a translation as shown, then a rotation about a line passing through P_1. Figure 12.14(b) shows that the same final position has been achieved via a rotation about point P_1 followed by a translation. Clearly the coordinates x_1, y_1 of P_1 and ϕ are apt choices to describe rigid-body motion in a plane.

12.13 VELOCITY AND ACCELERATION

It is fundamental to the development of rigid-body kinematics and dynamics that the distance between any two points on the body remains fixed. This imposes special restrictions with respect to the nature of the relative velocity and acceleration between any points on the body. Consider now the motion of point P, contained on a rigid body undergoing general translation and rotation with respect to XYZ, as shown in Fig. 12.16. Employing the general approach of Chapter 2, the xyz-coordinate system is fixed to the rigid body. The position, velocity, and acceleration of P with respect to XYZ are given by

$$\mathbf{r}_P = \mathbf{R} + \boldsymbol{\rho} \qquad (12.24)$$

$$\dot{\mathbf{r}}_P = \dot{\mathbf{R}} + \dot{\boldsymbol{\rho}} + \boldsymbol{\omega} \times \boldsymbol{\rho} \qquad (12.25)$$

$$\ddot{\mathbf{r}}_P = \ddot{\mathbf{R}} + 2\boldsymbol{\omega} \times \mathbf{v}_{rel} + \boldsymbol{\omega} \times (\boldsymbol{\omega} \times \boldsymbol{\rho}) + \mathbf{a}_{rel} + \dot{\boldsymbol{\omega}} \times \boldsymbol{\rho} \qquad (12.26)$$

where it is recalled that $\mathbf{v}_{rel} = \dot{\boldsymbol{\rho}}$ and $\mathbf{a}_{rel} = \ddot{\boldsymbol{\rho}}$ are relative quantities, measured with respect to the body. This leads us to the use of the rigid-body assumption: The

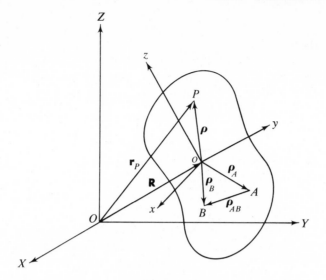

Fig. 12.16

relative velocity and relative acceleration terms as measured in the *xyz*-frame must be zero. The simplified versions of Eqs. (12.25) and (12.26) therefore become

$$\dot{\mathbf{r}}_P = \dot{\mathbf{R}} + \boldsymbol{\omega} \times \boldsymbol{\rho} \tag{12.27}$$

$$\ddot{\mathbf{r}}_P = \ddot{\mathbf{R}} + \boldsymbol{\omega} \times (\boldsymbol{\omega} \times \boldsymbol{\rho}) + \dot{\boldsymbol{\omega}} \times \boldsymbol{\rho} \tag{12.28}$$

The relative velocity and acceleration between any two points as viewed from *XYZ* are simply the vector differences between the respective quantities. Referring to Fig. 12.16,

$$\mathbf{v}_{BA} = \dot{\mathbf{r}}_B - \dot{\mathbf{r}}_A = (\dot{\mathbf{R}} + \boldsymbol{\omega} \times \boldsymbol{\rho}_B) - (\dot{\mathbf{R}} + \boldsymbol{\omega} \times \boldsymbol{\rho}_A)$$

$$= \boldsymbol{\omega} \times (\boldsymbol{\rho}_B - \boldsymbol{\rho}_A) = \boldsymbol{\omega} \times \boldsymbol{\rho}_{AB} \tag{12.29}$$

In examining the relative velocity of *B* to *A*, one can imagine the body experiencing rotation only about point *A*. This is because the only relative velocity possible is attributable to the rotation of the body. It is as though *xyz* were fixed to the body at point *A* rather than *O*. The relative velocity between any two points on a body in pure translation is therefore zero.

In a similar fashion, the relative acceleration of points *A* and *B* is

$$\mathbf{a}_{BA} = \ddot{\mathbf{r}}_B - \ddot{\mathbf{r}}_A = [\ddot{\mathbf{R}} + \boldsymbol{\omega} \times (\boldsymbol{\omega} \times \boldsymbol{\rho}_B) + \dot{\boldsymbol{\omega}} \times \boldsymbol{\rho}_B]$$

$$- [\ddot{\mathbf{R}} + \boldsymbol{\omega} \times (\boldsymbol{\omega} \times \boldsymbol{\rho}_A) + \dot{\boldsymbol{\omega}} \times \boldsymbol{\rho}_A]$$

$$= \boldsymbol{\omega} \times (\boldsymbol{\omega} \times \boldsymbol{\rho}_{AB}) + \dot{\boldsymbol{\omega}} \times \boldsymbol{\rho}_{AB} \tag{12.30}$$

Equations (12.29) and (12.30) may also be derived by applying Eqs. (2.34) and (2.44) and the definitions of relative velocity and acceleration.

Illustrative Example 12.2

Point 1 on the rod, Fig. 12.17, possesses a velocity v_1 and acceleration a_1 to the right, at the instant shown. Determine the instantaneous velocity of point 2 as well as the angular velocity and angular acceleration of the rod.

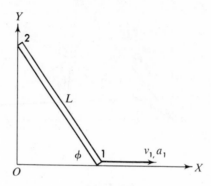

Fig. 12.17

SOLUTION

Applying Eq. (12.29), $\mathbf{v}_1 - \mathbf{v}_2 = \boldsymbol{\omega} \times \boldsymbol{\rho}_{21}$, where $\mathbf{v}_1 = v_1\hat{\mathbf{i}}$ (given), $\mathbf{v}_2 = v_2\hat{\mathbf{j}}$ (direction known because of constraint), $\boldsymbol{\rho}_{21} = L(\cos\phi\hat{\mathbf{i}} - \sin\phi\hat{\mathbf{j}})$, $\boldsymbol{\omega} = \omega_x\hat{\mathbf{i}} + \omega_y\hat{\mathbf{j}} + \omega_z\hat{\mathbf{k}}$. The angular velocity is presented as above even though it is clear that $\boldsymbol{\omega}$ has only a Z-component; let the results confirm this.

Expanding the relative velocity equation,

$$v_1\hat{\mathbf{i}} - v_2\hat{\mathbf{j}} = (\omega_x\hat{\mathbf{i}} + \omega_y\hat{\mathbf{j}} + \omega_z\hat{\mathbf{k}}) \times L(\cos\phi\hat{\mathbf{i}} - \sin\phi\hat{\mathbf{j}})$$

$$= -\omega_x L\sin\phi\hat{\mathbf{k}} - \omega_y L\cos\phi\hat{\mathbf{k}} + \omega_z L\cos\phi\hat{\mathbf{j}} + \omega_z L\sin\phi\hat{\mathbf{i}}$$

Equating the respective coefficients of the unit vectors:

(a) $\quad v_1 = \omega_z L\sin\phi$

(b) $\quad v_2 = \omega_z L\cos\phi$

(c) $\quad 0 = -\omega_x L\sin\phi - \omega_y L\cos\phi$

From Eq. (a),

$$\omega_z = \frac{v_1}{L\sin\phi}$$

Equation (c) is satisfied for all ϕ if $\omega_x = \omega_y = 0$. Consequently, the angular velocity of the rod is

$$\boldsymbol{\omega} = \omega_z\hat{\mathbf{k}} = \frac{v_1}{L\sin\phi}\,\hat{\mathbf{k}}$$

$$\mathbf{v}_2 = -\frac{v_1}{\sin\phi}\cos\phi\hat{\mathbf{j}} = -\frac{v_1}{\tan\phi}\,\hat{\mathbf{j}}$$

It is interesting to determine the components of $\mathbf{v}_1$ and $\mathbf{v}_2$ along the rod; these must clearly be equal if the rod is rigid. Forming the scalar product of velocity with a unit vector colinear with the rod, we have

$$\mathbf{v}_1 \cdot (\cos \phi \hat{\mathbf{i}} - \sin \phi \hat{\mathbf{j}}) = (\omega_z L \sin \phi \hat{\mathbf{i}}) \cdot (\cos \phi \hat{\mathbf{i}} - \sin \phi \hat{\mathbf{j}}) = \omega_z L \sin \phi \cos \phi$$

$$\mathbf{v}_2 \cdot (\cos \phi \hat{\mathbf{i}} - \sin \phi \hat{\mathbf{j}}) = (-\omega_z L \cos \phi \hat{\mathbf{j}}) \cdot (\cos \phi \hat{\mathbf{i}} - \sin \phi \hat{\mathbf{j}}) = \omega_z L \cos \phi \sin \phi$$

The acceleration of point 1 relative to point 2 is given by Eq. (12.30):

$$\mathbf{a}_1 - \mathbf{a}_2 = \boldsymbol{\omega} \times (\boldsymbol{\omega} \times \boldsymbol{\rho}_{21}) + \dot{\boldsymbol{\omega}} \times \boldsymbol{\rho}_{21}$$

$$a_1 \hat{\mathbf{i}} - a_2 \hat{\mathbf{j}} = \omega_z \hat{\mathbf{k}} \times [\omega_z \hat{\mathbf{k}} \times L(\cos \phi \hat{\mathbf{i}} - \sin \phi \hat{\mathbf{j}})] + (\dot{\omega}_x \hat{\mathbf{i}} + \dot{\omega}_y \hat{\mathbf{j}} + \dot{\omega}_z \hat{\mathbf{k}})$$

$$\times L(\cos \phi \hat{\mathbf{i}} - \sin \phi \hat{\mathbf{j}})$$

Expanding and equating coefficients of the unit vectors:

(a) $a_1 = -\omega_z^2 L \cos \phi + \dot{\omega}_z L \sin \phi$

(b) $a_2 = -\omega_z^2 L \sin \phi - \dot{\omega}_z L \cos \phi$

(c) $0 = -\dot{\omega}_x L \sin \phi - \dot{\omega}_y L \cos \phi$

From Eq. (a),

$$\dot{\omega}_z = \frac{a_1}{\sin L \phi} + \frac{\omega_z^2}{\tan \phi}$$

Equation (c) is satisfied for all ϕ if $\dot{\omega}_x = \dot{\omega}_y = 0$.

12.14 RIGID-BODY DYNAMICS — TRANSLATION OF A RIGID BODY

Rigid-body translation is described by the same expression introduced in connection with a system of particles:

$$\sum_{i=1}^{n} \mathbf{F}_i = \frac{d^2}{dt^2} (m\mathbf{r}_{CM}) = m\mathbf{a}_{CM} \tag{12.31}$$

Since we are dealing with a translating body, Eq. (12.31) is sufficient to completely describe the system and no consideration of rotation is relevant.

The fictitious particles of previous chapters were, of course, extended bodies in which a single point, the center of mass, played the role of the entire body. Such idealization has heretofore been limited to body shapes for which our intuition serves to locate the center of mass. Let us now discuss the determination of the mass center of bodies employing the results of multiparticle development. Assuming a continuous distribution of mass so that dm replaces m_i,

$$m = \int_m dm = \int_V \rho \, dV \tag{12.32}$$

where ρ is the mass density and dV the elemental volume, containing dm. From Eq. (11.9) we have, therefore,

$$\mathbf{r}_{CM} = \frac{\int_V \mathbf{r}\rho\, dV}{\int_V \rho\, dV} \qquad (12.33)$$

In calculations involving nonhomogeneous materials, as, for example, where a temperature distribution causes significant density variations, density is a function of position and must be treated as a variable.

Illustrative Example 12.3

Given the system shown in Fig. 12.18, determine the acceleration of masses m_1 and m_2, the cord tension, and the point at which the normal force between the horizontal plane and m_2 effectively acts to assure rotational equilibrium. Assume negligible friction between the massless, inextensible cord and guide G. The following data apply: $m_1 = 3$ kg, $m_2 = 10$ kg, and $\mu = 0.1$ between m_2 and the plane.

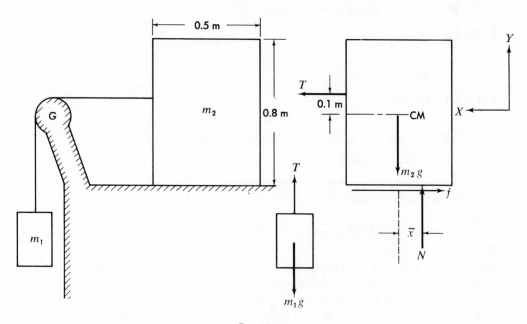

Fig. 12.18

SOLUTION

The free-body diagram of mass m_1 leads to the equations $\sum F_y = m_1 a_1$, $T - m_1 g = m_1 a_1$, and $T - 3(9.8) = 3a_1 = -3a_2$. From a free-body diagram of m_2, we obtain $\sum F_y = 0$, $N = m_2 g = 10(9.8) = 98$ N. (Therefore, $f = \mu N = 9.8$ N.) $\sum F_x = m_2 a_2$, $T - f = m_2 a_2$, and $T - 9.8 = 10a_2 = -10a_1$. Solving for a_2 and T, $a_2 = 1.51$ m/sec^2 and $T = 24.9$ N.

To ascertain the location of N for rotational equilibrium, moments about the mass center give $\sum M_{CM} = 0$, $\bar{x}N + 0.1T + 0.4f = 0$, and $\bar{x} = -0.065$ m. The negative sign indicates that $\bar{x}$ is located on the opposite side of the vertical line through the mass center. It is necessary to elect as a point about which moments are taken, one that agrees with the criteria developed in Chapter 11.

12.15 GENERAL MOTION OF A RIGID BODY AND THE TENSOR OF INERTIA

We now leave the dynamics of translation to delve into the rotational dynamics of a rigid body. First, the angular momentum will be developed employing the general approach of Chapter 11. From this will emerge the tensor of inertia and then, upon application of $\mathbf{M} = \dot{\mathbf{H}}$, the well-known Euler equations. Since in general motion, no point is fixed, it is most convenient to apply $\mathbf{M} = \dot{\mathbf{H}}$ to the center of mass.

The angular momentum about the center of mass of infinitesimal mass dm shown in Fig. 12.19 is formed by taking the moment about the CM of the linear

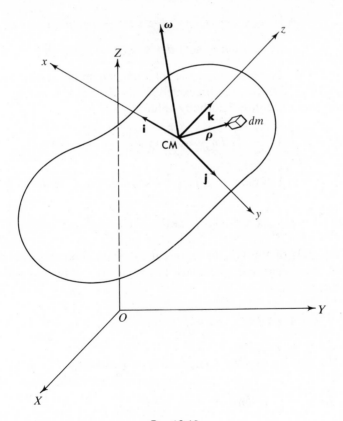

Fig. 12.19

momentum relative to the CM. The moment of linear momentum is thus $d\mathbf{H}_{CM} = \mathbf{r} \times dm\, \mathbf{v}$. Referring to this figure, the velocity of an arbitrary elemental mass relative to the center of mass is $\mathbf{v} = \boldsymbol{\omega} \times \mathbf{r}$, where $\boldsymbol{\omega}$ is the angular velocity of the body with respect to XYZ, an inertial frame of reference. In Fig. 12.19, the reader will note an xyz-coordinate system with origin coincident with the center of mass. For the time being we place no restrictions upon the orientation of xyz relative to either XYZ or the body. Temporarily, xyz may have any time-dependent orientation.

At any time, $d\mathbf{H}_{CM}$, $\mathbf{r}$ and $\boldsymbol{\omega}$ may each be resolved into components parallel to the xyz-axes. The angular momentum of dm about the mass center is

$$d\mathbf{H}_{CM} = (dH_{CM})_x\hat{\mathbf{i}} + (dH_{CM})_y\hat{\mathbf{j}} + (dH_{CM})_z\hat{\mathbf{k}} = \mathbf{r} \times dm(\boldsymbol{\omega} \times \mathbf{r})$$
$$= (x\hat{\mathbf{i}} + y\hat{\mathbf{j}} + z\hat{\mathbf{k}}) \times [(\omega_x\hat{\mathbf{i}} + \omega_y\hat{\mathbf{j}} + \omega_z\hat{\mathbf{k}}) \times (x\hat{\mathbf{i}} + y\hat{\mathbf{j}} + z\hat{\mathbf{k}})]\, dm$$

$$(12.34)$$

Expansion of Eq. (12.34) yields

$$d\mathbf{H}_{CM} = [\omega_x(y^2 + z^2)\, dm - \omega_y xy\, dm - \omega_z xz\, dm]\hat{\mathbf{i}}$$
$$+ [-\omega_x yx\, dm + \omega_y(x^2 + z^2)\, dm - \omega_z yz\, dm]\hat{\mathbf{j}}$$
$$+ [-\omega_x zx\, dm - \omega_y zy\, dm + \omega_z(x^2 + y^2)\, dm]\hat{\mathbf{k}} \qquad (12.35)$$

To obtain the components of the total angular momentum about the mass center, $H_{CM,x}, H_{CM,y}, H_{CM,z}$, the bracketed coefficients of $\hat{\mathbf{i}}$, $\hat{\mathbf{j}}$, and $\hat{\mathbf{k}}$, respectively, are integrated over the entire rigid-body volume:

$$H_{CM,x} = \omega_x \int_m (y^2 + z^2)\, dm - \omega_y \int_m xy\, dm - \omega_z \int_m xz\, dm$$

$$H_{CM,y} = -\omega_x \int_m yx\, dm + \omega_y \int_m (x^2 + z^2)\, dm - \omega_z \int_m yz\, dm$$

$$H_{CM,z} = -\omega_x \int_m zx\, dm - \omega_y \int_m zy\, dm + \omega_z \int_m (x^2 + y^2)\, dm \qquad (12.36)$$

The integrals of Eq. (12.36) represent the mass moments and products of inertia. It is convenient and customary to assign the following nomenclature to these quantities,

$$I_{xx} = \int_m (y^2 + z^2)\, dm \qquad I_{yy} = \int_m (x^2 + z^2)\, dm \qquad I_{zz} = \int_m (x^2 + y^2)\, dm$$

$$I_{xy} = I_{yx} = -\int_m xy\, dm \qquad I_{xz} = I_{zx} = -\int_m xz\, dm$$

$$I_{yz} = I_{zy} = -\int_m yz\, dm \qquad (12.37)$$

where terms such as I_{xx} and I_{yy} are the mass moments of inertia, and terms such as I_{xy} and I_{zx} are the mass products of inertia. A convenient and concise representation of the defining integrals given above is

$$I_{\alpha\beta} = \int_m (r^2 \hat{\mathbf{e}}_\alpha \cdot \hat{\mathbf{e}}_\beta - \alpha\beta)\, dm \tag{12.38}$$

where α and β each take on the value x, y, z and where $\hat{\mathbf{e}}_\alpha$ and $\hat{\mathbf{e}}_\beta$ represent unit vectors along the respective directions indicated by the subscripts. In Eq. (12.38) $r^2 = x^2 + y^2 + z^2$, that is, the square of the distance from the origin of xyz to an element of mass dm.

It is clear that the inertia integrals depend upon the location and orientation with respect to the body of the xyz-coordinate system. Indeed, this system need not even be attached to the body. In view of this dependence upon coordinate frame, it is interesting to demonstrate the following property: The sum of the mass moments of inertia with respect to a given coordinate system is an invariant, independent of the orientation of the coordinate axes. To demonstrate this, we need only add the moments of inertia:

$$\int_m (y^2 + z^2)\, dm + \int_m (x^2 + z^2)\, dm + \int_m (x^2 + y^2)\, dm$$

$$= \int_m (2x^2 + 2y^2 + 2z^2)\, dm = \int_m 2r^2\, dm = \text{constant} \tag{12.39}$$

The distance r, as shown in Fig. 12.20, is a function of the location of the origin of xyz, not of the orientation of the coordinate axes with respect to the body.

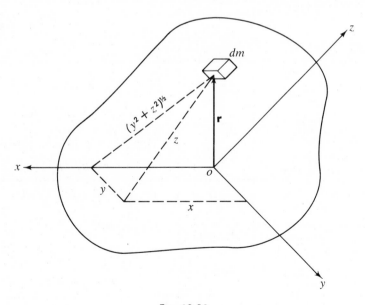

Fig. 12.20

The moments of inertia formed with respect to certain convenient reference axes are tabulated in Fig. 12.21. The reference axes are usually axes of rotation or axes passing through the center of mass. The inertia integrals are not difficult

Geometry	Mass Center	Moments of Inertia [*]
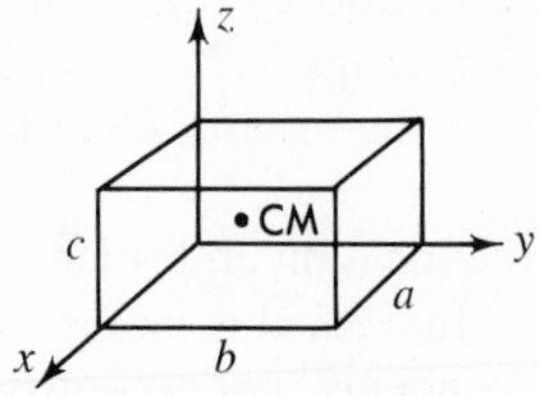 Rectangular Parallelepiped	$x_{cm} = a/2$ $y_{cm} = b/2$ $z_{cm} = c/2$	$I_{xx} = \frac{1}{3} m \left(b^2 + c^2 \right)$ $I_{x,cm} = \frac{1}{12} m \left(b^2 + c^2 \right)$ $m = \rho abc$
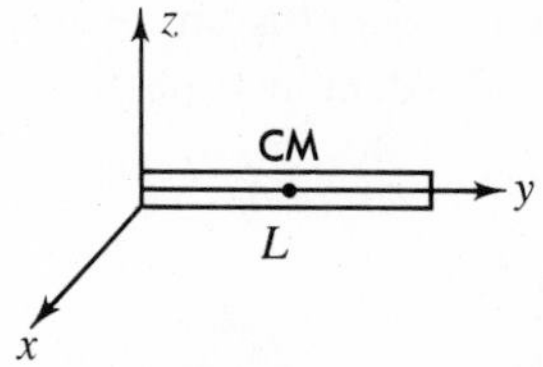 Thin Rod	$x_{cm} = 0$ $y_{cm} = L/2$ $z_{cm} = 0$	$I_{xx} = I_{zz} = \frac{1}{3} mL^2$ $I_{yy} = I_{y,cm} = 0$ $I_{x,cm} = I_{z,cm} = \frac{1}{12} mL^2$ $m = \rho_L L$
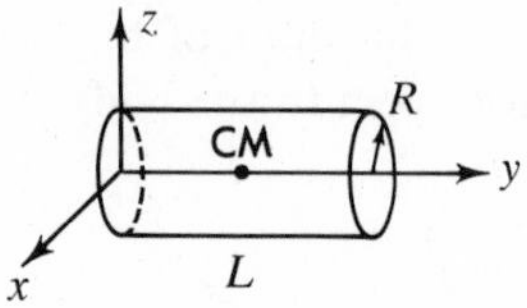 Circular Cylinder	$x_{cm} = 0$ $y_{cm} = L/2$ $z_{cm} = 0$	$I_{xx} = I_{zz} = \frac{m}{12} \left(3R^2 + 4L^2 \right)$ $I_{yy} = I_{y,cm} = \frac{1}{2} mR^2$ $I_{x,cm} = I_{z,cm} = \frac{m}{12} \left(3R^2 + L^2 \right)$ $m = \pi \rho R^2 L$
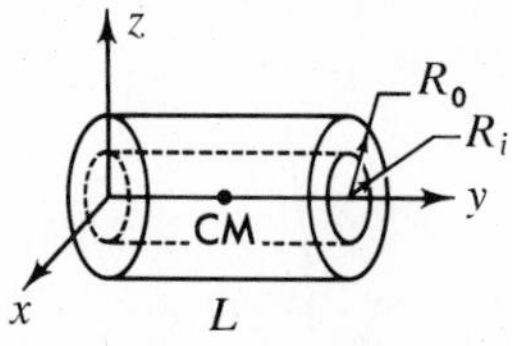 Hollow Circular Cylinder	$x_{cm} = 0$ $y_{cm} = L/2$ $z_{cm} = 0$	$I_{xx} = I_{zz} = \frac{m}{12} \left(3R_i^2 + 3R_0^2 + 4L^2 \right)$ $I_{yy} = I_{y,cm} = \frac{m}{2} \left(R_i^2 + R_0^2 \right)$ $I_{x,cm} = I_{z,cm} = \frac{m}{12} \left(3R_i^2 + 3R_0^2 + L^2 \right)$ $m = \pi \rho \left(R_0^2 - R_i^2 \right) L$
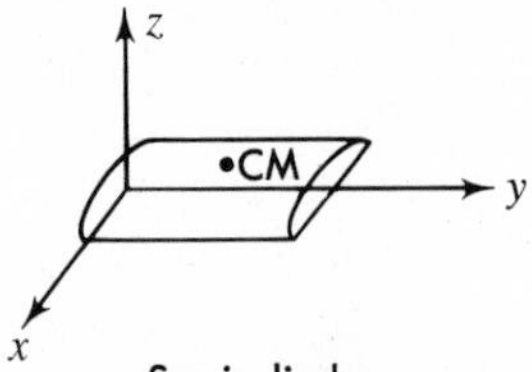 Semicylinder	$x_{cm} = 0$ $y_{cm} = L/2$ $z_{cm} = \frac{4R}{3\pi}$	$I_{xx} = I_{zz} = \frac{1}{4} mR^2 + \frac{1}{3} mL^2$ $I_{yy} = \frac{1}{2} mR^2$ $I_{x,cm} = I_{z,cm} = \frac{1}{4} mR^2 + \frac{1}{12} mL^2$ $m = \frac{1}{2} \pi \rho R^2 L$

Fig. 12.21(a)

to evaluate if the geometry is relatively simple, but considerable effort may be required for complex shapes. In this regard the digital computer is an invaluable tool, because it can quickly perform the summations implied by the defining inte-

Geometry	Mass Center	Moments of Inertia
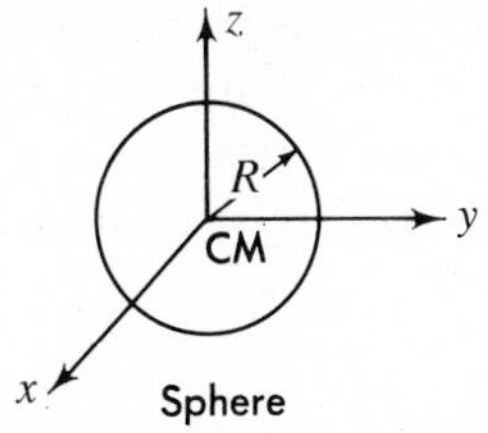 Sphere	$x_{cm} = 0$ $y_{cm} = 0$ $z_{cm} = 0$	$I_{xx} = I_{yy} = I_{zz} = \dfrac{2}{5} mR^2$ $I_{x,cm} = I_{y,cm} = I_{z,cm} = \dfrac{2}{5} mR^2$ $m = \dfrac{4}{3} \pi \rho R^3$
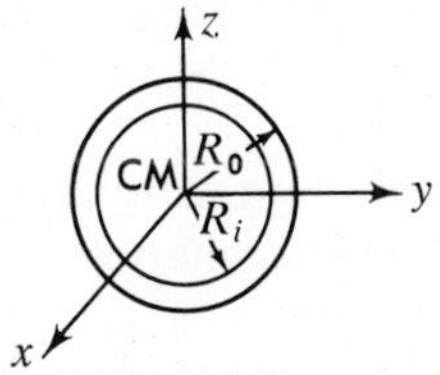 Hollow Sphere	$x_{cm} = 0$ $y_{cm} = 0$ $z_{cm} = 0$	$I_{xx} = I_{yy} = I_{zz} = \dfrac{2}{5} m \left(\dfrac{R_0^5 - R_i^5}{R_0^3 - R_i^3} \right)$ $I_{x,cm} = I_{y,cm} = I_{z,cm} = \dfrac{2}{5} m \left(\dfrac{R_0^5 - R_i^5}{R_0^3 - R_i^3} \right)$ $m = \dfrac{4}{3} \pi \rho \left(R_0^3 - R_i^3 \right)$
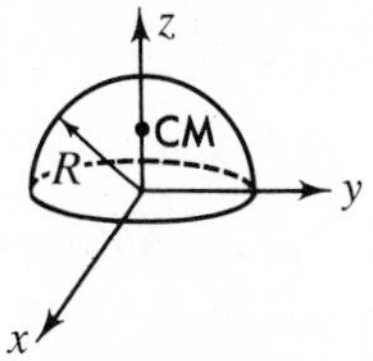 Hemisphere	$x_{cm} = 0$ $y_{cm} = 0$ $z_{cm} = \dfrac{3}{8} R$	$I_{xx} = I_{yy} = I_{zz} = \dfrac{2}{5} mR^2$ $m = \dfrac{2}{3} \pi \rho R^3$
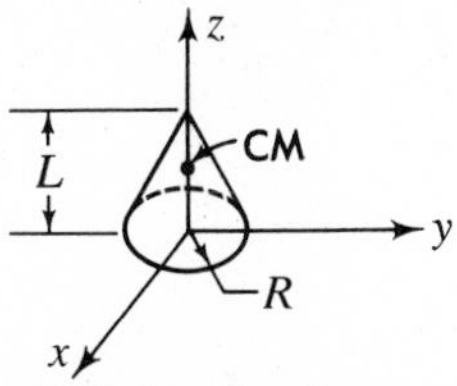 Right Circular Cone	$x_{cm} = 0$ $y_{cm} = 0$ $z_{cm} = h/4$	$I_{xx} = I_{yy} = \dfrac{1}{20} m \left(3R^2 + 2h^2 \right)$ $I_{zz} = I_{z,cm} = \dfrac{3}{10} mR^2$ $I_{x,cm} = I_{y,cm} = \dfrac{3}{80} m \left(4R^2 + h^2 \right)$ $m = \dfrac{1}{3} \pi \rho R^2 h$
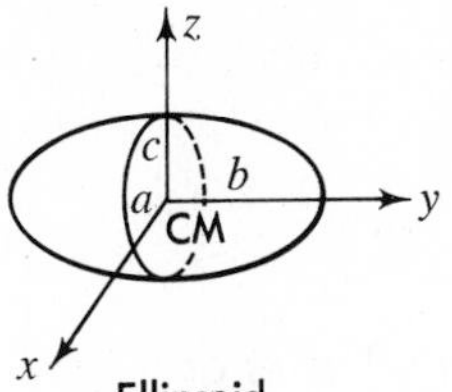 Ellipsoid	$x_{cm} = 0$ $y_{cm} = 0$ $z_{cm} = 0$	$I_{xx} = I_{x,cm} = \dfrac{m}{5} \left(b^2 + c^2 \right)$ $I_{yy} = I_{y,cm} = \dfrac{m}{5} \left(a^2 + c^2 \right)$ $I_{zz} = I_{z,cm} = \dfrac{m}{5} \left(a^2 + b^2 \right)$ $m = \dfrac{4}{3} \pi \rho\, abc$

Fig. 12.21(b)

grals once a given mass has been divided into an acceptable number of smaller masses. In any case, when the integrals are known for a given set of axes, transformations discussed in the sections to follow may be performed which enable the simple evaluation of the complete set of integrals with respect to any axes translated and rotated with respect to the original axes. Great savings in time and effort can thus be effected. In addition, evaluation of the integrals may often be simplified by dividing a complex geometry into constituent shapes for which the moments and products of inertia are already known, and adding the individual inertia terms. This, of course, requires that each term be referred to the same coordinate set.

When the nomenclature associated with the inertia integrals is introduced into Eq. (12.36), the angular momentum appears as follows:

$$H_{\text{CM},x} = I_{xx}\omega_x + I_{xy}\omega_y + I_{xz}\omega_z$$

$$H_{\text{CM},y} = I_{yx}\omega_x + I_{yy}\omega_y + I_{yz}\omega_z \qquad (12.40)$$

$$H_{\text{CM},z} = I_{zx}\omega_x + I_{zy}\omega_y + I_{zz}\omega_z$$

The coefficients of the angular velocity terms may be placed in an array in which they occupy the same relative positions as in Eq. (12.40). This array obeys transformations typical of a tensor and is termed the *tensor of inertia*.

$$\mathsf{I} = \begin{bmatrix} I_{xx} & I_{xy} & I_{xz} \\ I_{yx} & I_{yy} & I_{yz} \\ I_{zx} & I_{zy} & I_{zz} \end{bmatrix} \qquad (12.41)$$

Since $I_{xy} = I_{yx}$, and so on, as may be verified by examining the defining integrals, the inertia tensor is symmetrical with respect to the main diagonal.

We are now in a position to write the angular momentum as the product of two matrices:

$$\begin{bmatrix} H_x \\ H_y \\ H_z \end{bmatrix} = \begin{bmatrix} I_{xx} & I_{xy} & I_{xz} \\ I_{yx} & I_{yy} & I_{yz} \\ I_{zx} & I_{zy} & I_{zz} \end{bmatrix} \begin{bmatrix} \omega_x \\ \omega_y \\ \omega_z \end{bmatrix} \qquad (12.42)$$

or simply

$$\mathbf{H} = \mathsf{I}\boldsymbol{\omega} \qquad (12.43)$$

As we have stated, the rotational dynamics of a rigid body are developed from the expression

$$\mathbf{M} = \dot{\mathbf{H}} = \frac{d}{dt}(\mathsf{I}\boldsymbol{\omega}) \qquad (12.44)$$

where the moment of force may be taken about the center of mass or a fixed point and the tensor of inertia formed with respect to axes with origin at these respective points.

The development continues with motion referred to the center of mass. In determining the time derivative of the angular momentum, some caution is necessary. The derivative, although taken with respect to the inertial frame of reference, may nevertheless be expressed in terms of rectangular components in the xyz reference frame:

$$\mathbf{M}_{CM} = \frac{d}{dt}(H_{CM,x}\hat{\mathbf{i}}) + \frac{d}{dt}(H_{CM,y}\hat{\mathbf{j}}) + \frac{d}{dt}(H_{CM,z}\hat{\mathbf{k}}) \tag{12.45}$$

The unit vectors of the xyz-system must be treated as variables in Eq. (12.45), since their directions are continuously changing when viewed from XYZ.

$$\mathbf{M}_{CM} = (\dot{H}_{CM,x})\hat{\mathbf{i}} + (\dot{H}_{CM,y})\hat{\mathbf{j}} + (\dot{H}_{CM,z})\hat{\mathbf{k}}$$
$$+ (H_{CM,x})\dot{\hat{\mathbf{i}}} + (H_{CM,y})\dot{\hat{\mathbf{j}}} + (H_{CM,z})\dot{\hat{\mathbf{k}}} \tag{12.46}$$

It is now appropriate to reexamine our lack of restriction concerning the xyz-system. As matters stand, $\dot{H}_{CM,x}$ is expressed

$$\dot{H}_{CM,x} = I_{xx}\dot{\omega}_x + \dot{I}_{xx}\omega_x + I_{xy}\dot{\omega}_y + \dot{I}_{xy}\omega_y + I_{xz}\dot{\omega}_z + \dot{I}_{xz}\omega_z$$

Thus one is confronted with the rather uncomfortable prospect of dealing with time varying moments and products of inertia resulting from the rotation of xyz with respect to the body. Some simplification is achieved if xyz is so restricted as to always remain parallel to XYZ, thus eliminating the time derivatives of the unit vectors in Eq. (12.46). The problem of time-varying moments and products of inertias remains as long as xyz rotates relative to the body. Selecting the lesser of two complexities, one usually elects to fix xyz to the body, thus making the angular velocity of the body and that of xyz identical.

Substitution of Eqs. (12.40) together with $\dot{\hat{\mathbf{i}}} = \boldsymbol{\omega} \times \hat{\mathbf{i}}$, and so on, into Eq. (12.46) yields

$$(M_{CM,x})\hat{\mathbf{i}} + (M_{CM,y})\hat{\mathbf{j}} + (M_{CM,z})\hat{\mathbf{k}}$$
$$= [I_{xx}\dot{\omega}_x + (I_{zz} - I_{yy})\omega_y\omega_z + I_{xy}(\dot{\omega}_y - \omega_z\omega_x) + I_{xz}(\dot{\omega}_z + \omega_y\omega_x)$$
$$+ I_{yz}(\omega_y^2 - \omega_z^2)]\hat{\mathbf{i}}$$
$$+ [I_{yy}\dot{\omega}_y + (I_{xx} - I_{zz})\omega_z\omega_x + I_{yz}(\dot{\omega}_z - \omega_x\omega_y) + I_{yx}(\dot{\omega}_x + \omega_z\omega_y)$$
$$+ I_{zx}(\omega_z^2 - \omega_x^2)]\hat{\mathbf{j}}$$
$$+ [I_{zz}\dot{\omega}_z + (I_{yy} - I_{xx})\omega_x\omega_y + I_{zx}(\dot{\omega}_x - \omega_y\omega_z) + I_{zy}(\dot{\omega}_y + \omega_x\omega_z)$$
$$+ I_{yx}(\omega_x^2 - \omega_y^2)]\hat{\mathbf{k}} \tag{12.47}$$

Equation (12.47) is obviously quite cumbersome to apply. If the xyz reference frame can be so oriented with respect to the body as to cause the products of inertia to vanish, Eq. (12.47), written in scalar form, becomes

$$M_{CM,x} = I_{xx}\dot{\omega}_x + (I_{zz} - I_{yy})\omega_y\omega_z$$

$$M_{CM,y} = I_{yy}\dot{\omega}_y + (I_{xx} - I_{zz})\omega_z\omega_x \tag{12.48}$$

$$M_{CM,z} = I_{zz}\dot{\omega}_z + (I_{yy} - I_{xx})\omega_x\omega_y$$

These are the well-known Euler equations of motion.[4] Note that no generality has been lost in judiciously selecting the orientation of xyz. If conditions of symmetry exist such that $I_{yy} = I_{zz}$, for example, we have the familiar result of elementary physics, $M_{CM,x} = I_{xx}\dot{\omega}_x = I_{xx}\alpha_x$.

The foregoing derivation of the Euler equations is predicated upon the relationship of the net moment of force about the mass center to the time derivative with respect to an inertial frame of reference of the angular momentum relative to the center of mass. Recall from Chapter 2 that the derivative of a vector $\mathbf{A}$ with respect to XYZ is related to the time derivative of $\mathbf{A}$ as viewed from xyz by the expression

$$\left(\frac{d\mathbf{A}}{dt}\right)_{XYZ} = \left(\frac{d\mathbf{A}}{dt}\right)_{xyz} + \boldsymbol{\omega} \times \mathbf{A} \tag{12.49}$$

where $\boldsymbol{\omega}$ is the angular velocity of xyz relative to XYZ. Equations (12.47) and (12.48) would have also resulted, therefore, by starting with

$$\mathbf{M}_{CM} = \dot{\mathbf{H}}_{CM,XYZ} = \dot{\mathbf{H}}_{CM,xyz} + \boldsymbol{\omega} \times \mathbf{H}_{CM} \tag{12.50}$$

where the vector equations are expressed in terms of the instantaneous projections of $\boldsymbol{\omega}$, $\dot{\mathbf{H}}_{CM,xyz}$, and $\mathbf{H}_{CM}$ upon the xyz-axes.

12.16 KINETIC ENERGY OF A RIGID BODY

Referring to Eq. (11.86), derived for a system of discrete particles, and again noting that the velocity of an elemental mass relative to the center of mass is $\boldsymbol{\omega} \times \mathbf{r}$, the total kinetic energy of a rigid body in general motion is

$$T = \tfrac{1}{2}mv_{CM}^2 + \tfrac{1}{2}\int_m |\boldsymbol{\omega} \times \mathbf{r}|^2 \, dm \tag{12.51a}$$

or

$$T = \tfrac{1}{2}(\mathbf{v}_{CM}\cdot m\mathbf{v}_{CM}) + \tfrac{1}{2}\int_m (\boldsymbol{\omega} \times \mathbf{r})\cdot(\boldsymbol{\omega} \times \mathbf{r}) \, dm \tag{12.51b}$$

The first term represents the kinetic energy associated with the motion of the mass center and the second term is the kinetic energy of the body relative to the mass center, as viewed from XYZ. Referring to Fig. 12.19, we have a body-fixed xyz-coordinate system with origin at the mass center. Substitution for $\mathbf{r}$ and $\boldsymbol{\omega}$ in terms of this system into Eqs. (12.51) yields

[4] Euler, L., *Mechanica Sine Motus Scientia*, St. Petersburg, 1736.

$$T = \tfrac{1}{2}mv_{CM}^2 + \omega_x^2 \int_m (z^2 + y^2)\, dm + \omega_y^2 \int_m (x^2 + z^2)\, dm + \omega_z^2 \int_m (x^2 + y^2)\, dm$$

$$- 2\omega_x\omega_z \int_m xz\, dm - 2\omega_y\omega_z \int_m yz\, dm - 2\omega_x\omega_y \int_m xy\, dm \quad \textbf{(12.52)}$$

where the integrals are again the moments and products of inertia with respect to *xyz*. Employing the nomenclature introduced in connection with the angular momentum, Eq. (12.52) becomes

$$T = \tfrac{1}{2}mv_{CM}^2 + \tfrac{1}{2}(I_{xx}\omega_x^2 + I_{yy}\omega_y^2 + I_{zz}\omega_z^2 + 2\omega_x\omega_y I_{xy} + 2\omega_x\omega_z I_{xz}$$

$$+ 2\omega_y\omega_z I_{yz}) \quad \textbf{(12.53)}$$

where the terms involving the moments and products of inertia are easily shown to equal $\tfrac{1}{2}\boldsymbol{\omega} \cdot \mathbf{H}_{CM}$. Thus

$$T = \tfrac{1}{2}(\mathbf{v}_{CM} \cdot \mathbf{p}_{CM}) + \tfrac{1}{2}(\boldsymbol{\omega} \cdot \mathbf{H}_{CM}) \quad \textbf{(12.54)}$$

The rotational contribution to the kinetic energy may also be expressed by the following matrix operation:

$$T = \tfrac{1}{2}[\omega_x \quad \omega_y \quad \omega_z] \begin{bmatrix} I_{xx} & I_{xy} & I_{xz} \\ I_{yx} & I_{yy} & I_{yz} \\ I_{zx} & I_{zy} & I_{zz} \end{bmatrix} \begin{bmatrix} \omega_x \\ \omega_y \\ \omega_z \end{bmatrix} \quad \textbf{(12.55)}$$

or simply

$$T = \tfrac{1}{2}\boldsymbol{\omega}^T | \boldsymbol{\omega} \quad \textbf{(12.56)}$$

where $\boldsymbol{\omega}^T$ represents the transpose of the angular velocity.

If the *xyz*-axes coincide with the *principal axes* of the body, the products of inertia vanish and Eq. (12.53) reduces to

$$T = \tfrac{1}{2}mv_{CM}^2 + \tfrac{1}{2}(I_{xx}\omega_x^2 + I_{yy}\omega_y^2 + I_{zz}\omega_z^2) \quad \textbf{(12.57)}$$

Examination of the above expressions quickly points to the analogous roles played by the linear and angular velocities (and momenta), as well as the mass and tensor of inertia.

12.17 WORK-ENERGY EQUATION

The relationship between the work done by the forces acting on an aggregate of particles and the resulting change in kinetic energy was discussed in Chapter 11. For the rigid body, the work-energy equation expressed in terms of the motion of the mass center and that relative to the mass center is

$$\int_{t_1}^{t_2} \mathbf{F} \cdot \mathbf{v}_{CM}\, dt + \int_{t_1}^{t_2} \boldsymbol{\omega} \cdot \dot{\mathbf{H}}_{CM}\, dt = [\tfrac{1}{2}(\mathbf{v}_{CM} \cdot m\mathbf{v}_{CM})]_1^2 + [\tfrac{1}{2}\boldsymbol{\omega} \cdot \mathbf{H}_{CM})]_1^2 \quad \textbf{(12.58)}$$

where $\mathbf{F} \cdot \mathbf{v}_{CM}\, dt = \mathbf{F} \cdot \mathbf{dr}_{CM}$ and

$$\int_{t_1}^{t_2} \boldsymbol{\omega} \cdot \dot{\mathbf{H}}_{CM} \, dt = \int_{t_1}^{t_2} \boldsymbol{\omega} \cdot \mathbf{M}_{CM} \, dt = \int_{t_1}^{t_2} \frac{1}{2} \frac{d}{dt} (\boldsymbol{\omega} \cdot \mathbf{H}_{CM}) \, dt = [\tfrac{1}{2}(\boldsymbol{\omega} \cdot \mathbf{H}_{CM})]_1^2$$

(12.59)

If the net moment of force about the center of mass is zero, the scalar product $\frac{1}{2}\boldsymbol{\omega} \cdot \mathbf{H}_{CM}$ clearly remains constant; that is, the rotational kinetic energy is unchanged. This is analogous to a translating body on which no net force acts. In addition, the condition of zero torque dictates that $\dot{\mathbf{H}}_{CM} = \mathbf{0}$ or $\mathbf{H}_{CM}$ remains constant. Both conditions of constancy ($\boldsymbol{\omega} \cdot \mathbf{H}_{CM}$ and $\mathbf{H}_{CM}$) can be fulfilled simultaneously with a nonconstant angular velocity. To demonstrate this, resolve the angular velocity into components $\boldsymbol{\omega}_n$ and $\boldsymbol{\omega}_t$ normal and parallel to $\mathbf{H}_{CM}$. The normal component may vary, whereas the scalar product $(\boldsymbol{\omega}_n + \boldsymbol{\omega}_t) \cdot \mathbf{H}_{CM} = \boldsymbol{\omega}_t \cdot \mathbf{H}_{CM}$ remains constant.

12.18 IMPULSE AND MOMENTUM

The concepts of impulse and momentum as they relate to a system of particles have already been treated. Inasmuch as the internal constraints present in a rigid body do not invalidate any of the earlier conclusions, these relationships apply here as well.

The equation of impulse and momentum written for the mass center is identical with Eq. (11.56):

$$\int_{t_1}^{t_2} \mathbf{F} \, dt = [m\mathbf{v}_{CM}]_1^2$$

(12.60)

The rigid body may, of course, experience rotational motion as well as translation, and a second equation is required to express the effect of a net torque acting over a period of time:

$$\int_{t_2}^{t_2} \mathbf{M}_{CM} \, dt = [\mathbf{H}_{CM}]_1^2$$

(12.61)

12.19 IMPULSIVE FORCES AND ENERGY CONSIDERATIONS

In analyzing the effect of impulsive forces upon a rigid body, it is of interest to consider not only changes in momentum but also the associated energy changes and the work required to effect them.

The incremental change of kinetic energy which the body may be given manifests itself as changes in translational and rotational mechanical energy:

$$\Delta T = \Delta(\tfrac{1}{2}\mathbf{v}_{CM} \cdot \mathbf{p}_{CM} + \tfrac{1}{2}\boldsymbol{\omega} \cdot \mathbf{H}_{CM})$$

(12.62)

The change of kinetic energy may also be expressed in terms of the initial and final values of linear velocity, angular velocity, and angular momentum:

$$\Delta T = \tfrac{1}{2}[(\mathbf{v}_{CM} + \Delta\mathbf{v}_{CM})\cdot(\mathbf{p}_{CM} + \Delta\mathbf{p}_{CM}) - \mathbf{v}_{CM}\cdot\mathbf{p}_{CM}]$$

$$+ \tfrac{1}{2}[(\boldsymbol{\omega} + \Delta\boldsymbol{\omega})\cdot(\mathbf{H}_{CM} + \Delta\mathbf{H}_{CM}) - \boldsymbol{\omega}\cdot\mathbf{H}_{CM}]$$

$$= \Delta\mathbf{p}_{CM} \cdot \left(\mathbf{v}_{CM} + \frac{\Delta\mathbf{v}_{CM}}{2}\right) + \Delta\mathbf{H}_{CM} \cdot \left(\boldsymbol{\omega} + \frac{\Delta\boldsymbol{\omega}}{2}\right) \qquad (12.63)$$

The term $\Delta\mathbf{p}_{CM}$ is nothing more than the change in linear momentum associated with the motion of the mass center, equal to the linear impulse $\overline{\mathbf{F}} \Delta t$. Similarly, $\Delta\mathbf{H}_{CM}$ represents the increment of angular momentum (about the mass center) which we may equate to the angular impulse caused by moments of force about the mass center $\overline{\mathbf{M}}_{CM} \Delta t$. Making these substitutions into Eq. (12.63) we obtain

$$\Delta T = (\overline{\mathbf{F}} \Delta t) \cdot \left(\mathbf{v}_{CM} + \frac{\Delta\mathbf{v}_{CM}}{2}\right) + (\overline{\mathbf{M}}_{CM} \Delta t) \cdot \left(\boldsymbol{\omega} + \frac{\Delta\boldsymbol{\omega}}{2}\right) \qquad (12.64)$$

The right side of Eq. (12.64) represents the work done by $\overline{\mathbf{F}}$ and $\overline{\mathbf{M}}_{CM}$. We have thus derived the work energy relationship for the impulsive loading of a rigid body. Note that $\mathbf{v}_{CM} + (\Delta\mathbf{v}_{CM}/2)$ and $\boldsymbol{\omega} + (\Delta\boldsymbol{\omega}/2)$ are the average linear and angular velocities for the time interval Δt. Hence $[\mathbf{v}_{CM} + (\Delta\mathbf{v}_{CM}/2)] \Delta t$ and $[\boldsymbol{\omega} + (\Delta\boldsymbol{\omega}/2)] \Delta t$ represent the respective linear and angular displacements of the rigid body.

12.20 IMPACT

Two rigid bodies suffering a collision or impact are treated in a manner similar to point masses in Chapter 11. Recall that impact per se is not analyzed, only the motion immediately before and after. The similarity extends further to the conservation of linear momentum before and after impact, owing to equal and opposite collision forces.

Consider now the impact of two rigid bodies a and b undergoing general motion. Since linear momentum is conserved,

$$m_a\mathbf{v}_{ai} + m_b\mathbf{v}_{bi} = m_a\mathbf{v}_{af} + m_b\mathbf{v}_{bf} \qquad (12.65)$$

where the subscripts i and f refer to conditions before and after impact. The impulse on body a due to b is equal to the change in the linear momentum of a:

$$\int \mathbf{F}\, dt = \mathbf{I} = m_a(\mathbf{v}_{af} - \mathbf{v}_{ai}) \qquad (12.66)$$

The impulse on b is $-\mathbf{I}$:

$$-\mathbf{I} = m_b(\mathbf{v}_{bf} - \mathbf{v}_{bi}) \qquad (12.67)$$

The angular impulse on a during dt is

$$\int \mathbf{M}\, dt = \int (\boldsymbol{\rho}_a \times \mathbf{F})\, dt = \boldsymbol{\rho}_a \times \mathbf{I} = \mathbf{H}_{af} - \mathbf{H}_{ai} \qquad (12.68)$$

where ρ_a is a vector running from the mass center of a to the point of impact and H_a represents the angular momentum of a about the mass center.

Similarly,

$$\int [\rho_b \times (-F)] \, dt = \rho_b \times (-I) = H_{bf} - H_{bi} \qquad (12.69)$$

Assuming that the initial linear velocities and angular momenta are known, the linear impulse I, final linear velocities, and angular momenta represent 15 unknowns in 12 scalar equations. A relationship describing the conservation of energy provides a thirteenth equation,

$$\tfrac{1}{2}m_a v_{ai}^2 + \tfrac{1}{2}m_b v_{bi}^2 + \tfrac{1}{2}\omega_{ai} \cdot H_{ai} + \tfrac{1}{2}\omega_{bi} \cdot H_{bi} - E_L$$
$$= \tfrac{1}{2}m_a v_{af}^2 + \tfrac{1}{2}m_b v_{bf}^2 + \tfrac{1}{2}\omega_{af} \cdot H_{af} + \tfrac{1}{2}\omega_{bf} \cdot H_{bf} \qquad (12.70)$$

where E_L is the loss of mechanical energy associated with the impact. At this point the reader may have surmised that it is a good deal simpler to write the above set of equations than to determine the data required for their solution. For example, the solution requires a knowledge of the energy loss per impact. In addition, the direction cosines of the linear impulse must be ascertained to provide two additional known quantities. These can be deduced experimentally from examination of the deformation in the vicinity of the impact "point." Here, of course, we are utilizing information inconsistent with the assumption of rigidity.

12.21 TRANSLATION OF AXES

It is often the case that the moments and products of inertia are known with respect to a coordinate set, the origin of which coincides with the mass center. If the terms of the inertia tensor are required with respect to another coordinate system parallel to the first, we now demonstrate that it is not necessary to recompute the integrals, but merely to perform a simple transformation.

Referring to Fig. 12.22, the coordinates of the mass center in the xyz-system are 0,0,0 and in the $x_1 y_1 z_1$-frame, x_{CM}, y_{CM}, z_{CM}. The relationship between the coordinates in both sets is

$$x_1 = x + x_{CM} \qquad y_1 = y + y_{CM} \qquad z_1 = z + z_{CM} \qquad (12.71)$$

Employing the definition of the moment of inertia integral with respect to the y_1-axis, for example, we have

$$I_{y_1 y_1} = \int_m (x_1^2 + z_1^2) \, dm = \int_m [(x + x_{CM})^2 + (z + z_{CM})^2] \, dm \qquad (12.72)$$

Expanding Eq. (12.72) and rearranging,

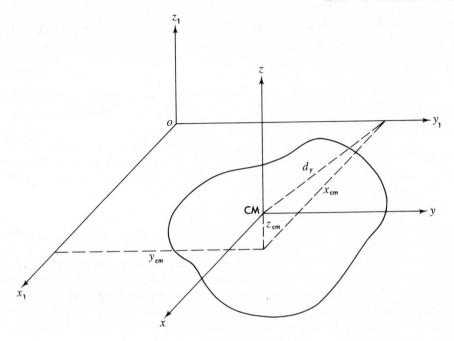

Fig. 12.22

$$I_{y_1y_1} = \int_m (x^2 + z^2)\,dm + (x_{CM}^2 + z_{CM}^2)\int_m dm + 2x_{CM}\int_m x\,dm$$

$$+ 2z_{CM}\int_m z\,dm \qquad (12.73)$$

The first term is the moment of inertia about the y-axis. The integral of the second term is the total mass and $x_{CM}^2 + z_{CM}^2$ is the square of the distance between the y- and y_1-axes. The integrals $\int_m x\,dm$ and $\int_m z\,dm$ are, by definition of the center of mass, both equal to zero. Consequently, we are left the extremely useful result

$$I_{y_1y_1} = I_{yy} + md_y^2 \qquad (12.74)$$

where $d_y = (x_{CM}^2 + z_{CM}^2)^{1/2}$. The *parallel-axis theorem* is thus: *To determine the moment of inertia with respect to a given axis, determine the moment of inertia with respect to a parallel axis passing through the center of mass and add the product of the body mass and the distance squared between the parallel axes involved.*

The product of inertia $I_{x_1y_1}$ may be determined in a similar fashion:

$$I_{x_1y_1} = \int_m x_1 y_1\,dm = \int_m [(x + x_{CM})(y + y_{CM})]\,dm \qquad (12.75)$$

Expanding and rearranging terms,

$$I_{x_1y_1} = \int_m xy\,dm + x_{CM}y_{CM}\int_m dm + x_{CM}\int_m y\,dm + y_{CM}\int_m x\,dm \qquad (12.76)$$

The first integral is the product of inertia I_{xy}. Since the last two integrals on the right side are zero, the following terms remain:

$$I_{x_1y_1} = I_{xy} + mx_{CM}y_{CM} \tag{12.77}$$

In applying Eq. (12.77), careful note of algebraic sign must be made. The parallel-axis theorem applied to both moments and products of inertia stated succinctly is

$$I_{\alpha_1\beta_1} = I_{\alpha\beta} + (mr^2_{CM}\hat{e}_{\alpha_1} \cdot \hat{e}_{\beta_1} - m\alpha_{CM}\beta_{CM}) \tag{12.78}$$

where α_1 and β_1 refer to the $x_1y_1z_1$-frame, α and β represent xyz in the center-of-mass system, α_{CM} and β_{CM} take on the values x_{CM}, y_{CM}, and z_{CM}, and $\hat{e}_{\alpha_1}$ and $\hat{e}_{\beta_1}$ represent the unit vectors of the $x_1y_1z_1$-system.

12.22 ROTATION OF AXES

Another important transformation enables us to determine the moments and products of inertia with respect to an xyz-coordinate set having a common origin with, and rotated relative to an $x'y'z'$ reference for which the moments and products of inertia are known, Fig. 12.23.

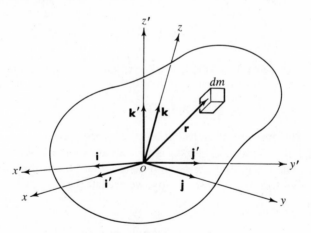

Fig. 12.23

Let us begin by noting that a radius vector $\mathbf{r}$ running from the origin to an arbitrary element of mass may be expressed in terms of its components and unit vectors in both coordinate sytems,

$$\mathbf{r} = x\hat{\mathbf{i}} + y\hat{\mathbf{j}} + z\hat{\mathbf{k}} = x'\hat{\mathbf{i}}' + y'\hat{\mathbf{j}}' + z'\hat{\mathbf{k}}' \tag{12.79}$$

and

$$r^2 = x^2 + y^2 + z^2 = x'^2 + y'^2 + z'^2 \tag{12.80}$$

We assume that $I_{x'x'}$ is known and I_{xx} is required. From the defining integral,

$$I_{xx} = \int_m (y^2 + z^2)\, dm = \int_m (r^2 - x^2)\, dm \qquad (12.81)$$

Substituting for r^2 in terms of primed quantities,

$$I_{xx} = \int_m (x'^2 + y'^2 + z'^2 - x^2)\, dm \qquad (12.82)$$

Since it is a transformation to the unprimed set which we require, it is necessary to express x^2 in terms of primed quantities. The unit vectors $\hat{\mathbf{i}}, \hat{\mathbf{j}}, \hat{\mathbf{k}}$ of the xyz set are related to $\hat{\mathbf{i}}', \hat{\mathbf{j}}', \hat{\mathbf{k}}'$ by the expressions

$$\hat{\mathbf{i}} = l_{xx'}\hat{\mathbf{i}}' + l_{xy'}\hat{\mathbf{j}}' + l_{xz'}\hat{\mathbf{k}}'$$

$$\hat{\mathbf{j}} = l_{yx'}\hat{\mathbf{i}}' + l_{yy'}\hat{\mathbf{j}}' + l_{yz'}\hat{\mathbf{k}}' \qquad (12.83)$$

$$\hat{\mathbf{k}} = l_{zx'}\hat{\mathbf{i}}' + l_{zy'}\hat{\mathbf{j}}' + l_{zz'}\hat{\mathbf{k}}'$$

where $l_{xx'}$ represents the cosine of the angle between Ox and Ox', and so on. The l terms are thus the nine direction cosines which describe the angular relationship between the primed and unprimed systems. We now write

$$x = \mathbf{r}\cdot\hat{\mathbf{i}} = (x'\hat{\mathbf{i}}' + y'\hat{\mathbf{j}}' + z'\hat{\mathbf{k}}')\cdot(l_{xx'}\hat{\mathbf{i}}' + l_{xy'}\hat{\mathbf{j}}' + l_{xz'}\hat{\mathbf{k}}') \qquad (12.84a)$$

or

$$x = l_{xx'}x' + l_{xy'}y' + l_{xz'}z' \qquad (12.84b)$$

which together with corresponding y and z equations may be expressed in matrix form as follows:

$$\begin{bmatrix} x \\ y \\ z \end{bmatrix} = \begin{bmatrix} l_{xx'} & l_{xy'} & l_{xz'} \\ l_{yx'} & l_{yy'} & l_{yz'} \\ l_{zx'} & l_{zy'} & l_{zz'} \end{bmatrix} \begin{bmatrix} x' \\ y' \\ z' \end{bmatrix} \qquad (12.85)$$

or

$$\mathbf{r} = \mathbf{l}\mathbf{r}' \qquad (12.86)$$

Substituting for x in Eq. (12.82), the moment of inertia with respect to the unprimed x-axis is now expressed in terms of primed quantities and direction cosines:

$$I_{xx} = \int_m [(x'^2 + y'^2 + z'^2)(l_{xx'}^2 + l_{xy'}^2 + l_{xz'}^2) - (l_{xx'}x' + l_{xy'}y' + l_{xz'}z')^2]\, dm \qquad (12.87)$$

where the term $l_{xx'}^2 + l_{xy'}^2 + l_{xz'}^2$ is equal to unity and does not alter the expression. Expanding Eq. (12.87) and employing the definition of the various inertia terms,

$$I_{xx} = l_{xx'}^2 I_{x'x'} + l_{xy'}^2 I_{y'y'} + l_{xz'}^2 I_{z'z'} + 2l_{xx'}l_{xy'}I_{x'y'} + 2l_{xx'}l_{xz'}I_{x'z'} + 2l_{xy'}l_{xz'}I_{y'z'} \qquad (12.88)$$

where the algebraic manipulation is left to the reader. A similar procedure yields the transformation for the product of inertia terms. Consider, for example:

$$-I_{xy} = \int_m xy\, dm = \int_m (l_{xx'}x' + l_{xy'}y' + l_{xz'}z')(l_{yx'}x' + l_{yy'}y' + l_{yz'}z')\, dm$$

(12.89)

Expanding Eq. (12.89) and utilizing the definition of the product of inertia,

$$I_{xy} = (l_{yy'}l_{xx'} + l_{yx'}l_{xy'})I_{x'y'} + (l_{yz'}l_{xx'} + l_{yx'}l_{xz'})I_{x'z'} + (l_{yy'}l_{xz'} + l_{yz'}l_{xy'})I_{y'z'}$$

$$- \int_m l_{xx'}l_{yx'}x'^2\, dm - \int_m l_{yy'}l_{xy'}y'^2\, dm - \int_m l_{yz'}l_{xz'}z'^2\, dm$$

(12.90)

Substituting,

$$x'^2 = r^2 - (y'^2 + z'^2) \qquad y'^2 = r^2 - (x'^2 + z'^2) \qquad z'^2 = r^2 - (x'^2 + y'^2)$$

the last three terms of Eq. (12.90) become

$$l_{xx'}l_{yx'}I_{x'x'} + l_{yy'}l_{xy'}I_{y'y'} + l_{yz'}l_{xz'}I_{z'z'} - \int_m r^2(l_{xx'}l_{yx'} + l_{xy'}l_{yy'} + l_{xz'}l_{yz'})\, dm$$

The term above in parentheses is the scalar product $\hat{\mathbf{i}} \cdot \hat{\mathbf{j}}$, equal to zero. Finally, we have

$$I_{xy} = (l_{yy'}l_{xx'} + l_{yx'}l_{xy'})I_{x'y'} + (l_{yz'}l_{xx'} + l_{yx'}l_{xz'})I_{x''} + (l_{yy'}l_{xz'}$$

$$+ l_{yz'}l_{xy'})I_{y'z'} + l_{xx'}l_{yx'}I_{x'x'} + l_{yy'}l_{xy'}I_{y'y'} + l_{yz'}l_{xz'}I_{z'z'}$$

(12.91)

The transformations for both moments and products of inertia may be expressed in concise form as follows:

$$I_{\alpha\beta} = \sum_{\alpha'} \sum_{\beta'} l_{\alpha\beta'}l_{\beta\alpha'}I_{\alpha'\beta'}$$

(12.92)

The interpretation of terms such as α' and β' has already been treated. The procedure to be followed in connection with Eq. (12.92) requires first summing with respect to the inner or β' index, yielding three terms. Summation over α' then produces the nine required terms of the inertia tensor. Quantities whose elements transform according to an expression such as Eq. (12.92) are termed tensors of second rank.

It is of interest to explore briefly an alternative and more expeditious approach to the derivation of Eq. (12.92). Begin with the expression for rotational kinetic energy expressed in terms of primed quantities (Fig. 12.23),

$$T = \tfrac{1}{2}\boldsymbol{\omega}'^T \mathbf{l}' \boldsymbol{\omega}'$$

(12.93)

In the unprimed system, $\boldsymbol{\omega} = \mathbf{l}\boldsymbol{\omega}'$, where the transformation is identical with that given by Eq. (12.85). Solving for $\boldsymbol{\omega}'$ we have

$$\mathbf{l}^T\boldsymbol{\omega} = \mathbf{l}^T\mathbf{l}\boldsymbol{\omega}' = \boldsymbol{\omega}'$$

(12.94)

Since l is an orthogonal matrix, $l^T l = 1$. Substituting for ω' in Eq. (12.93),

$$T = \tfrac{1}{2}\omega^T l I' l^T \omega \tag{12.95}$$

where $\omega^T l$ has replaced ω'^T. The kinetic energy is identical whether expressed in terms of primed or unprimed quantities and consequently

$$\tfrac{1}{2}\omega^T l I' l^T \omega = \tfrac{1}{2}\omega^T I \omega \tag{12.96}$$

which yields the required transformation, equivalent to Eq. (12.92),

$$I = l I' l^T \tag{12.97}$$

Illustrative Example 12.4

Derive an expression for the mass moment of inertia of a homogeneous circular cylinder of length L and radius R_0, with respect to its longitudinal axis (axis z-z), which passes through the center of mass.

SOLUTION
From the definition,

$$I_{zz} = \int_V \rho(x^2 + y^2)\, dV = \int_V \rho R^2\, dV$$

where the elemental volume dV may be expressed in cylindrical polar coordinates, $dV = (R\,d\phi)\,dR\,L$. Therefore,

$$I_{zz} = \int_0^{2\pi} \int_0^{R_0} \rho L R^3\, dR\, d\phi = 2\pi\rho L \int_0^{R_0} R^3\, dR = \frac{\pi\rho L R_0^4}{2}$$

Since the total cylinder mass $m = \rho V = \rho\pi R_0^2 L$, we have $I_{zz} = \tfrac{1}{2}mR_0^2$.

Illustrative Example 12.5

Determine the mass moment of inertia of the cylinder of the previous example about a $z_1 z_1$-axis parallel to the geometric zz-axis, tangent to the surface of the cylinder.

SOLUTION
Here the parallel-axis theorem may be used to good advantage since I_{zz} (through the mass center) has already been determined,

$$I_{z_1 z_1} = I_{zz} + md_z^2$$

where d_z is the perpendicular distance between axes. Therefore,

$$I_{z_1 z_1} = \tfrac{1}{2}mR_0^2 + mR_0^2 = \tfrac{3}{2}mR_0^2$$

Illustrative Example 12.6

For the rectangular parallelepiped shown in Fig. 12.24, determine I_{xx} and I_{yz}. The xyz-system has its origin at the center of mass. Employing the parallel-axis theorem, determine $I_{x_1x_1}$ and $I_{y_1z_1}$. The $x_1y_1z_1$-coordinate system is located along the edges of the body as shown.

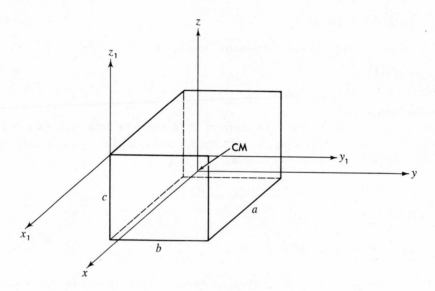

Fig. 12.24

SOLUTION

From the definition of the moment of inertia,

$$I_{xx} = \int_V \rho(y^2 + z^2)\, dV = \int_V \rho y^2\, dV + \int_V \rho z^2\, dV$$

Substituting $dV = ac(dy)$ and $dV = ab(dz)$, respectively, in the last two integrals of the above equation, we have

$$I_{xx} = \int_{-b/2}^{b/2} \rho ac y^2\, dy + \int_{-c/2}^{c/2} \rho ab z^2\, dz = \frac{m}{12}(b^2 + c^2)$$

since the total mass $m = \rho abc$.

The product of inertia I_{yz} is defined as

$$I_{yz} = -\int_V \rho yz\, dV = -\int_{-c/2}^{c/2} \int_{-b/2}^{b/2} \int_{-a/2}^{a/2} \rho yz\, dx\, dy\, dz$$

$$= -\int_{-c/2}^{c/2} \int_{-b/2}^{b/2} \rho xyz\, dy\, dz \Big]_{-a/2}^{a/2} = -\int_{-c/2}^{c/2} \rho az\, dz \frac{y^2}{2}\Big]_{-b/2}^{b/2} = 0$$

This is the expected result, since yz is a plane of symmetry. Employing the parallel-axis theorem,

$$I_{x_1 x_1} = I_{xx} + md_x^2 \quad \text{where } d_x = \left[\left(\frac{b}{2}\right)^2 + \left(\frac{c}{2}\right)^2 \right]^{1/2}$$

Therefore,

$$I_{x_1 x_1} = \frac{m}{12}(b^2 + c^2) + m\left(\frac{b^2}{4} + \frac{c^2}{4}\right) = \frac{m}{3}(b^2 + c^2)$$

Similarly,

$$I_{y_1 z_1} = I_{yz} + m y_{CM} z_{CM} = 0 + m\left(\frac{b}{2}\right)\left(\frac{c}{2}\right) = \frac{mbc}{4}$$

12.23 PRINCIPAL AXES OF INERTIA

It is clear that if the xyz-axes are oriented relative to the body in such a way as to cause all the products of inertia to vanish, the solutions of problems in rigid-body dynamics are vastly simplified. What we seek then is an inertia tensor with zeros for all terms not on the main diagonal. There are nine direction cosines which relate the original coordinate axes to that orthogonal set leading to a diagonalized inertia tensor. We therefore seek nine independent equations, three of which represent the statement that the products of inertia are zero, $I_{xy} = I_{yx} = I_{zx} = I_{xz} = I_{zy} = I_{yz} = 0$. The remaining six equations are the general relationships involving the direction cosines:

$$\hat{\mathbf{i}} \cdot \hat{\mathbf{i}} = 1 \qquad = l_{xx'}^2 + l_{xy'}^2 + l_{xz'}^2$$

$$\hat{\mathbf{j}} \cdot \hat{\mathbf{j}} = 1 \qquad = l_{yx'}^2 + l_{yy'}^2 + l_{yz'}^2 \qquad (12.98)$$

$$\hat{\mathbf{k}} \cdot \hat{\mathbf{k}} = 1 \qquad = l_{zx'}^2 + l_{zy'}^2 + l_{zz'}^2$$

$$\hat{\mathbf{i}} \cdot \hat{\mathbf{j}} = \hat{\mathbf{j}} \cdot \hat{\mathbf{i}} = 0 = l_{xx'} l_{yx'} + l_{xy'} l_{yy'} + l_{xz'} l_{yz'}$$

$$\hat{\mathbf{i}} \cdot \hat{\mathbf{k}} = \hat{\mathbf{k}} \cdot \hat{\mathbf{i}} = 0 = l_{xx'} l_{zx'} + l_{xy'} l_{zy'} + l_{xz'} l_{zz'} \qquad (12.99)$$

$$\hat{\mathbf{j}} \cdot \hat{\mathbf{k}} = \hat{\mathbf{k}} \cdot \hat{\mathbf{j}} = 0 = l_{yx'} l_{zx'} + l_{yy'} l_{zy'} + l_{yz'} l_{zz'}$$

where the unit vectors in the primed and unprimed sets are related in Eq. (12.83).

The matter of determining the orientation of the principal axes of inertia relative to an arbitrary set of body-fixed axes may also be stated: Locate those axes of rotation for which the angular momentum vector and angular velocity vector are parallel. As may readily be observed from Eq. (12.40), this requires that the products of inertia vanish. For principal axes therefore, Eq. (12.40) takes the form

$$\mathbf{H} = I_1 \omega_1 \hat{\mathbf{e}}_1 + I_2 \omega_2 \hat{\mathbf{e}}_2 + I_3 \omega_3 \hat{\mathbf{e}}_3$$

where $\hat{e}_1$, $\hat{e}_2$, $\hat{e}_3$ are the unit vectors of the principal-axis system. As in the case of Eq. (12.40), **H** in the above expression is not parallel to **ω**. Only if rotation occurs entirely about one of the principal axes will **ω** and **H** be parallel. For this case, $\mathbf{H} = I\boldsymbol{\omega} = I(\omega_x\hat{\mathbf{i}} + \omega_y\hat{\mathbf{j}} + \omega_z\hat{\mathbf{k}})$, where I represents a scalar and **H**, lying along a principal axis, has been written in terms of components parallel to the *nonprincipal* x, y, z axes. The three components of angular momentum may thus be expressed

$$H_x = I_{xx}\omega_x + I_{xy}\omega_y + I_{xz}\omega_z = I\omega_x$$

$$H_y = I_{yx}\omega_x + I_{yy}\omega_y + I_{yz}\omega_z = I\omega_y \qquad (12.100)$$

$$H_z = I_{zx}\omega_x + I_{zy}\omega_y + I_{zz}\omega_z = I\omega_z$$

leading to the following linear homogeneous equations:

$$(I_{xx} - I)\omega_x + \qquad I_{xy}\omega_y + \qquad I_{xz}\omega_z = 0$$

$$I_{yx}\omega_x + (I_{yy} - I)\omega_y + \qquad I_{yz}\omega_z = 0 \qquad (12.101)$$

$$I_{zx}\omega_x + \qquad I_{zy}\omega_y + (I_{zz} - I)\omega_z = 0$$

where it is to be recalled that I_{xx}, I_{xy}, and so on, are known.

Nontrivial solutions of Eqs. (12.101) are obtained by equating the determinant of the coefficients of the components of angular velocity to zero. This yields a cubic equation in I, having three real roots, and consequently three principal moments of inertia. Assume now that these roots are I_1, I_2, and I_3, and that they have already been determined. The reader will recall that it was our initial task to ascertain the orientation of principal axes, and this remains to be done. Return now to Eqs. (12.101) and divide through by $\omega_x \neq 0$, to obtain

$$(I_{xx} - I) + I_{xy}\frac{\omega_y}{\omega_x} + I_{xz}\frac{\omega_z}{\omega_x} = 0$$

$$I_{yx} + (I_{yy} - I)\frac{\omega_y}{\omega_x} + I_{yz}\frac{\omega_z}{\omega_x} = 0 \qquad (12.102)$$

$$I_{zx} + I_{zy}\frac{\omega_y}{\omega_x} + (I_{zz} - I)\frac{\omega_z}{\omega_x} = 0$$

A good deal more information is now available than when Eqs. (12.101) were first encountered, for now there are three values of I for substitution into Eqs. (12.102). These, together with the known values of the moments and products of inertia with respect to the nonprincipal coordinate axes xyz, means that Eqs. (12.102) represent three equations in two unknown velocity component ratios ω_y/ω_x and ω_z/ω_x. We must now relate these ratios to the direction cosines of the desired principal axes. Since $\omega_x = l_x\omega$, $\omega_y = l_y\omega$, and $\omega_z = l_z\omega$,

$$\frac{\omega_y}{\omega_x} = \frac{l_y}{l_x} \qquad \frac{\omega_z}{\omega_x} = \frac{l_z}{l_x} \qquad (12.103)$$

Substituting Eqs. (12.103) into Eqs. (12.102) together with one of the principal moments of inertia, I_1, for example, we have

$$(I_{xx} - I_1) + I_{xy}\frac{l_{y1}}{l_{x1}} + I_{xz}\frac{l_{z1}}{l_{x1}} = 0$$

$$I_{yx} + (I_{yy} - I_1)\frac{l_{y1}}{l_{x1}} + I_{yz}\frac{l_{z1}}{l_{x1}} = 0$$

$$I_{zx} + I_{zy}\frac{l_{y1}}{l_{x1}} + (I_{zz} - I_1)\frac{l_{z1}}{l_{x1}} = 0$$

which with $l_{x1}^2 + l_{y1}^2 + l_{z1}^2 = 1$ leads to the solution of the direction cosines of $\hat{e}_1$. In a similar manner, the direction cosines of $\hat{e}_2$ and $\hat{e}_3$ are determined. It is left as an exercise to verify the orthogonality of the principal axes.

12.24 THE ELLIPSOID OF INERTIA

Return now to the expression for the kinetic energy of a rigid body, Eq. (12.53). The rotational contribution may be written in terms of ω and the direction cosines which relate the inclination of the body-fixed xyz-axes to the angular velocity vector ($\omega_x = l_x\omega$, and so on),

$$T = \tfrac{1}{2}\omega^2(I_{xx}l_x^2 + I_{yy}l_y^2 + I_{zz}l_z^2 + 2I_{xy}l_xl_y + 2I_{yz}l_yl_z + 2I_{zx}l_zl_x) \tag{12.104}$$

or simply

$$T = \tfrac{1}{2}I\omega^2 \tag{12.105}$$

where I is an effective moment of inertia about the axis of rotation, determined from a knowledge of the products and moments of inertia relative to xyz and the direction cosines of ω. The following relationship thus transforms the tensor of inertia to the moment of inertia about a given axis of rotation.

$$I = I_{xx}l_x^2 + I_{yy}l_y^2 + I_{zz}l_z^2 + 2I_{xy}l_xl_y + 2I_{yz}l_yl_z + 2I_{zx}l_zl_x \tag{12.106}$$

Now consider a vector ρ colinear with the axis of rotation and of magnitude equal to $1/\sqrt{I}$. The direction cosines of ρ are clearly identical with those of ω, and therefore $l_x = \omega_x/\omega = \rho_x/\rho$, and so on, and

$$\rho = \frac{1}{\sqrt{I}}\frac{\omega}{\omega} = \frac{1}{k\sqrt{m}}\frac{\omega}{\omega} \tag{12.107}$$

where k represents the *radius of gyration* of a rigid body of mass m about a particular point in the body through which passes the axis of rotation. Substituting for the direction cosines and dividing by I, Eq. (12.106) becomes

$$I_{xx}\rho_x^2 + I_{yy}\rho_y^2 + I_{zz}\rho_z^2 + 2I_{xy}\rho_x\rho_y + 2I_{yz}\rho_y\rho_z + 2I_{zx}\rho_z\rho_x = 1 \tag{12.108}$$

The above expression is an ellipsoidal surface centered at the reference point from which ρ originates. It describes the locus of the end point of ρ as the axis of rotation (which is colinear with ρ) is varied in space. The *ellipsoid of inertia*, as it is termed, is thus attached to and rotates with the body. In general, there is a different ellipsoid associated with each reference point. Varying the angle between the reference axes and $\boldsymbol{\omega}$ results in different intersections with the inertia ellipsoid and consequently different magnitudes of ρ and $1/\sqrt{I}$. The ellipsoid of inertia thus represents the locus of I as the axis of rotation is varied.

Beginning with the general ellipsoid given by Eq. (12.108) it is always possible to determine a set of axes, centered at the same reference point, for which the ellipsoid takes the form $I_{xx}\rho_x^2 + I_{yy}\rho_y^2 + I_{zz}\rho_z^2 = 1$, which is exactly the same form as one obtains when the analysis is performed in a coordinate system in which the inertia tensor contains no off-diagonal terms. To proceed from Eq. (12.108) to the above, a transformation identical with that required to obtain the principal axes is employed. It follows that the axes of symmetry of the inertia ellipsoid and the principal axes are identical, and that since there are at least three axes of symmetry or principal axes of the ellipsoid, there must be at least three principal axes of inertia for any rigid body. We shall return to the inertia ellipsoid in discussing the torque-free motion of a rigid body.

12.25 RIGID-BODY MOTION ABOUT A FIXED AXIS

Since the expression $\mathbf{M} = \dot{\mathbf{H}}$ applies to motion about a fixed point, as well as about the center of mass, the derivation leading to Eqs. (12.47) and (12.48) for motion about the mass center pertains equally to the motion about a fixed point. The results previously obtained need only be modified by appropriately changing the subscripts.

Consider a body rotating about the space fixed Z-axis. If the body fixed z-axis is colinear with Z, $\omega_x = \omega_y = 0$, $\dot{\omega}_x = \dot{\omega}_y = 0$, and Eqs. (12.47) thus reduce to

$$\mathbf{M}_0 = M_{0,x}\hat{\mathbf{i}} + M_{0,y}\hat{\mathbf{j}} + M_{0,z}\hat{\mathbf{k}} = (I_{xz}\dot{\omega}_z - I_{yz}\omega_z^2)\hat{\mathbf{i}}$$

$$+ (I_{zx}\omega_z + I_{yz}\dot{\omega}_z)\hat{\mathbf{j}} + I_{zz}\dot{\omega}_z\hat{\mathbf{k}} \tag{12.109}$$

If the products of inertia are zero (as is the case when z is a principal axis or xy a plane of symmetry),

$$\mathbf{M}_0 = I_{zz}\dot{\omega}_z\hat{\mathbf{k}} \tag{12.110}$$

The three scalar moment equations provide more information than necessary to describe the motion about a fixed axis for only the z-moment equation is required to accomplish this. Together with the expressions describing the acceleration of the center of mass, the two remaining moment equations provide the means by which the forces of constraint, responsible for maintenance of the axis of rotation in its space-fixed orientation, may be determined.

12.26 DYNAMIC BALANCING AND SUPPORT REACTIONS

The material of Section 12.25 is directly applicable to the dynamic balancing of rotating machine components. Consider in this regard a mass m rotating about the z-axis as shown in Fig. 12.25(a).

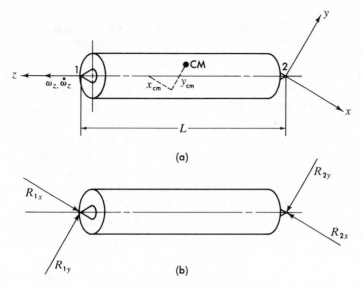

Fig. 12.25

Assume the rotating body to have its mass center displaced from the axis of rotation and to have nonzero products of inertia. Assume further that the reaction forces R_{1x}, R_{1y}, R_{2x}, R_{2y} do not include the contribution required to support the static weight of the rotating member. The latter is added to the dynamic reactions if the *total* reaction forces are required. The dynamic reactions rotate with the rotating *xyz*-axes at the angular velocity of the mass, ω_z. These periodic forces can be quite undesirable, as they may excite structural vibration, result in damage to supporting bearings, and lead to high noise levels.

Consider first the motion of the mass center,

$$\sum F_x = R_{1x} + R_{2x} = m\ddot{x}_{CM} = m(-x_{CM}\omega_z^2 - y_{CM}\dot{\omega}_z) \tag{12.111a}$$

$$\sum F_y = R_{1y} + R_{2y} = m\ddot{y}_{CM} = m(x_{CM}\dot{\omega}_z - y_{CM}\omega_z^2) \tag{12.111b}$$

Where $-x_{CM}\omega_z^2$ and $-y_{CM}\omega_z^2$ represent the components of centripetal acceleration and $x_{CM}\dot{\omega}_z$ and $-y_{CM}\dot{\omega}_z$ are the tangential accelerations. Next, the moment equations are

$$-R_{1y}L = I_{xz}\dot{\omega}_z - I_{yz}\omega_z^2 \tag{12.112a}$$

$$R_{1x}L = I_{xz}\omega_z^2 + I_{yz}\dot{\omega}_z \tag{12.112b}$$

where moments have been taken about fixed point 2.

To eliminate the reactions R_{1x} and R_{1y}, Eqs. (12.112) indicate that the products of inertia I_{yz} and I_{xz} must be zero. If the axis of rotation is a principal axis, this requirement is satisfied   There are, however, dynamic reactions at support 2 with which to contend. From Eqs. (12.111) it is clear that these too can be eliminated if the mass center lies on the axis of rotation ($x_{CM} = y_{CM} = 0$), and thus we have a second criterion for dynamic balance. Let us consider the necessity of satisfying both.

If the center of mass lies on the axis of rotation, it is clear that no gravity-induced torque can exist, and the rotor is in static balance. This fact alone, according to Eqs. (12.111), assures only that $R_{1x} = -R_{2x}$ and $R_{1y} = -R_{2y}$. Thus couples may indeed act, owing to these reactions, as shown in Fig. 12.25(b), from which it is apparent that mass must be deployed in such a way as to cancel the dynamic reactions and, equivalently, cause the products of inertia to vanish.

There are a number of practical devices which facilitate the dynamic balancing of rotors. These are often based upon the addition of mass in two planes normal to the axis of rotation, as shown in Fig. 12.26. The rotating reaction forces are measured by the small displacement of the rotor supports; mass is added to null these forces.

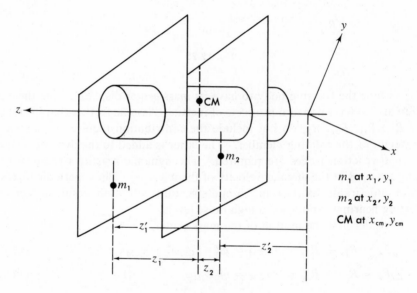

Fig. 12.26

For the mass center of the new mass system (including the balance masses) to lie on the axis of rotation, the moments of mass about the axis of rotation, must add to zero; that is, the new x and y coordinates of the mass center (x'_{CM} and y'_{CM}) must be zero,

$$m_1x_1 + m_2x_2 + mx_{CM} = (m_1 + m_2 + m)x'_{CM} = 0 \qquad (12.113a)$$

$$m_1y_1 + m_2y_2 + my_{CM} = (m_1 + m_2 + m)y'_{CM} = 0 \qquad (12.113b)$$

where m_1 and m_2 are the balance masses placed at x_1y_1 and x_2y_2, respectively, and m is the total system mass before balancing, located at the mass center of the unbalanced system.

The first condition of dynamic balance requires zero products of inertia. Referring to the rotating xyz-system,

$$I'_{xz} = I_{xz} - m_1 x_1 z'_1 - m_2 x_2 z'_2 = 0 \qquad \text{(12.114a)}$$

$$I'_{yz} = I_{yz} - m_1 y_1 z'_1 - m_2 y_2 z'_2 = 0 \qquad \text{(12.114b)}$$

where I'_{xz} and I'_{yz} are the products of inertia for the balanced system of mass and I_{xz} and I_{xy} apply to the unbalanced system.

Since z'_1 and z'_2 locate the balance planes and are known, the four equations above contain six unknowns: $m_1, m_2, x_1, x_2, y_1, y_2$. Two of these must therefore be preselected to determine the remaining quantities which assure dynamic balance.

12.27 CENTER OF PERCUSSION

As an illustration of the application of Eqs. (12.60) and (12.61), consider the effect of an impulsive force upon a rigid bar constrained to rotate in a plane about fixed point O. The force, acting for a short time interval Δt, causes an impulse $\int_0^{\Delta t} F \, dt$. Because in such situations it is difficult to determine the actual time history of the force, it convenient to replace $F(t)$ with a constant average force $\bar{F}$ acting throughout Δt. Applying Eq. (12.61) (for motion about a fixed point rather than the mass center),

$$\int_0^{\Delta t} M \, dt = \bar{M} \, \Delta t = \bar{F} a \, \Delta t = H_1 - H_0 = I\omega - I\omega_0 \qquad \text{(12.115)}$$

where a is the perpendicular distance from O to the point where $\bar{F}$ is applied, $\bar{M}$ the average moment of force associated with $\bar{F}$, and I the moment of inertia about the fixed axis of rotation passing through point O. Solving for the final angular velocity we have

$$\omega = \frac{\bar{F} a \, \Delta t}{I} + \omega_0 \qquad \text{(12.116)}$$

Since the rod is constrained to rotate about a fixed axis, the velocity of the mass center (located at the geometric center of a uniform bar) is

$$v_{\text{CM}} = \omega \frac{L}{2} \qquad \text{(12.117)}$$

where L is the length of the bar. Substituting for ω [Eq. (12.116)], the linear velocity of the center of mass is

$$v_{\text{CM}} = \left(\frac{\bar{F} a \, \Delta t}{I} + \omega_0 \right) \frac{L}{2} \qquad \text{(12.118)}$$

We now apply the equation of linear impulse and momentum, bearing in mind that all forces acting on the rod, *including reactions*, must be considered,

$$(\bar{F} + \bar{R}) \, \Delta t = m v_{\text{CM}} - m v_{\text{CM},0} \qquad \text{(12.119)}$$

where $v_{CM,0}$, the initial velocity of the center of mass, is equal to $\omega_0 L/2$. Substituting for v_{CM} [Eq. (12.118)],

$$(\bar{F} + \bar{R}) \Delta t = \frac{\bar{F} a \, \Delta t \, Lm}{2I} \tag{12.120}$$

For a uniform rod of length L, $I_0 = \frac{1}{3} m L^2$ and Eq. (12.120) becomes

$$\bar{F} + \bar{R} = \frac{3}{2} \frac{\bar{F} a}{L} \tag{12.121}$$

The reaction force at the point of support is therefore

$$\bar{R} = \bar{F} \left(\frac{3}{2} \frac{a}{L} - 1 \right) \tag{12.122}$$

For the specific geometry under consideration it is clear that $\bar{R} = 0$, provided $a = \frac{2}{3} L$. Consequently, when an impulsive force is applied at a point two thirds from the point of support of a *uniform rod*, it causes no momentum or impulse transfer to the support. This point of application is termed the *center of percussion*.

12.28 THE PHYSICAL PENDULUM

Recall from the study of the simple pendulum that the force of gravitational attraction on an initially displaced point mass suspended by means of a massless rod results in oscillatory motion of period $T = 2\pi\sqrt{L/g}$.

Now consider the physical or compound pendulum shown in Fig. 12.27, a further example of rigid-body motion about a fixed axis. The pendulum is free to rotate about a horizontal axis through point O. All the mass is assumed concentrated at the mass center. For moments about point O,

$$M_0 = \dot{H}_0 = I_0 \dot{\omega} = I_0 \ddot{\phi} \tag{12.123}$$

With respect to point O, the only moment of force is $-mgL_1 \sin \phi$, and for motion restricted to small ϕ,

$$\ddot{\phi} + \frac{mgL_1}{I_0} \phi = 0 \tag{12.124a}$$

The above equation has a solution of the form $\phi = \phi_0 \cos(\omega_0 t + \theta)$ where ϕ_0 is the amplitude of the motion (restricted to the limits of the small-angle approximation) and where the natural circular frequency

$$\omega_0 = \left(\frac{mgL_1}{I_0} \right)^{1/2} \tag{12.124b}$$

Consequently, the natural cyclic frequency and period are, respectively,

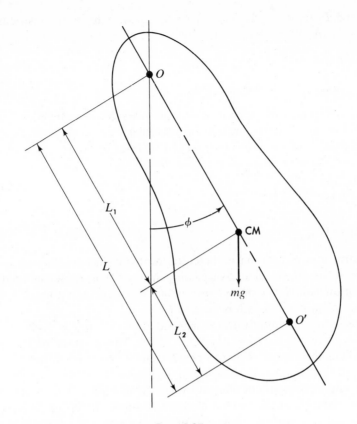

Fig. 12.27

$$f_0 = \frac{1}{2\pi} \left(\frac{mgL_1}{I_0} \right)^{\frac{1}{2}} \qquad (12.125a)$$

$$T = \frac{1}{f_0} = 2\pi \left(\frac{I_0}{mgL_1} \right)^{\frac{1}{2}} \qquad (12.125b)$$

Replacing I_0/m by the square of the radius of gyration, k^2,

$$T = 2\pi \left(\frac{k^2}{gL_1} \right)^{\frac{1}{2}} \qquad (12.126)$$

The period of a physical pendulum is thus identical with that of a simple pendulum of length $L = k^2/L_1$.

It is interesting to explore the question of what other axes of suspension of the physical pendulum result in the same period as given by Eq. (12.126). To satisfy this condition (referring to Fig. 12.27),

$$2\pi \left(\frac{I_0}{mgL_1} \right)^{\frac{1}{2}} = 2\pi \left(\frac{I_0'}{mgL_2} \right)^{\frac{1}{2}} \qquad (12.127)$$

where L_1 and L_2 represent distances from the mass center to the respective points of suspension. Applying the parallel-axis theorem,

$$I_0 = I_{CM} + mL_1^2 = mk_{CM}^2 + mL_1^2 \qquad (12.128a)$$

$$I_0' = I_{CM} + mL_2^2 = mk_{CM}^2 + mL_2^2 \qquad (12.128b)$$

where k_{CM}, the radius of gyration about the mass center, equals $(I_{CM}/m)^{\frac{1}{2}}$ and I_{CM} is the moment of inertia about the mass center.

Substitution of the above in Eq. (12.127) leads to

$$\frac{k_{CM}^2 + L_1^2}{L_1} = \frac{k_{CM}^2 + L_2^2}{L_2} \qquad (12.129a)$$

or

$$L_1 L_2 = k_{CM}^2 \qquad (12.129b)$$

An alternative axis of rotation is therefore located a distance $L_2 = k_{CM}^2/L_1$ from the mass center as shown. This axis passing through O' is located at what is called the *center of oscillation* or *center of suspension*, with respect to point O. Similarly, O is a center of suspension with respect to O'.

In the case of a uniform rod suspended at an end, $I_{CM} = \frac{1}{12}mL^2$ and $k_{CM}^2 = L^2/12$. Since $L_1 = L/2$, $L_2 = k_{CM}^2/L_1 = L/6$. The location of the center of oscillation O' is therefore $L/2 + L/6$ or $\frac{2}{3}L$ from O, the point of support. This is identical with the location of the center of percussion.

The physical pendulum offers a means for making a fairly accurate experimental determination of the gravitational acceleration. Referring to Eq. (12.125b) in which $I_0 = I_{CM} + mL_1^2$ has been substituted, we have

$$T = 2\pi \left(\frac{k_{CM}^2 + L_1^2}{gL_1} \right)^{\frac{1}{2}} \qquad (12.130)$$

If the pendulum is suspended from some other point (not necessarily the center of oscillation), the period of oscillation is

$$T' = 2\pi \left(\frac{k_{CM}^2 + L_2^2}{gL_2} \right)^{\frac{1}{2}} \qquad (12.131)$$

To minimize error associated with the introduction of a mathematical or experimental determination of k_{CM}, this term is eliminated between Eqs. (12.130) and (12.131), yielding

$$\frac{4\pi^2}{g} = \frac{L_1 T^2 - L_2 T'^2}{L_1^2 - L_2^2} \qquad (12.132)$$

Finally, the working form of Eq. (12.132) is obtained by applying the method of partial fractions:

$$\frac{1}{g} = \frac{1}{4\pi^2}\left[\frac{T^2 + T'^2}{2(L_1 + L_2)} + \frac{T^2 - T'^2}{2(L_1 - L_2)}\right] \tag{12.133}$$

The term $L_1 + L_2$ represents the distance between the two points of suspension and can be ascertained quite accurately. The individual distances L_1 and L_2 may be determined approximately by balancing the entire pendulum on a knife-edge. If T is close in value to T', the first term in the brackets of Eq. (12.133) predominates over the second, and an accurate finding of g may be made. The foregoing analysis serves as the basis of *Kater's pendulum*, a device for determining the acceleration of gravity or anomalies in the g field.

12.29 THE RIGID BODY IN PLANE MOTION

Another special case considers the rigid body subject to such constraints as to assure motion in a plane, although not necessarily about a fixed axis. This requires the specification of three coordinates for complete description of system configuration at any time. Considering rotation about the body-fixed z-axis only, we formally define plane motion by stipulating that the xy body-fixed plane and the space-fixed XY-plane must maintain a constant separation at all times; that is, they are parallel, as are the z- and Z-axes.

Since it is inconvenient to relate the rotational motion to a space-fixed point, we apply $\mathbf{M} = \dot{\mathbf{H}}$ referred to the mass center. The origin of the xyz-system is therefore located at the center of mass. Since no rotation takes place about the x- and y-axes, Eqs. (12.47) reduce to

$$M_{\mathrm{CM},x} = I_{xz}\dot{\omega}_z - I_{yz}\omega_z^2$$

$$M_{\mathrm{CM},y} = I_{yz}\dot{\omega}_z + I_{xz}\omega_z^2 \tag{12.134}$$

$$M_{\mathrm{CM},z} = I_{zz}\dot{\omega}_z$$

Note that these relations are identical in form with Eq. (12.109), which is referred to a fixed axis. Equations (12.134), together with those which treat the motion of the mass center, describe the total motion of the rigid body. The ensuing illustrative examples demonstrate further simplifications of Eqs. (12.134) made possible by considerations of symmetry.

Ilustrative Example 12.7
Study the dynamics of a uniform cylinder shown in Fig. 12.28(a) undergoing plane motion down an inclined plane subject to the following conditions: (1) the cylinder does not slip, and (2) the cylinder experiences slip.

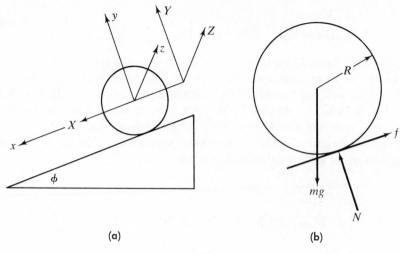

(a)   (b)

Fig. 12.28

SOLUTION

The motion is general inasmuch as no point is fixed; simplification is attributable to the constraints associated with maintaining plane motion and the condition that the mass center of the cylinder remains a fixed distance from the inclined plane. Thus the mass center undergoes rectilinear motion while the elements of mass comprising the cylinder rotate about the mass center.

Equations (12.134) are employed in considerably simplified form since $I_{xz} = I_{yz} = 0$. The governing equations are $\sum F_x = m\ddot{x}_{CM}$, $\sum F_y = m\ddot{y}_{CM}$, and $\sum M_{CM,z} = I_{zz}\dot{\omega}_z$.

The free-body diagram, Fig. 12.28(b), applies whether or not slip occurs. Summing forces,

$$\sum F_x = mg \sin \varphi - f = m\ddot{x}_{CM}$$

$$\sum F_y = N - mg \cos \varphi = m\ddot{y}_{CM} = 0 \quad \text{(since } y_{CM} = \text{constant)}$$

The frictional force causes the only torque about the mass center:

$$M_{CM,z} = -fR = I_{zz}\dot{\omega}_z \qquad f = \frac{-I_{zz}\dot{\omega}_z}{R}$$

The treatment to this point has been general. Further development depends upon the conditions relating to slip.

(1) If no slip occurs, that is, if there is no relative velocity at the point of contact between the cylinder and the inclined surface, the X-displacement of the mass center is the developed circumference of the cylinder, $x_{CM} = -R\theta$, where θ represents the angular displacement of an arbitrary line on the cylinder. Consequently, the velocity and acceleration of the mass center are, respectively,

$$\dot{x}_{\text{CM}} = -R\dot{\theta} = -R\omega_z$$

$$\ddot{x}_{\text{CM}} = -R\ddot{\theta} = -R\dot{\omega}_z$$

Equating the force of friction in the force and moment equations,

$$f = mg \sin \varphi - m\ddot{x}_{\text{CM}} = -\frac{I_{zz}\dot{\omega}_z}{R} = \frac{I_{zz}\ddot{x}_{\text{CM}}}{R^2}$$

where $-\ddot{x}_{\text{CM}}/R$ has been substituted for $\dot{\omega}_z$. Solving for $\ddot{x}_{\text{CM}}$, we have

$$\ddot{x}_{\text{CM}} = \frac{mg \sin \varphi}{m + I_{zz}/R^2}$$

Since every term in the above expression is constant, it follows that $\ddot{x}_{\text{CM}}$ and therefore $\dot{\omega}_z$ are constant. Substituting $I_{zz} = \frac{1}{2}mR^2$ for a uniform cylinder, $\ddot{x}_{\text{CM}} = \frac{2}{3}g \sin \varphi$.

Observe that the acceleration of the mass center is less than that of a non-rotating mass sliding without friction down the incline. This is because of the I_{zz}/R^2 term in the denominator which increases the effective mass being accelerated. The kinetic energy associated with the mass center must be less for the case of rotation because a portion of the initial system energy is transformed into rotational kinetic energy.

For the "no-slip" case, relative sliding does not occur and therefore no work is done in overcoming friction. Since we have a conservative system, the sum of the gravitational potential energy relative to some datum (for example, the base of the incline) and the kinetic energy remains constant:

$$\tfrac{1}{2}m\dot{x}_{\text{CM}}^2 + \tfrac{1}{2}I_{zz}\omega_z^2 - mgx_{\text{CM}} \sin \varphi = E$$

Substituting $\omega_{z,\text{CM}} = -\dot{x}_{\text{CM}}/R$ and differentiating with respect to time,

$$m\dot{x}_{\text{CM}}\ddot{x}_{\text{CM}} + I_{zz}\dot{x}_{\text{CM}}\frac{\ddot{x}_{\text{CM}}}{R^2} - mg\dot{x}_{\text{CM}} \sin \varphi = 0$$

Dividing by $\dot{x}_{\text{CM}}$ and solving for $\ddot{x}_{\text{CM}}$, we derive the same result as before.

The frictional force required for zero slip can also be determined by making the appropriate substitutions for $\ddot{x}_{\text{CM}}$:

$$f = \frac{-I_{zz}\dot{\omega}_z}{R} = \frac{-\frac{1}{2}mR^2}{R}\left(\frac{-\ddot{x}_{\text{CM}}}{R}\right) = \tfrac{1}{3}mg \sin \varphi$$

Thus the *minimum* coefficient of friction required for zero slip (pure rolling) is

$$\mu = \frac{f}{N} = \frac{\tfrac{1}{3}mg \sin \varphi}{mg \cos \varphi} = \tfrac{1}{3} \tan \varphi$$

(2) Assuming now that relative motion occurs between the line of contact of the cylinder and inclined plane, we may apply $f = \mu_k N$. The X-equation thus becomes

$$mg \sin \varphi - \mu_k N = m\ddot{x}_{\text{CM}}$$

Substituting $N = mg \cos \varphi$ and solving for the acceleration of the mass center, we have

$$\ddot{x}_{CM} = g \sin \varphi - \mu_k g \cos \varphi$$

and it is observed that for the case of slip as well as for no slip, the acceleration of the mass center remains constant. The angular acceleration is given by

$$\dot{\omega}_z = -\frac{fR}{I_{zz}} = -\frac{\mu_k mgR \cos \varphi}{I_{zz}}$$

If the cylinder begins its travel with zero translational and angular velocities, integration of $\ddot{x}_{CM}$ and $\dot{\omega}_z$ with respect to time yields

$$\dot{x}_{CM} = (g \sin \varphi - \mu_k g \cos \varphi)t$$

$$\omega_z = -\frac{\mu_k mgR \cos \varphi}{I_{zz}}t$$

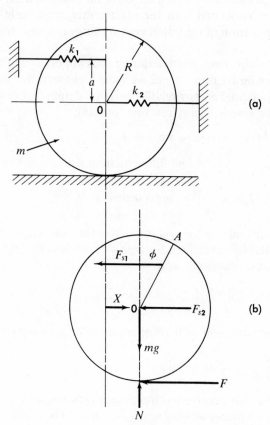

Fig. 12.29

While the ratio $-\dot{x}_{CM}/\omega_z R$ is equal to unity for the zero-slip case, it exceeds unity if slip occurs. This ratio can therefore serve as a slip parameter for this specific example:

$$\frac{-\dot{x}_{CM}}{\omega_z R} = \frac{g \sin \varphi - \mu_k g \cos \varphi}{\mu_k mgR^2 \cos \varphi / I_{zz}} = \frac{\tan \varphi}{\mu_k/(mR^2/I_{zz})} - 1$$

Substituting $I_{zz} = \frac{1}{2}mR^2$,

$$\frac{-\dot{x}_{CM}}{\omega_z R} = \frac{\tan \varphi}{2\mu_k} - \frac{1}{2}$$

Clearly, for $\tan \phi > 3\mu_k$, $-x_{CM}/\omega_z R > 1$, and slip occurs.

Illustrative Example 12.8

Determine the natural frequency of oscillation of the spring-mass system shown in Fig. 12.29(a) by two methods, force and energy. Assume pure rolling between the cylinder and the plane.

SOLUTION

Referring to the free-body diagram, Fig. 12.29(b), in which the cylinder is displaced a small amount to the right, Newton's second law written for the mass center (O) is

$$-k_1(x + a\phi) - k_2 x - F = m\ddot{x}$$

where F_{s1}, the spring force due to spring 1, is based upon a total deformation $(x = a\phi)$, and F represents the force of the plane upon the cylinder.

Summing moments of force about the mass center and equating to $I\ddot{\phi}$, we have

$$FR - k_1(x + a\phi)a = I\ddot{\phi}$$

Multiplying the first equation by R and adding to the second, F is eliminated:

$$-k_1(x + a\phi)(R + a) - k_2 xR = m\ddot{x}R + I\ddot{\phi}$$

Since we have pure rolling, $x = R\phi$ and $\ddot{x} = R\ddot{\phi}$, and these may be substituted in the above expression to yield an equation in x or ϕ alone. Thus x and ϕ are not independent, and the system possesses only one degree of freedom. In terms of ϕ,

$$-k_1(R\phi + a\phi)(R + a) - k_2 R^2\phi = mR^2\ddot{\phi} + I\ddot{\phi}$$

or,

$$\ddot{\phi} + \frac{[k_1(R + a)^2 + k_2 R^2]\phi}{mR^2 + I} = 0$$

which has the familiar form of an undamped single degree of freedom oscillator. The natural circular frequency is, by inspection,

$$\omega_0 = \left[\frac{k_1(R + a)^2 + k_2 R^2}{mR^2 + I} \right]^{1/2}$$

Identical results can also be obtained if moments are taken about the point of contact between cylinder and plane. The term $mR^2 + I$ is clearly the moment of inertia about this point.

Employing an energy approach, the kinetic- and potential-energy terms are

$$T = \tfrac{1}{2}m\dot{x}^2 + \tfrac{1}{2}I\dot{\phi}^2 = \tfrac{1}{2}mR^2\dot{\phi}^2 + \tfrac{1}{2}I\dot{\phi}^2$$

$$V = \tfrac{1}{2}k_1(x + a\phi)^2 + \tfrac{1}{2}k_2 x^2 = \tfrac{1}{2}k_1(R\phi + a\phi)^2 = \tfrac{1}{2}k_2 R^2\phi^2$$

The force of friction does no work (because there is zero slip) and consequently the energy is constant:

$$\frac{d}{dt}[\tfrac{1}{2}mR^2\dot{\phi}^2 + \tfrac{1}{2}I\dot{\phi}^2 + \tfrac{1}{2}k_1(R\phi + a\phi)^2 + \tfrac{1}{2}k_2 R^2\phi^2] = 0$$

Performing the differentiation and dividing through by $\dot{\phi}$, the resulting differential equation is identical with the one previously derived.

12.30 THE GYROSCOPE — TORQUE-FREE MOTION ABOUT A POINT

Consider a rigid body subject to no external forces whatever or external forces having lines of action that pass through the center of rotation. In either situation, the components of torque about the center of rotation are zero. If a set of xyz-coordinate axes is fixed to the rigid body, and if these are the principal axes, Euler's equations become

$$I_{xx}\dot{\omega}_x + (I_{zz} - I_{yy})\omega_y\omega_z = 0$$

$$I_{yy}\dot{\omega}_y + (I_{xx} - I_{zz})\omega_z\omega_x = 0 \qquad\qquad \text{(12.135)}$$

$$I_{zz}\dot{\omega}_z + (I_{yy} - I_{xx})\omega_x\omega_y = 0$$

For a body under the influence of a uniform gravitational field, the point about which rotations occur must be the center of mass in order to satisfy the second condition given above.

If now a further restriction is applied that the mass be a body of revolution, symmetrical about the body-fixed z-axis, it follows that $I_{xx} = I_{yy}$. Examples of such a rigid body are the symmetrical top containing a fixed point on the axis of symmetry, and the symmetrical gyroscope, shown in Fig. 12.30. Here we see the familiar Cardan suspension, which permits rotation about axes 1, 2, and 3, each of which passes through the center of mass. Thus torque-free motion is assured. For the symmetrical body of revolution, Euler's equations reduce to

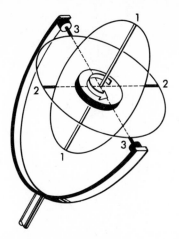

Fig. 12.30

$$I_{xx}\dot{\omega}_x + (I_{zz} - I_{xx})\omega_y\omega_z = 0 \tag{12.136a}$$

$$I_{xx}\dot{\omega}_y + (I_{xx} - I_{zz})\omega_z\omega_x = 0 \tag{12.136b}$$

$$I_{zz}\dot{\omega}_z = 0 \tag{12.136c}$$

Integration of Eq. (12.136c) gives the result that $I_{zz}\omega_z = H_{CM,z}$ is constant. Therefore, not only is the angular momentum $\mathbf{H}_{CM}$ constant (on the basis of $\mathbf{M}_{CM} = \mathbf{0}$), but the component of $\mathbf{H}_{CM}$ along the body-fixed symmetry axis Oz is also constant. Clearly, the angular relationship between $\mathbf{H}_{CM}$ and $H_{CM,z}$ must remain fixed, and the motion of the figure axis relative to the constant direction of the total angular momentum, termed the precession, describes a cone as shown in Fig. 12.31.

We must now determine ω_x and ω_y, the components of $\boldsymbol{\omega}$ along the body-fixed x- and y-axes as functions of time. First, Eqs. (12.136a) and (12.136b) are expressed

$$\dot{\omega}_x + \lambda\omega_y = 0 \tag{12.137a}$$

$$\dot{\omega}_y - \lambda\omega_x = 0 \tag{12.137b}$$

where $\lambda = \omega_z[(I_{zz} - I_{xx})/I_{xx}]$ = constant, since ω_z is constant.

Equation (12.137b), when differentiated with respect to time and combined with Eq. (12.137a), results in a second-order differential equation in ω_y alone:

$$\ddot{\omega}_y + \lambda^2\omega_y = 0 \tag{12.138}$$

An equation in ω_x is similarly obtained:

$$\ddot{\omega}_x + \lambda^2\omega_x = 0 \tag{12.139}$$

The solution of Eqs. (12.138) and (12.139) may be written

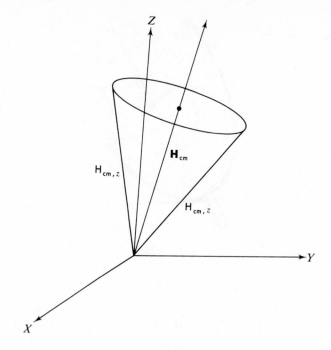

Fig. 12.31

$$\omega_x = A \cos(\lambda t + \gamma) \tag{12.140a}$$

$$\omega_y = A \sin(\lambda t + \gamma) \tag{12.140b}$$

where the sine and cosine solutions are selected to satisfy the original equations [Eqs. (12.137)].

We have already observed that the projection of $\boldsymbol{\omega}$ along the symmetry axis of the body, ω_z is constant. The angular velocity $\boldsymbol{\omega}$ is also of constant magnitude, as is now demonstrated. On the basis of Eqs. (12.140), adding $\omega_z \hat{\mathbf{k}}$, the total angular velocity is

$$\boldsymbol{\omega} = \omega_x \hat{\mathbf{i}} + \omega_y \hat{\mathbf{j}} + \omega_z \hat{\mathbf{k}} = [A \cos(\lambda t + \gamma)]\hat{\mathbf{i}} + [A \sin(\lambda t + \gamma)]\hat{\mathbf{j}} + \omega_z \hat{\mathbf{k}} \tag{12.141}$$

Since $[A^2 \cos^2(\lambda t + \gamma) + A^2 \sin^2(\lambda t + \gamma)]^{1/2} = \text{constant} = A$, and ω_z is constant, it is clear that $\omega = (A^2 + \omega_z^2)^{1/2}$ is likewise constant. Referring to Fig. 12.32, one can visualize the vector $\omega_x \hat{\mathbf{i}} + \omega_y \hat{\mathbf{j}}$ (parallel to the xy body-fixed plane) rotating about the geometric Oz-axis of the body at angular speed λ. The angular velocity $\boldsymbol{\omega}$ thus precesses about the axis of symmetry of the rigid body, and in so doing traces a *body cone* of radius A as shown in Fig. 12.32.

The angular speed of precession, λ, is dependent directly upon ω_z and the ratio involving the moments of inertia. If λ is positive, the direction of this pre-

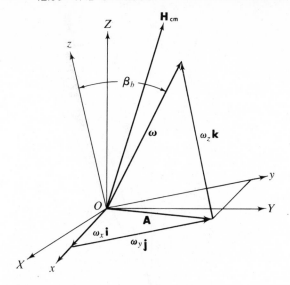

Fig. 12.32

cession is the same as ω_z. This is the case in which $I_{zz} > I_{xx}$. One speaks of recession or retrograde precession if $I_{xx} > I_{zz}$, in which case ω_z and the precessional velocity are of opposite sense. A long cylindrical object spinning about its longitudinal axis exemplifies such a case. On the other hand, the usual mass distribution of the gyroscope and flywheel are such that $I_{zz} > I_{xx}$, and the resulting precession is direct.

The angle between the axis of symmetry and the angular velocity vector, from Fig. 12.32, is given by

$$\tan \beta_b = \frac{A}{\omega_z} \qquad (12.142)$$

and is constant, because A and ω_z are constant.

To relate $\boldsymbol{\omega}$ to a space-fixed direction (the angular momentum vector being an ideal candidate in view of the torque-free motion), consider the definition of the scalar product

$$\boldsymbol{\omega} \cdot \mathbf{H}_{CM} = \omega H_{CM} \cos \beta_s \qquad (12.143)$$

where β_s represents the angle between $\boldsymbol{\omega}$ and $\mathbf{H}_{CM}$, Fig. 12.33. Since the kinetic energy of rotation, $T = \frac{1}{2}\boldsymbol{\omega} \cdot \mathbf{H}_{CM}$,

$$\cos \beta_s = \frac{\boldsymbol{\omega} \cdot \mathbf{H}_{CM}}{\omega H_{CM}} = \frac{2T}{\omega H_{CM}} \qquad (12.144)$$

Since ω, $\mathbf{H}_{CM}$, and T are all constant, it is clear that β_s is constant, and we may speak of a *space cone*, formed as $\boldsymbol{\omega}$ precesses about the fixed direction of

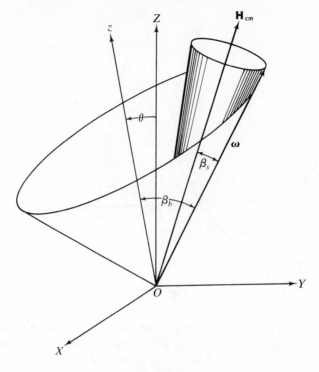

Fig. 12.33

H_{CM}. We thus have a most interesting representation of torque-free rigid-body motion in terms of the rolling of the body cone on the space cone, where the angular velocity vector is an instantaneous line of contact between the two.

12.31 POINSOT'S CONSTRUCTION [5]

An interesting construction, due to Poinsot, connects the inertia ellipsoid with the characteristics of torque-free motion. We begin by recalling that the vector ρ from a reference point to the surface of the inertia ellipsoid is given by

$$\rho = \frac{\omega}{\omega \sqrt{I}} \tag{12.145}$$

Since the torque is zero, the angular momentum vector is fixed in space, as we have discussed. In addition, the rotational kinetic energy is constant. Thus

$$T = \tfrac{1}{2} I \omega^2 = \tfrac{1}{2} \omega \cdot H = \text{constant} \tag{12.146}$$

or in view of Eq. (12.145), $T = \tfrac{1}{2} \omega \sqrt{I}\, \rho \cdot H$ or

$$\rho \cdot H = \frac{2T}{\omega \sqrt{I}} \tag{12.147}$$

---

[5] Poinsot, L., *Theorie nouvelle de la rotation des corps*, Paris, 1834.

From Eq. (12.147) it is clear that the projection of ρ upon the angular momentum vector (the *invariable line*) is also constant, and therefore the tangent plane containing the end points of ρ is a single space-fixed plane, the *invariable plane*, perpendicular to the invariable line as shown in Fig. 12.34. The ellipsoid of inertia thus moves on the invariable plane during the torque-free motion of a rigid body about point O. That the motion is pure rolling can be demonstrated by observing that the angular velocity vector also passes through the point of contact of the invariable plane and the ellipsoid, ω and ρ being colinear. Thus, through the point of contact passes an instantaneous axis of rotation of the ellipsoid with respect to the invariable plane.

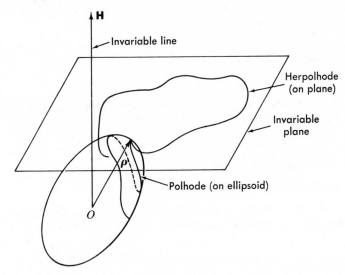

Fig. 12.34

The trace of the point of contact upon the invariable plane is called the *herpolhode*, shown in Fig. 12.34. For a body of symmetry, the herpolhode is a circle, as the reader has probably already concluded from the previous section. The trace of the point of contact on the ellipsoid of inertia is called the *polhode*. The polhode of a symmetrical body is also a circle.

12.32 STABILITY ANALYSIS — TORQUE-FREE MOTION

It is useful to explore a technique by which the stability of a particular configuration of a rigid body may be predicted. Given, for example, an asymmetrical body ($I_{xx} \neq I_{yy} \neq I_{zz}$) rotating about a principal axis, under what circumstances will the motion be unstable? We apply the same general definition of stability here as in Chapter 5: A body is stable when disturbances causing departures from equilibrium are counteracted by correcting influences.

Consider the body to be subject to the action of no forces or torques. The motion is described by Eqs. (12.135). Suppose that rotation takes place about the x-axis only, with $\omega_x = $ constant $= \omega_c$; $\omega_y = \omega_z = 0$. If now a disturbance in the angular velocity components occurs, we have

$$\omega_x = \omega_c + \delta_{\omega x} \qquad \dot{\omega}_x = \dot{\delta}_{\omega x} \tag{12.148a}$$

$$\omega_y = 0 + \delta_{\omega y} \qquad \dot{\omega}_y = \dot{\delta}_{\omega y} \tag{12.148b}$$

$$\omega_z = 0 + \delta_{\omega z} \qquad \dot{\omega}_z = \dot{\delta}_{\omega z} \tag{12.148c}$$

where $\delta_{\omega x}$, $\delta_{\omega y}$, $\delta_{\omega z}$ are the small increments in the angular velocity resulting from the disturbance. Rewriting the equations of motion in terms of Eqs. (12.148),

$$I_{xx}\dot{\delta}_{\omega x} + \delta_{\omega y}\delta_{\omega z}(I_{zz} - I_{yy}) = 0 \tag{12.149a}$$

$$I_{yy}\dot{\delta}_{\omega y} + \delta_{\omega z}(\omega_c + \delta_{\omega x})(I_{xx} - I_{zz}) = 0 \tag{12.149b}$$

$$I_{zz}\dot{\delta}_{\omega z} + (\omega_c + \delta_{\omega x})\delta_{\omega y}(I_{yy} - I_{xx}) = 0 \tag{12.149c}$$

Discarding all terms involving products of the increments, such as $\delta_{\omega y}\delta_{\omega z}$, as being small relative to the remaining terms, we have

$$I_{xx}\dot{\delta}_{\omega x} = 0 \tag{12.150a}$$

$$I_{yy}\dot{\delta}_{\omega y} + \omega_c\delta_{\omega z}(I_{xx} - I_{zz}) = 0 \tag{12.150b}$$

$$I_{zz}\dot{\delta}_{\omega z} + \omega_c\delta_{\omega y}(I_{yy} - I_{xx}) = 0 \tag{12.150c}$$

It is a simple matter to combine Eqs. (12.150b) and (12.150c) to produce a single second-order equation in one variable, $\delta_{\omega y}$ or $\delta_{\omega z}$. For example, from Eq. (12.150c),

$$\dot{\delta}_{\omega z} = \frac{\omega_c\delta_{\omega y}(I_{xx} - I_{yy})}{I_{zz}} \tag{12.151}$$

When Eq. (12.150b) is differentiated once with respect to time, Eq. (12.151) may be substituted to yield the following expression in $\delta_{\omega y}$ alone:

$$\ddot{\delta}_{\omega y} + \left[\frac{\omega_c^2(I_{xx} - I_{yy})(I_{xx} - I_{zz})}{I_{yy}I_{zz}}\right]\delta_{\omega y} = 0 \tag{12.152}$$

which is of the form $\ddot{\delta}_{\omega y} + m^2\,\delta_{\omega y} = 0$ where

$$m^2 = \frac{\omega_c^2(I_{xx} - I_{yy})(I_{xx} - I_{zz})}{I_{yy}I_{zz}}$$

The foregoing relationships apply not to the angular speed but to disturbances or departures from that speed. From the theory of differential equations, we know that $\delta_{\omega y}$ increases with time if $m^2 \leq 0$, and we therefore can avail ourselves of the following condition for stability:

$$\frac{\omega_c^2(I_{xx} - I_{yy})(I_{xx} - I_{zz})}{I_{yy}I_{zz}} > 0 \tag{12.153}$$

If I_{xx} is a maximum or minimum moment of inertia, it is clear that the above quotient will be positive. Stable rotation about a principal axis therefore requires that the moment of inertia about that axis be a maximum or a minimum.

12.33 THE GYROCOMPASS

We now demonstrate the useful fact that a rigid symmetric body spinning about an axis constrained to move in a horizontal plane, in the absence of applied torques, experiences oscillatory motion. The mean direction of the rotor axis throughout this oscillation is along a north-south meridian. A device operating in accordance with the principles detailed below is called a gyrocompass.

Consider now such a rotor with spin axis in a horizontal plane as shown in Fig. 12.35 and a coordinate system with origin at the mass center of the rotor but

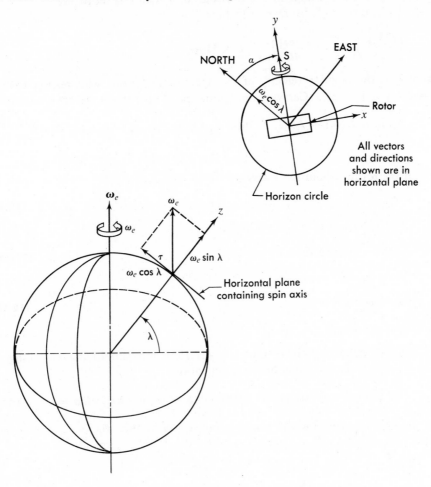

Fig. 12.35

not rotating with the rotor. The z-axis of the coordinate system is vertical, the y-axis coincides with the spin axis, and the x-axis completes the right-handed triad and is not fixed to the rotor.

We have already derived equations for treating the dynamics of a rigid body rotating about its mass center. One approach to this problem leads to the development of the Euler equations, which are written in terms of body-fixed coordinates. In the case at hand, the coordinates are not fixed to the spinning body. In writing the angular momentum of the rotor one must include not only the spin, but appropriate components of the earth's angular velocity as well.

For this case, the products of inertia are zero and the angular momentum is

$$\mathbf{H_{CM}} = I_{xx}\omega_x\hat{\mathbf{i}} + I_{yy}(\omega_y + s)\hat{\mathbf{j}} + I_{zz}\omega_z\hat{\mathbf{k}} \tag{12.154}$$

where s represents the angular speed of the rotor with respect to xyz.

Differentiating with respect to time,

$$\left(\frac{d\mathbf{H_{CM}}}{dt}\right)_{XYZ} = \left(\frac{d\mathbf{H_{CM}}}{dt}\right)_{xyz} + \boldsymbol{\omega} \times \mathbf{H_{CM}}$$

where $\boldsymbol{\omega}$ is the angular velocity of the xyz rotating frame with respect to XYZ, $\boldsymbol{\omega} = \omega_x\hat{\mathbf{i}} + \omega_y\hat{\mathbf{j}} + \omega_z\hat{\mathbf{k}}$. Carrying out the indicated operations, we have

$$\mathbf{M_{CM}} = I_{xx}\dot{\omega}_x\hat{\mathbf{i}} + I_{yy}(\dot{\omega}_y + \dot{s})\hat{\mathbf{j}} + I_{zz}\dot{\omega}_z\hat{\mathbf{k}} + [I_{zz}\omega_y\omega_z - I_{yy}\omega_z(\omega_y + s)]\hat{\mathbf{i}}$$
$$+ [I_{xx}\omega_x\omega_z - I_{zz}\omega_x\omega_z]\hat{\mathbf{j}} + [I_{yy}\omega_x(\omega_y + s) - I_{xx}\omega_x\omega_y]\hat{\mathbf{k}} \tag{12.155}$$

so that the torque about the z-axis is

$$M_{CM,z} = I_{zz}\dot{\omega}_z + (I_{yy} - I_{xx})\omega_x\omega_y + I_{yy}s\omega_x = 0 \tag{12.156}$$

Referring to Fig. 12.35 we obtain

$$\omega_x = -\omega_e \cos\lambda \sin\alpha \tag{12.157a}$$

$$\omega_y = \omega_e \cos\lambda \cos\alpha \tag{12.157b}$$

$$\omega_z = \omega_e \sin\lambda - \dot{\alpha} \tag{12.157c}$$

$$\dot{\omega}_z = -\ddot{\alpha} \tag{12.157d}$$

Substituting the above into Eq. (12.156) and simplifying (recalling that $I_{zz} = I_{xx}$) yields the following nonlinear equation:

$$-I_{xx}\ddot{\alpha} - \omega_e^2 \cos^2\lambda(I_{yy} - I_{xx})\sin\alpha\cos\alpha - s\omega_e\cos\lambda I_{yy}\sin\alpha = 0 \tag{12.158}$$

Assuming that the rotor spin is much greater than the contribution in the z-direction of the earth's angular velocity, the middle term above may be neglected. If, in addition, α is confined to small angles,

$$\ddot{\alpha} + \left(\frac{I_{yy}s\omega_e\cos\lambda}{I_{xx}}\right)\alpha = 0 \tag{12.159}$$

which is the equation of a linear oscillator. Thus the spin axis oscillates about a meridian line with period

$$T = 2\pi \left(\frac{I_{xx}}{I_{yy} S \omega_e \cos \lambda} \right)^{\frac{1}{2}}$$ (12.160)

Although the theory is not complex, note must be taken of the simplifications made not only in the linearization of the differential equation but in the assumption regarding zero torque. The reduction of theory into a properly operating device requires complex design and careful assembly. Frictional torques do indeed exist. In the horizontal or xy-plane they serve to damp out the oscillation given by Eq. (12.159), thus causing the spin axis to come to rest along a north-south line. That frictional torques exist about the spin axis is made abundantly clear by the fact that in practice a motor is required to drive the rotor.

12.34 MOTION ABOUT A FIXED POINT — THE TOP

Consider the top shown in Fig. 12.36, the configuration of which must be specified with respect to an inertial frame of reference. Examination of the figure yields the following "natural" coordinates:

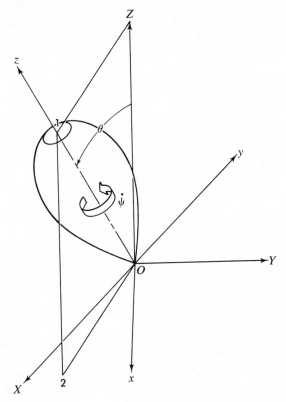

Fig. 12.36

a. Angle θ between the vertical Z-axis of the inertial frame of reference and Oz, the axis of symmetry of the body.

b. Angle ϕ, which measures the rotation of vertical plane 1-2-0 containing Oz.

c. Angle of rotation ψ with respect to vertical plane 1-2-0.

A Cartesian coordinate system $x'y'z'$ based upon the *Euler angles* [6] θ, ϕ, ψ may readily be determined. This coordinate frame is neither fixed to the body (as is the xyz-system shown) nor fixed in space (as is XYZ). Instead, $x'y'z'$ rotates with, and is attached to, the vertical 1-2-0 plane. The angular velocity of $x'y'z'$ about Z is the precessional velocity of the top. The $x'y'z'$-set is determined as follows:

a. Location of Oz'. This axis is identical with Oz and is thus the axis of symmetry of the body.

b. Location of Ox'. This axis is defined by a plane normal to Oz' through point O and its intersection with the XY-plane. Note that since Oz and Oz' coincide, the normal plane must be the body-fixed xy-plane. The axis thus formed must also always lie in a horizontal plane, because XY is horizontal. This axis is called the *line of nodes*.

c. Location of Oy'. Axis Oy' completes the right-handed set $x'y'z'$. The relationship between the Euler angles and the $x'y'z'$-coordinate system may be seen in Fig. 12.37.

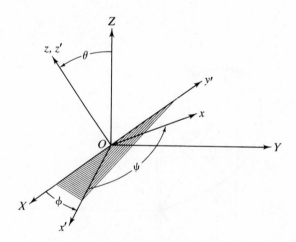

Fig. 12.37

What is the relationship between the body-fixed xyz-axes and the Euler angles? If a body is initially so located that its body-fixed axes coincide with the axes of an inertial frame of reference, any desired final orientation of the body may be effected by successive rotations through the Euler angles. A specific scheme

---

[6] See footnote on page 434.

for performing the rotations is given in the following development, and it should be understood that the sense of rotation, whether clockwise or counterclockwise, is applied when viewing along a given axis toward the origin.

First Rotation

Through a counterclockwise rotation ϕ about OZ, a plane is formed containing OX, Ox_1, OY, Oy_1 as shown in Fig. 12.38. The body-fixed triad is now designated x_1, y_1, z_0, indicating that the x- and y-axes have experienced their first change in orientation while that of the z-axis remains unchanged.

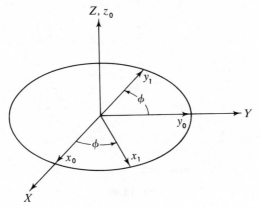

Fig. 12.38

Second Rotation

The plane formed by the first rotation is now brought into a new orientation through a counterclockwise rotation θ about Ox_1, as shown in Fig. 12.39. The body-fixed axes are now designated $x_1 y_2 z_1$.

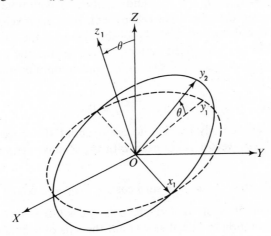

Fig. 12.39

Third and Final Rotation

Oz_1 represents the final axis of rotation, about which a counterclockwise angular displacement ψ is effected as shown in Fig. 12.40. The body-fixed axes are now shown as x, y, z at an arbitrary orientation with respect to the starting orientation.

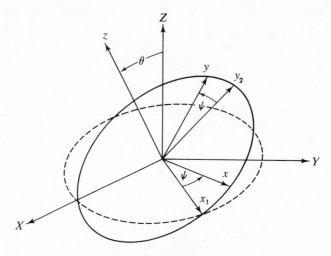

Fig. 12.40

It should be emphasized that since finite angular displacements are not vector quantities, the sequence described must be strictly adhered to; that is, the final orientation of xyz is dependent upon the order of operations.

We have carried the analysis to a point where the issue of the most suitable coordinate system must be broached: Shall xyz or $x'y'z'$ be selected? For an object symmetrical about Oz (Oz'), the moments of inertia are identical in either coordinate system, that is, $I_{xx} = I_{yy} = I_{x'x'} = I_{y'y'}$, and do not vary with time. Thus this factor does not enter into our decision. The components of angular velocity do, of course, differ. We shall first determine the relationship between the Euler angles and the components of ω in the body-fixed system. Imagine that the total body rotation is about the line of nodes: $\omega = \dot{\theta}$. The x-, y-, and z-components of ω are, for this case,

$$\omega_x = \dot{\theta} \cos \psi \qquad \omega_y = -\dot{\theta} \sin \psi \qquad \omega_z = 0 \tag{12.161}$$

Now consider the entire body rotation to occur about the space fixed Z-axis; that is, the angular velocity vector is colinear with the Z-axis. The x-, y-, and z-components are now

$$\omega_x = \dot{\phi} \sin \theta \sin \psi \qquad \omega_y = \dot{\phi} \sin \theta \cos \psi \qquad \omega_z = \dot{\phi} \cos \theta \tag{12.162}$$

since the projection of $\dot{\phi}$ on the xy-plane is $\dot{\phi} \cos (90 - \theta) = \dot{\phi} \sin \theta$. Finally, imagine the body rotation to be entirely about the axis of symmetry (Oz, Oz'):

$$\omega_x = 0 \qquad \omega_y = 0 \qquad \omega_z = \dot{\psi} \tag{12.163}$$

For a rigid body experiencing general rotation, rotations about the line of nodes, Z-axis, and the symmetry axis may occur simultaneously, and therefore the x-, y-, and z-components of $\boldsymbol{\omega}$ must include contributions caused by all three rotations above:

$$\omega_x = \dot{\theta} \cos \psi + \dot{\phi} \sin \theta \sin \psi \tag{12.164a}$$

$$\omega_y = -\dot{\theta} \sin \psi + \dot{\phi} \sin \theta \cos \psi \tag{12.164b}$$

$$\omega_z = \dot{\phi} \cos \theta + \dot{\psi} \tag{12.164c}$$

These are the components which should be employed in the Euler equations if a body-fixed coordinate system is selected.

If, on the other hand, one selects the rotating $x'y'z'$ system, the components of $\boldsymbol{\omega}$, the angular velocity of $x'y'z'$ relative to XYZ are

$$\omega_{x'} = \dot{\theta} \qquad \omega_{y'} = \dot{\phi} \sin \theta \qquad \omega_{z'} = \dot{\phi} \cos \theta \tag{12.165}$$

Since the Euler equations require derivatives of the components of $\boldsymbol{\omega}$ as well as the components themselves, the primed coordinate system provides the simpler expressions inasmuch as they are themselves simpler than the corresponding terms written for the body-fixed axes.

The angular momentum of the top in terms of components in the primed set is written

$$\mathbf{H} = I_{xx}\omega_{x'}\hat{\mathbf{i}}' + I_{xx}\omega_{y'}\hat{\mathbf{j}}' + I_{zz}(\omega_{z'} + \dot{\psi})\hat{\mathbf{k}}' \tag{12.166}$$

Note that the top spin $\dot{\psi}$ has been added to obtain the total z'-component of the angular velocity of the top. This is because the angular velocity given by Eq. (12.165) is not that of the top but rather of the primed system with respect to XYZ. The time derivative of the angular momentum may now be determined:

$$\dot{\mathbf{H}} = I_{xx}\dot{\omega}_{x'}\hat{\mathbf{i}}' + I_{xx}\dot{\omega}_{y'}\hat{\mathbf{j}}' + I_{zz}(\dot{\omega}_{z'} + \ddot{\psi})\hat{\mathbf{k}}' + [I_{zz}\omega_{y'}(\omega_{z'} + \dot{\psi}) - I_{xx}\omega_{z'}\omega_{y'}]\hat{\mathbf{i}}'$$

$$+ [I_{xx}\omega_{z'}\omega_{x'} - I_{zz}\omega_{x'}(\omega_{z'} + \dot{\psi})]\hat{\mathbf{j}}' + [I_{xx}\omega_{x'}\omega_{y'} - I_{xx}\omega_{y'}\omega_{x'}]\hat{\mathbf{k}}' \tag{12.167}$$

The only torque acting on the top is that attributable to its own weight $mgL \sin \theta$, L representing the distance from O to the center of mass (which lies on Oz'). This torque has a component about the rotating x'-axis only. The Euler equations for the top are thus

$$\dot{H}_{x'} = I_{xx}\dot{\omega}_{x'} + (I_{zz} - I_{xx})\omega_{y'}\omega_{z'} + I_{zz}\omega_{y'}\dot{\psi} = mgL \sin \theta \tag{12.168}$$

$$\dot{H}_{y'} = I_{xx}\dot{\omega}_{y'} + (I_{xx} - I_{zz})\omega_{x'}\omega_{z'} - I_{zz}\omega_{x'}\dot{\psi} = 0 \tag{12.169}$$

$$\dot{H}_{z'} = I_{zz}(\dot{\omega}_{z'} + \ddot{\psi}) = 0 \tag{12.170}$$

Defining $S = \dot{\psi} + \omega_{x'}$, Eqs. (12.168), (12.169), and (12.170) become

$$I_{xx}\dot{\omega}_{x'} - I_{xx}\omega_{y'}\omega_{z'} + I_{zz}S\omega_{y'} = mgL \sin \theta \qquad (12.171)$$

$$I_{xx}\omega_{y'} + I_{xx}\omega_{x'}\omega_{z'} - I_{zz}S\omega_{x'} = 0 \qquad (12.172)$$

$$I_{zz}\dot{S} = 0 \qquad (12.173)$$

The total mechanical energy of the top is obtained by simply adding to the kinetic energy, the potential energy relative to a convenient datum such as the XY-plane,

$$E = \tfrac{1}{2}I_{xx}\omega_{x'}^2 + \tfrac{1}{2}I_{xx}\omega_{y'}^2 + \tfrac{1}{2}I_{zz}S^2 + mgL \cos \theta \qquad (12.174)$$

where $L \cos \theta$ represents the vertical distance of the center of mass above the XY-plane. Equation (12.174) may also be derived by multiplying Eq. (12.171) by $\omega_{x'}$ and Eq. (12.172) by $\omega_{y'}$ and adding, thereby eliminating the second and third terms. If to the expression thus obtained we add Eq. (12.173) multiplied by S, we have

$$I_{xx}\omega_{x'}\dot{\omega}_{x'} + I_{xx}\omega_{y'}\dot{\omega}_{y'} + I_{zz}S\dot{S} = mgL \sin \theta\omega_{x'} \qquad (12.175)$$

where $\omega_{x'} = \dot{\theta}$. When integrated with respect to time, Eq. (12.175) yields Eq. (12.174), where the constant of integration represents the system energy.

Equation (12.173) when integrated tells us that $I_{zz}S$ is a constant or S is constant; that is, the z'-component of the top's angular velocity and angular momentum are constant. A similar result was obtained for the gyroscope in a Cardan suspension.

The solution of the foregoing rather formidable set of equations is now obtained only for several special situations. For example, for the case in which the total spin S dwarfs the other angular-speed terms (usually because of large $\dot{\psi}$), Eq. (12.171) becomes

$$I_{zz}S\omega_{y'} = mgL \sin \theta \qquad (12.176)$$

Substituting $\omega_{y'} = \dot{\phi} \sin \theta$,

$$\dot{\phi} = \frac{mgL}{SI_{zz}} \qquad (12.177)$$

which is the result obtained from the elementary theory of the gyroscope. In this connection, it is now appropriate to review the elementary gyroscope, Fig. 12.41.

Neglecting the effect caused by rotation of the spin axis about OO, the total angular momentum of the gyroscope rotor is $\mathbf{H} = I\omega$. The torque vector, caused entirely by the weight mg, must be directed parallel to the change of the angular momentum vector in accordance with $\mathbf{M} = \dot{\mathbf{H}} = d/dt(I\omega)$. The angle between $\mathbf{H}$ at any time and $\mathbf{H}$ at some subsequent infinitesimal time is $d\phi$ and from Fig. 12.41 is given by $d\phi = dH/H$; therefore,

$$\dot{\phi} = \frac{d\phi}{dt} = \frac{1}{H}\frac{dH}{dt} = \frac{M}{H} = \frac{mgL}{I\omega} \qquad (12.178)$$

as in Eq. (12.177).

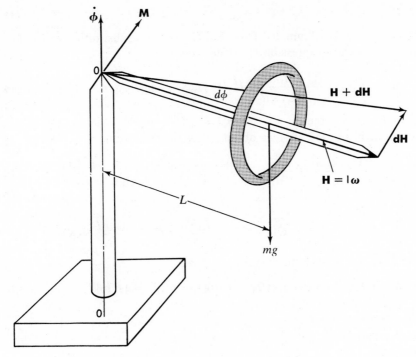

Fig. 12.41

Although Fig. 12.41 leads to the impression that the angle θ which the symmetry axis makes with the space-fixed Z-axis remains constant, this is in general not true, and the top is usually observed to *nutate*. *Nutation*, the periodic variation in the angle θ between certain limits θ_1 and θ_2, is now explored.

Since the gravity force acts parallel to the Z-axis, any moment ascribable to this force about the Z-axis must be zero and therefore $M_Z = \dot{H}_Z = 0$. As a consequence, the component of the angular momentum vector along the space-fixed Z-direction equals a constant, C, leading to the statement

$$I_{xx}\omega_{y'} \sin \theta + I_{zz}S \cos \theta = C \tag{12.179}$$

Substituting $\omega_{y'} = \dot{\phi} \sin \theta$, Eq. (12.179) becomes

$$I_{xx}\dot{\phi} \sin^2 \theta + I_{zz}S \cos \theta = C \tag{12.180}$$

and we thus have a new expression for the precessional velocity

$$\dot{\phi} = \frac{C - I_{zz}S \cos \theta}{I_{xx} \sin^2 \theta} \tag{12.181}$$

Substituting into the energy expression, Eq. (12.174), for $\omega_{x'}$, $\omega_{y'}$, $\omega_{z'}$, Eqs. (12.165), we have

$$E = \tfrac{1}{2}I_{xx}\dot{\theta}^2 + \tfrac{1}{2}I_{xx}\dot{\phi}^2 \sin^2\theta + \tfrac{1}{2}I_{zz}S^2 + mgL\cos\theta \qquad (12.182)$$

The term $\dot{\phi}^2$ may be eliminated from Eq. (12.182) by substitution of Eq. (12.181), yielding yet another expression for system energy:

$$E = \tfrac{1}{2}I_{xx}\dot{\theta}^2 + \frac{(C - I_{zz}S\cos\theta)^2}{2I_{xx}\sin^2\theta} + \tfrac{1}{2}I_{zz}S^2 + mgL\cos\theta \qquad (12.183)$$

If we set $p = \cos\theta$, then $\dot{p} = -(\sin\theta)\dot{\theta} = -(1 - p^2)^{1/2}\dot{\theta}$. Making the appropriate substitutions in Eq. (12.183) we obtain, after simplification,

$$\dot{p}^2 = \frac{2E - I_{zz}S^2}{I_{xx}}(1 - p^2) - \frac{2mgLp(1 - p^2)}{I_{xx}} - \left(\frac{C}{I_{xx}} - \frac{I_{zz}Sp}{I_{xx}}\right)^2 \qquad (12.184)$$

Setting

$$\frac{C}{I_{xx}} = C_1 \qquad \frac{I_{zz}S}{I_{xx}} = C_2 \qquad \frac{2E - I_{zz}S^2}{I_{xx}} = C_3 \qquad \frac{2mgL}{I_{xx}} = C_4$$

Eq. (12.184) now reads, after factoring $1 - p^2$,

$$p = [-(C_1 - C_2 p)^2 + (C_3 - C_4 p)(1 - p^2)]^{1/2} = [f(p)]^{1/2} \qquad (12.185)$$

or

$$\int_{p_1}^{p_2} \frac{dp}{[f(p)]^{1/2}} = t_2 - t_1 \qquad (12.186)$$

Although no simple solutions to the elliptic integral Eq. (12.186) are available, some insight into the time variation of p (and hence θ) may be gained without actually integrating. The following analysis will enable us to piece together a plausible description of $f(p)$:

a. $f(p)$ is cubic and therefore may have no more than three real roots, p_1, p_2, p_3, that is, a maximum of three intersections with the p-axis, Fig. 12.42.

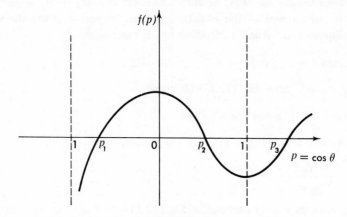

Fig. 12.42

b. For motion to occur, the integrand of Eq. (12.186) must be real and therefore $f(p)$ may not be negative.

c. Since $p = \cos\theta$, acceptable values of p may not lie outside the range 1 and -1.

d. At $p = 1, f(p) = -(C_1 - C_2)^2$, meaning that $f(p)$ can be zero or negative only. The same conclusion is drawn if one substitutes $p = -1$, for now $f(p) = -(C_1 + C_2)^2$.

e. For large positive values of p, the cubic term in $f(p)$ is positive. For large negative values of p, the cubic term is negative and $f(p)$ is negative.

In general, p_3 exceeds unity and represents an unacceptable root. The roots p_1 and p_2 indicate that a periodic variation in θ occurs between the limits $\theta_1 = \arccos p_1$ and $\theta_2 = \arccos p_2$. The motion associated with this variation in θ is as we have noted termed *nutation*, and the limiting angles for such motion are termed the *libration limits*. The nutation is shown in Fig. 12.43(a), in which the fixed point of motion is at O. The path traced on the surface of a sphere of unit radius is that of the symmetry axis of the top.

The trace of the axis of symmetry upon the sphere of Fig. 12.43 can be studied on the basis of the relationship between the precessional frequency $\dot\phi$ and the angle of inclination θ. These quantities are connected in Eq. (12.181), written below in slightly modified form:

$$\dot\phi = \frac{SI_{zz}}{I_{xx}\sin^2\theta}\left(\frac{C}{SI_{zz}} - \cos\theta\right) \tag{12.187}$$

Consider the case in which the initial conditions cause $C/SI_{zz} > \cos\theta_2$. It follows that C/SI_{zz} also exceeds $\cos\theta_1$, and therefore $\dot\phi$ does not alter its algebraic sign as θ proceeds from θ_1 to θ_2 and back to θ_1 again. In other words, if one substitutes a positive p_2 and a negative p_1 into Eq. (12.187), or any value in between the allowable roots shown in Fig. 12.42, $\dot\phi$ will be positive (hence never changing sign). Since $\dot\theta$ must be zero at θ_1 and θ_2, the trace shown in Fig. 12.43(a) is fully consistent with $\dot\phi$ positive for all θ. If C/SI_{zz} should lie between $\cos\theta_1$ and $\cos\theta_2$, it is clear that $\dot\phi_2 < 0$ and $\dot\phi_1 > 0$. The precessional velocity is thus of different sign at the two circles shown in Fig. 12.43(b). $\dot\phi$ must pass through a zero at some intermediate θ, and therefore displays a characteristic *looping* trace. The average precession is not zero, however, and the symmetry axis does exhibit a net positive or negative precession.

For conditions under which C/SI_{zz} corresponds to a root of $f(p)$, the precessional velocity at that root is zero, from Eq. (12.187). At $p_2 = \cos\theta_2$, for example, if $C/SI_{zz} = p_2$, both $\dot\theta$ and $\dot\phi$ are zero. The other values assumed by $\dot\phi$ are positive only, according to Eq. (12.187). Consequently, the *cuspidal motion* of Fig. 12.43(c) is observed.

The trace of Fig. 12.43(c) can be obtained by fixing a spinning top at an initial angle θ_i and releasing it with initial conditions $\dot\theta = \dot\phi = 0$. Thus θ_i is actually θ_2. As the top begins to fall (θ increasing), it exchanges some of the potential

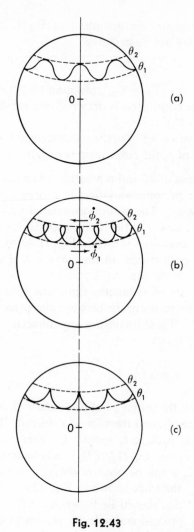

Fig. 12.43

energy it had at $\theta_i = \theta_2$ for precessional kinetic energy. When θ_1 is reached, the top can go no further on the basis of the allowable roots of $f(p)$, and it returns to θ_2, precessing as before. That the stationary point can exist at θ_2 only is clear from the energy equation rearranged as follows:

$$E - \tfrac{1}{2}I_{zz}S^2 = \tfrac{1}{2}I_{xx}\dot\theta^2 + \tfrac{1}{2}I_{xx}\dot\phi^2\sin^2\theta + mgL\cos\theta \qquad (12.188)$$

where the left side is constant [recall Eq. (12.173)]. At a stationary point, the terms in $\dot\theta$ and $\dot\phi$ are zero and the potential energy is therefore a maximum, meaning that θ is a minimum corresponding to θ_2. In the event a double root of $f(p)$ exists, $\cos\theta_1 = \cos\theta_2 = p_1 = p_2$ and there is no nutation, only a steady precession of the top.

12.35 CONDITIONS FOR NO NUTATION

Under what conditions can we expect the top to exhibit no nutational motion? Clearly this requires that $\theta = \theta_c$, a constant, and $\dot{\theta} = 0$. Equations (12.165) thus become $\omega_{x'} = 0$, $\omega_{y'} = \dot{\phi} \sin \theta_c$, $\omega_{z'} = \dot{\phi} \cos \theta_c$. Substituting the foregoing into Eqs. (12.171), (12.172), and (12.173), we have

$$S\dot{\phi}I_{zz} - \dot{\phi}^2 \cos \theta_c I_{xx} = mgL \tag{12.189}$$

$$I_{xx} \sin \theta_c \ddot{\phi} = 0 \tag{12.190}$$

$$I_{zz} \dot{S} = 0 \tag{12.191}$$

The last two equations relate that the precessional velocity $\dot{\phi}$ and the total spin S are constant. Equation (12.189) is quadratic in $\dot{\phi}$ possessing roots given by

$$\dot{\phi} = \frac{SI_{zz} \pm [(SI_{zz})^2 - 4mgLI_{xx} \cos \theta_c]^{1/2}}{2I_{xx} \cos \theta_c} \tag{12.192}$$

Since the radical must be real, $(SI_{zz})^2 \geq 4mgLI_{xx} \cos \theta_c$. For each value of $\cos \theta_c = p_c$ there are two possible values of precessional velocity $\dot{\phi}$ corresponding to the positive and negative values of the square root.

Rewriting Eq. (12.192) in the form

$$\dot{\phi} = \frac{SI_{zz} \pm SI_{zz}[1 - 4mgLI_{xx} \cos \theta_c(SI_{zz})^{-2}]^{1/2}}{2I_{xx} \cos \theta_c} \tag{12.193}$$

and expanding the radical,

$$\dot{\phi} = \frac{SI_{zz} \pm SI_{zz}[1 - 2mgLI_{xx} \cos \theta_c(SI_{zz})^{-2} + \cdots]}{2I_{xx} \cos \theta_c} \tag{12.194a}$$

or

$$\dot{\phi} = \frac{SI_{zz} \pm [SI_{zz} - (SI_{zz})^{-1}2mgLI_{xx} \cos \theta_c + \cdots]}{2I_{xx} \cos \theta_c} \tag{12.194b}$$

Dropping the second term in the brackets, on the basis of large S, the higher of the two precessional velocities is

$$\dot{\phi}_2 = \frac{SI_{zz}}{I_{xx} \cos \theta_c} \tag{12.195}$$

which we observe to depend solely upon the total spin S and the constant angle which the spin axis makes with the space-fixed vertical. The smaller value of $\dot{\phi}$ is determined by retaining the second term in the brackets. This is the precession usually observed:

$$\dot{\phi}_1 = \frac{mgL}{SI_{zz}} \tag{12.196}$$

Here the precession is inversely proportional to the total spin S, a result previously obtained in a simplified analysis of the gyroscope.

12.36 THE SLEEPING TOP

As a further application of the top dynamics already developed, consider the sleeping top, that is, a top initiated and remaining with its axis of symmetry vertical.

The locus of the axis of the top is described by the function $f(p)$ in Eq. (12.185), now modified to correspond to the conditions $\theta = 0$, $\dot{\theta} = 0$, representing the special case of zero nutation. In addition, the system energy is assumed to remain constant:

$$E = \tfrac{1}{2} I_{zz} S^2 + mgL \tag{12.197}$$

From Eq. (12.179), the component along space-fixed Z of the angular momentum is now $C = I_{zz} S$, and the constants in $f(p)$ are therefore

$$C_1 = C_2 = \frac{I_{zz} S}{I_{xx}}$$

$$C_3 = \frac{2E - I_{zz} S^2}{I_{xx}} = \frac{I_{zz} S^2 + 2mgL - I_{zz} S^2}{I_{xx}} = \frac{2mgL}{I_{xx}} = C_4$$

When $C_1 = C_2$ and $C_3 = C_4$ are substituted into $f(p)$, we obtain

$$f(p) = (1 - p)^2 [-C_1^2 + (1 + p) C_3] \tag{12.198}$$

which has a double root at $p = 1$, corresponding to zero nutation at the vertical position of the symmetry axis. The remaining root is obtained by equating the bracketed term to zero. This yields

$$p = \frac{C_1^2}{C_3} - 1 = \frac{I_{zz}^2 S^2 I_{xx}}{I_{xx}^2 2mgL} - 1 \tag{12.199a}$$

or

$$\cos \theta_3 = \frac{I_{zz}^2 S^2}{2 I_{xx} mgL} - 1 \tag{12.199b}$$

Assume that $\cos \theta_3$ can be realized by the top, that is, represents an allowable angle of inclination in accordance with the previous discussion of the roots of $f(p)$. The motion will then be unstable in the vertical position, because the top can also exist in a configuration of potential energy lower than that corresponding to $\theta = 0$.

The reader may find it interesting to examine the influence of friction upon the top. Even if the spin is so great as to preclude p_3 as an allowable root initially, eventually friction will cause S to diminish, leading to an acceptable θ_3 at a lower system potential energy, to which the top will fall.

12.37 TORQUE-FREE MOTION (continued)

In analyzing the torque-free motion of a symmetrical body of revolution, the Euler equations were written in terms of body-fixed coordinates. Next we observed that the Euler angles led to a rotating coordinate system (not fixed to the body) which simplified the equations describing the motion of a body of revolution under the action of a gravity-induced torque. More importantly, the precessional and nutational motions were related to a fixed direction in space, the Z-axis, which proved particularly useful for purposes of visualization. We now return to torque-free motion, employing the techniques of the analysis of the top.

To simplify our analysis somewhat, the space-fixed Z-axis is taken as colinear with the direction of the fixed angular momentum vector **H**, as shown in Fig. 12.44. The components of the angular momentum in the $x'y'z'$-system are

$$H_{x'} = 0 \tag{12.200a}$$

(since x' lies in the XY-plane, and **H** is normal to this plane)

$$H_{y'} = H \cos\left(\frac{\pi}{2} - \theta\right) = H \sin\theta \tag{12.200b}$$

$$H_{z'} = H \cos\theta \tag{12.200c}$$

Since we are treating a body of revolution, $I_{x'x'} = I_{xx} = I_{y'y'} = I_{yy}$ and the primed notation will, as before, be dropped in writing the moments of inertia. The angular momenta are therefore

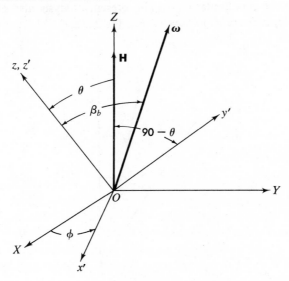

Fig. 12.44

$$H_{x'} = I_{xx}\omega_{x'} = 0 \qquad \omega_{x'} = 0 \qquad\qquad\qquad (12.201a)$$

$$H_{y'} = I_{yy}\omega_{y'} \qquad\qquad\qquad (12.201b)$$

$$H_{z'} = I_{zz}\omega_{z'} \qquad\qquad\qquad (12.201c)$$

Since $\omega_{x'} = 0$, $\boldsymbol{\omega}$ must lie entirely within the $y'z'$-plane and is therefore in the same plane as z' and $\mathbf{H}$. In terms of β_b, the angle between the angular velocity and the symmetry axis $Oz' = Oz$,

$$\omega_{y'} = \omega \sin \beta_b \qquad\qquad\qquad (12.202a)$$

$$\omega_{z'} = \omega \cos \beta_b \qquad\qquad\qquad (12.202b)$$

The Euler angle θ, between OZ (also $\mathbf{H}$ in this case) and Oz', is related to β_b as follows:

$$\frac{H_{y'}}{H_{z'}} = \tan \theta = \frac{I_{yy}\omega_{y'}}{I_{zz}\omega_{z'}} = \frac{I_{yy}}{I_{zz}} \tan \beta_b \qquad\qquad\qquad (12.203)$$

Consequently the relative location of $\mathbf{H}$, z', and $\boldsymbol{\omega}$ is a function of the ratio I_{yy}/I_{zz}. For $I_{yy} > I_{zz}$, which applies to elongated bodies, θ exceeds β_b as shown in Fig. 12.45. For mass distributions in which $I_{zz} > I_{yy}$, typical of flywheels, β_b is larger than θ, as in Fig. 12.46. The vectors shown in Fig. 12.46 have already been shown to form the basis of space and body cone diagrams. The counterpart of Fig. 12.33, corresponding to $I_{zz} > I_{yy}$, is depicted in Fig. 12.47.

The reader will recall that the angles θ and β_s are constant, thus, of course, permitting the cone representations. The present analysis also confirms that the angle $\beta_b = \theta - \beta_s$ is constant in view of Eq. (12.203).

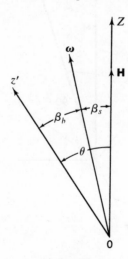

Fig. 12.45

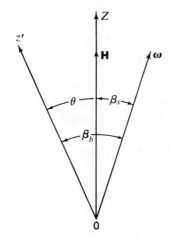

Fig. 12.46

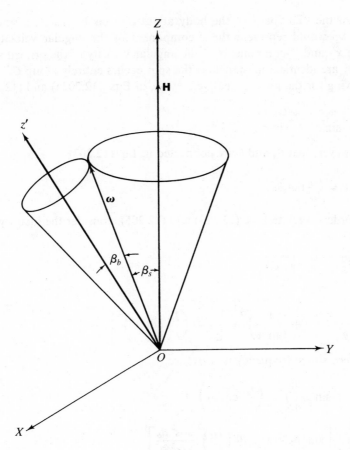

Fig. 12.47

12.38 THE PRECESSIONAL FREQUENCY — TORQUE-FREE MOTION

Examination of Fig. 12.37 reveals that rotation of the $y'z'$-plane is identical with the angular displacement ϕ and that, inasmuch as this plane contains ω, the angular velocity, $\dot\phi$ is the precessional velocity or angular frequency of ω with respect to the fixed direction of the angular momentum vector.

Referring to Eqs. (12.165), the angular velocity components of the *body* along the $x'y'z'$-axes are

$$\omega_{x'} = 0 \tag{12.204a}$$

$$\omega_{y'} = \dot\phi \sin\theta \tag{12.204b}$$

$$\omega_{z'} = \dot\phi \cos\theta + \dot\psi \tag{12.204c}$$

where again the spin speed of the body about Oz' has been added since without this term, $\omega_{z'}$ would represent the z'-component of the angular velocity of $x'y'z'$ only. The x'- and y'-components of the angular velocity of the primed system and of the body are identical inasmuch as the spin occurs entirely along Oz'.

Solving for the precessional velocity $\dot\phi$, in Eqs. (12.202a) and (12.204b),

$$\dot\phi = \frac{\omega_{y'}}{\sin\theta} = \frac{\omega \sin\beta_b}{\sin\theta} \tag{12.205}$$

Recall, however, that β_b and θ are connected in Eq. (12.203):

$$\tan\theta = \frac{I_{yy}}{I_{zz}} \tan\beta_b$$

In order to relate Eqs. (12.203) and (12.205), consider the identity

$$\frac{1}{\sin^2\theta} = 1 + \frac{1}{\tan^2\theta}$$

Then

$$\frac{1}{\sin\theta} = \left(1 + \frac{1}{\tan^2\theta}\right)^{\frac12} = \left[1 + \left(\frac{I_{zz}}{I_{yy}}\cot\beta_b\right)^2\right]^{\frac12} \tag{12.206}$$

and the precessional frequency is therefore,

$$\dot\phi = \omega \sin\beta_b \left[1 + \left(\frac{I_{zz}}{I_{yy}}\cot\beta_b\right)^2\right]^{\frac12}$$

$$= \omega \left[\sin^2\beta_b + \sin^2\beta_b \left(\frac{I_{zz}}{I_{yy}}\right)^2 \frac{\cos^2\beta_b}{\sin^2\beta_b}\right]^{\frac12} \tag{12.207}$$

Adding and subtracting $\cos^2 \beta_b$ within the radical,

$$\dot{\phi} = \omega \left\{ 1 + \cos^2 \beta_b \left[\left(\frac{I_{zz}}{I_{yy}} \right)^2 - 1 \right] \right\}^{\frac{1}{2}} \tag{12.208}$$

In the event that the angle between the axis of symmetry and the angular velocity is very small, as is true for the earth, for example, $\cos \beta_b$ is approximately unity, and Eq. (12.208) becomes

$$\dot{\phi} = \omega \frac{I_{zz}}{I_{xx}}$$

References

Christie, D. E., *Vector Mechanics*, Second Edition, McGraw-Hill, Inc., New York, 1964.

Fowles, G. R., *Analytical Mechanics*, Holt, Rinehart and Winston, Inc., New York, 1962.

Goldstein, H., *Classical Mechanics*, Addison-Wesley Publishing Company, Inc., Reading, Mass., 1950.

Greenwood, D. T., *Principles of Dynamics*, Prentice-Hall, Inc., Englewood Cliffs, N.J., 1965.

Wangsness, R. K., *Introduction to Theoretical Physics*, John Wiley & Sons, Inc., New York, 1963.

Whittaker, E. T., *Analytical Dynamics*, Fifth Edition, Dover Publications, Inc., New York, 1944.

EXERCISES

12.1 Determine at the origin and at $(1, 1, 1)$, the resultant of the following system of forces and couple moments:

$$\mathbf{F}_1 = -3\hat{\mathbf{i}} + 4\hat{\mathbf{j}} + 2\hat{\mathbf{k}} \text{ at } (3, 3, 2)$$

$$\mathbf{F}_2 = 2\hat{\mathbf{i}} + 2\hat{\mathbf{j}} - 2\hat{\mathbf{k}} \text{ at } (-1, 6, 4)$$

$$\mathbf{F}_3 = 6\hat{\mathbf{i}} + 2\hat{\mathbf{j}} + 7\hat{\mathbf{k}} \text{ at } (1, 1, 1)$$

$$\mathbf{C}_1 = 2\hat{\mathbf{i}} + 6\hat{\mathbf{j}}$$

$$\mathbf{C}_2 = 3\hat{\mathbf{i}} + 3\hat{\mathbf{j}} + 6\hat{\mathbf{k}}$$

12.2 Reduce the following system to a wrench:

$$\mathbf{F} = 3\hat{\mathbf{i}} + 2\hat{\mathbf{j}} - 8\hat{\mathbf{k}} \text{ N at } (3, 3, 3) \text{ m}$$

$$\mathbf{C} = 2\hat{\mathbf{i}} + 2\hat{\mathbf{j}} - 2\hat{\mathbf{k}} \text{ N-m}$$

12.3 Locate the center of mass of a triangle having sides a, b, c and constant thickness t. Assume in this problem, as well as those following, that matter is uniformly distributed.

12.4 Locate the center of mass of a sector θ of a circle of radius R_0 and constant thickness t.

12.5 Locate the center of mass of a right circular cone of height h and base radius R_0.

12.6 Locate the center of mass of a hemisphere of radius R_0.

12.7 Show that a maximum of six independent equations of statics may be written for a general system of forces, and that, of these, a maximum of three may be force equations, but as many as six may be moment equations.

12.8 Demonstrate that no more than three independent equations of statics may be written for a coplanar system of forces, and that no more than two of these may be force equations.

12.9 Prove that only one independent equation of statics may be written for a two-force system.

12.10 Prove that in a coplanar parallel system of forces, only two independent equations of statics may be written, and that these may be a force equation plus a moment equation, or two moment equations.

12.11 A cantilever beam is loaded as shown in Fig. P12.11. Write the equations for shear and bending moment for all representative sections of the beam and draw the shear and moment diagrams.

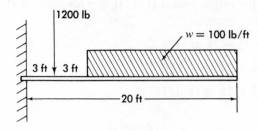

Fig. P12.11

12.12 For the simply supported beam described in Fig. P12.12, write the equations for shear and moment at representative sections, draw the shear and moment diagrams, and calculate the point of maximum bending moment.

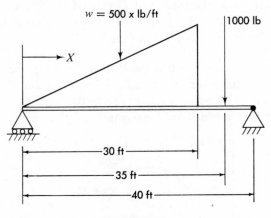

Fig. P12.12

12.13 Replace the system of applied forces (Fig. P12.13) with a single force and determine its point of application.

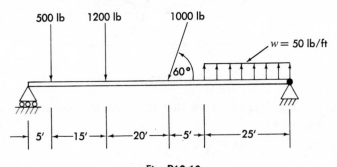

Fig. P12.13

12.14 A rigid body rotates about fixed point O, coincident with the origin of coordinate system XYZ. A point P_1 on the body is located with respect to O by the vector

$$\boldsymbol{\rho}_1 = 2\hat{\mathbf{i}} - 3\hat{\mathbf{j}} + 6\hat{\mathbf{k}} \text{ m}$$

If the angular velocity of the body with respect to XYZ is constant, given by

$$\boldsymbol{\omega} = 3\hat{\mathbf{i}} + 3\hat{\mathbf{j}} - 2\hat{\mathbf{k}} \text{ rad/sec}$$

what are the velocity and acceleration of P_1?

Show that the velocity is perpendicular to the angular velocity and position vectors by forming the appropriate scalar products.

12.15 Referring to Exercise 12.14, a second point P_2 is located by

$$\rho_2 = 7\hat{i} + 7\hat{j} - 6\hat{k} \text{ m}$$

What are the velocity and acceleration of P_2 relative to P_1? Assume the constant angular velocity given in Exercise 12.14. What is the relative acceleration if the instantaneous angular acceleration is $3\hat{k}$ rad/sec²?

12.16 A rigid body undergoes motion parallel to the XY-plane as shown in Fig. P12.16. Demonstrate that given the velocities of points A and B, point C is at rest at the instant shown. This point is referred to as the *instantaneous center of rotation*. It is that point about which one body rotates relative to another at any instant.

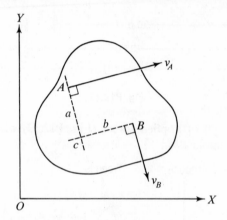

Fig. P12.16

12.17 Referring to Fig. P12.16, prove that the instantaneous angular velocity of the rigid body is given by

$$\omega = \frac{v_A}{a} = \frac{v_B}{b}$$

12.18 Given n rigid bodies in plane relative motion, how many instantaneous centers of rotation exist at any time?

12.19 An Atwood's machine consists of masses m_1 and m_2 ($m_1 > m_2$) attached to a cord which passes over a pulley of mass m and radius R_0. Derive an expression for the acceleration of each mass and determine the tension on each side of the pulley. What torque does the system of mass apply to the pulley? Assume no slip occurs between the cord and the pulley surface.

12.20 A massless, inextensible rope, attached to a mass of 2000 g, passes over frictionless surface a and is wrapped around pulley b, as shown in Fig. P12.20. The friction in the pulley bearings may be neglected. If the system is initially at

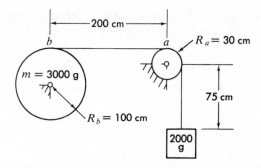

Fig. P12.20

rest and then released, determine, at $t = 2$ sec, the velocity and acceleration of the mass, the angular velocity and acceleration of the large pulley, the rope tension, and the reactions at the pulley bearings. Assume the pulley to be a homogeneous cylinder.

12.21 This problem is similar to Exercise 12.20, except that the cord mass of 10 g/cm of length is to be taken into account. Determine the same quantities as in Exercise 12.20. The system is released from rest in the position shown in Fig. P12.20.

12.22 The cylinder shown in Fig. P12.22 is connected by means of a linear spring to a rigid wall. If the cylinder rolls without slipping, describe the motion which results from an initial displacement x_0 and an initial velocity v_0.

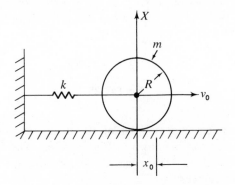

Fig. P12.22

12.23 A cylinder is given an initial velocity v_0 up an inclined plane making an angle θ with the horizontal. If the cylinder does not slip, how far up the incline will it travel before stopping?

12.24 Determine the natural frequencies of the system described in Fig. P12.24, in which a rigid bar of mass m is supported by two identical linear springs.

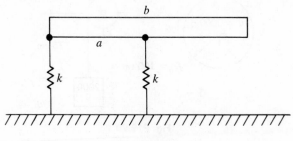

Fig. P12.24

12.25 Assuming no slip occurs between the belt and the pulley, determine the natural frequency of oscillation of the system described in Fig. P12.25.

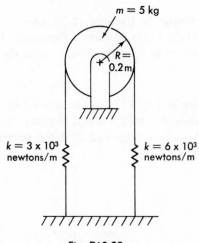

Fig. P12.25

12.26 A homogeneous circular cylinder of 10 cm diameter and 150 cm length is free to rotate about point O as shown in Fig. P12.26. The density of the cylinder material is 2 g/cm³. If released from rest at $\phi = 30$ degrees, determine the following quantities at $\phi = 150$ degrees:

    a. the velocity of the center of mass.
    b. the acceleration of the center of mass.
    c. the angular velocity.
    d. the angular acceleration.
    e. the force acting on the support.

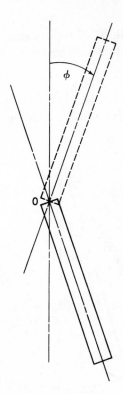

Fig. P12.26

12.27 Determine the moment of inertia of a rectangular parallelepiped of sides a, b, c as shown in Fig. P12.27 about diagonal 1-2 and diagonal 3-4.

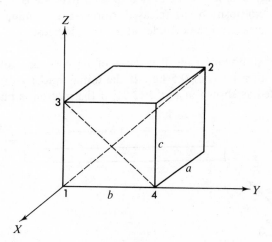

Fig. P12.27

12.28 If the rigid body of Exercise 12.27 rotates about axis 1-2 with angular velocity ω, what is its angular momentum with respect to XYZ?

12.29 A circular cylinder of mass 2500 g and radius 10 cm rolls without slipping on a cylindrical surface, as shown in Fig. P12.29. Employing energy considerations, derive the differential equation for small oscillations about $\phi = 0$. What is the period of oscillation?

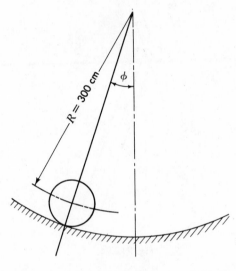

Fig. P12.29

12.30 If the cylinder of Exercise 12.29 is displaced to angle $\phi = 45$ degrees to the left of the equilibrium position and released from rest, determine the force exerted by the cylindrical surface on the cylinder at $\phi = 30$ degrees.

12.31 An initially stationary rod, 1 m long of mass 2 kg, suffers a completely inelastic impact with a mass of 0.5 kg. If the initial velocity of the smaller mass is 10 m/sec directed as shown in Fig. P12.31, if the motion is planar, and if there

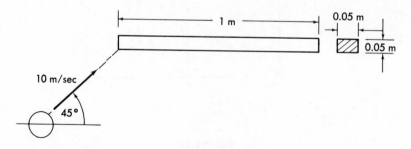

Fig. P12.31

is negligible friction between both objects and the plane on which they move, determine:

    a. the velocity of the mass center of the rod immediately after impact.
    b. the angular velocity of the rod immediately after impact.
    c. the velocity of the impact point on the rod, immediately after impact.

12.32 An assembly consisting of a block of mass 200 kg and two rods each of mass 20 kg rotates in a vertical plane as shown in Fig. P12.32. Determine the forces at the support as well as the angular acceleration at the instant shown if the assembly is released from rest in a vertical position.

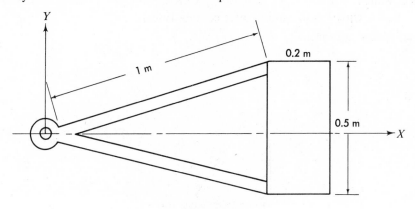

Fig. P12.32

12.33 Determine the components of angular velocity parallel to the principal axes of a plane object free to rotate about fixed point O, following the application of an impulse I at point $x, 0, z$. As shown in Fig. P12.33, the x- and z-axes are in the

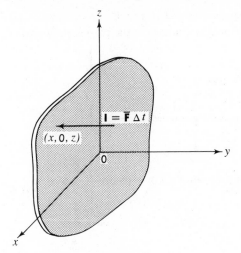

Fig. P12.33

plane of the body. Demonstrate that the component of angular velocity parallel to the xz-plane remains constant.

12.34 A semicylinder is held at an angle ϕ_0 with respect to a frictionless plane, as shown in Fig. P12.34, and then released. Determine the following quantities:

   a. the reaction of the vertical wall on the cylinder for all ϕ until contact is lost.

   b. the value of ϕ at which contact is lost.

   c. the velocity of the mass center when the cylinder loses contact with the vertical wall.

   d. the equation of motion after contact is lost.

Use an energy approach.

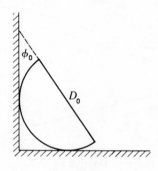

Fig. P12.34

12.35 For the system shown in Fig. P12.35, rotating at constant angular velocity $\omega = 2$ rad/sec, determine:

   a. the bearing reactions at A and B.

   b. a system of balance masses in the planes shown to effect dynamic balance.

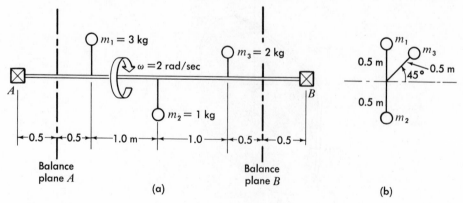

Fig. P12.35

12.36 Determine the center of percussion of the hammer shown in Fig. P12.36, which is composed of a sphere mounted on a circular cylinder. The hammer is pivoted at point O, and fabricated of a metal of density 7.5 g/cm^3.

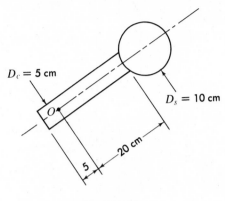

$D_c = 5$ cm

$D_s = 10$ cm

20 cm

5

Fig. P12.36

12.37 Prove that the principal axes of inertia of an asymmetrical body are mutually perpendicular.

12.38 The following array describes the inertia tensor of a rigid body:

$$| = \begin{bmatrix} 300 & 50 & 100 \\ 50 & 400 & 20 \\ 100 & 20 & 500 \end{bmatrix}$$

What are the principal moments of inertia referred to the same point as the above matrix?

12.39 Determine the principal moments of inertia as well as the orientation of the principal axes, given the following matrix:

$$| = \begin{bmatrix} 200 & 0 & 50 \\ 0 & 500 & 0 \\ 50 & 0 & 300 \end{bmatrix}$$

12.40 Because the thrust vector does not pass through the mass center, as shown in Fig. P12.40, a constant moment about the body-fixed y-axis occurs, and rotational motion results. Describe the angular velocity and acceleration of the rocket. Take the initial conditions to be $\omega_x = \omega_y = \omega_z = 0$. The following data apply:

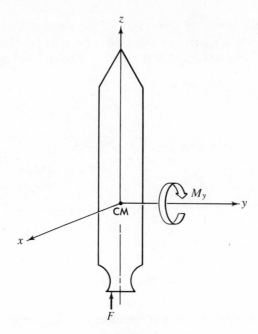

Fig. P12.40

$$I_{zz} = 3 \times 10^4 \text{ kg-m}^2$$

$$I_{xx} = I_{yy} = 1.5 \times 10^6 \text{ kg-m}^2$$

$$M_y = 5000 \text{ N-m}$$

12.41 Repeat Exercise 12.40 with the following initial conditions: $\omega_x = \omega_y = 0$, $\omega_z = 2$ rad/sec.

12.42 Determine the frequency of precession of a circular disk in torque-free motion.

12.43 The angle between the symmetry axis of the earth and the earth's angular velocity is approximately 0.2 sec of arc. For the earth, $(I_{zz} - I_{xx})/I_{xx} = 0.00327$. Verify (approximately) that the period of precession of the angular velocity vector about the symmetry axis is 427 days. To what do you attribute the difference between your solution and the observed value?

12.44 Verify that the precession of the earth's axis of symmetry about the angular momentum vector is 0.997 day.

12.45 Demonstrate that for a symmetrical body, the polhode represents the curve of intersection of the body cone with the ellipsoid of inertia.

12.46 Demonstrate that for a symmetrical body, the herpolhode represents the curve of intersection of the space cone with the invariable plane.

In Exercises 12.47 through 12.51, describe the geometry of the ellipsoids of inertia corresponding to the figure geometries given. The reference point is the center of mass.

12.47 A rectangular parallelepiped of sides a, b, c.

12.48 A cube.

12.49 A sphere.

12.50 An ellipsoid.

12.51 A right-circular cylinder of radius R_0 and length L.

12.52 The disk shown in Fig. P12.52 has an angular speed of 50 rad/sec about axis A-A, and a moment of inertia about the same axis of 120,000 g-cm². Assuming the shaft inertia negligible, determine the following:

    a. the precessional velocity for $m = 600$ g.
    b. the bearing reactions for $m = 600$ g.
    c. the precessional velocity for $m = 3600$ g.

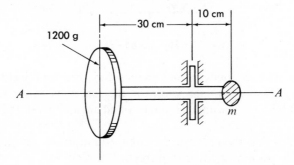

Fig. P12.52

12.53 Investigate the dynamics of a top started at a particular angle θ_i, with zero precession ($\dot\phi = 0$). In particular, develop an expression for the precession rate as a function of the initial inclination θ_i and the instantaneous value of θ. For what values of θ is the nutational speed zero?

12.54 A top is composed of three circular disks of equal thickness t and radii R_0, $2R_0$, and R_0 as shown in Fig. P12.54. A thin, massless rod joins the disks. What spin velocity is required for the top to sleep?

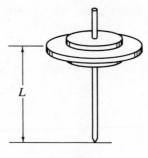

Fig. P12.54

12.55 At the instant shown in Fig. P12.55, point A on the cube has a velocity of 30 cm/sec in the positive Y-direction. What are its kinetic energy and angular momentum with respect to XYZ? The density of the cube material is 4 g/cm³.

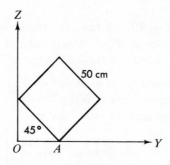

Fig. P12.55

12.56 A uniform cylinder rotates about an axis displaced from its longitudinal axis by an amount e. Derive an expression for the reactions at supports located at each end of the cylinder. The angular velocity and angular acceleration are ω and $\dot{\omega}$.

12.57 Demonstrate that the roots of the cubic equation introduced in connection with the diagonalization of the tensor of inertia are real.

12.58 A rectangular plate suspended from a corner experiences small oscillations in the vertical plane. Determine the period of the motion.

12.59 Assume that the plate of the previous exercise experiences oscillation normal to a vertical plane. Determine the period of the motion as well as the length of an equivalent simple pendulum.

CHAPTER 13

MECHANICS OF DEFORMABLE CONTINUA

13.1 INTRODUCTION

The fundamental structure of matter is discrete and not uniform or continuous; the subdivisions or building blocks of matter include molecules and atoms, and subatomic particles such as electrons, neutrons, and protons. When describing phenomena involving a great many particles, such as the flow of a liquid or the vibration of a string, it is greatly to our advantage to replace a system of discrete particles with a continuous distribution of matter as we have done in Chapter 12 for rigid continua.

In this chapter our concern is with *deformable continua*. To state that matter is continuous and deformable provides in itself an incomplete picture for there are important distinctions within this general grouping. We have the three basic states of matter: gases, liquids, and solids. A fixed mass of gas fills any shape or volume into which it is placed. A liquid is similar to a gas in its indefinite shape but differs in having a definite volume and in often displaying a distinct surface. A liquid ordinarily undergoes only very small volumetric changes, whereas a gas subject to similar ranges of pressure undergoes relatively large changes in volume. Gases and liquids are often classified together as fluids, sharing an inability to withstand the action of shearing stresses. As distinct from a fluid, a solid maintains its shape without need of a container, and displays an ability to resist continuous shear without accelerating. A solid that experiences creep or slow flow under the action of a stress is termed *viscoelastic*. A so-called "fourth" state of matter is the plasma or ionized gas. In addition to the usual properties associated with the gaseous state, a plasma also manifests strong electric and magnetic interactions, which are its distinguishing features.

The necessity for going beyond the simple idealizations presented in this chapter rests entirely with the extent to which one is satisfied with the results obtained through their use. More complete or elaborate description of the behavior of a given material is the subject of continuing research, resting in large measure upon a more secure knowledge of the structure of matter.

13.2 DEFINITION OF STRESS

Consider a body, such as shown in Fig. 13.1(a), on which is imposed an arbitrary system of forces. A free body of the portion shown in Fig. 13.1(b) requires that the force resultant $\mathbf{F}$ act at surface A. In general, $\mathbf{F}$ will be neither normal nor tangent to the surface. The average stress acting on area A is

$$\mathbf{S} = \frac{\mathbf{F}}{A} \tag{13.1}$$

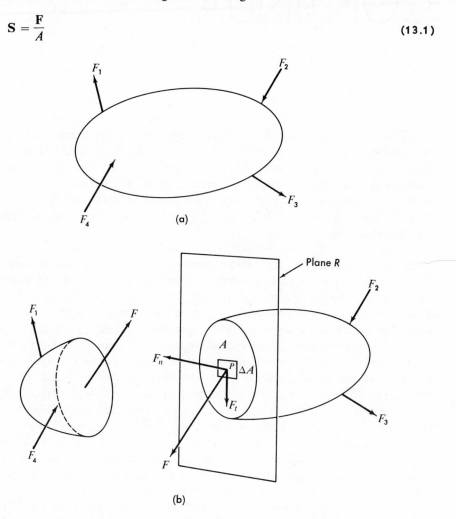

Fig. 13.1

If now attention is focused on the force resultant acting on smaller and smaller areas, ΔA, enclosing point P on the surface, a definite limit of the ratio of force to area is found to exist, as the area approaches zero. The stress at point P is thus defined as

$$\mathbf{S} = \lim_{\Delta A \to 0} \frac{\Delta \mathbf{F}}{\Delta A} = \frac{d\mathbf{F}}{dA} \tag{13.2}$$

Clearly, the magnitude of $\mathbf{S}$ depends upon the orientation of the plane selected, and it is particularly significant that an infinite number of planes may be drawn through point P. In addition, the reader is reminded that a continuum has replaced the actual discrete mass distribution, but we need not hesitate permitting the area to approach very small values unless the actual surface dimensions are of the same order of magnitude as interatomic distances.

Resolving $\mathbf{S}$ into components normal and tangent to plane R, the *normal stress* σ is defined as

$$\sigma = \lim_{\Delta A \to 0} \frac{\Delta F_n}{\Delta A} = \frac{dF_n}{dA} \tag{13.3a}$$

Similarly, the *shear stress*,

$$\tau = \lim_{\Delta A \to 0} \frac{\Delta F_t}{\Delta A} = \frac{dF_t}{dA} \tag{13.3b}$$

13.3 SIMPLE STRESS ANALYSIS

A particle, subject to zero net force, experiences no displacement if initially at rest. A distribution of solid matter of finite extent, however, acted on by a system of forces and torques adding to zero, retains its shape only if perfectly rigid. Otherwise a change of shape occurs which depends upon the magnitude and location of the external forces and reactions as well as the nature of the material.

Consider a rod of length L_0 and cross-sectional area A subject to tensile forces F. The average force per unit area throughout the rod, normal to A is given by

$$\sigma = \frac{F}{A} \tag{13.4}$$

The accompanying elongation is $\Delta L = L - L_0$, where L represents the final length and L_0 the initial length. The fractional change of length ϵ, a dimensionless quantity known as the *strain*, is defined as

$$\epsilon = \frac{\Delta L}{L_0} = \frac{L - L_0}{L_0} \tag{13.5}$$

If a linear relationship exists between the applied force and the change in length, that is, if the material obeys Hooke's law,

$$F = k \, \Delta L \tag{13.6}$$

where k represents the same spring constant or stiffness introduced in Chapter 3.

Dividing Eq. (13.6) by the area normal to the applied force,

$$\frac{F}{A} = \frac{k \, \Delta L}{A} = \frac{k L_0}{A} \frac{\Delta L}{L_0}$$

or

$$\sigma = E\epsilon \tag{13.7}$$

where E, called Young's modulus[1] or the *modulus of elasticity*, represents a constant of proportionality relating stress and strain. Since strain is dimensionless, the units of E $(= \sigma/\epsilon)$ are those of stress. We assume that E is the same in tension and compression, a property of a *bilateral* material.

In order that material properties be presented in a general manner, it is usual to depict load-deformation characteristics in terms of a stress-strain relation. Consider part of such a curve for a brittle type of material shown in Fig. 13.2. All stresses at or below point A are linearly proportional to strain, hence the term *proportional limit*. If the material is stressed beyond point B, the *elastic limit*, some *permanent* deformation occurs. Point B is therefore the point beyond which the stress curve is no longer reversible. The elastic and proportional limits are identical, or nearly so, for many materials. It is customary to base stress curves such as Fig. 13.2 on the original cross-sectional area. This has the effect of distorting the curve, particularly in those regions in which the lateral contraction

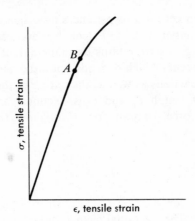

Fig. 13.2

[1] Young, T., *A Course of Lectures on Natural Philosophy and the Mechanical Arts*, Volume II, 1807, page 46.

is significant. Since the stress thus calculated is force divided by a *constant*, the stresses indicated are actually indicative of the force rather than the instantaneous ratio of force to true area.

In this connection it is of interest to define *true strain* as the integral of the ratio dL/L, where dL represents an infinitesimal amount of elongation and L is the actual instantaneous length of the specimen. Thus

$$\epsilon_{\text{true}} = \int_{L_0}^{L_f} \frac{dL}{L} = \ln \frac{L_f}{L_0} \tag{13.8}$$

where L_f is the final length of the specimen. The above expression may be written

$$\epsilon_{\text{true}} = \ln \frac{L_0 + \Delta L}{L_0} = \frac{\Delta L}{L_0} + \text{higher-order terms}$$

which to first order is the result for *small strain*.

Another type of stress is produced by a pair of equal and opposite forces not directed along the same line of action, as shown in Fig. 13.3. This *shearing* action causes distortion but no changes of length. It thus appears as though the material were composed of layers attempting to slide over one another. The shearing stress is the ratio of shear force to area of application. The shearing strain is the tangent of the angle γ, or for small angles, simply γ. Assuming that

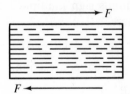

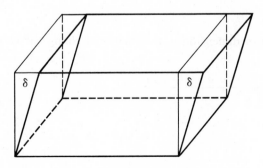

Fig. 13.3

Hooke's law is applicable to shear, the relation between shearing stress and shearing strain is

$$G = \frac{\tau}{\gamma} \tag{13.9}$$

where G is the *shear modulus*.

A stress acting in one direction results in transverse as well as parallel strains. The ratio of the lateral or transverse strain to the longitudinal strain in a simple tension test is termed the *Poisson's ratio*,[2] ν, usually assumed constant throughout the elastic range of a material:

$$\nu = \frac{\epsilon_{lat}}{\epsilon_{long}} = \frac{\epsilon_t}{\epsilon_l} \tag{13.10}$$

Poisson's ratio has the following approximate values: 0.28 for wrought iron and steel, 0.5 for rubber, and 0.25 for glass.

Consider now a volume of material subjected to a normal compressive stress acting over its *entire* surface. The stress is related to the fractional change in volume (a volumetric strain) by the equation

$$\sigma = -k \frac{\Delta V}{V} \tag{13.11}$$

where k, the *bulk modulus*, not to be confused with the stiffness, possesses dimensions of stress. The minus sign multiplied by the negative ΔV assures a positive value of k. For a sphere,

$$\Delta V = V_{final} - V_{initial} = \tfrac{4}{3}\pi(R - \Delta R)^3 - \tfrac{4}{3}\pi R^3$$

where ΔR represents the radial contraction. Discarding terms in $(\Delta R)^2$ and $(\Delta R)^3$,

$$\Delta V = -4\pi R^2 \, \Delta R \tag{13.12}$$

Thus

$$\sigma = -k \frac{-4\pi R^2 \, \Delta R}{\tfrac{4}{3}\pi R^3} = 3k \frac{\Delta R}{R} \tag{13.13a}$$

or

$$k = \frac{R}{3 \, \Delta R} \sigma \tag{13.13b}$$

For a cube of dimension a, the fractional change of volume, to first order in Δa, is

$$\frac{\Delta V}{V} = \frac{(a - \Delta a)^3 - a^3}{a^3} = \frac{-3 \, \Delta a}{a} \tag{13.14}$$

so that

$$\sigma = 3k \frac{\Delta a}{a} \tag{13.15}$$

---

[2] Poisson, S. D., *Mémoires de l'Institut Français* (1814).

The reader should not draw the hasty conclusion that $3k$ equals Young's modulus, since the results obtained for a cube loaded uniaxially (for which E is defined) are quite different from those described by Eq. (13.15). The strain in one direction, associated with application of load on each face of the cube, is the sum of the strain due to the longitudinal load ϵ_l plus that associated with loads on the remaining faces, ϵ_t'. The total strain (along a given load direction) is

$$\epsilon = \epsilon_l + \epsilon_t' \tag{13.16}$$

or

$$\epsilon = \frac{\sigma}{E} - \frac{2\sigma\nu}{E} \tag{13.17}$$

where the minus sign appears because the strains are oppositely directed; see Eq. (13.10). The total strain given above is equal to that in Eq. (13.15), $\Delta a/a$, so that

$$\frac{\sigma}{E} - \frac{2\sigma\nu}{E} = \frac{\sigma}{3k} \tag{13.18a}$$

Therefore,

$$k = \frac{E}{3(1 - 2\nu)} \tag{13.18b}$$

Equation (13.18b) thus relates the modulus of elasticity, Poisson's ratio, and the bulk modulus for homogeneous and isotropic materials, that is, uniform materials displaying identical elastic properties in all directions.

13.4 STATIC DEFLECTION OF A BEAM

The determination of the shear and bending moments in a beam subjected to various concentrated and distributed forces has been treated in Chapter 12. The relationship between the moment thus determined and the associated deformation and stress is now explored.

Consider a beam, initially straight, of constant cross-sectional area, subject to pure bending, as shown in Fig. 13.4. By pure bending, an assumption of elementary beam theory, is meant that shearing forces are absent, and hence deformation due to shear is neglected. Another important assumption relates to the plane

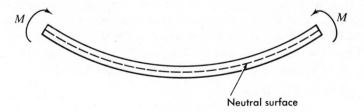

Neutral surface

Fig. 13.4

sections of the undeformed beam, shown by the dashed lines of Fig. 13.5(a). According to the elementary theory, these sections remain plane after bending, as in Fig. 13.5(b). Thus if one considers all the points lying in a plane normal to the axis of the beam before bending, these points still lie within a single plane after bending. In addition, it is assumed that these plane sections remain perpendicular to the beam axis, and that the shape to which the beam is deformed over the differential length under study is circular.

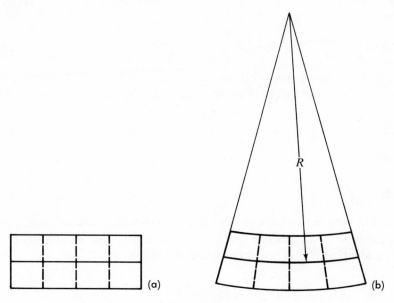

Fig. 13.5

Let the beam under study be restricted to one possessing a plane of symmetry along the longitudinal axis. A differential element before and after bending is shown in Fig. 13.6.

Imagining the beam to be composed of longitudinal fibers, it is clear that bending causes the upper fibers to be compressed or shortened, and the lower fibers to be lengthened. It is then reasonable to assume that there exists a surface that retains its initial or undeformed length though the beam is deformed. This *neutral surface*, as it is termed, marks the boundary between compressive and tensile straining. If it is further assumed that nowhere is the beam inelastically stressed, the distribution of deformation above and below the neutral surface is as shown in Fig. 13.6. On the basis of the foregoing, the strain of an arbitrary differential segment of longitudinal fiber of an initially straight beam is

$$\epsilon = \frac{(R + y)\, d\phi - R\, d\phi}{R\, d\phi} = \frac{y}{R} \tag{13.19}$$

where $R\, d\phi$ represents the initial or undeformed length of any differential segment

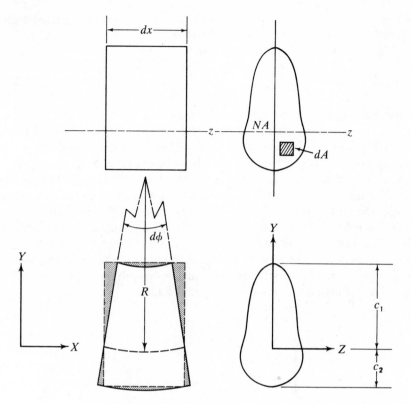

Fig. 13.6

of the beam.   It is measured at the neutral surface because this is the only location at which the beam is undistorted.   The term $(R + y) \, d\phi$ is the deformed length at an arbitrary distance y from the neutral surface.

The strain thus obtained is related to the stress by Hooke's law,

$$\epsilon = \frac{\sigma}{E} = \frac{y}{R} \tag{13.20a}$$

or

$$\sigma = \epsilon E = \left(\frac{y}{R}\right) E \tag{13.20b}$$

In order for each section of the beam to be in equilibrium, the total force acting over a cross section must be zero:

$$F = \int_A dF = \int_A \sigma \, dA = \frac{E}{R} \int_A y \, dA = 0 \tag{13.21a}$$

Note that the cross-sectional area of the section in bending is assumed identical with that of the unstressed section.   The assumption that the normal stresses do not result in transverse strains and consequent distortion of the cross section is equivalent to taking Poisson's ratio to be zero.

Since E/R is not zero, it follows that

$$\int_A y\, dA = 0 \tag{13.21b}$$

The above integral is the first moment of area about the *neutral axis*, labeled *NA*. This is the line of intersection of the neutral surface and the plane of the cross section *A*, Fig. 13.6. It is clear from Eq. (13.21b) that the centroid of area *A* lies on the neutral axis. This is true for initially straight beams only.

Recall from Chapter 12 that equilibrium requires that the beam supply a couple. It is this couple which was determined as a function of the distance measured along the longitudinal axis. This may be readily seen from an examination of a cantilever beam, loaded as shown in Fig. 13.7. Unless the beam furnishes couple F', the couple F will cause rotation. The mechanism by which F' is produced is now quite clear in view of Fig. 13.6, in which the forces corresponding to the tensile and compressive stresses are equal in magnitude, oppositely directed, and have lines of action separated by a finite distance.

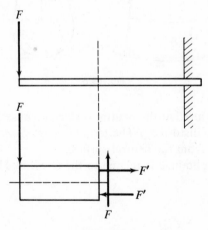

Fig. 13.7

The moment of the internal force about the neutral axis is

$$M_z = \int_A y\, dF = \frac{E}{R}\int_A y^2\, dA = \frac{EI_A}{R} \tag{13.22}$$

where the elastic moduli in tension and compression have been assumed equal, and $I_A = \int_A y^2\, dA$ represents the second moment of area about the neutral axis or, more commonly, the area moment of inertia about axis z-z.

Recall from the calculus that the radius of curvature R is related to the first and second derivatives of beam deflection with respect to the beam length variable by the following expression:

$$R = \frac{[1 + (dy/dx)^2]^{3/2}}{d^2y/dx^2} \tag{13.23}$$

If the slope of the beam is small, $(dy/dx)^2 \ll 1$ and

$$\frac{1}{R} \cong \frac{d^2y}{dx^2} \tag{13.24}$$

The moment is therefore

$$M_z = EI_A \frac{d^2y}{dx^2} \tag{13.25}$$

or

$$\frac{M_z}{EI_A} = \frac{d^2y}{dx^2} = \frac{1}{R} \tag{13.26}$$

To apply this relationship, $M = M(x)$ is substituted (as was obtained in Chapter 12). Integration then provides the slope of the deflection curve (dy/dx) and the bending deflection at each point, $y = y(x)$. The two required boundary conditions are provided from information regarding y and dy/dx.

13.5 BENDING STRESS

The normal stress associated with bending is given by Eq. (13.20b). Eliminating R by substitution of Eq. (13.26), we have

$$\sigma = \frac{Ey}{R} = \frac{M_z y}{I_A} \tag{13.27}$$

Maximum values of σ occur in the extreme beam fibers, at $y = c_1$ and c_2, Fig. 13.6. The stress is thus related directly to the bending moment and distance from the neutral axis to the fiber of interest and inversely to the area moment of inertia. It is therefore desirable, in applications requiring resistance to bending, to arrange the beam material in such a way as to maximize I_A while retaining an ability to resist shear as in an I-beam.

Illustrative Example 13.1

A simply supported beam of span L is loaded by a concentrated force at $L/2$. The beam is of rectangular cross section: width b, height h. Determine the equation of beam deformation due to bending, as well as the maximum bending stress. Neglect the weight of the beam.

SOLUTION

Applying the techniques of Chapter 12, the bending moment as a function of x for the region $0 \leq x < L/2$ is $M = Px/2$ and therefore

$$\frac{M}{EI_A} = \frac{Px}{2EI_A} = \frac{d^2y}{dx^2}$$

Integration yields

$$\frac{dy}{dx} = \frac{Px^2}{4EI_A} + C_1$$

which is, for the prescribed region, the slope of the beam as a function of x. From considerations of symmetry, $dy/dx = 0$ at $L/2$. Hence $C_1 = -PL^2/16EI_A$ and

$$\frac{dy}{dx} = \frac{Px^2}{4EI_A} - \frac{PL^2}{16EI_A} \qquad (0 \leq x < L/2)$$

A second integration gives

$$y = \frac{Px^3}{12EI_A} - \frac{PL^2 x}{16EI_A} + C_2$$

Applying the condition $y = 0$ at $x = 0$, $C_2 = 0$. Note that the condition $y = 0$ at $x = L$ was not applied because the above expression is inapplicable outside the region $0 \leq x < L/2$. The deflection is thus

$$y = \frac{Px^3}{12EI_A} - \frac{PL^2 x}{16EI_A} \qquad (0 \leq x < L/2)$$

For the region to the right of the concentrated force P,

$$M = \frac{PL}{2} - \frac{Px}{2}$$

$$\frac{dy}{dx} = \frac{PLx}{2EI_A} - \frac{Px^2}{4EI_A} + C_3 \qquad C_3 = -\frac{3}{16}\frac{PL^3}{EI_A}$$

$$y = \frac{PLx^2}{4EI_A} - \frac{Px^3}{12EI_A} - \frac{3}{16}\frac{PL^2 x}{EI_A} + C_4 \qquad C_4 = \frac{1}{48}\frac{PL^3}{EI_A}$$

The maximum deflection may be found from either expression for $y(x)$. It occurs when $dy/dx = 0$, $x = L/2$:

$$y_{max} = -\frac{PL^3}{48EI_A}$$

To determine the stress, the second moment of area about the neutral axis must be determined:

$$I_A = \int_{-h/2}^{h/2} y^2 \, dA = \int_{-h/2}^{h/2} y^2 b \, dy = \frac{by^3}{3}\bigg]_{-h/2}^{h/2} = \frac{bh^3}{12}$$

The maximum bending stress occurs at the extreme fibers of the beam, and at that location where the bending moment is a maximum. In this example, M_{max} occurs at $x = L/2$:

$$M_{max} = PL/4$$

and

$$\sigma_{\text{tensile max}} = \sigma_{\text{compressive max}} = \frac{(PL/4)(h/2)}{bh^3/12} = \frac{3}{2}\frac{PL}{bh^2}$$

13.6 STATICALLY INDETERMINATE BEAMS

The beams discussed in Chapter 12 were such that the number of unknown reactions was equal to the number of equations of equilibrium. These were instances of statically determinate beams. If the number of unknown reactions exceeds that which may be determined from statics, the beam is *statically indeterminate*, and additional information, based upon geometric considerations related to deformation, must be applied.

Illustrative Example 13.2
Determine the deformation $y = y(x)$ of the beam shown in Fig. 13.8(a).

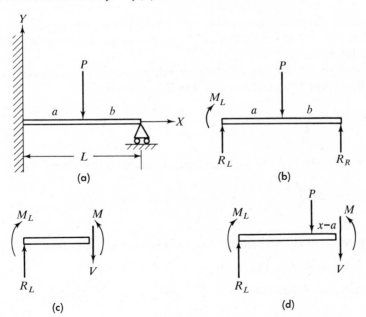

Fig. 13.8

SOLUTION

From the free-body diagram of the entire beam, Fig. 13.8(b), it is clear that there are three unknown reactions, R_R, R_L, and M_L and only two equations of statics.

We begin by writing the moment equation for the representative beam section ($0 \leq x < a$), Fig. 13.8(c):

$$M = M_L + R_L x = EI_A \frac{d^2y}{dx^2} \tag{a}$$

The equations for slope and deformation are, therefore:

$$EI_A \frac{dy}{dx} = M_L x + R_L \frac{x^2}{2} + C_1 \tag{b}$$

$$EI_A y = M_L \frac{x^2}{2} + R_L \frac{x^3}{6} + C_1 x + C_2 \tag{c}$$

For the remaining representative section ($a \leq x \leq L$), Fig. 13.8(d):

$$M = M_L + R_L x - P(x - a) \tag{d}$$

$$EI_A \frac{dy}{dx} = M_L x + R_L \frac{x^2}{2} - \frac{P}{2}(x - a)^2 + C_3 \tag{e}$$

$$EI_A y = M_L \frac{x^2}{2} + R_L \frac{x^3}{6} - \frac{P}{6}(x - a)^3 + C_3 x + C_4 \tag{f}$$

Because the beam cannot rotate in a built-in support, $dy/dx = 0$ at $x = 0$. Substituting into Eq. (b), we obtain $C_1 = 0$. From the condition $y = 0$ at $x = 0$, Eq. (c) yields $C_2 = 0$.

The slope of the beam deformation (elastic curve) is continuous, and consequently dy/dx may be equated at $x = a$ in Eqs. (b) and (e):

$$M_L a + R_L \frac{a^2}{2} = M_L a + R_L \frac{a^2}{2} - \frac{P}{2}(a - a)^2 + C_3$$

Thus $C_3 = 0$.

Similarly, the deflections given by Eqs. (c) and (f) are equated at $x = a$:

$$M_L \frac{a^2}{2} + R_L \frac{a^3}{6} = M_L \frac{a^2}{2} + R_L \frac{a^3}{6} - \frac{P}{6}(a - a)^3 + C_4$$

Thus $C_4 = 0$.

The moment at the right-hand support is zero, and therefore Eq. (d) becomes

$$0 = M_L + R_L L - P(L - a)$$

Finally, the deflection at $x = L$ is zero, and Eq. (f) yields

$$0 = M_L \frac{L^2}{2} + R_L \frac{L^3}{6} - \frac{P}{6}(L - a)^3$$

Simultaneous solution of the last two equations provides the following expressions for the reactions at the left end:

$$R_L = \frac{Pa(3L^2 - a^2)}{2L^3}$$

$$M_L = \frac{Pa}{2L^2}(a^2 - L^2)$$

Substitution of R_L and M_L into Eqs. (c) and (f) gives the required beam deflection.

13.7 SHEARING STRESS — TWISTING OF A CIRCULAR BAR

An expression, analogous to Eq. (13.27), relates the shearing stress associated with the twisting of a circular bar to the shearing strain. We begin by considering an elemental volume of such a bar, subjected to torque, as in Fig. 13.9(a).

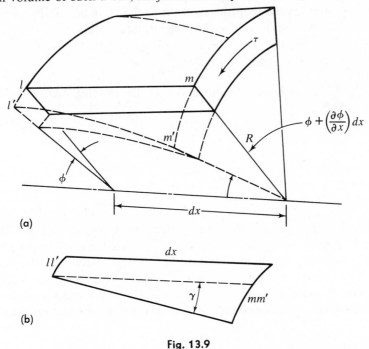

Fig. 13.9

The angle of twist at x is ϕ, and at $x + dx$, $\phi + (\partial\phi/\partial x)dx$. Therefore, the length ll' may be expressed,

$$ll' = R\phi \tag{13.28}$$

and mm',

$$mm' = R\left(\phi + \frac{\partial\phi}{\partial x}dx\right) \tag{13.29}$$

The shearing strain, as already noted, is simply

$$\tan \gamma = \frac{mm' - ll'}{dx} \tag{13.30}$$

as depicted in Fig. 13.9(b). Substituting for mm' and ll' we have

$$\tan \gamma = \frac{R[\phi + (\partial\phi/\partial x)dx] - R\phi}{dx} = R\,\frac{\partial\phi}{\partial x} \tag{13.31}$$

where for small angles $\tan \gamma = \gamma$.

The shearing stress associated with this strain is given by Eq. (13.9):

$$\tau = G\gamma = GR\frac{\partial\phi}{\partial x} \tag{13.32}$$

The mechanism by which twisting is resisted by the bar, that is, the means by which the bar supplies the requisite moment to maintain equilibrium, is provided through the collective action of the shearing stresses. Thus

$$M = \int_A R\tau\, dA = \int_A GR^2\frac{\partial\phi}{\partial x}\, dA = G\,\frac{\partial\phi}{\partial x}\int_A R^2\, dA \tag{13.33}$$

where Eq. (13.32) has been substituted for τ. Eliminating $\partial\phi/\partial x$ between Eqs (13.32) and (13.33),

$$\tau = \frac{MR}{I_p} \tag{13.34}$$

where

$$I_p = \int_A R^2\, dA \tag{13.35}$$

represents the polar moment of inertia of the cross-sectional area A. The above expression is clearly analogous to Eqs. (13.27) in that the stress varies directly as the distance from the longitudinal axis of the bar and inversely as the moment of inertia. Note that as in the case of pure bending of a beam, plane sections have been assumed to remain plane; that is, no warping of the cross section has been taken into account. This assumption is not valid for noncircular cross sections.

13.8 THE STRESS TENSOR

It has been noted that to specify the stress at a point it is necessary to prescribe the orientation of a plane passing through the point. The normal and shear stresses are then determined by resolving the stress tangent and normal to the plane.

One method of completely specifying the stress at a point is to define the stress on three planes passing through the point and, most conveniently, by three mutually perpendicular planes, as shown in Fig. 13.10. In the figure the planes are taken to be the sides of a cube of dimensions dx, dy, and dz. To prove that the

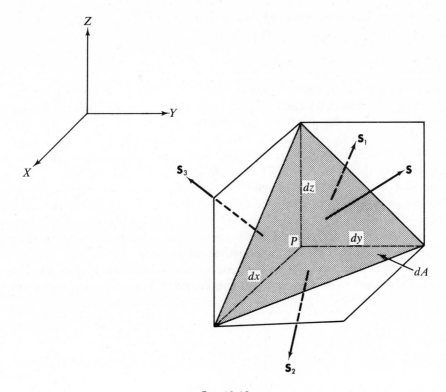

Fig. 13.10

stress associated with any plane passing through P can now be ascertained, begin with stresses S_1, S_2, S_3, arbitrarily oriented relative to the areas on which they act: $dy\,dz/2$, $dx\,dy/2$, and $dx\,dz/2$, respectively. Consider the plane of interest to be parallel to dA. Since the tetrahedron is in equilibrium, the sum of all forces must be zero:

$$\mathbf{S}_1\,\frac{dy\,dz}{2} + \mathbf{S}_2\,\frac{dx\,dy}{2} + \mathbf{S}_3\,\frac{dx\,dz}{2} + \mathbf{S}\,dA = \mathbf{0} \tag{13.36}$$

The areas in Eq. (13.36) are related by the following expressions:

$$\frac{dy\,dz}{2} = dA\,\cos(\hat{\mathbf{e}}_n,\hat{\mathbf{i}}) = dA\,l_{nx} \tag{13.37a}$$

$$\frac{dx\,dz}{2} = dA\,\cos(\hat{\mathbf{e}}_n,\hat{\mathbf{j}}) = dA\,l_{ny} \tag{13.37b}$$

$$\frac{dx\,dy}{2} = dA\,\cos(\hat{\mathbf{e}}_n,\hat{\mathbf{k}}) = dA\,l_{nz} \tag{13.37c}$$

where $\hat{\mathbf{e}}_n$ is a unit vector normal to the surface dA, and l_{nx}, l_{ny}, and l_{nz} are the respective cosines of the angles between $\hat{e}_n$ and the positive X-, Y-, and Z-directions. Substitution of Eqs. (13.37) into Eq. (13.36) yields

$$\mathbf{S} = -\mathbf{S}_1 l_{nx} - \mathbf{S}_2 l_{ny} - \mathbf{S}_3 l_{nz} \qquad (13.38)$$

The stress on dA is thus determined on the basis of the known stresses $\mathbf{S}_1$, $\mathbf{S}_2$, $\mathbf{S}_3$ and a complete knowledge of the orientation of dA. In the limit, as dx, dy, dz each approach zero, the plane, dA, contains the point P.

In terms of the components of $\mathbf{S}_1$, $\mathbf{S}_2$, and $\mathbf{S}_3$ parallel to the X-, Y-, and Z-coordinate directions, it is clear that nine such scalar stresses exist, defining the stress at P, as shown in Fig. 13.11. The stresses σ_{xx}, σ_{xy}, σ_{xz}, σ_{yx}, σ_{yy}, σ_{yz}, σ_{zx}, σ_{zy}, σ_{zz} provide a complete and unambiguous description of the stress at a point. The stresses in Fig. 13.11 obey the following scheme: σ_{ij} represents the ith component of stress on a plane, a normal to which is in the jth direction. That is, $\sigma_{ij} = dF_i/dA_j$. Thus σ_{xy} is the X-component of stress on the plane normal to Y, that is, the XZ-plane. This is a shear stress, as depicted.

Consider the case of two-dimensional stress, depicted in Fig. 13.12, and assume the stresses to be represented as continuous functions of the space variables.

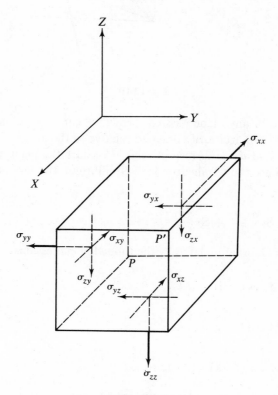

Fig. 13.11

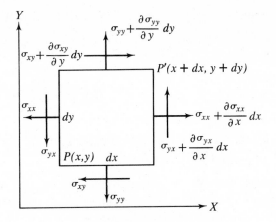

Fig. 13.12

Application of Taylor's theorem in two dimensions, expanding about $P(x, y)$, leads to the following stresses at $P'(x + dx, y + dy)$:

$$\sigma_{xx}(x + dx) = \sigma_{xx}(x) + \frac{\partial \sigma_{xx}}{\partial x} dx + \cdots$$

$$\sigma_{yy}(y + dy) = \sigma_{yy}(y) + \frac{\partial \sigma_{yy}}{\partial y} dy + \cdots$$

$$\sigma_{yx}(x + dx) = \sigma_{yx}(x) + \frac{\partial \sigma_{yx}}{\partial x} dx + \cdots$$

$$\sigma_{xy}(y + dy) = \sigma_{xy}(y) + \frac{\partial \sigma_{xy}}{\partial y} dy + \cdots$$

Applying the equation of equilibrium to the X-direction (assuming unit depth),

$$\left(\sigma_{xx} + \frac{\partial \sigma_{xx}}{\partial x} dx \right) dy - \sigma_{xx}\, dy + \left(\sigma_{xy} + \frac{\partial \sigma_{xy}}{\partial y} dy \right) dx - \sigma_{xy}\, dx = 0 \qquad \textbf{(13.39)}$$

or

$$\left(\frac{\partial \sigma_{xx}}{\partial x} + \frac{\partial \sigma_{xy}}{\partial y} \right) dx\, dy = 0$$

Since $dx\, dy \neq 0$,

$$\frac{\partial \sigma_{xx}}{\partial x} + \frac{\partial \sigma_{xy}}{\partial y} = 0 \qquad \textbf{(13.40a)}$$

Similarly, for Y-equilibrium,

$$\frac{\partial \sigma_{yy}}{\partial y} + \frac{\partial \sigma_{yx}}{\partial x} = 0 \qquad \textbf{(13.40b)}$$

Summing moments about $P(x, y)$ and equating to zero, we have

$$-\sigma_{yy}\,dx\left(\frac{dx}{2}\right) + \left(\sigma_{yy} + \frac{\partial\sigma_{yy}}{\partial y}\,dy\right)dx\left(\frac{dx}{2}\right) - \left(\sigma_{xx} + \frac{\partial\sigma_{xx}}{\partial x}\,dx\right)dy\left(\frac{dy}{2}\right)$$

$$+ \sigma_{xx}\,dy\left(\frac{dy}{2}\right) - \left(\sigma_{xy} + \frac{\partial\sigma_{xy}}{\partial y}\,dy\right)dx\,dy + \left(\sigma_{yx} + \frac{\partial\sigma_{yx}}{\partial x}\,dx\right)dy\,dx = 0$$

$$(13.41)$$

where the normal forces have been taken to act at the center of each face. Dropping all higher-order terms, Eq. (13.41) is written

$$(-\sigma_{xy} + \sigma_{yx})\,dy\,dx = 0$$

or

$$\sigma_{xy} = \sigma_{yx} \tag{13.42}$$

The equations of equilibrium, for three-dimensional stress yield the following equations:

$$\sigma_{yx} = \sigma_{xy}$$

$$\sigma_{zx} = \sigma_{xz} \tag{13.43}$$

$$\sigma_{zy} = \sigma_{yz}$$

and

$$\frac{\partial\sigma_{xx}}{\partial x} + \frac{\partial\sigma_{xy}}{\partial y} + \frac{\partial\sigma_{xz}}{\partial z} = 0$$

$$\frac{\partial\sigma_{yx}}{\partial x} + \frac{\partial\sigma_{yy}}{\partial y} + \frac{\partial\sigma_{yz}}{\partial z} = 0 \tag{13.44}$$

$$\frac{\partial\sigma_{zx}}{\partial x} + \frac{\partial\sigma_{zy}}{\partial y} + \frac{\partial\sigma_{zz}}{\partial z} = 0$$

Equations (13.43) thus state that shear stresses contained in perpendicular planes are equal. The number of different stress components required to describe the state of stress at a point is therefore reduced from nine to six. In addition, there are the three *compatibility equations* which must be satisfied [Eqs. (13.44)].

As in the cases of simple stress involving the extension of a bar, the bending of beams, and shaft torsion, it is necessary to relate the external effects such as force and torque to the resulting internal stresses. Consider now the general relationship between a force **dF** applied to an element of external area, arbitrarily oriented with respect to the area, and the internal state of stress. Applying the equations of equilibrium,

$$dF_x = \sigma_{xx}dA_x + \sigma_{xy}dA_y + \sigma_{xz}dA_z$$

$$dF_y = \sigma_{yx}dA_x + \sigma_{yy}dA_y + \sigma_{yz}dA_z \tag{13.45}$$

$$dF_z = \sigma_{zx}dA_x + \sigma_{zy}dA_y + \sigma_{zz}dA_z$$

where dA_x represents the component of **dA**, the normal to which is parallel to the X-direction, and so on. Treating the differential area as a vector colinear with a unit normal to the surface as shown, we have

$$\mathbf{dA} = dA_x\hat{\mathbf{i}} + dA_y\hat{\mathbf{j}} + dA_z\hat{\mathbf{k}}$$

$$= dA(l_{nx}\hat{\mathbf{i}} + l_{ny}\hat{\mathbf{j}} + l_{nz}\hat{\mathbf{k}}) \tag{13.46}$$

where, as before, l_{nx}, l_{ny}, l_{nz} are the cosines corresponding to the unit normal and the coordinate directions. Dividing Eqs. (13.45) by dA,

$$\frac{dF_x}{dA} = S_x = \sigma_{xx}l_{nx} + \sigma_{xy}l_{ny} + \sigma_{xz}l_{nz}$$

$$\frac{dF_y}{dA} = S_y = \sigma_{yx}l_{nx} + \sigma_{yy}l_{ny} + \sigma_{yz}l_{nz} \tag{13.47}$$

$$\frac{dF_z}{dA} = S_z = \sigma_{zx}l_{nx} + \sigma_{zy}l_{ny} + \sigma_{zz}l_{nz}$$

The above set of equations is identical in form to the expressions relating the angular momentum and angular velocity of a rigid body. Equations (13.47) may be succinctly expressed

$$S_i = \sigma_{ij}l_{nj} \tag{13.48}$$

In Eq. (13.48) σ_{ij} represents the stress tensor, analogous to the *tensor of inertia*.

$$\sigma = \begin{bmatrix} \sigma_{xx} & \sigma_{xy} & \sigma_{xz} \\ \sigma_{yx} & \sigma_{yy} & \sigma_{yz} \\ \sigma_{zx} & \sigma_{zy} & \sigma_{zz} \end{bmatrix} \tag{13.49}$$

Note, as in the case of the inertia tensor, that the corresponding off-diagonal terms (shear stresses) are equal:

$$\sigma_{ij} = \sigma_{ji} \tag{13.50}$$

For $i = j$, the components are the normal stresses, $\sigma_{xx}, \sigma_{yy}, \sigma_{zz}$.

The transformation characteristics of symmetric tensors of second rank have been discussed in connection with the properties of the inertia tensor, and apply equally here. We should expect, therefore, that three orthogonal axes, termed principal axes, exist, for which the off-diagonal terms are all zero, leaving three *principal stresses*.

13.9 STRAIN

In the case of the simple extension of a rod, the strain was obtained directly through the application of Hooke's law. When normal and shear stresses act simultaneously, the resulting strain is now of mixed variety, longitudinal and transverse to an arbitrary direction. In addition, normal stresses contribute to shearing strains. Let us first examine strain and then relate it to the stresses which cause it.

A general strain exists in a body when a relative displacement between points on the body occurs. Let us consider the cube of Fig. 13.13 for an elastic material. Suppose that a stress deforms the cube in such a way as to cause point A to move to A', and so on. What strain occurs?

The change of length of line OA is $\Delta L = OA' - OA$ and the strain is

$$\frac{\Delta L}{OA} = \frac{OA' - OA}{OA}$$

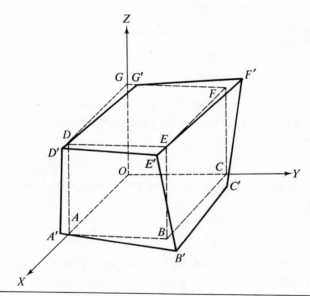

Coordinates		
	Before Deformation	After Deformation
A	x, o, o	$x + u, v, w$
B	x, y, o	$x + u, y + v, w$
C	o, y, o	$u, y + v, w$
D	x, o, z	$x + u, v, z + w$
E	x, y, z	$x + u, y + v, z + w$
F	o, y, z	$u, y + v, z + w$
G	o, o, z	$u, v, z + w$

Fig. 13.13

From Fig. 13.13,

$$OA' = [(x + u)^2 + v^2 + w^2]^{1/2}$$

and

$$\frac{\Delta L}{OA} = \frac{[(x + u)^2 + v^2 + w^2]^{1/2} - x}{x} = \left[1 + \frac{2u}{x} + \frac{u^2 + v^2 + w^2}{x^2}\right]^{1/2} - 1$$

Since u, v, w are necessarily small compared with lengths x, y, z, we are justified in retaining only terms of first order, and the strain is consequently

$$\frac{\Delta L}{OA} = \frac{u}{x}$$

and

$$\lim_{x \to 0} \frac{u}{x} = \frac{\partial u}{\partial x} = \epsilon_x \tag{13.51a}$$

where ϵ_x is the normal strain in the X-direction at the point to which the cube shrinks as x, y, z approach zero. By a similar analysis,

$$\epsilon_y = \frac{\partial v}{\partial y} \tag{13.51b}$$

$$\epsilon_z = \frac{\partial w}{\partial z} \tag{13.51c}$$

Consequently, paralleling stress at a point, is the concept of strain at a point.

The shear strain is determined by considering the change of an angle resulting from deformation. For simplicity, let us examine the change of the angle AOC. Before deformation AOC is a right angle in the XY-plane. The changes of position of A and C perpendicular to OA and OC in the XY-plane are $\Delta A = v$ and $\Delta C = u$, and the total change of angle is $(\Delta A/x) + (\Delta C/y)$, the shear strain. Thus

$$\frac{\Delta A}{x} + \frac{\Delta C}{y} = \frac{v}{x} + \frac{u}{y}$$

In the limit as x and y approach zero, we have the shear strain at a point:

$$\gamma_{xy} = \frac{\partial v}{\partial x} + \frac{\partial u}{\partial y} \tag{13.52a}$$

The shear strains γ_{xy} and γ_{yx} are identical as may be verified by deriving the change of angle OAB.

Similarly, for the shear strains in the XZ- and YZ-planes,

$$\gamma_{xz} = \gamma_{zx} = \frac{\partial u}{\partial z} + \frac{\partial w}{\partial x} \tag{13.52b}$$

$$\gamma_{yz} = \gamma_{zy} = \frac{\partial v}{\partial z} + \frac{\partial w}{\partial y} \tag{13.52c}$$

There are thus six independent strain components at a point. These components must satisfy certain differential equations which are termed the *compatibility equations*. For a first set, consider the following:

$$\frac{\partial^2}{\partial y\,\partial z}\gamma_{yz} = \frac{\partial^2}{\partial y\,\partial z}\left(\frac{\partial v}{\partial z} + \frac{\partial w}{\partial y}\right) = \frac{\partial^2}{\partial z^2}\frac{\partial v}{\partial y} + \frac{\partial^2}{\partial y^2}\frac{\partial w}{\partial z} \qquad (13.53)$$

where the interchange of the order of partial differentiation is permissible if the strain is continuous and single-valued. From Eqs. (13.51b) and (13.51c),

$$\frac{\partial^2}{\partial y\,\partial z}\gamma_{yz} = \frac{\partial^2\epsilon_y}{\partial z^2} + \frac{\partial^2\epsilon_z}{\partial y^2} \qquad (13.54a)$$

By a similar process for γ_{xy} and γ_{xz}, we have

$$\frac{\partial^2}{\partial x\,\partial y}\gamma_{xy} = \frac{\partial^2\epsilon_y}{\partial x^2} + \frac{\partial^2\epsilon_x}{\partial y^2} \qquad (13.54b)$$

$$\frac{\partial^2}{\partial x\,\partial z}\gamma_{xz} = \frac{\partial^2\epsilon_z}{\partial x^2} + \frac{\partial^2\epsilon_x}{\partial z^2} \qquad (13.54c)$$

Another useful relationship may be obtained by considering the terms

$$\frac{\partial^2}{\partial x\,\partial y}\gamma_{yz} + \frac{\partial^2}{\partial y\,\partial z}\gamma_{xy} = \frac{\partial^2}{\partial x\,\partial y}\left(\frac{\partial v}{\partial z} + \frac{\partial w}{\partial y}\right) + \frac{\partial^2}{\partial y\,\partial z}\left(\frac{\partial u}{\partial y} + \frac{\partial v}{\partial x}\right)$$

$$= \frac{\partial^2}{\partial x\,\partial z}\left(\frac{\partial v}{\partial y} + \frac{\partial v}{\partial y}\right) + \frac{\partial^2}{\partial y^2}\left(\frac{\partial w}{\partial x} + \frac{\partial u}{\partial z}\right)$$

$$= 2\frac{\partial^2\epsilon_y}{\partial x\,\partial z} + \frac{\partial^2\gamma_{xy}}{\partial y^2} \qquad (13.55)$$

or

$$\frac{\partial^2\epsilon_y}{\partial x\,\partial z} = \frac{1}{2}\frac{\partial}{\partial y}\left(\frac{\partial\gamma_{yz}}{\partial x} + \frac{\partial\gamma_{xy}}{\partial z} - \frac{\partial\gamma_{xz}}{\partial y}\right) \qquad (13.56a)$$

Similarly, by permutation of the coordinates,

$$\frac{\partial^2\epsilon_x}{\partial y\,\partial z} = \frac{1}{2}\frac{\partial}{\partial x}\left(\frac{\partial\gamma_{xz}}{\partial y} + \frac{\partial\gamma_{xy}}{\partial z} - \frac{\partial\gamma_{yz}}{\partial x}\right) \qquad (13.56b)$$

$$\frac{\partial^2\epsilon_z}{\partial x\,\partial y} = \frac{1}{2}\frac{\partial}{\partial z}\left(\frac{\partial\gamma_{yz}}{\partial x} + \frac{\partial\gamma_{xz}}{\partial y} - \frac{\partial\gamma_{xy}}{\partial z}\right) \qquad (13.56c)$$

There are thus six compatibility equations relating strains within a continuous material.

13.10 STRESS-STRAIN RELATED

Having derived a number of equations involving stress and strain separately, we are now prepared to relate them to one another. Only certain materials can be described by the methods to be developed. To begin with, the material must be strained within its elastic limit so that the principle of superposition applies. The strains are thus independent of any previous condition of stress.

Based upon these considerations, Hooke's law may be generalized by assuming each component of stress proportional to each component of strain. If we develop the stress equations in a principal-axis system, then there are only three stress equations, one each for σ_{11}, σ_{22}, σ_{33}. When describing the stress and strain in an isotropic material along principal axes,

$$\epsilon_1 = \frac{1}{E}(\sigma_{11} - \nu\sigma_{22} - \nu\sigma_{33})$$

$$\epsilon_2 = \frac{1}{E}(-\nu\sigma_{11} + \sigma_{22} - \nu\sigma_{33}) \tag{13.57}$$

$$\epsilon_3 = \frac{1}{E}(-\nu\sigma_{11} - \nu\sigma_{22} + \sigma_{33})$$

where ν is Poisson's ratio and E is Young's modulus. The inverse of these equations, determined by solving for stress, is

$$\sigma_{11} = \frac{E}{(1 + \nu)(1 - 2\nu)}[(1 - \nu)\epsilon_1 + \nu\epsilon_2 + \nu\epsilon_3]$$

$$\sigma_{22} = \frac{E}{(1 + \nu)(1 - 2\nu)}[\nu\epsilon_1 + (1 - \nu)\epsilon_2 + \nu\epsilon_3] \tag{13.58}$$

$$\sigma_{33} = \frac{E}{(1 + \nu)(1 - 2\nu)}[\nu\epsilon_1 + \nu\epsilon_2 + (1 - \nu)\epsilon_3]$$

The matrix form of Eqs. (13.57) is

$$\begin{bmatrix} \epsilon_1 \\ \epsilon_2 \\ \epsilon_3 \end{bmatrix} = \frac{1}{E} \begin{bmatrix} 1 & -\nu & -\nu \\ -\nu & 1 & -\nu \\ -\nu & -\nu & 1 \end{bmatrix} \begin{bmatrix} \sigma_{11} \\ \sigma_{22} \\ \sigma_{33} \end{bmatrix} \tag{13.59}$$

with the inverse relationship

$$\begin{bmatrix} \sigma_{11} \\ \sigma_{22} \\ \sigma_{33} \end{bmatrix} = \frac{E}{(1 + \nu)(1 - 2\nu)} \begin{bmatrix} 1 - \nu & \nu & \nu \\ \nu & 1 - \nu & \nu \\ \nu & \nu & 1 - \nu \end{bmatrix} \begin{bmatrix} \epsilon_1 \\ \epsilon_2 \\ \epsilon_3 \end{bmatrix} \tag{13.60}$$

These equations apply for homogeneous, isotropic, continuous, and bilateral materials. Most typically used materials fall approximately into these categories.

We have taken for granted that values of E and ν can be determined. In practice, this is accomplished experimentally. Solid-state physics and metallurgy, whose province it is to relate intermolecular binding forces to E and ν, have not yet been able to accurately predict these properties.

13.11 ENERGY IN STRESS AND STRAIN

The forces that cause stress in a body produce displacements in the form of strains and hence do work. Because the body is assumed to return to its original shape when the forces are removed, the work done in establishing the internal condition of strain is stored in the form of elastic potential energy.

To ascertain the energy associated with strain, consider a rod in tension. Since it is in equilibrium, we may treat one end as fixed and ask how much work is done in causing the total displacement in the direction of the force at the free end. From the definition of work,

$$dW = \mathbf{F}' \cdot \mathbf{ds} = F' \, d(\Delta L)'$$

where $\mathbf{F}'$ represents the force required to change the length of the rod by $d(\Delta L)'$. The total work done on the rod as it stretches from 0 to ΔL is therefore

$$W = \int_0^{\Delta L} F' \, d(\Delta L)' \tag{13.61}$$

If the force F' were constant as the rod is displaced, the work would simply be $(\Delta L)F$. However, the force itself varies linearly from zero to a maximum, and consequently, from Eq. (13.7),

$$F' = \frac{EA}{L_0} (\Delta L)'$$

and

$$W = \int_0^{\Delta L} \frac{EA}{L_0} (\Delta L)' \, d(\Delta L)' = \frac{EA(\Delta L)^2}{2L_0} \tag{13.62a}$$

or

$$W = \frac{1}{2} \frac{EA \, \Delta L}{A L_0} \frac{\Delta L}{L_0} A L_0 = \tfrac{1}{2}\sigma\epsilon V \tag{13.62b}$$

where AL_0 is the volume of the rod, V. Thus the work done per unit volume equals the potential energy per unit volume stored within the rod, which in turn is one half the product of stress and strain. The strain energy per unit volume of a three-dimensional solid may be written

$$\frac{W}{V} = \tfrac{1}{2}(\sigma_{xx}\epsilon_x + \sigma_{yy}\epsilon_y + \sigma_{zz}\epsilon_z) \tag{13.63}$$

since a stress, in the X-direction, for example, does no work in producing a strain in the Y- or Z-direction.

In a similar manner we may evaluate the strain energy of a body subject to shear. For the one-dimensional case of Fig. 13.3, the work done to produce the strain is

$$W = \int_0^\gamma G\gamma' A L_0 \, d\gamma' = \frac{G\gamma^2}{2} A L_0 = \tfrac{1}{2}\tau\gamma V \tag{13.64}$$

Thus the work done per unit volume in shear, the shear strain potential energy per unit volume, is one half the product of the shear stress and shear strain. When three-dimensional shear is considered,

$$\frac{W}{V} = \tfrac{1}{2}(\sigma_{xy}\gamma_{xy} + \sigma_{xz}\gamma_{xz} + \sigma_{yz}\gamma_{yz}) \tag{13.65}$$

For an elastic body subject to both normal and shear stress,

$$\frac{W}{V} = \tfrac{1}{2}(\sigma_{xx}\epsilon_x + \sigma_{yy}\epsilon_y + \sigma_{zz}\epsilon_z + \sigma_{xy}\gamma_{xy} + \sigma_{yz}\gamma_{yz} + \sigma_{xz}\gamma_{xz}) \tag{13.66}$$

Mixed stress–strain terms do not appear because only those stresses in the direction of strain do work.

13.12 CABLES

The methods of statics will now be applied to a simple structural element: the inextensible, flexible, planar cable. Such cables are characteristically capable of sustaining only tensile loading. The direction of such loads is everywhere tangent to the cable.

Consider Fig. 13.14(a) where A and B represent rigid supports. Note that the cable shape in the analyses which follow is the final shape, resulting from the conditions of loading.

In Fig. 13.14(b), τ_0 represents the cable tension at the lowest point, τ is

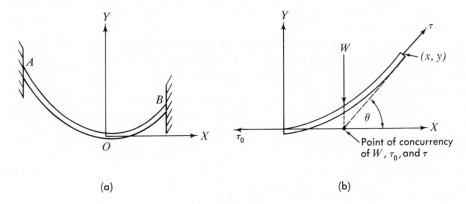

(a)   (b)

Fig. 13.14

the tension at any other point (x, y), and W is the equivalent of the external distributed cable loading. The load W acts at the center of gravity of the distributed loading which it represents.

Employing $\sum F_x = 0$, we find that $-\tau_0 + \tau \cos \theta = 0$. Thus

$$\tau_0 = \tau \cos \theta \tag{13.67}$$

Since $\sum F_y = 0$, $-W + \tau \sin \theta = 0$, or

$$W = \tau \sin \theta \tag{13.68}$$

Dividing W by τ_0 yields

$$\frac{W}{\tau_0} = \tan \theta = \left(\frac{dy}{dx}\right)_{x,y} \tag{13.69}$$

which is the basic differential equation used to solve a variety of cable problems.

Since the free-body diagram reveals that the cable section can be reduced to a three-force member, equilibrium requires that τ_0, τ, and W have a common point of intersection. Adding the three forces at this point of concurrency, of course, produces the same result as above.

Consider now a cable in which the vertical loading is a function of x, the horizontal distance from the origin. This may be a reasonable assumption in the case of a suspension bridge. Assume further that the loading is uniform, so that $W = wx$, where w is the constant loading rate. The differential equation now becomes

$$\frac{dy}{dx} = \frac{wx}{\tau_0} \tag{13.70}$$

which, when integrated, yields a parabolic variation of y as a function of x:

$$y = \frac{wx^2}{2\tau_0} + C_1 \tag{13.71}$$

Since $y = 0$ at $x = 0$, $C_1 = 0$.

A frequently encountered instance of uniform loading of the type $W = wx$ occurs in the symmetric cable in which L is the cable span and f the sag. At $x = \pm L/2$ we have $y = f$, so that, from Eq. (13.71),

$$f = \frac{wL^2}{8\tau_0} \tag{13.72}$$

or

$$\tau_0 = \frac{wL^2}{8f} \tag{13.73}$$

From Fig. 13.14(b) observe that at any point

$$\tau^2 = \tau_0^2 + W^2 \tag{13.74}$$

Thus, where loading is represented by $W = wx$,

$$\tau = [\tau_0^2 + (wx)^2]^{\frac{1}{2}} \tag{13.75}$$

Note that τ is a minimum at $x = 0$ and a maximum at $x = \pm L/2$. Hence the minimum tension in the cable is given by

$$\tau_{min} = \tau_0 = \frac{W_t L}{8f} \tag{13.76}$$

where $W_t = wL$, the total cable weight. The maximum tension is

$$\tau_{max} = \left[\left(\frac{wL^2}{8f}\right)^2 + w^2 \left(\frac{L}{2}\right)^2 \right]^{\frac{1}{2}} = \frac{W_t}{2} \left(\frac{L^2}{16f^2} + 1\right)^{\frac{1}{2}} \tag{13.77}$$

An important consideration in the use and analysis of cables is the total cable length, S. The relationship for differential length of arc dS in terms of differential Cartesian coordinates is

$$dS = [(dx)^2 + (dy)^2]^{\frac{1}{2}} = \left[1 + \left(\frac{dy}{dx}\right)^2 \right]^{\frac{1}{2}} dx \tag{13.78}$$

Because of the symmetry condition, S is expressed

$$S = \int_{-L/2}^{L/2} dS = 2 \int_0^{L/2} dS \tag{13.79}$$

Substituting for dS,

$$S = 2 \int_0^{L/2} \left[1 + \left(\frac{dy}{dx}\right)^2 \right]^{\frac{1}{2}} dx \tag{13.80}$$

But dy/dx may be expressed as a function of x from the relationship of Eq. (13.70), so that

$$S = 2 \int_0^{L/2} \left[1 + \left(\frac{wx}{\tau_0}\right)^2 \right]^{\frac{1}{2}} dx \tag{13.81}$$

Integration of Eq. (13.81) gives

$$S = \frac{L}{2} \left[1 + \left(\frac{wL}{2\tau_0}\right)^2 \right]^{\frac{1}{2}} + \frac{\tau_0}{w} \sinh^{-1} \left(\frac{wL}{2\tau_0}\right) \tag{13.82}$$

Another commonly used approximation for the solution of cable problems is the use of loading described by $W = wS$, where $S = S(x, y)$ is the length measured along the cable from $x = 0$, $y = 0$ to the point (x, y). This loading condition corresponds to a constant load per unit length of cable rather than the constant load per unit length in the X-direction previously employed, and is therefore suited to a description of a cable loaded principally by its own weight. The differential equation is now

$$\frac{dy}{dx} = \frac{W}{\tau_0} = \frac{wS(x, y)}{\tau_0} \tag{13.83}$$

Recall, from Eq. (13.78),

$$\frac{dS}{dx} = \left[1 + \left(\frac{dy}{dx}\right)^2\right]^{\frac{1}{2}} \tag{13.84a}$$

or

$$dx = \frac{dS}{[1 + (dy/dx)^2]^{\frac{1}{2}}} \tag{13.84b}$$

Substituting Eq. (13.83) into Eq. (13.84b) and integrating,

$$x = \int \frac{dS}{[1 + (wS/\tau_0)^2]^{\frac{1}{2}}} = \frac{\tau_0}{w} \sinh^{-1}\left(\frac{wS}{\tau_0}\right) + C_2 \tag{13.85}$$

where $C_2 = 0$ on the basis of $S = 0$ at $x = 0$. Thus

$$S = \frac{\tau_0}{w} \sinh\left(\frac{wx}{\tau_0}\right) \tag{13.86}$$

In order to eliminate S and obtain y as a function of x, Eq. (13.86) is substituted into the original differential equation:

$$\frac{dy}{dx} = \frac{wS}{\tau_0} = \sinh\left(\frac{wx}{\tau_0}\right) \tag{13.87}$$

Hence

$$y = \int \sinh\left(\frac{wx}{\tau_0}\right) dx = \frac{\tau_0}{w} \cosh\left(\frac{wx}{\tau_0}\right) + C_3 \tag{13.88}$$

From the boundary condition, $x = 0$ at $y = 0$, we obtain $C_3 = -\tau_0/w$. Substituting the value of C_3 into Eq. (13.88) we obtain the well-known equation of the *catenary*.

$$y = \frac{\tau_0}{w}\left(\cosh\frac{wx}{\tau_0} - 1\right) \tag{13.89}$$

13.13 VIBRATION OF A STRING

Consider now a uniform string of mass per unit length ρ_L and total length L, fixed at each end. As the tensile force τ is increased, the slack is removed, and the curve described is *essentially* a straight line. We now study the motion of such a string, following an initial displacement from its rest position, subject to the condition that the force of gravity $(\rho_L g L)$ is quite small relative to the tension, and may therefore be neglected.

In Fig. 13.15 is shown a small length Δx of the string. For small displacements, the tension may be assumed constant in magnitude throughout, and therefore the X- and Y-components of string tension are simply

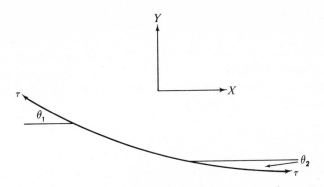

Fig. 13.15

$$\tau \cos \theta_2 - \tau \cos \theta_1 \quad \text{and} \quad \tau \sin \theta_1 - \tau \sin \theta_2$$

If θ_1 and θ_2 are small angles, $\cos \theta_1 \cong \cos \theta_2$, and there is no net horizontal force. The small-angle assumption also leads to

$$\sin \theta_1 \cong \tan \theta_1 \quad \text{and} \quad \sin \theta_2 \cong \tan \theta_2$$

For small motions of the string, no longitudinal displacement occurs. This is a consequence of insignificant changes in wire tension throughout. Perpendicular to the string, however, there is considerable motion. Applying Newton's second law in the Y-direction,

$$\sum F_y = ma_y = m \frac{\partial^2 y}{\partial t^2} \tag{13.90}$$

or

$$\tau \tan \theta_1 - \tau \tan \theta_2 = \rho_L \, \Delta x \, \frac{\partial^2 y}{\partial t^2} \tag{13.91}$$

where the partial derivative is employed because the transverse displacement is a function of x and t. Since $\tan \theta_1 = (\partial y / \partial x)_1$ and $\tan \theta_2 = (\partial y / \partial x)_2$, the slope of the string may be conveniently expanded in a Taylor series about point 2,

$$\left(\frac{\partial y}{\partial x}\right)_1 = \left(\frac{\partial y}{\partial x}\right)_2 + \left(\frac{\partial^2 y}{\partial x^2}\right)_2 \Delta x + \cdots$$

Substituting this result into the equation of motion,

$$\rho_L \, \Delta x \, \frac{\partial^2 y}{\partial t^2} = \tau(\tan \theta_1 - \tan \theta_2) = \tau \left[\left(\frac{\partial y}{\partial x}\right)_2 + \left(\frac{\partial^2 y}{\partial x^2}\right)_2 \Delta x - \left(\frac{\partial y}{\partial x}\right)_2\right] \tag{13.92}$$

resulting in the *one-dimensional* wave equation,

$$\tau \frac{\partial^2 y}{\partial x^2} = \rho_L \frac{\partial^2 y}{\partial t^2} \tag{13.93}$$

applicable only where it is valid to disregard gravitational forces and where the assumption of small displacements relative to string length is acceptable.

The term τ/ρ_L has dimensions of speed squared, leading to the natural substitution, $\tau/\rho_L = v^2$ in Eq. (13.93):

$$\frac{\partial^2 y}{\partial x^2} = \frac{1}{v^2} \frac{\partial^2 y}{\partial t^2} \tag{13.94}$$

It is appropriate to demonstrate that any function of $x + vt$ or $x - vt$ is a solution. Setting $z = x \pm vt$, and assuming $y = y(x \pm vt)$, we have

$$\frac{\partial y}{\partial x} = \frac{\partial}{\partial x}[y(x \pm vt)] = \frac{\partial y}{\partial z}\frac{\partial z}{\partial x} = \frac{\partial y}{\partial z}$$

$$\frac{\partial^2 y}{\partial x^2} = \frac{\partial^2}{\partial x^2}[y(x \pm vt)] = \frac{\partial}{\partial x}\frac{\partial y}{\partial x} = \frac{\partial}{\partial x}\frac{\partial y}{\partial z} = \frac{\partial^2 y}{\partial z^2}\frac{\partial z}{\partial x} = \frac{\partial^2 y}{\partial z^2}$$

$$\frac{\partial y}{\partial t} = \frac{\partial y}{\partial z}\frac{\partial z}{\partial t} = \pm v \frac{\partial y}{\partial z}$$

$$\frac{\partial^2 y}{\partial t^2} = \frac{\partial^2 y}{\partial z^2}\left(\frac{\partial z}{\partial t}\right)^2 = v^2 \frac{\partial^2 y}{\partial z^2}$$

Substituting the appropriate terms above into Eq. (13.94),

$$\frac{\partial^2 y}{\partial z^2} = \frac{1}{v^2}\left(v^2 \frac{\partial^2 y}{\partial z^2}\right) \tag{13.95}$$

Thus it is proved that *any* function of $x + vt$ or $x - vt$ solves Eq. (13.94). Since this equation is linear in y, a constant multiplied by any solution is also a solution, and hence the most general representation of $y(x, t)$ appears as follows:

$$y = C_1 y_1(x + vt) + C_2 y_2(x - vt) \tag{13.96}$$

where C_1 and C_2 are constants and y_1 and y_2 are not necessarily of the same functional form.

Consider a *particular* displacement of the string represented by

$$y = y_2\left(\frac{x - vt}{\lambda}\right)$$

where λ is a constant length, and the argument of y_2 is, by virtue of λ, dimensionless. Furthermore, let y_2 equal zero everywhere, except in the small region between x_0 and $x_0 + \Delta x_0$, as shown in Fig. 13.16 at the instant T. We define the phase of the string displacement as the argument of y_2, $(x - vt)/\lambda$. Thus there is a nonzero displacement at time T, only for phase between $(x_0 - vT)/\lambda$ and $(x_0 + \Delta x_0 - vT)/\lambda$.

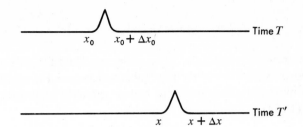

Fig. 13.16

Let us examine the progress of displacement y_2 as time increases. Suppose T changes to T', the phase of y_2 is now $(x - vT')/\lambda$. The position at which the displacement differs from zero is found by equating the phase at time T to that at T': $(x - vT')/\lambda = (x_0 - vT)/\lambda$ for the beginning of the disturbance and

$$\frac{x + \Delta x - vT'}{\lambda} = \frac{x_0 + \Delta x_0 - vT}{\lambda}$$

for the end of the disturbance. The limits of the disturbance have moved from x_0 and $x_0 + \Delta x_0$ to

$$x_0 + v(T' - T) \qquad \text{and} \qquad x_0 + \Delta x_0 + v(T' - T)$$

Hence the disturbance travels in the positive X-direction with velocity v, termed the *phase velocity*.

A disturbance similar to $y_2(x - vt)$ described by $y_1(x + vt)$ travels in the negative X-direction. Any general disturbance of a string can thus be decomposed into waves traveling in the positive and negative X-directions by applying the principle of superposition.

A very important wave type satisfying Eq. (13.94) is a *plane wave*, represented by

$$y = A_+ e^{i[(x-vt)/\lambda]} \tag{13.97a}$$

for motion in the $+X$-direction, and

$$y = A_- e^{i[(x+vt)/\lambda]} \tag{13.97b}$$

for motion in the $-X$-direction, where A_+ and A_- are the wave amplitudes. Since we eventually require the real part of y, it is clear that these expressions describe cosinusoidal waves on the string with phase velocities $\pm v$.

The possibility of *standing waves*, that is, composite or resultant waves which do not appear to propagate, occurs when the displacement is given by

$$y = A\left[\cos\left(\frac{x + vt}{\lambda}\right) - \cos\left(\frac{x - vt}{\lambda}\right)\right] \tag{13.98}$$

Consider a point of zero displacement, *a node*, requiring that $y = 0$, or

$$\frac{x + vt}{\lambda} = \frac{x - vt}{\lambda} + 2n\pi \qquad (n = 0, \pm 1, \pm 2, \cdots) \tag{13.99}$$

This expression applies at any time. The nodes are fixed for all time at locations given by $x = n\pi\lambda$, as may be verified by substitution into Eq. (13.98). Points of maximum (or minimum) amplitude are called *antinodes* and may be found by expanding Eq. (13.98),

$$y = -2A \sin \frac{x}{\lambda} \sin \frac{vt}{\lambda}$$

Clearly, if

$$\frac{x}{\lambda} = (n - \tfrac{1}{2})\pi \qquad (n = 0, \pm 1, \pm 2, \cdots) \tag{13.100}$$

these points experience a maximum displacement at a given time. Thus a standing wave is composed of two plane waves displaying the same magnitude of phase velocity, traveling in opposite directions with identical amplitudes and λ.

Although the form of the solutions of Eq. (13.94) is known, the method of separation of variables, will be applied to ascertain its analytical solution. The partial differential equation of motion is linear, containing two independent variables whose derivatives are each of second order. We therefore anticipate four constants of integration, and require two boundary and two initial conditions. The boundary conditions are immediately apparent from considerations of end fixity:

$$y(0, t) = 0 \tag{13.101a}$$

$$y(L, t) = 0 \tag{13.101b}$$

Two other conditions, related to time, are the initial displacement $y(x, 0)$ and the initial velocity $\partial y / \partial t|_{x,0}$ for the entire string.

Although there are no *general* techniques for solving a partial differential equation, the method of separation of variables is quite often applicable. Assume that $y(x, t)$ is the product of two functions, $X(x)$ and $T(t)$, and that these functions are, as implied in the way they are written, each dependent upon one variable only. Thus

$$y = X(x)T(t) \tag{13.102}$$

The terms $\partial^2 y / \partial x^2$ and $\partial^2 y / \partial t^2$ are required:

$$\frac{\partial y}{\partial x} = T \frac{\partial X}{\partial x} \qquad \frac{\partial^2 y}{\partial x^2} = T \frac{\partial^2 X}{\partial x^2}$$

$$\frac{\partial y}{\partial t} = X \frac{\partial T}{\partial t} \qquad \frac{\partial^2 y}{\partial t^2} = X \frac{\partial^2 T}{\partial t^2}$$

Substitution into Eq. (13.94) yields

$$T\frac{d^2X}{dx^2} = \frac{1}{v^2}\left(X\frac{d^2T}{dt^2}\right) \tag{13.103}$$

where total derivatives have been used because X and T are functions only of x and t, respectively. Dividing each side by $y = XT$,

$$\frac{1}{X}\frac{d^2X}{dx^2} = \frac{1}{v^2T}\frac{d^2T}{dt^2} \tag{13.104}$$

Clearly, the left side of Eq. (13.104) is a function only of x, and the right side only of t. Since x and t are each independent, Eq. (13.104) can be satisfied only if each side is equal to a constant for all values of the variables. This may be demonstrated more formally by assuming each side to equal a function of x and t, say $\gamma(x, t)$. Thus

$$\frac{1}{X}\frac{d^2X}{dx^2} = \gamma \tag{13.105a}$$

and

$$\frac{1}{v^2T}\frac{d^2T}{dt^2} = \gamma \tag{13.105b}$$

The partial derivative of Eq. (13.105a) with respect to time is $\partial\gamma/\partial t = 0$; the partial derivative of Eq. (13.105b) with respect to x is $\partial\gamma/\partial x = 0$. Therefore, γ is neither a function of x nor t, and must be a constant, which we call β^2. Equation (13.105a) now becomes

$$\frac{1}{X}\frac{d^2X}{dx^2} = \beta^2 \tag{13.106a}$$

or

$$\frac{d^2X}{dx^2} = \beta^2 X \tag{13.106b}$$

The general solution of this equation is

$$X(x) = A_1 e^{\beta x} + A_2 e^{-\beta x} \tag{13.107}$$

Applying the boundary conditions given by Eqs. (13.101), the end displacements must equal zero at all times and consequently from Eq. (13.102), $X(0) = X(L) = 0$. Hence

$$A_1 + A_2 = 0$$

and

$$A_1 e^{\beta L} + A_2 e^{-\beta L} = 0$$

Only if the determinant of the coefficients of A_1 and A_2 vanishes can there be a nontrivial solution of the simultaneous homogeneous equations above:

$$\begin{vmatrix} 1 & 1 \\ e^{\beta L} & e^{-\beta L} \end{vmatrix} = 0$$

or

$$e^{-\beta L} - e^{\beta L} = 0$$

The only possible solution (for real β) is $\beta = 0$. Thus no real value of β except $\beta = 0$ satisfies Eq. (13.106). That $\beta = 0$ results in a trivial solution is verified by solving the equation

$$\frac{d^2 X}{dx^2} = 0$$

or

$$X = a_1 + a_2 x$$

At $x = 0$, $X = 0$, and $a_1 = 0$. At $x = L$, $X = 0$, and $a_2 = 0$. Consequently, for $\beta = 0$, $X(x) = 0$, and there is no motion.

A nontrivial solution exists only if β is complex, and we therefore substitute $\beta^2 = -\alpha^2$ in Eq. (13.106b),

$$\frac{d^2 X}{dx^2} = -\alpha^2 X \tag{13.108}$$

where α is real. The general solution of this familiar equation is

$$X = B_1 \sin \alpha x + B_2 \cos \alpha x \tag{13.109}$$

From the boundary condition $X(0) = 0$, $B_2 = 0$. From $X(L) = 0$, $B_1 \sin \alpha L = 0$. Since $B_1 = 0$ also results in a trivial solution, we seek therefore the values of α such that $\sin \alpha L = 0$. That is, $\alpha L = n\pi$ or

$$\alpha_n = \frac{n\pi}{L} \tag{13.110}$$

where n is any positive or negative integer.

Subject to the boundary conditions given, we now have the solution:

$$X(x) = B_1 \sin n\pi \frac{x}{L} \tag{13.111}$$

That $(1/X)(d^2 X/dx^2)$ is constant is verified as follows:

$$\frac{B_1}{B_1 \sin n\pi(x/L)}\left(-\frac{n^2\pi^2}{L^2}\right) \sin n\pi \frac{x}{L} = -\frac{n^2\pi^2}{L^2}$$

and Eq. (13.105b) is thus

$$\frac{1}{v^2 T} \frac{d^2 T}{dt^2} = -\frac{n^2 \pi^2}{L^2}$$ (13.112a)

or

$$\frac{d^2 T}{dt^2} = -\frac{n^2 \pi^2 v^2}{L^2} T$$ (13.112b)

Hence

$$T(t) = C_1 \sin\left(\frac{n\pi v}{L}\right) t + C_2 \cos\left(\frac{n\pi v}{L}\right) t$$ (13.113)

The solution, $y(x, t)$, is consequently

$$y = B_1 C_1 \sin\left(\frac{n\pi x}{L}\right) \sin\left(\frac{n\pi v t}{L}\right) + B_1 C_2 \sin\left(\frac{n\pi x}{L}\right) \cos\left(\frac{n\pi v t}{L}\right)$$ (13.114a)

or

$$y = D_1' \sin\left(\frac{n\pi x}{L}\right) \sin\left(\frac{n\pi v t}{L}\right) + D_2' \sin\left(\frac{n\pi x}{L}\right) \cos\left(\frac{n\pi v t}{L}\right)$$ (13.114b)

Employing the trigonometric identities,

$$\sin a \sin b = \tfrac{1}{2}[\cos(a - b) - \cos(a + b)]$$

$$\sin a \cos b = \tfrac{1}{2}[\sin(a - b) + \sin(a + b)]$$

Equation (13.114b) is written

$$y = D_1 \left[\cos \frac{n\pi}{L}(x - v_t) - \cos \frac{n\pi}{L}(x + vt) \right]$$
$$+ D_2 \left[\sin \frac{n\pi}{L}(x - vt) + \sin \frac{n\pi}{L}(x + vt) \right]$$ (13.115)

and it is further verified that y is a function of $x - vt$ and $x + vt$. Actually, $y(x, t)$ is more involved than appears above, for n may be any positive or negative integer. Because the original differential equation is linear, the principle of superposition applies and the complete solution is the sum of all solutions corresponding to integral n:

$$y = \sum_{n=1}^{\infty} \left\{ D_{1n} \left[\cos \frac{n\pi}{L}(x - vt) - \cos \frac{n\pi}{L}(x + vt) \right] \right.$$
$$\left. + D_{2n} \left[\sin \frac{n\pi}{L}(x - vt) + \sin \frac{n\pi}{L}(x + vt) \right] \right\}$$ (13.116)

We begin the summation at $n = 1$, inasmuch as it has been demonstrated that $n = 0$ yields a trivial solution. Negative values of n do not lead to linearly independent solutions. D_{1n} and D_{2n} are evaluated from a knowledge of the initial displacement and velocity.

To gain some insight into the motion, assume only one frequency present. This is equivalent to stating that all D_{1n} are zero except one. The same applies for D_{2n}. Labeling this specific value of n, m, we have

$$y = D_{1m} \left[\cos \frac{m\pi}{L} (x - vt) - \cos \frac{m\pi}{L} (x + vt) \right]$$

$$+ D_{2m} \left[\sin \frac{m\pi}{L} (x - vt) + \sin \frac{m\pi}{L} (x + vt) \right] \qquad (13.117)$$

Substituting the following identities in Eq. (13.117),

$$\sin \theta = \frac{1}{2i} (e^{i\theta} - e^{-i\theta})$$

$$\cos \theta = \frac{1}{2} (e^{i\theta} + e^{-i\theta})$$

we have

$$y = \left(\frac{D_{1m}}{2} + \frac{D_{2m}}{2i} \right) e^{(im\pi/L)(x - vt)} + \left(\frac{D_{1m}}{2} - \frac{D_{2m}}{2i} \right) e^{(i - m\pi/L)(x - vt)}$$

$$+ \left(\frac{D_{2m}}{2i} - \frac{D_{1m}}{2} \right) e^{(im\pi/L)(x + vt)} + \left(\frac{D_{2m}}{2i} + \frac{D_{1m}}{2} \right) e^{(-im\pi/L)(x + vt)}$$

$$(13.118)$$

The first two terms represent a plane wave traveling in the positive X-direction; the second two terms, a plane wave in the negative X-direction. The two waves combine at all points to produce a standing wave. It is a simple matter to verify that the amplitude is always zero at $x = 0$ and $x = L$. Cancellation of the two waves at points such as $x = 0$ and $x = L$ is referred to as destructive interference.

According to the familiar definition, wavelength is the distance measured, at a given time, between any adjacent corresponding positions of a sinusoidal wave. The total string length must equal an integral number of *half*-wavelengths, and as a consequence the following wavelengths are possible: $2L, L, \frac{2}{3}L, \frac{1}{2}L, \cdots$. This is generalized as $2L/(p - 1)$, where p represents the number of nodes, including those at the walls. Several cases are shown in Fig. 13.17. For any integer p, the frequency ν and wavelength λ are related by

$$\lambda \nu = v \qquad (13.119)$$

where v is the velocity of the wave.

Consider now a long string fixed only at $x = L$, and assume that nothing is known about the other end. Let a disturbance Δy_t be transmitted toward $x = L$, represented as

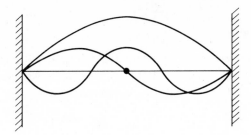

Fig. 13.17

$$\Delta y_t = \Delta y_t(x - vt) \tag{13.120}$$

When Δy_t reaches $x = L$ it can proceed no further, and is reflected toward $x = 0$ as a traveling wave, Δy_r:

$$\Delta y_r = \Delta y_r(x + vt) \tag{13.121}$$

If the wave were not reflected it would disappear, and the energy associated with it would vanish (see Section 13.17).

The net deflection of the string at any point is the sum of the transmitted and reflected disturbances. At $x = L$ these cancel, and hence $\Delta y_r(L + vt) = \Delta y_t(L - vt)$. If the disturbances are represented cosinusoidally,

$$\Delta y_t = A \cos \frac{2\pi n}{L}(L - vt) \qquad (x = L)$$

and

$$\Delta y_r = A \cos \frac{2\pi n}{L}(L + vt + \phi) \qquad (x = L)$$

where the reader may verify that the phase angle ϕ must equal π rad. We thus observe that a reflected wave experiences a phase change of π at the wall.

Let us now complete the solution of the vibrating string, proceeding from Eq. (13.116), written in the form

$$y(x, t) = \sum_{n=1}^{\infty} a_n \cos\left(\frac{n\pi vt}{L}\right) \sin\left(\frac{n\pi x}{L}\right) + \sum_{n=1}^{\infty} b_n \sin\left(\frac{n\pi vt}{L}\right) \sin\left(\frac{n\pi x}{L}\right) \tag{13.122}$$

There are $2n$ constants to be evaluated from the initial conditions $y(x, 0)$ and $\partial y/\partial t|_{x,0}$. To accomplish this, we apply Fourier's theorem,[3] which states that an arbitrary function, defined in a finite interval and having a finite number of maxima and minima and a finite number of discontinuities, can be expanded in a doubly infinite series of sinusoidal and cosinusoidal functions, for example, Eq. (13.122).

---

[3] Fourier, J., *The Analytical Theory of Heat*, Dover Publications Inc., New York, 1955.

13.14 FOURIER SERIES

Consider a function $g(u)$ defined in the interval $0 \leq u \leq 2T$ and having only a finite number of discontinuities located at u_1, u_2, $\cdots$, u_l.[4] Assume that the function may be adequately represented by the Fourier expansion,

$$g(u) = \frac{a_0}{2} + \sum_{n=1}^{\infty} \left[a_n \cos\left(\frac{n\pi u}{T}\right) + b_n \sin\left(\frac{n\pi u}{T}\right) \right] \qquad (13.123)$$

where a_0, a_n, and b_n are constants to be evaluated. We do not prove that the expansion is a complete set, that is, includes *all* the necessary terms, or that the series converges for all u. Such proofs are found in mathematical treatises.[5]

To ascertain a_0, multiply Eq. (13.123) by du and integrate from 0 to $2T$,

$$\int_0^{2T} g(u)\,du = \int_0^{u_1} g(u)\,du + \int_{u_1}^{u_2} g(u)\,du + \cdots + \int_{u_l}^{2T} g(u)\,du$$

$$= \frac{a_0}{2}\int_0^{2T} du + \int_0^{2T} \sum_{n=1}^{\infty}\left[a_n \cos\left(\frac{n\pi u}{T}\right) + b_n \sin\left(\frac{n\pi u}{T}\right) \right] du$$

$$(13.124)$$

where it is noted that the integration is piecewise to avoid the discontinuities.

Consider first the integrals on the right side of the above expression,

$$\frac{a_0}{2}\int_0^{2T} du = a_0 T \qquad (13.125)$$

$$\int_0^{2T} a_n \cos\left(\frac{n\pi u}{T}\right) du = \frac{a_n T}{n\pi}\left[\sin\left(\frac{n\pi u}{T}\right) \right]_0^{2T} = 0 \qquad (13.126)$$

$$\int_0^{2T} b_n \sin\left(\frac{n\pi u}{T}\right) du = -\frac{b_n T}{n\pi}\left[\cos\left(\frac{n\pi u}{T}\right) \right]_0^{2T} = 0 \qquad (13.127)$$

for all integral values of n. The result is therefore

$$\int_0^{2T} g(u)\,du = a_0 T \qquad (13.128a)$$

or

$$a_0 = \frac{1}{T}\int_0^{2T} g(u)\,du \qquad (13.128b)$$

To determine the a_n, multiply Eq. (13.123) by $\cos(m\pi u/T)\,du$, where m is an integer, and integrate from 0 to $2T$:

---

[4] Of course, a string would have no discontinuities in displacement.
[5] Churchill, R. V., *Fourier Series and Boundary Value Problems*, McGraw-Hill, Inc., New York, 1952.

$$\int_0^{2T} g(u) \cos\left(\frac{m\pi u}{T}\right) du = \int_0^{2T} \frac{a_0}{2} \cos\left(\frac{m\pi u}{T}\right) du$$

$$+ \int_0^{2T} \sum_{n=1}^{\infty} a_n \cos\left(\frac{n\pi u}{T}\right) \cos\left(\frac{m\pi u}{T}\right) du$$

$$+ \int_0^{2T} \sum_{n=1}^{\infty} b_n \sin\left(\frac{n\pi u}{T}\right) \cos\left(\frac{m\pi u}{T}\right) du \quad \textbf{(13.129)}$$

The first term above, as before, is zero for all m. A representative term of the second integral may be written

$$\int_0^{2T} a_n \cos\left(\frac{n\pi u}{T}\right) \cos\left(\frac{m\pi u}{T}\right) du$$

$$= \int_0^{2T} \frac{a_n}{2}\left[\cos(n+m)\frac{\pi u}{T} + \cos(n-m)\frac{\pi u}{T}\right] du \quad \textbf{(13.130)}$$

where the trigonometric identity $\cos a \cos b = \frac{1}{2}[\cos(a+b) + \cos(a-b)]$ has been applied. Integrating, we have

$$\int_0^{2T} a_n \cos\left(\frac{n\pi u}{T}\right) \cos\left(\frac{m\pi u}{T}\right) du$$

$$= \frac{a_n}{2}\left[\frac{T}{\pi} \frac{\sin(n+m)(\pi u/T)}{n+m} + \frac{T}{\pi} \frac{\sin(n-m)(\pi u/T)}{n-m}\right]_0^{2T}$$

The first term is zero for all n and m, while the second term is zero *except* when $n = m$. For this case, integration yields the indeterminate form $0/0$. Referring to the final term of Eq. (13.130) for $n = m$,

$$\frac{a_n}{2}\int_0^{2T} \cos 0 \, du = a_n T$$

Consider now the final integral of Eq. (13.129),

$$\int_0^{2T} b_n \sin\left(\frac{n\pi u}{T}\right) \cos\left(\frac{m\pi u}{T}\right) du$$

$$= \int_0^{2T} \frac{b_n}{2}\left[\sin(n+m)\frac{\pi u}{T} + \sin(n-m)\frac{\pi u}{T}\right] du \quad \textbf{(13.131)}$$

where the trigonometric identity $\sin a \cos b = \frac{1}{2}[\sin(a+b) + \sin(a-b)]$ has been applied. Integrating,

$$\int_0^{2T} b_n \sin\left(\frac{n\pi u}{T}\right) \cos\left(\frac{m\pi u}{T}\right) du$$

$$= \frac{b_n}{2}\left[\frac{-\cos(n+m)(\pi u/T)}{(\pi/T)(n+m)} + \frac{-\cos(n-m)(\pi u/T)}{(\pi/T)(n-m)}\right]_0^{2T} \quad \textbf{(13.132)}$$

Clearly the integrals are zero for all n except $n = m$. When $n = m$, Eq. (13.131) becomes

$$\frac{b_n}{2} \int_0^{2T} \sin 0 \; du = 0$$

and Eq. (13.132) is zero for $n = m$ as well. The result is therefore

$$\int_0^{2T} g(u) \cos \left(\frac{n\pi u}{T}\right) du = a_n T \qquad (13.133a)$$

or

$$a_n = \frac{1}{T} \int_0^{2T} g(u) \cos \left(\frac{n\pi u}{T}\right) du \qquad (13.133b)$$

A similar analysis yields the b_n. Multiply each side of Eq. (13.123) by $\sin (m\pi u/T) \, du$, where m is an integer, and integrate from 0 to $2T$:

$$\int_0^{2T} g(u) \sin \left(\frac{m\pi u}{T}\right) du = \int_0^{2T} \frac{a_0}{2} \sin \left(\frac{m\pi u}{T}\right) du$$

$$+ \int_0^{2T} \sum_{n=1}^{\infty} a_n \cos \left(\frac{n\pi u}{T}\right) \sin \left(\frac{m\pi u}{T}\right) du$$

$$+ \int_0^{2T} \sum_{n=1}^{\infty} b_n \sin \left(\frac{n\pi u}{T}\right) \sin \left(\frac{m\pi u}{T}\right) du \qquad (13.134)$$

According to Eq. (13.127),

$$\int_0^{2T} \frac{a_n}{2} \sin \left(\frac{m\pi u}{T}\right) du = 0$$

From Eq. (13.131),

$$\int_0^{2T} \sum_{n=1}^{\infty} a_n \cos \left(\frac{n\pi u}{T}\right) \sin \left(\frac{m\pi u}{T}\right) du = 0$$

for all integral values of n and m.

Consider now a term of the final integral:

$$\int_0^{2T} b_n \sin \left(\frac{n\pi u}{T}\right) \sin \left(\frac{m\pi u}{T}\right) du$$

Employing the trigonometric identity, $\sin a \sin b = \frac{1}{2}[\cos (a - b) - \cos (a + b)]$

$$\int_0^{2T} b_n \sin \left(\frac{n\pi u}{T}\right) \sin \left(\frac{m\pi u}{T}\right) du$$

$$= \frac{b_n}{2} \int_0^{2T} \left[\cos (n - m)\frac{\pi u}{T} - \cos (n + m)\frac{\pi u}{T} \right] du \qquad (13.135)$$

The above integral is zero except when $n = m$, in which case,

$$\int_0^{2T} b_n \sin\left(\frac{n\pi u}{T}\right) \sin\left(\frac{n\pi u}{T}\right) du = b_n T$$

Thus

$$\int_0^{2T} g(u) \sin\left(\frac{n\pi u}{T}\right) du = b_n T \qquad (13.136a)$$

or

$$b_n = \frac{1}{T} \int_0^{2T} g(u) \sin\left(\frac{n\pi u}{T}\right) du \qquad (13.136b)$$

Substituting the results of Eqs. (13.128b), (13.133b), and (13.136b) into (13.123),

$$g(u) = \frac{1}{2}\left[\frac{1}{T}\int_0^{2T} g(u')\, du'\right] + \sum_{n=1}^{\infty}\left[\frac{1}{T}\int_0^{2T} g(u') \cos\left(\frac{n\pi u'}{T}\right) du'\right] \cos\left(\frac{n\pi u}{T}\right)$$

$$+ \sum_{n=1}^{\infty}\left[\frac{1}{T}\int_0^{2T} g(u') \sin\left(\frac{n\pi u'}{T}\right) du'\right] \sin\left(\frac{n\pi u}{T}\right) \qquad (13.137)$$

and we have therefore an expansion for $g(u)$, in an infinite series of sinusoidal and cosinusoidal functions.

Illustrative Example 13.3
Expand the function

$$g(u) = 1 \qquad (0 \leq u < T)$$
$$g(u) = -1 \qquad (T < u \leq 2T)$$

in a Fourier series.

SOLUTION
Either Eqs. (13.128b), (13.133b), and (13.136b), or Eq. (13.137) may be used to evaluate the coefficients. From Eq. (13.128b),

$$a_0 = \frac{1}{T} \int_0^{2T} g(u)\, du = \frac{1}{T} \int_0^T du + \frac{1}{T} \int_T^{2T} (-du) = \frac{1}{T}(T - 2T + T) = 0$$

From Eq. (13.133b),

$$a_n = \frac{1}{T} \int_0^{2T} g(u) \cos\left(\frac{n\pi u}{T}\right) du = \frac{1}{T}\left[\int_0^T \cos\left(\frac{n\pi u}{T}\right) du + \int_T^{2T} - \cos\left(\frac{n\pi u}{T}\right) du\right]$$

$$= \frac{1}{T}\left[\frac{T}{n\pi} \sin\left(\frac{n\pi u}{T}\right)\right]_0^T - \frac{1}{T}\left[\frac{T}{n\pi} \sin\left(\frac{n\pi u}{T}\right)\right]_T^{2T} = 0$$

From Eq. (13.136b),

$$b_n = \frac{1}{T} \int_0^{2T} g(u) \sin\left(\frac{n\pi u}{T}\right) du = \frac{1}{T}\left[\int_0^T \sin\left(\frac{n\pi u}{T}\right) du + \int_T^{2T} - \sin\left(\frac{n\pi u}{T}\right) du\right]$$

$$= \frac{1}{n\pi}(-\cos n\pi + 1 + \cos 2n\pi - \cos n\pi) = \frac{2}{n\pi}(1 - \cos n\pi)$$

since $\cos 2n\pi = 1$.

Recall that $\cos n\pi = 1$ if n is positive and $\cos n\pi = -1$ if n is negative. Thus $b_n = (2/n\pi)[1 - (-1)^n]$. Only when n is odd is b_n different from zero; on the other hand, a_0 and a_n are zero for all n.

For the given function,

$$g(u) = \sum_{n=1}^{\infty} \frac{2}{n\pi}[1 - (-1)^n]\sin\left(\frac{n\pi u}{T}\right)$$

If we set $2l - 1 = n$,

$$g(u) = \frac{2}{\pi} \sum_{l=1}^{\infty} \frac{1}{2l-1}\sin\left[\frac{(2l-1)\pi u}{T}\right]$$

This expression describes $g(u)$ in the interval 0 to $2T$.

13.15 VIBRATING STRING (continued)

Returning to the vibrating string, we assume the initial position and velocity of the string given by

$$y(x, 0) = F(x) \tag{13.138a}$$

and

$$\left.\frac{\partial y}{\partial t}\right|_{x,0} = G(x) \tag{13.138b}$$

To determine the coefficients a_n and b_n we employ the same techniques as with the general Fourier series. From Eq. (13.122),

$$y(x, 0) = \sum_{n=1}^{\infty} a_n \sin\left(\frac{n\pi x}{L}\right) = F(x) \tag{13.139}$$

From the condition on the transverse velocity at time zero,

$$\left.\frac{\partial y}{\partial t}\right|_{x,0} = \sum_{n=1}^{\infty} \frac{n\pi v}{L} a_n\left[-\sin\frac{n\pi v(0)}{L}\right]\sin\left(\frac{n\pi x}{L}\right)$$

$$+ \sum_{n=1}^{\infty} \frac{n\pi v}{L} b_n \cos\left[\frac{n\pi v(0)}{L}\right]\sin\left(\frac{n\pi x}{L}\right) = G(x) \tag{13.140a}$$

or

$$G(x) = \sum_{n=1}^{\infty} \frac{n\pi v}{L} b_n \sin\left(\frac{n\pi x}{L}\right) \tag{13.140b}$$

For the values of a_n, multiply Eq. (13.139) by $\sin(m\pi x/L)\,dx$ and integrate from 0 to L:

$$\int_0^L F(x) \sin\left(\frac{m\pi x}{L}\right) dx = \int_0^L \sum_{n=1}^{\infty} a_n \sin\left(\frac{n\pi x}{L}\right) \sin\left(\frac{m\pi x}{L}\right) dx \tag{13.141}$$

Recall that

$$\int_0^L \sin\left(\frac{n\pi x}{L}\right) \sin\left(\frac{m\pi x}{L}\right) dx = 0 \qquad (n \neq m)$$

and consequently,

$$\int_0^L \sin\left(\frac{n\pi x}{L}\right) \sin\left(\frac{m\pi x}{L}\right) dx = \int_0^L \sin^2\left(\frac{n\pi x}{L}\right) dx = \frac{L}{2}$$

Thus

$$a_n = \frac{2}{L} \int_0^L F(x) \sin\left(\frac{n\pi x}{L}\right) dx \tag{13.142}$$

For b_n, multiply Eq. (13.140b) by $\sin(m\pi x/L)\,dx$ and integrate from 0 to L:

$$\int_0^L G(x) \sin\left(\frac{m\pi x}{L}\right) = \int_0^L \sum_{n=1}^{\infty} \frac{n\pi v}{L} b_n \sin\left(\frac{n\pi x}{L}\right) \sin\left(\frac{m\pi x}{L}\right) dx$$

This leads to

$$b_n = 2 \frac{L}{n\pi v} \int_0^L G(x) \sin\left(\frac{n\pi x}{L}\right) dx \tag{13.143}$$

Thus, $y(x, t)$ may be expressed

$$y(x, t) = \sum_{n=1}^{\infty} \left\{ \frac{2}{L} \left[\int_0^L F(x') \sin\left(\frac{n\pi x'}{L}\right) dx' \right] \cos\left(\frac{n\pi vt}{L}\right) \right.$$

$$\left. + 2\frac{L}{n\pi v} \left[\int_0^L G(x') \sin\left(\frac{n\pi x'}{L}\right) dx' \right] \sin\left(\frac{n\pi vt}{L}\right) \right\} \sin\left(\frac{n\pi x}{L}\right) \tag{13.144}$$

Except in the rare case in which $F(x)$ and $G(x)$ are both simple sinusoidal functions, $y(x, t)$ is composed of an infinite number of frequencies, each of which is called a *normal frequency*.

Illustrative Example 13.4
A string of mass m, length L, and tension τ, fixed at both ends is distorted into the shape of the parabola, $y = c(L - x)x$ and released from rest. Determine the motion.

SOLUTION

The differential equation for the motion of the string is

$$\frac{\partial^2 y}{\partial x^2} = \frac{m}{LT}\frac{\partial^2 y}{\partial t^2} = \frac{1}{v^2}\frac{\partial^2 y}{\partial t^2}$$

The boundary conditions are $y(0, t) = y(L, t) = 0$; the initial conditions are $y(x, 0) = c(L - x)x$ and $\partial y/\partial t$ at $(x, 0) = 0$.

Equation (13.144) applies:

$$y(x, t) = \sum_{n=1}^{\infty}\frac{2}{L}\Bigg[\int_0^L\bigg(cLx'\sin\frac{n\pi x'}{L}$$

$$-cx'^2\sin\frac{n\pi x'}{L}\bigg)dx'\cos\frac{n\pi vt}{L}\sin\frac{n\pi x}{L}$$

$$+\frac{L^2}{n\pi v}\int_0^L(0)\sin\frac{n\pi x'}{L}dx'\sin\frac{n\pi vt}{L}\sin\frac{n\pi x}{L}\Bigg]$$

There are two integrals to evaluate:

$$\int_0^L x'\sin\left(\frac{n\pi x'}{L}\right)dx' = -\frac{L^2}{n\pi}\cos n\pi = -\frac{L^2}{n\pi}(-1)^n$$

and

$$\int_0^L x'^2\sin\left(\frac{n\pi x'}{L}\right)dx' = \frac{L^3}{n\pi}\left[\frac{2(\cos n\pi - 1)}{n^2\pi^2} - \cos n\pi\right]$$

Thus

$$y(x, t) = \sum_{n=1}^{\infty}\frac{2c}{L}\left\{-\frac{L^3}{n\pi}(-1)^n - \frac{L^3}{n\pi}\left[\frac{2(\cos n\pi - 1)}{n^2\pi^2} - (-1)^n\right]\right\}$$

$$\times\cos\left(\frac{n\pi vt}{L}\right)\sin\left(\frac{n\pi x}{L}\right)$$

or

$$y(x, t) = 4cL^2\sum_{n=1}^{\infty}\frac{1 - (-1)^n}{n^3\pi^3}\cos\left(\frac{n\pi vt}{L}\right)\sin\left(\frac{n\pi x}{L}\right)$$

By setting $n = 2l - 1$, only terms involving odd n appear, and we have

$$y(x, t) = 8cL^2\sum_{l=1}^{\infty}\frac{1}{(2l - 1)^3\pi^3}\cos\left[\frac{(2l - 1)\pi vt}{L}\right]\sin\left[\frac{(2l - 1)\pi x}{L}\right]$$

There are two nodes, $x = 0$ and $x = L$. The maximum displacement of y, at any time, occurs at $x = L/2$.

The wave equation for the string has been derived neglecting nonlinearities in the differential equation as well as dissipative mechanisms. Where these cannot be neglected, the wave shape or wave packet can still be decomposed into a series of harmonic waves, but now the velocity of propagation is found to be wavelength-dependent. Consequently, the shape of the wave changes with time and we speak of *dispersion*. The propagation of water waves and the decomposition of white light into its constituent colors by a prism are examples of dispersion. The phase velocity does not describe the progress of a dispersive wave packet. For such phenomena, a more meaningful quantity is the *group velocity*, the velocity of a maximum of a given wave packet.

Returning to the nondissipative case, the magnitude of the phase velocity of each pair of harmonic waves constituting a standing wave is v. Since their directions are opposite, the maximum of the wave does not move, and hence its group velocity is zero.

13.16 ADDITIONAL EXAMPLES OF THE WAVE EQUATION

Presented here briefly are two additional examples of the wave equation. These relate to longitudinal vibrations of a beam and electrical oscillations in a transmission line.

Consider first a uniform beam of length L and mass m, as in Fig. 13.18.

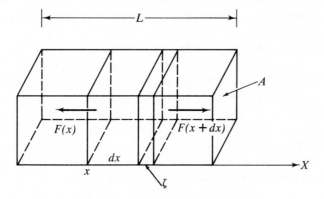

Fig. 13.18

If the axial force at $x + dx$ is $F(x + dx)$ and that at x is $F(x)$, the net force acting on a differential element dx is $F(x + dx) - F(x)$. Equating the net force to the product of mass and acceleration, we have, to first order,

$$F(x + dx) - F(x) = F(x) + \frac{\partial F}{\partial x} dx - F(x) = \frac{m}{L} dx \frac{\partial^2 \zeta}{\partial t^2} \qquad (13.145)$$

where m/L is the mass per unit length of the uniform beam and ζ the change of length of dx associated with elastic distortion. The strain at any point in the beam is

$$\epsilon = \frac{\partial \zeta}{\partial x} \tag{13.146}$$

and is related by Hooke's law to the stress:

$$\frac{F}{A} = E\epsilon = E\frac{\partial \zeta}{\partial x} \tag{13.147}$$

Equation (13.145) requires $\partial F/\partial x$, which is obtained by differentiating Eq. (13.147) with respect to x:

$$\frac{\partial F}{\partial x} = EA\frac{\partial^2 \zeta}{\partial x^2} \tag{13.148}$$

When this is substituted into Eq. (13.145) we have

$$EA\frac{\partial^2 \zeta}{\partial x^2} dx = \frac{m}{L} dx \frac{\partial^2 \zeta}{\partial t^2} \tag{13.149a}$$

or

$$\frac{\partial^2 \zeta}{\partial x^2} = \frac{m}{EAL}\frac{\partial^2 \zeta}{\partial t^2} = \frac{\rho}{E}\frac{\partial^2 \zeta}{\partial t^2} \tag{13.149b}$$

where ρ is the mass density. The term E/ρ possesses dimensions of velocity squared, and in fact represents the square of the velocity of propagation of longitudinal waves through the beam.

Consider now a two-wire electrical transmission line. The wires are assumed ideal, that is, they possess only distributed capacitance *between* the wires, for example, C farads per meter, and distributed inductance *along* each wire, for example, L henries per meter. Series resistance and parallel conductance are neglected in this idealization.

To derive the differential equations of current and voltage, consider a differential length of wire, small enough so that the inductance and capacitance can be treated as *lumped parameters*, as in Fig. 13.19. The entire line is assumed to be composed of elements of this type, connected in series.

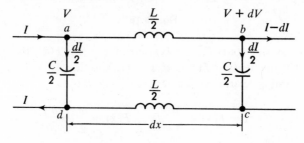

Fig. 13.19

Node d is taken as a point of zero potential, with node a at potential V. Current I enters node a, as shown, and leaves node d. The voltage difference between nodes a and b is equal to the electromotive force induced between these points:

$$(V + dV) - V = -L\,dx\,\frac{d}{dt}\left(I - \frac{dI}{2}\right) \tag{13.150}$$

To first order,

$$dV = -L\,dx\,\frac{dI}{dt} \tag{13.151a}$$

or

$$\frac{\partial V}{\partial x} = -L\frac{\partial I}{\partial t} \tag{13.151b}$$

Since the voltage and current are functions of both position and time, partial derivatives have been employed.

Another expression connecting current and voltage relates the current passing through the capacitor between nodes a and d to the time rate of change of voltage:

$$\frac{dI}{2} = -\frac{C}{2}\,dx\,\frac{\partial V}{\partial t} \tag{13.152a}$$

or

$$\frac{\partial I}{\partial x} = -C\frac{\partial V}{\partial t} \tag{13.152b}$$

Differentiating Eq. (13.151b) with respect to time, and Eq. (13.152b) with respect to x, and combining in order to decouple the circuit equations,

$$\frac{\partial^2 V}{\partial t\,\partial x} = -L\frac{\partial^2 I}{\partial t^2}$$

and

$$\frac{\partial^2 I}{\partial x^2} = -C\frac{\partial^2 V}{\partial x\,\partial t}$$

Combining we have

$$\frac{\partial^2 I}{\partial x^2} = LC\frac{\partial^2 I}{\partial t^2} \tag{13.153}$$

Reversing the order of differentiation prescribed above, the result is

$$\frac{\partial^2 V}{\partial x^2} = LC\frac{\partial^2 V}{\partial t^2} \tag{13.154}$$

Equations (13.153) and (13.154) are wave equations. The phase velocity, identical for each, is $(1/LC)^{1/2}$. In electromagnetic theory it is demonstrated that the quantity LC is independent of configuration.

13.17 ENERGY RELATIONSHIPS FOR A STRING

The kinetic energy of a string is simply the sum of the kinetic energy of each differential length of which it is composed. Each infinitesimal of kinetic energy is thus

$$dT = \tfrac{1}{2}\rho_L \, dx \left(\frac{\partial y}{\partial t}\right)^2 \tag{13.155}$$

The work done on the string by the tension F is the force multiplied by those components of displacement parallel to the force. It is necessary, therefore, to determine the amount by which the string lengthens, for this represents the required displacement. For an element dx, the change of length along F is

$$dS - dx = \left[1 + \left(\frac{\partial y}{\partial x}\right)^2\right]^{1/2} dx - dx \tag{13.156}$$

Since $\partial y/\partial x$ is small compared with unity,

$$dS - dx \cong \frac{1}{2}\left(\frac{\partial y}{\partial x}\right)^2 dx$$

to first order. The work done on a string element by the tension is, therefore,

$$dW = F\left[\frac{1}{2}\left(\frac{\partial y}{\partial x}\right)^2\right] dx \tag{13.157}$$

because the work done on the string is stored as elastic potential energy, $dW = dV$,[6] where the zero of potential energy occurs at the unstretched string configuration $(dS = dx)$. The total energy per unit length is

$$\frac{dW}{dx} = \tfrac{1}{2}\rho_L \left(\frac{\partial y}{\partial t}\right)^2 + \tfrac{1}{2}F\left(\frac{\partial y}{\partial x}\right)^2 \tag{13.158}$$

13.18 TRANSVERSE BEAM VIBRATION — RAYLEIGH'S METHOD

Since the mechanics of deformable continua treats systems possessing an infinite number of degrees of freedom, there are associated with such systems an infinite number of natural frequencies. This is, of course, true of beams as well as the other configurations discussed in this chapter. As in the case of the simple two-degree-of-freedom systems of Chapter 11, the total vibration of continuous media

---

[6] Note that the sign is opposite to that in Eq. (5.31). Since dV is equal to the *negative* of the work done by a conservative force, it is clearly equal to the work done by the reaction to this force.

is the sum of the individual modes. The *Rayleigh method*,[7] now applied to the *transverse* vibration of a beam of constant section and uniform mass, provides a simple means by which the fundamental frequency (first mode) may be determined. The method is restricted to conservative systems and assumes the maximum kinetic energy associated with a given mode equal to the maximum potential energy in the same mode. The energy, as we shall see, is dependent upon the deflection of the beam under dynamic conditions, and since this is presumably unknown, an assumption is required. In general, it is quite difficult to assume the curves of beam deformation associated with the higher modes, and consequently the Rayleigh method is applied primarily for determining the fundamental frequency. If the assumed beam deflection differs from the actual curve, it is equivalent to imposing additional constraints to the beam, thereby increasing the calculated natural frequency. Since the static deflection is quite often selected as the assumed dynamic curve of beam deformation, the fundamental frequency obtained by the Rayleigh method is higher than the actual value.

We begin by deriving an expression for the elastic potential energy associated with the bending of a beam, for it is this energy which is used in the Rayleigh method. Consider an infinitesimal length of beam subject to pure bending, as in Fig. 13.20. Just as in the case of a linear spring in which the force is proportional to the deformation, the angle θ is linearly proportional to the bending moment.

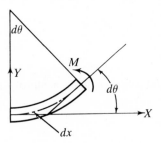

Fig. 13.20

The energy associated with the straining of dx can only have resulted from the work done by moment M (neglecting the influence of shearing forces). Consequently,

$$dW = M \frac{d\theta}{2} = dV \tag{13.159}$$

where V is the elastic potential energy of bending. Since $d\theta$ is simply the beam slope as shown in Fig. 13.20,

[7] Rayleigh, J. W. S., *The Theory of Sound* (Second Edition, 1894), Dover Publications, Inc., New York, 1945.

$$d\theta = d\frac{dy}{dx} = \frac{M}{EI}\,dx \tag{13.160}$$

which, when substituted in Eq. (13.159) yields

$$dV = \frac{M^2\,dx}{2EI}$$

or

$$V = \int_0^L \frac{M^2\,dx}{2EI} = \int_0^L \frac{[EI(d^2y/dx^2)]^2\,dx}{2EI} = \frac{EI}{2}\int_0^L \left(\frac{d^2y}{dx^2}\right)^2 dx \tag{13.161}$$

where EI has been assumed constant, owing to constant beam properties and cross section.

Returning to the Rayleigh method, the maximum kinetic and potential elastic energies associated with the fundamental mode are equated: $V_{max} = T_{max}$. The maximum kinetic energy,

$$T_{max} = \tfrac{1}{2}mv_{max}^2 = \tfrac{1}{2}\rho_L\int_0^L (y\omega_0)^2\,dx = \tfrac{1}{2}\rho_L\omega_0^2\int_0^L y^2\,dx \tag{13.162}$$

where ω_0 represents the fundamental circular frequency of the beam and $\omega_0 y$ the maximum velocity at any point. Equating energies and solving for ω_0^2,

$$\omega_0^2 = \frac{EI\int_0^L (d^2y/dx^2)^2\,dx}{\rho_L\int_0^L y^2\,dx} \tag{13.163}$$

Illustrative Example 13.5
Determine the fundamental frequency of a simply supported uniform beam of length L.

SOLUTION
The question of dynamic deflection immediately arises, for $y = y(x)$ must be assumed in order for Eq. (13.163) to be integrated. One may select, for example, the static deflection associated with a uniformly distributed force. We instead guess at a sinusoid to illustrate the method:

$$y = y_{max}\sin\frac{\pi x}{L}$$

$$\frac{dy}{dx} = \frac{\pi}{L}y_{max}\cos\frac{\pi x}{L}$$

$$\frac{d^2y}{dx^2} = -\left(\frac{\pi}{L}\right)^2 y_{max}\sin\frac{\pi x}{L}$$

Substituting into Eq. (13.163),

$$\omega_0^2 = \frac{EI y_{max}^2 (\pi^4/L^4) \int_0^L \sin^2 \pi x/L \, dx}{\rho_L y_{max}^2 \int_0^L \sin^2 \pi x/L \, dx} = \frac{EI}{\rho_L} \frac{\pi^4}{L^4} \tag{13.164}$$

Note that the amplitude y_{max} plays no role in the final expression for the fundamental frequency.

13.19 TRANSVERSE BEAM VIBRATION — GENERAL ANALYSIS

The case of small transverse vibration of a beam subject to the conditions of simple beam theory is now treated. Consider a uniformly loaded, simply supported beam. The moment and elastic curve, for a beam of constant section, are related by

$$EI \frac{d^2 y}{dx^2} = M$$

which, when differentiated, yields

$$EI \frac{d^3 y}{dx^3} = \frac{dM}{dx} = V \tag{13.165a}$$

and

$$EI \frac{d^4 y}{dx^4} = \frac{dV}{dx} = - w \tag{13.165b}$$

where V now represents the shear and w is the loading rate or load intensity (force per unit length). For a beam in transverse free vibration, external forcing is absent, and application of Newton's second law leads to the expression

$$- EI \frac{\partial^4 y}{\partial x^4} dx = ma_y = \rho_L \, dx \frac{\partial^2 y}{\partial t^2} \tag{13.166}$$

where ρ_L represents the beam mass per unit length. It is important to note that the above differential equation carries with it the same assumptions as elementary beam theory. The vibration it represents is transverse to the longitudinal axis and includes no shear or rotational effects. The latter becomes significant when the beam cross section is no longer small relative to beam length, and rotary inertia must then be taken into account. It is also assumed that internal friction is negligible.

Rearranging Eq. (13.166),

$$\frac{EI}{\rho_L} \frac{\partial^4 y}{\partial x^4} = - \frac{\partial^2 y}{\partial t^2} \tag{13.167}$$

Again applying the method of separation of variables, a solution of the form $y(x, t) = X(x)T(t)$ is assumed. Upon substitution of the appropriate derivatives, Eq. (13.167) is written

$$T \frac{EI}{\rho_L} \frac{d^4X}{dx^4} = -X \frac{d^2T}{dt^2}$$

Dividing by $y = XT$,

$$\frac{EI}{\rho_L} \frac{1}{X} \frac{d^4X}{dx^4} = -\frac{1}{T} \frac{d^2T}{dt^2} \tag{13.168}$$

Since the left and right sides are functions solely of x and t, respectively, both of which are independent variables, each side must equal the same constant, which we label λ^4. Therefore, Eq. (13.168) is separated into two equations,

$$\frac{EI}{\rho_L} \frac{1}{X} \frac{d^4X}{d^4x} = \lambda^4 \tag{13.169a}$$

$$-\frac{1}{T} \frac{d^2T}{dt^2} = \lambda^4 \tag{13.169b}$$

It is convenient to rewrite Eq. (13.169a) as follows:

$$\frac{d^4X}{dx^4} - \frac{\lambda^4 \rho_L}{EI} X = 0 \tag{13.170}$$

The solution of Eq. (13.169b) is

$$T(t) = C_1 \cos \lambda^2 t + C_2 \sin \lambda^2 t \tag{13.171a}$$

or

$$T(t) = C \cos (\lambda^2 t - \alpha) \tag{13.171b}$$

where α is a phase angle.

For the fourth-order equation, the following exponential solution $X(x) = C'e^{sx}$ is assumed, yielding, after substitution into Eq. (13.170),

$$C's^4 e^{sx} - \frac{C'\lambda^4 \rho_L e^{sx}}{EI} = 0 \tag{13.172a}$$

or

$$s^4 = \frac{\lambda^4 \rho_L}{EI} = \beta^4 \tag{13.172b}$$

The four roots of Eq. (13.172b) are $s = \pm\beta$, $s = \pm i\beta$ and therefore

$$X(x) = C_3' e^{\beta x} + C_4' e^{-\beta x} + C_5' e^{i\beta x} + C_6' e^{-i\beta x}$$

which, upon substitution of the trigonometric and hyperbolic identities, is written

$$X(x) = C_3'' \cosh \beta x + C_4'' \sinh \beta x + C_5'' \cos \beta x + C_6'' \sin \beta x \tag{13.173}$$

The solutions $T(t)$ and $X(x)$ are now combined to yield

$$y(x, t) = (C_3 \cosh \beta x + C_4 \sinh \beta x + C_5 \cos \beta x + C_6 \sin \beta x)[\cos (\lambda^2 t - \alpha)] \tag{13.174}$$

To continue beyond this point, a specific beam must be introduced. Returning to the simply supported beam previously discussed in connection with the Rayleigh method, the following express the condition of zero deflection and zero moment at $x = 0$ and $x = L$:

$$y(0, t) = 0 \qquad\qquad (13.175a)$$

$$y(L, t) = 0 \qquad\qquad (13.175b)$$

$$\frac{M}{EI} = \frac{\partial^2 y}{\partial x^2}\bigg|_{0,t} = 0 \qquad\qquad (13.175c)$$

$$\frac{M}{EI} = \frac{\partial^2 y}{\partial x^2}\bigg|_{L,t} = 0 \qquad\qquad (13.175d)$$

In order to apply the boundary conditions given by Eqs. (13.175c) and (13.175d), $\partial^2 y/\partial x^2$ is required. Differentiating Eq. (13.174) twice, we have

$$\frac{\partial^2 y}{\partial x^2} = (C_3\beta^2 \cosh \beta x + C_4\beta^2 \sinh \beta x - C_5\beta^2 \cos \beta x - C_6\beta^2 \sin \beta x)$$

$$\times [\cos (\lambda^2 t - \alpha)] \qquad\qquad (13.176)$$

Substituting Eqs. (13.175a) and (13.175c) into Eqs. (13.174) and (13.176), respectively,

$$0 = (C_3 + C_5) \cos (\lambda^2 t - \alpha)$$

$$0 = (C_3 - C_5)\beta^2 \cos (\lambda^2 t - \alpha)$$

Thus $C_3 = C_5 = 0$. Similarly, when the conditions of Eqs. (13.175b) and (13.175d) are substituted,

$$0 = C_4 \sinh \beta L + C_6 \sin \beta L \qquad \text{and} \qquad 0 = C_4 \sinh \beta L - C_6 \sin \beta L$$

For the solution to be nontrivial, the determinant of the coefficients of C_4 and C_6 must vanish:

$$-\sinh \beta L \sin \beta L - \sinh \beta L \sin \beta L = 0 \qquad \text{or} \qquad \sinh \beta L \sin \beta L = 0$$

This is satisfied in a nontrivial way if $\sin \beta L = 0$, or

$$\beta L = n\pi \qquad \beta_n = \frac{n\pi}{L} \qquad (n = 1, 2, \cdots)$$

Equation (13.174) is now written

$$y(x, t) = \left(A_n \sinh \frac{n\pi x}{L} + B_n \sin \frac{n\pi x}{L} \right) \cos (\lambda_n^2 t - \alpha_n) \qquad\qquad (13.177)$$

where the relationship of λ_n and β_n has previously been noted. If $\beta_n = n\pi/L$ is substituted into Eq. (13.175b), $0 = C_4 \sinh n\pi + C_6 \sin n\pi$. Inasmuch as $\sin n\pi$ is zero and $\sinh n\pi$ is not, we conclude that $C_4 = 0$, and Eq. (13.177) becomes

$$y(x, t) = \sum_{n=1}^{\infty} B_n \sin\left(\frac{n\pi x}{L}\right) \cos\left(\lambda_n^2 t - \alpha_n\right) \qquad \textbf{(13.178)}$$

where B_n, λ_n, and α_n correspond to each integer. B_n and α_n are evaluated on the basis of conditions such as the shape of the elastic curve at $t = 0$ and the initial beam velocity.

Returning now to Eq. (13.172b) it is clear that

$$\lambda_n^2 = \beta_n^2 \left(\frac{EI}{\rho_L}\right)^{\frac{1}{2}}$$

represents the natural circular frequencies of the beam:

$$\omega_n = \frac{n^2 \pi^2}{L^2} \left(\frac{EI}{\rho_L}\right)^{\frac{1}{2}}$$

For the fundamental mode,

$$\omega_0 = \frac{\pi^2}{L^2} \left(\frac{EI}{\rho_L}\right)^{\frac{1}{2}}$$

which is identical with the equation derived in Illustrative Example 13.5.

13.20 ELASTIC INSTABILITY — COLUMNS

We have so far discussed simple structural elements for which failure is related to the attainment of some stress within the material. In the analysis below, due to L. Euler,[8] it is clear that failure may be initiated at stress levels within the elastic regime.

Consider a column, pinned (hinged) at both ends, as in Fig. 13.21(a), and imagine that the connection permits one end to move axially, enabling the member to deflect laterally. The column, defined as an axially compressed slender rod, is distinguished from a short compression member in that it fails by a mechanism resulting from *elastic instability* in which a lateral disturbance results in *buckling*. The short compression member, on the other hand, fails by *crushing*.

If the forces P are exactly colinear and there is no deviation between the column axis and a straight line, we may anticipate no buckling failure, but rather failure by yielding in compression. Since such a situation is not likely to occur, we seek a value of the axial load P which is just sufficient to maintain the column in a slightly deflected configuration.

Again, referring to Fig. 13.21(a) and to the accompanying free-body diagram, rotational equilibrium requires that

$$M = -Py \qquad \textbf{(13.179)}$$

---

[8] Euler, L., *Mémoires de l'Académie de Berlin* **XIII**, 252 (1759).

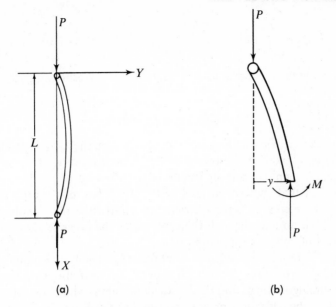

Fig. 13.21

Assuming that the deflection is small, the differential equation for the bending of beams is applicable here:

$$M = EI\frac{d^2y}{dx^2} = -Py \qquad \text{(13.180a)}$$

or

$$\frac{d^2y}{dx^2} + \frac{P}{EI}y = 0 \qquad \text{(13.180b)}$$

Making the usual substitution, $k^2 = P/EI$, Eq. (13.180b) is written

$$\frac{d^2y}{dx^2} + k^2y = 0 \qquad \text{(13.181)}$$

and the familiar solution is

$$y = C_1 \cos kx + C_2 \sin kx \qquad \text{(13.182)}$$

Applying the boundary condition $y(0) = 0$, we obtain $C_1 = 0$. Similarly, from the condition $y(L) = 0$, $C_2 \sin kL = 0$. Since $C_2 = 0$ leads to the trivial solution corresponding to no deflection whatever, the result we seek must be $\sin kL = 0$, or $kL = n\pi$:

$$k = k_n = \frac{n\pi}{L} \qquad (n = 1, 2, \cdots) \qquad \text{(13.183)}$$

Recalling that $k^2 = P/EI$, the critical loads are

$$P_{cr} = k^2 EI = \frac{n^2 \pi^2}{L^2} EI$$

Our primary concern is with the lowest value of P_{cr}, corresponding to $n = 1$:

$$P_{cr} = \frac{\pi^2}{L^2} EI$$

If P is less than P_{cr}, no buckling occurs. Note that the critical load is directly proportional to the product EI, known as the *flexural rigidity*. It is this combination which plays a similar role in beams in determining the deflection. It is of particular interest that the critical load is *not* a function of material strength. If the deflection of the column is not restricted to a particular plane, failure will occur by buckling about an axis through the centroid of the cross section for which the second moment of area is a minimum.

We have determined a value of axial force for which the column will sustain a small lateral deformation. In fact, in view of the restrictions on the expression relating the bending moment and the elastic curve upon which the foregoing development is based, the deflection must be so small as to assure elastic stressing of the column. If the applied force is less than P_{cr}, after a lateral disturbing influence is removed, the column assumes a straight configuration. If P exceeds P_{cr}, the column continues to deflect until total collapse.

In terms of an energy concept, $P = P_{cr}$ represents a boundary between stable $(P < P_{cr})$ and unstable $(P > P_{cr})$ states. An energy analysis of the pin-ended column is left to the reader as an exercise.

Substituting for k_n in Eq. (13.182), the column deflection is found to be

$$y = C_2 \sin\left(\frac{n \pi x}{L}\right) \tag{13.184}$$

From Eqs. (13.182) and (13.183) it is observed that according to the linearized theory from which Eq. (13.184) is developed, the amplitude C_2 of the elastic curve is not determined, and the critical load and amplitude are independent. P_{cr} will thus sustain any small lateral deflection. More exact analysis does reveal some dependence of P_{cr} upon C_2, however.

As to the meaning of n in Eq. (13.184), we may state that all modes of deflection are theoretically possible, but all but the first mode require the assistance of lateral constraints.

The average stress associated with the critical state of the column (just prior to buckling) is given by

$$\sigma_{cr} = \frac{P_{cr}}{A} = \frac{\pi^2 EI}{AL^2} \tag{13.185}$$

Defining the radius of gyration of a plane area, $r = (I/A)^{1/2}$, and substituting into Eq. (13.185) for I, we have

$$\sigma_{cr} = \frac{\pi^2 E r^2}{L^2} = \frac{\pi^2 E}{(L/r)^2} \tag{13.186}$$

where L/r, the *slenderness ratio*, represents an important yardstick for judging whether columns are short, intermediate, or long. The Euler formula, Eq. (13.186), gives good agreement with experiment for columns having a slenderness ratio higher than 100.

We have touched upon a single example of elastic instability. There are, of course, many complex problems involving columns with different end constraints, initial curvature, and inelastic action. The buckling of shells, plates, and bars under several combined loads are additional examples of elastic instability.

13.21 FLUIDS

A fluid is a continuum of matter which undergoes finite displacement under the action of infinitesimal shear stress. It is characterized by physical and mechanical properties such as density, temperature, velocity, and pressure, all of which are time-dependent fields. A small fluid volume can be treated as a continuum if it contains a statistically significant number of molecules so that the average distance traveled by a representative molecule between collisions is small compared with the smallest dimension of the volume. Random fluctuations of fluid properties are thus disregarded. We are then concerned with the average properties of the volume but not individual molecular motion.

Fluids that experience motion are said to flow. When the flow is time-independent, it is described as *steady*. *Uniform* fluid flow occurs when the fluid properties (except pressure) are constant in the direction of motion. *Laminar* or *streamlined flow* takes place when layers of fluid move past each other with no mixing or *turbulence*. Such flow is not necessarily steady.

We concern ourselves first with a fluid at rest for which Newton's first law is applicable. Then we consider the motion of ideal or frictionless fluids, and finally that of real fluids.

13.22 FLUID STATICS

In a static fluid each elemental volume is at rest in an inertial frame of reference. Thus the velocity characterizing each point in the fluid is zero. In a gas, for example, where each molecule is in rapid random motion, the velocity we refer to is the molecular velocity averaged over a significant volume.

Consider a prismatic element of fluid in equilibrium as shown in Fig. 13.22. If any forces act which are tangent to the prism surfaces, then in accordance with the definition of a fluid, motion must ensue. Since this contradicts the assumption of a static condition, it follows that the forces acting on the prism surfaces must be normal.

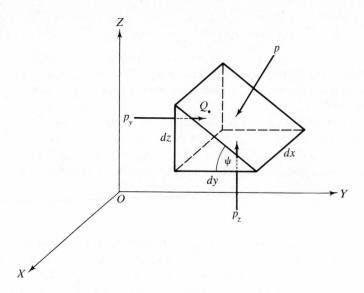

Fig. 13.22

The only forces acting result from pressure (normal force per unit area) and the weight of the fluid. The sum of forces in the Y- and Z-directions are

$$\sum F_y = p_y \, dx \, dz - p \, dx \frac{dz}{\sin \psi} \sin \psi = 0 \qquad (13.187)$$

$$\sum F_z = p_z \, dx \, dy - p \, dx \frac{dz}{\cos \psi} \cos \psi - \rho g \frac{dx \, dy \, dz}{2} = 0 \qquad (13.188)$$

where ρ is the mass density and $dx \, dy \, dz/2$ the prism volume. From Eq. (13.187) $p_y = p$. In the limit as dx, dy, dz approach zero, the weight term in Eq. (13.188) is of higher order than the pressure terms, and thus $p_z = p$. By rotating the prism about Q, it is clear that p is independent of direction and may be regarded as a scalar.

Let us now determine the manner in which pressure varies with position in a static fluid. Consider an infinitesimal volume of fluid, Fig. 13.23. The forces acting in the Z-direction must add to zero:

$$\sum F_z = p(z) \, dx \, dy - p(z + dz) \, dx \, dy - \rho g \, dx \, dy \, dz = 0 \qquad (13.189)$$

Expanding the pressure in a Taylor series about z to first order,

$$p(z) \, dx \, dy - p(z) \, dx \, dy - \frac{dp}{dz} \, dz \, dx \, dy - \rho g \, dx \, dy \, dz = 0 \qquad (13.190)$$

or

$$\frac{dp}{dz} = -\rho g \qquad (13.191)$$

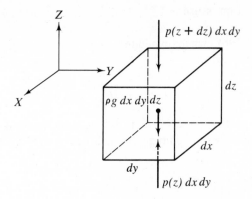

Fig. 13.23

For an incompressible fluid, integration of Eq. (13.191) yields

$$p = p_0 - \rho g z \tag{13.192}$$

where the pressure is p_0 at $z = 0$.

Since z is measured positively upward, pressure increases as z decreases. We may thus interpret the total pressure as the weight of a column of fluid of unit area added to the pressure at the top of the column.

We have assumed that fluid density is constant in order to integrate Eq. (13.191). If the density is a known function of vertical position, this equation may also be simply integrated. In the "isothermal" atmosphere, for example, density may be approximately represented by the expression

$$\rho = \rho_0 e^{-z/H} \tag{13.193}$$

where ρ_0 is the density at $z = 0$ (sea level) and H is the atmospheric scale height, the distance in which the density decreases by $1/e$ of its initial value. The scale height of the standard atmosphere varies from 5.5 to 8.1 km, primarily because the temperature is not actually constant.

The differential equation for atmospheric pressure, substituting Eq. (13.193) into Eq. (13.191), is therefore

$$dp = -\rho_0 g e^{-z/H} \, dz \tag{13.194}$$

Integrating,

$$p = C_1 + \rho_0 g H e^{-z/H} \tag{13.195}$$

From the condition that $p = 0$ for infinite z, $C_1 = 0$. At $z = 0$, $p = p_0 = \rho_0 g H$, and therefore

$$p = p_0 e^{-z/H} \tag{13.196}$$

Standard atmospheric pressure is 1.013×10^6 dyn/cm^2 or 14.7 lb/in.^{2}, also called 1 atmosphere.

Equation (13.196), when expanded in powers of z/H for z/H much less than unity, gives Eq. (13.192):

$$p = p_0\left(1 - \frac{z}{H} + \cdots\right) \cong p_0 - \rho_0 g H \frac{z}{H} = p_0 - \rho_0 g z \qquad \textbf{(13.197)}$$

This linear approximation is accurate to 0.5 percent if $z < 0.1H \approx 500$ m. The variation of g for this altitude change is much less than 0.5 percent.

13.23 ARCHIMEDES' PRINCIPLE

Consider a body immersed in a static fluid. What net force does the fluid exert on the body? To answer this question, we begin with a differential cube within the fluid. The net force exerted by the fluid on the cube acts in the Z-direction and is given by

$$dF_z = p(z)\, dx\, dy - p(z + dz)\, dx\, dy \qquad \textbf{(13.198)}$$

To first order in dz,

$$dF_z = -\frac{dp}{dz}\, dx\, dy\, dz \qquad \textbf{(13.199)}$$

Substituting Eq. (13.191),

$$dF_z = \rho g\, dx\, dy\, dz = \rho g\, dV \qquad \textbf{(13.200)}$$

The force on the cube is thus equal to the weight of the fluid displaced and in a direction opposite the force of gravity. For a body of finite dimension, we have only to integrate over that portion of the body which is immersed. The point of application of this force is the center of gravity of the *displaced fluid*.

The foregoing derivation is simply a development of *Archimedes' principle*[9]: *A body immersed in a fluid is buoyed up by a force equal to the weight of the fluid volume displaced.*

13.24 KINEMATICS OF FLUIDS

There are two approaches commonly employed to describe fluid motion, both attributable to the work of Euler,[10] although one is called the Lagrange system. In the latter, the motions of individual fluid particles which start from a known position at a given time are followed. This method represents a direct application of Newton's laws to a system of particles. Time is the only independent variable

---

[9] *The Works of Archimedes*, Edited by T. L. Heath, Dover Publications, Inc., New York, 1953.
[10] Euler, L., *Hist. de l'Academie de Berlin* (1755).

and the motion evolves from the initial conditions. Because of the problem of dividing the fluid into a manageable number of particles, this approach is not frequently applied. More often the Eulerian system is employed, in which kinematic quantities and the associated fluid properties are expressed as functions of position and time. The coordinates are thus *independent* variables equivalent to time, in contrast with the Lagrangian system. Observers may be thought of as fixed at every point in space. Quantities such as fluid velocity, pressure, and temperature are in this way time-dependent *fields* measured by the observers. Individual particles of fluid are thus not followed. Consequently, it is reasonable to describe the velocity vector as follows:

$$\mathbf{v} = \mathbf{v}(x, y, z, t) \tag{13.201}$$

The total change in $\mathbf{v}$ is therefore written

$$d\mathbf{v} = \frac{\partial \mathbf{v}}{\partial x} dx + \frac{\partial \mathbf{v}}{\partial y} dy + \frac{\partial \mathbf{v}}{\partial z} dz + \frac{\partial \mathbf{v}}{\partial t} dt$$

For the scalar pressure, $p(x, y, z, t)$,

$$dp = \frac{\partial p}{\partial x} dx + \frac{\partial p}{\partial y} dy + \frac{\partial p}{\partial z} dz + \frac{\partial p}{\partial t} dt$$

The total time derivatives of $\mathbf{v}$ and p are

$$\frac{d\mathbf{v}}{dt} = \frac{\partial \mathbf{v}}{\partial x} \frac{dx}{dt} + \frac{\partial \mathbf{v}}{\partial y} \frac{dy}{dt} + \frac{\partial \mathbf{v}}{\partial z} \frac{dz}{dt} + \frac{\partial \mathbf{v}}{\partial t} \tag{13.202}$$

and

$$\frac{dp}{dt} = \frac{\partial p}{\partial x} \frac{dx}{dt} + \frac{\partial p}{\partial y} \frac{dy}{dt} + \frac{\partial p}{\partial z} \frac{dz}{dt} + \frac{\partial p}{\partial t} \tag{13.203}$$

The quantities dx/dt, dy/dt, and dz/dt are the components of the fluid velocity at point (x, y, z) in the X-, Y-, Z-directions, respectively, at time t. Concise forms of Eqs. (13.202) and (13.203) are

$$\frac{d\mathbf{v}}{dt} = (\mathbf{v} \cdot \nabla)\mathbf{v} + \frac{\partial \mathbf{v}}{\partial t} \tag{13.204}$$

and

$$\frac{dp}{dt} = (\mathbf{v} \cdot \nabla)p + \frac{\partial p}{\partial t} \tag{13.205}$$

The above manner of representation enables us to deduce the operator

$$\frac{d}{dt} = \mathbf{v} \cdot \nabla + \frac{\partial}{\partial t} \tag{13.206}$$

known as the *substantial derivative*, applicable to either vector or scalar quantities.

For comparison, the velocity of a particle in the Lagrangian system is given by

$$\mathbf{v} = \mathbf{v}(\dot{x}_0, \dot{y}_0, \dot{z}_0, t) \tag{13.207}$$

where $\dot{x}_0, \dot{y}_0, \dot{z}_0$ are the known components of velocity at time t_0. The acceleration,

$$\frac{d\mathbf{v}}{dt} = \frac{\partial \mathbf{v}}{\partial t} \tag{13.208}$$

since $\dot{x}_0, \dot{y}_0, \dot{z}_0$ are constants.

The principle of conservation of mass is now applied to an Eulerian system to develop the equation of continuity. Consider a small differential volume $dx\, dy\, dz$ surrounding point (x, y, z). Referring to Fig. 13.24, the mass flow across face I in time dt is

$$\rho(x, y, z, t)v_x(x, y, z, t)\, dy\, dz\, dt$$

where v_x is the X-component of velocity, normal to area $dy\, dz$, and ρ is the mass density. The mass flow across side II is

$$\rho(x + dx, y, z, t)v_x(x + dx, y, z, t)\, dy\, dz\, dt$$

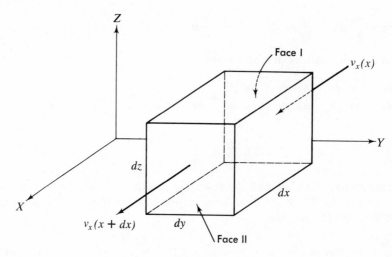

Fig. 13.24

The net mass leaving the volume in the X-direction is therefore

$$-\rho(x, y, z, t)v_x(x, y, z, t)\, dy\, dz\, dt + \rho(x + dx, y, z, t)v_x(x + dx, y, z, t)\, dy\, dz\, dt$$

or

$$\frac{\partial \rho}{\partial x}v_x\, dx\, dy\, dz\, dt + \rho \frac{\partial v_x}{\partial x}\, dx\, dy\, dz\, dt$$

The same procedure applied to the remaining faces yields the net mass leaving the elemental volume during dt:

$$\left(\frac{\partial \rho}{\partial x} v_x + \frac{\partial \rho}{\partial y} v_y + \frac{\partial \rho}{\partial z} v_z + \rho \frac{\partial v_x}{\partial x} + \rho \frac{\partial v_y}{\partial y} + \rho \frac{\partial v_z}{\partial z} \right) dx \, dy \, dz \, dt$$

The net mass leaving must, of course, equal the decrease of mass within the volume, $-(\partial \rho / \partial t) \, dx \, dy \, dz \, dt$, lest matter be destroyed. We are thus led to the continuity equation,

$$-\frac{\partial \rho}{\partial t} = \frac{\partial \rho}{\partial x} v_x + \frac{\partial \rho}{\partial y} v_y + \frac{\partial \rho}{\partial z} v_z + \rho \frac{\partial v_x}{\partial x} + \rho \frac{\partial v_y}{\partial y} + \rho \frac{\partial v_z}{\partial z} \tag{13.209}$$

or, in vector notation,

$$\nabla \cdot \rho \mathbf{v} + \frac{\partial \rho}{\partial t} = 0 \tag{13.210}$$

Equations (13.209) and (13.210) merely represent affirmations of the nondestructibility of mass. The quantity $\rho \mathbf{v}$ is the mass flux, mass per unit time per unit area leaving the volume in question. Equation (13.210) may be interpreted as follows: The divergence of the mass flux leaving a volume equals the rate at which the mass density decreases.

For steady flow, $\partial \rho / \partial t$ vanishes, and the mass entering is exactly accounted for by mass leaving. The mass flow can be determined by integrating Eq. (13.210) over the entire volume:

$$\int_V (\nabla \cdot \rho \mathbf{v}) \, dV + \int_V \frac{\partial \rho}{\partial t} \, dV = 0 \tag{13.211}$$

Applying Gauss' divergence theorem for steady conditions,

$$\int_A \rho \mathbf{v} \cdot d\mathbf{A} = 0 \tag{13.212}$$

If we attribute to the fluid constant density and constant velocity at flow areas A_1 and A_2 of a conduit, integration of Eq. (13.212) gives

$$\rho_1 A_1 v_1 = \rho_2 A_2 v_2 \tag{13.213}$$

Another special case occurs for incompressible flow (ρ constant). For this situation, the continuity equation becomes

$$\nabla \cdot \mathbf{v} = 0 \tag{13.214}$$

Since the divergence of the curl of any vector is zero, $\mathbf{v}$ is derivable from a vector potential Φ. Thus

$$\mathbf{v} = \nabla \times \Phi \tag{13.215}$$

and

$$\nabla \cdot (\nabla \times \Phi) = 0 \tag{13.216}$$

Equations (13.214) and (13.216) are reminiscent of the equations for the magnetic field and the associated vector potential.

When the curl of **v** is zero, **v** is derivable from a scalar potential ϕ, because the curl of the gradient of a scalar is zero. Thus $\nabla \times \mathbf{v} = \mathbf{0}$ implies $\mathbf{v} = -\nabla\phi$. The differential equation for ϕ is

$$\nabla \times \nabla\phi = \mathbf{0} \tag{13.217}$$

This type of flow is referred to as *irrotational*.

13.25 EULER'S EQUATION OF MOTION — INVISCID FLOW

A fluid must not only satisfy the continuity equation but Newton's second law as well. In order to derive the equations of motion, both internal and external forces are assumed functions of space variables and time. In Fig. 13.23 we observed the influence of two types of forces, one acting over an area, the other acting throughout the entire volume (a *body force*). Consider now such a volume, for which the net force is not zero, subject to a general body force **f** per unit volume. Applying Newton's second law to the X-direction,

$$f_x \, dx \, dy \, dz + p(x, y, z, t) \, dy \, dz - p(x + dx, y, z, t) \, dy \, dz = \frac{d}{dt} (pv_x) \, dx \, dy \, dz \tag{13.218}$$

There are, of course, similar equations for the Y- and Z-directions.

Expanding the pressure to first order in $dx \, dy \, dz$ and dividing by $dx \, dy \, dz$, we have

$$f_x - \frac{\partial p}{\partial x} = \frac{d}{dt} \rho v_x \tag{13.219a}$$

$$f_y - \frac{\partial p}{\partial y} = \frac{d}{dt} \rho v_y \tag{13.219b}$$

$$f_z - \frac{\partial p}{\partial z} = \frac{d}{dt} \rho v_z \tag{13.219c}$$

In vector notation,

$$\mathbf{f} - \nabla p = \frac{d}{dt} \rho \mathbf{v} = (\mathbf{v} \cdot \nabla)\rho \mathbf{v} + \frac{\partial}{\partial t} \rho \mathbf{v} = \rho \frac{d\mathbf{v}}{dt} \tag{13.220}$$

The final form is obtained by employing Eqs. (13.206) and (13.210). The above expression, in which fluid friction is neglected, is Euler's equation of motion[11] for an ideal fluid. When the body force **f** is given, there are five unknowns: pressure, density, and the three components of velocity. At this juncture, there are available

---

[11] Euler, L., *Hist. de l'Academie de Berlin* (1755).

only four scalar equations, however: the continuity equation and Euler's equation of motion.

An incompressible fluid, from its definition, leads to a fifth equation, $\rho =$ constant: For *compressible* fluid flow, thermodynamic considerations are usually applied such as the equation of state for a perfect gas,

$$\frac{p}{\rho} = nR_0T \tag{13.221}$$

where R_0 is the universal gas constant, n the number of moles, and T the absolute temperature. This equation unfortunately introduces an additional variable, temperature, and additional restrictions are necessary. For example, if circumstances permit, one may assume idealized isothermal (constant temperature) or adiabatic (zero heat transfer) processes to occur. For the former p/ρ is constant, whereas for the latter, $p(1/\rho)^\gamma$ (where γ is the ratio of the specific heat at constant pressure to that at constant volume) remains unchanged. It is clear that incompressible fluids are simpler to treat; therefore, where the approximation is not too unrealistic, incompressible flow is often assumed.

When the fluid velocity is zero, it is readily demonstrated that the basic equation of fluid statics is obtained. Let $\mathbf{f}$ be the gravitational force per unit volume $-\rho g\hat{\mathbf{k}}$. From Eq. (13.220),

$$-\rho g\hat{\mathbf{k}} - \frac{\partial p}{\partial x}\hat{\mathbf{i}} - \frac{\partial p}{\partial y}\hat{\mathbf{j}} - \frac{\partial p}{\partial z}\hat{\mathbf{k}} = 0$$

or

$$-\frac{\partial p}{\partial x} = -\frac{\partial p}{\partial y} = 0$$

and

$$-\rho g = \frac{\partial p}{\partial z}$$

The pressure is thus independent of x and y and we may write

$$\frac{dp}{dz} = -\rho g$$

which corresponds to Eq. (13.191).

13.26 BERNOULLI'S EQUATION

Under special conditions, Euler's equation of motion may be integrated. Consider the scalar product of Eq. (13.220) with $\mathbf{v}$:

$$\mathbf{f}\cdot\mathbf{v} - \nabla p\cdot\mathbf{v} = \rho\frac{d\mathbf{v}}{dt}\cdot\mathbf{v} \tag{13.222}$$

The first term, the dot product of force per unit volume and velocity, is the power per unit volume supplied by $\mathbf{f}$. The second term may be manipulated as follows:

$$-\nabla p \cdot \mathbf{v} = -\frac{\partial p}{\partial x}\frac{dx}{dt} - \frac{\partial p}{\partial y}\frac{dy}{dt} - \frac{\partial p}{\partial z}\frac{dz}{dt} = -\frac{dp}{dt} + \frac{\partial p}{\partial t}$$

Finally,

$$\rho\frac{d\mathbf{v}}{dt} \cdot \mathbf{v} = \rho\frac{d}{dt}\tfrac{1}{2}v^2 = \frac{d}{dt}\tfrac{1}{2}\rho v^2 - \tfrac{1}{2}v^2\frac{d\rho}{dt}$$

Collecting terms, we have

$$\mathbf{f}\cdot\mathbf{v} - \frac{dp}{dt} + \frac{\partial p}{\partial t} = \frac{d}{dt}\tfrac{1}{2}\rho v^2 - \tfrac{1}{2}v^2\frac{d\rho}{dt}$$

For steady incompressible flow, $\partial p/\partial t$ and $d\rho/dt$ are zero and, consequently,

$$\mathbf{f}\cdot\mathbf{v} - \frac{dp}{dt} - \frac{d}{dt}\tfrac{1}{2}\rho v^2 = 0 \tag{13.223}$$

Multiplying by dt and integrating,

$$\int (\mathbf{f}\cdot\mathbf{v})\,dt - p - \tfrac{1}{2}\rho v^2 = \text{constant} \tag{13.224}$$

The term $\int (\mathbf{f}\cdot\mathbf{v})\,dt = \int \mathbf{f}\cdot\mathbf{dr}$ represents the work done by the body force per unit volume. If $\mathbf{f}$ is derivable from a scalar potential, $\mathbf{f} = -\nabla U$, and Eq. (13.224) becomes

$$U + p + \tfrac{1}{2}\rho v^2 = \text{constant} \tag{13.225}$$

where U is the potential energy per unit volume. For a body force of the gravitational type, $U = \rho g z$ and we have

$$\rho g z + p + \tfrac{1}{2}\rho v^2 = \text{constant} \tag{13.226}$$

This expression, known as *Bernoulli's equation*,[12] applies to a fluid in steady motion. The pressure term represents the work per unit volume done by the fluid and $\tfrac{1}{2}\rho v^2$ is the kinetic energy per unit volume of the fluid. This equation is thus a statement of conservation of energy for an ideal fluid in steady motion.

Illustrative Example 13.6

Determine the flow velocity of an incompressible fluid emanating from a small hole (b) in the side of a container open to the atmosphere.

---

[12] Bernoulli, D., *Hydrodynamics*, Argentorati, 1738.

SOLUTION

If the hole size is very small compared with the diameter of the container, the velocity of the fluid at the top (a) may be neglected relative to the velocity at exit. The liquid flows into the atmosphere which has virtually the same pressure at a and b. Bernoulli's equation between these points is written

$$p_a + \tfrac{1}{2}\rho_a v_a^2 + \rho_a g z_a = p_b + \tfrac{1}{2}\rho_b v_b^2 + \rho_b g z_b$$

Since the fluid may be regarded as incompressible, $\rho_a = \rho_b$:

$$\frac{p_{\text{atm}}}{\rho} + 0 + g z_a = \frac{p_{\text{atm}}}{\rho} + \tfrac{1}{2}v_b^2 + g z_b$$

or

$$v_b = [2g(z_a - z_b)]^{\frac{1}{2}}$$

The expression is known as *Torricelli's theorem* governing the velocity of efflux from a small hole. We observe that this is exactly the speed acquired by an object falling freely (in the absence of frictional effects) through a vertical distance $z_a - z_b$. Since it appears as though the fluid at the top is disappearing, this rationale is readily justified.

13.27 SOUND WAVES IN A FLUID

To this point, we have been primarily concerned with the steady flow of incompressible fluids. As an illustration of compressible, unsteady flow, the propagation of sound waves in a fluid is now treated.

Previous sections of this chapter have demonstrated the transmissibility of longitudinal and transverse waves within a solid medium. For a fluid, we shall treat longitudinal waves, to the exclusion of transverse waves which occur at the interface of a fluid with its surroundings.

For a fluid at rest, the continuity equation is simply

$$\frac{\partial \rho_0}{\partial t} = 0 \tag{13.227}$$

where ρ_0 is the density prior to the initiation of any disturbance. Euler's equation of motion is

$$\mathbf{f} - \nabla p_0 = 0 \tag{13.228}$$

If a small disturbance not associated with $\mathbf{f}$ occurs, ρ_0, p_0, and the fluid velocity change from their static values to $\rho_0 + \delta\rho$, $p_0 + \delta p$, and $\delta\mathbf{v}$. The continuity equation and Euler's equation written subsequent to the disturbance are, respectively,

$$\frac{\partial}{\partial t}(\rho_0 + \delta\rho) + \nabla \cdot [(\rho_0 + \delta\rho)\,\delta\mathbf{v}] = 0 \tag{13.229}$$

and

$$\mathbf{f} - \nabla(p_0 + \delta p) = \frac{d}{dt}[(\rho_0 + \delta\rho)\,\delta\mathbf{v}] \tag{13.230}$$

To first order in small quantities, we have

$$\frac{\partial\rho_0}{\partial t} + \frac{\partial}{\partial t}\delta\rho + \nabla \cdot \rho_0\,\delta\mathbf{v} = 0 \tag{13.231}$$

and

$$\mathbf{f} - \nabla p_0 - \nabla(\delta p) = \rho_0 \frac{d}{dt}\delta\mathbf{v} = \rho_0 \frac{\partial}{\partial t}\delta\mathbf{v} \tag{13.232}$$

Substituting Eqs. (13.227) and (13.228),

$$\frac{\partial}{\partial t}\delta\rho + \nabla \cdot \rho_0\,\delta\mathbf{v} = 0 \tag{13.233}$$

$$-\nabla(\delta p) = \rho_0 \frac{\partial}{\partial t}\delta\mathbf{v} \tag{13.234}$$

Taking the time derivative of Eq. (13.233) and the divergence of Eq. (13.234),

$$\frac{\partial^2}{\partial t^2}\delta\rho + \frac{\partial}{\partial t}[\nabla \cdot (\rho_0\,\delta\mathbf{v})] = \frac{\partial^2}{\partial t^2}\delta\rho + \nabla \cdot \left(\rho_0 \frac{\partial}{\partial t}\delta\mathbf{v}\right) = 0 \tag{13.235}$$

and

$$(-\nabla \cdot \nabla)\,\delta p = \nabla \cdot \left(\rho_0 \frac{\partial}{\partial t}\delta\mathbf{v}\right) \tag{13.236}$$

where the order of partial differentiation has been changed, permitted because all physical quantities are assumed continuous. Combining the two equations above, we obtain

$$\frac{\partial^2}{\partial t^2}\delta\rho = (\nabla \cdot \nabla)\,\delta p \tag{13.237}$$

This partial differential equation relates the density change to the pressure change. For a liquid these are related by the bulk modulus

$$\delta p = k\frac{\delta\rho}{\rho_0} \tag{13.238}$$

Substitution into Eq. (13.237) yields

$$\frac{\partial^2}{\partial t^2}\delta\rho = \frac{k}{\rho_0}(\nabla \cdot \nabla)\,\delta\rho = \frac{k}{\rho_0}\left(\frac{\partial^2}{\partial x^2}\delta\rho + \frac{\partial^2}{\partial y^2}\delta\rho + \frac{\partial^2}{\partial z^2}\delta\rho\right) \tag{13.239}$$

which is a three-dimensional wave equation. The phase velocity, $\sqrt{k/\rho_0}$, should not be confused with $\delta \mathbf{v}$, the velocity of the fluid.

This equation is valid for liquids only, because small pressure changes in a liquid cause negligible temperature changes. For a gas the situation is somewhat different, because even small pressure changes cause moderate temperature variations. Rapid changes are closely approximated by adiabatic processes[13] for which $p(1/\rho)^\gamma$ remains constant. Using this expression, an adiabatic bulk modulus may be developed:

$$k' = \rho_0 \frac{dp}{d\rho} = \gamma k \qquad (13.240)$$

The phase velocity is thus increased to $\sqrt{\gamma k/\rho_0}$, which agrees with experiment.

To derive a differential equation for velocity disturbance, Eq. (13.240) and the gradient of Eq. (13.233) are substituted into the time derivative of Eq. (13.234):

$$\frac{\gamma k}{\rho_0} (\nabla \cdot \nabla) \, \delta \mathbf{v} = \frac{\partial^2}{\partial t^2} \, \delta \mathbf{v} \qquad (13.241)$$

This expression is a *vector* wave equation for the velocity disturbance. To interpret its characteristics we examine the pressure.

The direction of the maximum change of the pressure disturbance is given by the gradient of δp, Eq. (13.234), from which it can be ascertained that the acceleration is also in this direction. Since the pressure disturbance satisfies a wave equation [see Eqs. (13.238) and (13.239)], we may represent it by

$$\delta p = \delta p_0 e^{i(k_x x + k_y y + k_z z - \omega t)} \qquad (13.242)$$

where k_x, k_y, k_z are constants and ω is the natural circular frequency. Equation (13.242) may be written in terms of a single spatial variable in the direction of the gradient of δp. Hence the pressure gradient and acceleration are in the direction of propagation of the wave.

Integrating the acceleration and recalling that the undisturbed velocity is zero, the fluid velocity and phase velocity of the waves are readily demonstrated to be colinear. Thus the pressure wave (the propagation of a disturbance), which is called a sound wave, is longitudinal.

13.28 VISCOSITY AND VISCOUS FLOW

The foregoing analyses have neglected the effects of friction within the fluid and are, strictly speaking, valid for perfect (nonviscous) fluids only. Since there are no perfect fluids, of what value are the results of the theory? The answer lies in the great mathematical complexity resulting from the introduction of fluid friction into the equations of motion. It is usual to avoid the application of the more general

---

[13] Redding, J. L., *Am. J. Phys.* **34**, 626 (1966).

Navier–Stokes equations[14] in favor of the Euler equations when only rough answers are required or when other dynamical influences overshadow those of fluid friction.

Viscosity is easily demonstrated in an experiment in which a flat plate is pulled over a fluid at constant velocity v, as shown in Fig. 13.25. A force F is required to maintain a constant velocity indicating the existence of opposing forces within the fluid. Under ordinary circumstances, the fluid in contact with the moving plate adheres to it and consequently assumes its velocity. Similarly, the fluid contacting the stationary wall experiences zero velocity. There being zero relative velocity at both solid-fluid interfaces, one speaks of conditions of *zero slip* at these surfaces.

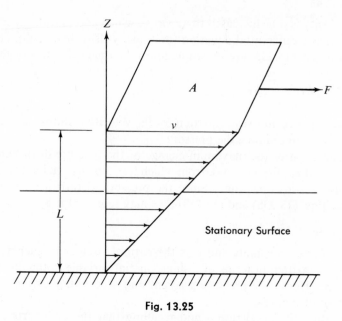

Fig. 13.25

The coefficient of viscosity may be defined by the expression

$$\eta = \frac{F/A}{v/L} \tag{13.243}$$

where F/A is recognized as a shearing stress. The gap L is necessarily small, because laminar (smooth, layered) flow must be assured. The quantity v/L, for an assumed linear velocity variation, is equal to the velocity gradient dv/dz. Making this substitution, we are led to Newton's law of viscosity:

$$\tau_{yz} = \eta \frac{\partial v}{\partial z} \tag{13.244}$$

where τ_{yz} represents a Y-directed shearing stress acting on a surface, a normal to

---

[14] Longwell, P.A., *Mechanics of Fluid Flow*, McGraw-Hill, Inc., New York, 1966.

which is parallel to the Z-axis. The fact that the velocity distributes itself as shown is a manifestation of the shearing forces acting on each of the layers which we may assume to comprise the fluid.

Fluids governed by expressions of the form of Eq. (13.244), in which the shear is proportional to the first power of the velocity gradient, are referred to as Newtonian. Non-Newtonian fluids include those for which the coefficient of viscosity is a function of time or of velocity gradient, even at constant temperature and pressure. For low pressures, the coefficient of viscosity is essentially a function of temperature only. The viscosity of liquids decreases with temperature, whereas the reverse is true of gases.

In principle, all the quantities of Eq. (13.243) may be measured and the viscosity determined. To minimize edge effects, the area of the parallel surfaces should be quite large, introducing the difficulties associated with the large fluid volumes and high forces required. This practical difficulty is avoided by employing an apparatus similar to that shown in Fig. 13.26, in which the cylinder and cylindrical container are coaxial. The volume between cylinders is filled with the liquid to be tested. The outer cylinder is rotated at constant angular velocity ω, the inner cylinder restrained from moving by a calibrated torsion spring or an electromagnetically supplied torque. Assuming the fluid Newtonian and the annular spacing quite small, the velocity gradient is simply the difference in fluid velocity at R_2 and R_1 divided by the gap width:

$$\frac{dv}{dR} = \frac{R_2\omega - 0}{R_2 - R_1} \tag{13.245}$$

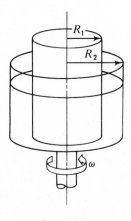

Fig. 13.26

Therefore,

$$\eta \frac{R_2\omega}{R_2 - R_1} = \frac{F}{A} \tag{13.246}$$

where A represents the cylindrical area in contact with the fluid, $2\pi R_1 L$. Note that the flat area of contact at the base of the cylinder has been neglected. When only the torque arising from shearing stresses acting on the vertical cylinder walls is considered,

$$M = R_1 F \tag{13.247}$$

and the coefficient of viscosity is determined by combining Eqs. (13.246) and (13.247):

$$\eta = \frac{(R_2 - R_1)M}{2\pi R_1^2 R_2 L \omega} \tag{13.248}$$

The quantity $R_2 - R_1$ must be small compared with either R_1 or R_2 if this method is to yield satisfactory results. If the gap is too large, the flow may not be laminar.

When the torque associated with the fluid at the bottom cannot be safely ignored, it is accounted for as follows. The force dF acting on a differential ring of the stationary cylinder base between R and $R + dR$ is related to the velocity gradient between the rotating cylinder and the ring by first noting that

$$\frac{dv}{dR} = \frac{\omega R}{a} \tag{13.249}$$

where a is the gap at the bottom of the apparatus. Thus

$$\eta \frac{dv}{dR} = \frac{dF}{dA} \tag{13.250}$$

or

$$dF = \eta \frac{dv}{dR} dA = \eta \frac{\omega R}{a} 2\pi R \, dR \tag{13.251}$$

The differential torque due to dF is

$$dM' = R \, dF = \eta \frac{\omega}{a} 2\pi R^3 \, dR \tag{13.252}$$

and consequently,

$$M' = \int dM' = \frac{2\pi\eta\omega}{a} \int_0^{R_1} R^3 \, dR = \frac{\pi\eta\omega}{2a} R_1^4 \tag{13.253}$$

The torque indicated by the sensing device is the sum of the two moments

$$M_T = M + M' = \frac{\eta 2\pi R_1^2 R_2 L \omega}{R_2 - R_1} + \frac{\pi\eta\omega R_1^4}{2a} \tag{13.254}$$

and therefore

$$\eta = \frac{2a(R_2 - R_2)M_T}{\omega\pi R_1^2(4aR_2L + R_1^2 R_2 - R_1^3)} \tag{13.255}$$

If the experiment is performed with two different liquid levels, the viscosity may be determined from the following expression in which the effect of the bottom surface cancels:

$$M_{T,1} - M_{T,2} = M_1 + M' - M_2 - M' = M_1 - M_2$$

$$= \frac{\eta 2\pi R_1^2 R_2 \omega (L_1 - L_2)}{R_2 - R_1} \tag{13.256}$$

The need for applying the mathematical expression for M' is thereby obviated.

To observe, in a simple analysis, the effect of viscosity, consider an element of a viscous fluid flowing at constant velocity in a circular pipe. Since the fluid does not accelerate, the forces associated with pressure and fluid friction cancel. Referring to Fig. 13.27 we have

$$p\pi r^2 - \left[p + \left(\frac{dp}{dx}\right) dx\right] \pi r^2 - \tau 2\pi r \, dx = 0 \tag{13.257}$$

where it has been assumed that the pressure varies with x only. From Eq. (13.257),

$$\frac{dp}{dx} = -\frac{2\tau}{r} \tag{13.258}$$

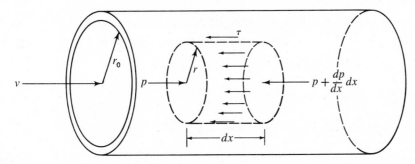

Fig. 13.27

For laminar flow of a Newtonian fluid, the shearing stress is related to the velocity gradient by the expression

$$\tau = -\eta \frac{dv}{dr} \tag{13.259}$$

where the minus sign indicates that the velocity decreases with increasing radius, $r = 0$ corresponding to maximum fluid velocity. Substituting Eq. (13.258) into Eq. (13.259), the pressure gradient becomes

$$\frac{dp}{dx} = \frac{2}{r} \eta \frac{dv}{dr} \tag{13.260}$$

Since dv/dr is inherently negative, v decreasing with r, Eq. (13.260) indicates that the pressure decreases with increasing x. Note that in Eq. (13.260), v is regarded as a function of r alone, indicating not only steady-state conditions but *fully developed* flow as well. The latter means that changes in velocity with x do not occur in that section of pipe under analysis, and consequently the velocity profile $v(r)$ is independent of x.

The difference in pressure between two axial locations of the pipe is determined by integrating Eq. (13.260) with respect to x:

$$\int_1^2 dp = \frac{2}{r} \eta \frac{dv}{dr} \int_1^2 dx \tag{13.261}$$

$$p_2 - p_1 = \frac{2}{r} \eta \frac{dv}{dr} (x_2 - x_1) \tag{13.262}$$

To determine the variation of velocity, we have only to integrate Eq. (13.262) with respect to the radius:

$$\int dv = \frac{p_2 - p_1}{2(x_2 x - {}_1)\eta} \int r \, dr \tag{13.263}$$

$$v = \frac{p_2 - p_1}{4(x_2 - x_1)\eta} r^2 + C_1 \tag{13.264}$$

The velocity distribution is thus parabolic; this is a characteristic profile for laminar flow. The constant C_1 is found by applying the condition $v = 0$ at $r = r_0$, yielding

$$C_1 = -\frac{(p_2 - p_1)r_0^2}{4(x_2 - x_1)\eta}$$

and therefore

$$v = \frac{p_2 - p_1}{4(x_2 - x_1)\eta} (r^2 - r_0^2) = \frac{p_1 - p_2}{4(x_2 - x_1)\eta} (r_0^2 - r^2)$$

$$= \frac{\Delta p}{4\eta L} (r_0^2 - r^2) \tag{13.265}$$

where the substitutions $\Delta p = p_1 - p_2$ and $L = x_2 - x_1$ have been made.

The maximum velocity clearly corresponds to $r = 0$ $(dv/dr = 0)$:

$$v_{max} = \frac{\Delta p r_0^2}{4\eta L} \tag{13.266a}$$

or

$$\Delta p = \frac{4\eta L v_{max}}{r_0^2} \tag{13.266b}$$

To determine the average velocity, consider the equality

$$\rho \bar{v} A = \rho \int v \, dA = \text{mass flow} \tag{13.267}$$

where $\bar{v}$ represents the velocity, averaged over the flow cross section A, and ρ the mass density, is regarded as a constant. Substituting $A = \pi r_0^2$ and $dA = 2\pi r\, dr$ together with Eq. (13.265) into the above expression, we have

$$\bar{v} = \frac{\Delta p}{2\eta L r_0^2} \int_0^{r_0} (r_0^2 - r^2) r\, dr \tag{13.268}$$

or

$$\bar{v} = \frac{\Delta p r_0^2}{8\eta L} \tag{13.269}$$

and therefore the pressure drop and average velocity are connected by the expression

$$\Delta p = \frac{8\eta L \bar{v}}{r_0^2} = \frac{32\eta L \bar{v}}{d_0^2} \tag{13.270}$$

When the flow is not as simple as that described by the foregoing development, as, for example, under turbulent flow conditions, it is usual to plot the experimental data employing the parameters

$$f = \frac{\Delta p}{(L/d_0)\bar{v}^2/2} = \text{friction factor}$$

and $\mathrm{Re} = \rho \bar{v} d_0/\eta = $ Reynolds number. Equation (13.270) may be rewritten as follows to introduce the above dimensionless variables:

$$\frac{\Delta p}{\rho(\bar{v}^2/2)(L/d_0)} = \frac{32\eta L \bar{v}}{d_0 d_0 \rho (\bar{v}^2/2)(L/d_0)} = 64\left[\frac{1}{\rho(\bar{v}_0 d/\eta)}\right] = \frac{64}{\mathrm{Re}} \tag{13.271}$$

or, in general, $f = $ function of Re. Equation (13.271) agrees well with experiment for $\mathrm{Re} < 2000$.

13.29 REYNOLDS NUMBER

A full and exact mathematical treatment of the flow of real fluids is virtually impossible. For one thing, the Navier–Stokes equations are nonlinear partial differential equations in several dependent variables. For another, the boundary conditions are often unknown or quite difficult to formulate. One approach to the solution of such complex equations is to discard as many terms of lesser importance as possible. This may be accomplished through order of magnitude calculations. The resulting solutions, necessarily approximate, are often modified by empirical factors to account for neglected effects.

In physical phenomena governed by many variables, it is often advantageous, particularly in fluid mechanics, to represent the differential equations, their solutions, and experimental data in terms of dimensionless groupings. These are exemplified by such ratios t/τ, ω/ω_0, b/b_{cr}, and x/L. In less obvious circumstances, the intuitive approach to the formation of these ratios may not succeed, and systematic methods of dimensional analysis[15] are employed to determine these groupings.

---

[15] Buckingham, E., *Phys. Rev.* **4**, 345 (1914); Bridgman, P. W., *Dimensional Analysis*, Yale University Press, New Haven, Conn., 1922.

Dimensional analysis also enables the determination of the maximum number of independent dimensionless parameters which may be formed for a given physical situation. Such methods require that all the influencing factors be known. Indeed, it is important to recognize that dimensional analysis per se does not shed light upon the physics but does serve to organize the variables so as to reduce the clutter of representation, particularly when there are experimental data to be analyzed.

There are dangers in dimensional analysis associated with an incomplete knowledge of the pertinent variables or the inclusion of more variables than are actually significant. Under such circumstances the dimensionless groupings determined may be inconsequential, incorrect, or misleading.

Whether the dimensionless groupings have been ascertained through manipulation of the differential equation or by formal dimensional analysis, experimental studies serve to verify and modify the analytical results, and in the absence of analysis the dimensionless variables provide the framework for coherently representing the data. The use of dimensionless parameters also permits the representation, on a single graph, for example, of the behavior of similar physical systems which may exhibit wide differences as to size and characteristic physical properties.

The work of O. Reynolds[16] serves as an historic example of the development of a most important dimensionless grouping named in his honor, to which we have already referred. In his experimental studies of the flow of liquids through glass tubes, Reynolds deduced that the ratio $\rho v d / \eta$ provides a key for determining whether flow is laminar or turbulent; for values of $\mathrm{Re} < 2000$, flow is laminar, whereas for $\mathrm{Re} > 4000$, turbulent flow occurs. A low value of Reynolds number is indicative of the predominance of viscous forces which tend to damp out noticeable fluctuations. High Reynolds numbers characterize flows in which viscous effects are insignificant. In the intermediate or transition region, factors such as surface roughness, pressure, and surface tension are important in establishing the nature of flow.

A rationale for the Reynolds number can be found by considering the ratio of the product of mass and acceleration to the forces due to viscosity:

$$\frac{\rho V | \mathbf{v} \nabla \cdot \mathbf{v}|}{A | \tau |} = \frac{\rho V | \mathbf{v} \nabla \cdot \mathbf{v}|}{A \eta | \nabla \mathbf{v}|}$$

where a general expression has been substituted for τ and where V represents a volume of fluid and A the cross-sectional area of flow. To the same order of magnitude, $|\nabla \mathbf{v}| = |\nabla \cdot \mathbf{v}|$, and the above ratio assumes the form of the Reynolds number:

$$\mathrm{Re} = \frac{\rho V v}{A \eta} = \frac{\rho v L}{\eta}$$

The characteristic length L has been here taken to be the ratio of volume to area. Other significant lengths are often used in the correlation of experimental data.

[16] Reynolds, O., *Phil. Trans. Roy. Soc. London* (1883); *Scientific Papers*, Cambridge University Press, New York, 1900–1903.

References

Becker, R. A., *Introduction to Theoretical Mechanics*, McGraw-Hill, Inc., New York, 1954.

Churchill, R. V., *Fourier Series and Boundary Value Problems*, McGraw-Hill, Inc., New York, 1952.

Cottrell, A. H., *The Mechanical Properties of Matter*, John Wiley & Sons, Inc., New York, 1964.

Den Hartog, J. P., *Advanced Strength of Materials*, McGraw-Hill, Inc., New York, 1952.

Halliday, D., and Resnick, R., *Physics*, Part 2, John Wiley & Sons, Inc., New York, 1962.

Janke, E., and Emde, F., *Tables of Functions*, Dover Publications, Inc., New York, 1943.

Johnson, D. E., and Johnson, J. R., *Mathematical Methods in Engineering and Physics*, The Ronald Press Company, New York, 1965.

Lamb, H., *Hydrodynamics*, Dover Publications, Inc., New York, 1945.

Li, W. H., and Lam, S. H., *Principles of Fluid Mechanics*, Addison-Wesley Publishing Company, Inc., Reading, Mass., 1964.

Longwell, P. A., *Mechanics of Fluid Flow*, McGraw-Hill, Inc., New York, 1966.

Margenau, H., and Murphy, G. M., *The Mathematics of Physics and Chemistry*, Second Edition, D. Van Nostrand Company, Inc., Princeton, N.J., 1956.

Matveyev, A. N., *Principles of Electrodynamics*, Translated by L. F. Landovitz, Reinhold Publishing Corporation, New York, 1966.

Nadeau, G., *Introduction to Elasticity*, Holt, Rinehart and Winston, Inc., New York, 1964.

Osgood, W. F., *Advanced Calculus*, The Macmillan Company, New York, 1925.

Pao, R. H. F., *Fluid Mechanics*, John Wiley & Sons, Inc., New York, 1961.

Rogers, G. L., *Mechanics of Solids*, John Wiley & Sons, Inc., New York, 1964.

Sokolnikoff, I. S., *Mathematical Theory of Elasticity*, McGraw-Hill, Inc., New York, 1956.

Symon, K. R., *Mechanics*, Second Edition, Addison-Wesley Publishing Company, Inc., Reading, Mass., 1960.

Timoshenko, S. P., and Young, D. H., *Theory of Structures*, McGraw-Hill, Inc., New York, 1945.

Von Kármán, T., and Biot, M. A., *Mathematical Methods in Engineering*, McGraw-Hill, Inc., New York, 1940.

Wang, C., *Applied Elasticity*, McGraw-Hill, Inc., New York, 1953.

Wangsness, R. K., *Introduction to Theoretical Physics*, John Wiley & Sons, Inc., New York, 1963.

EXERCISES

13.1 A steel bar of constant cross-sectional area 0.2 ft^2 and length 30 ft is suspended from a ceiling. Calculate the total change of length as a result of its weight as well as the stress as a function of vertical distance. $E = 30 \times 10^6$ lb/in.^{2}.

13.2 Given the stresses at a point,

$$\sigma_{xx} = 200, \qquad \sigma_{yy} = 100, \qquad \sigma_{zz} = 0$$
$$\sigma_{xy} = -50 \qquad \sigma_{xz} = -20 \qquad \sigma_{yz} = 40$$

determine the orientation of the principal axes relative to the original set as well as the principal stresses.

13.3 Demonstrate that the sum of the normal stresses at a point is an invariant, independent of orthogonal coordinate system.

13.4 A beam of mass m and length L is built-in at both ends. In addition to its own weight, the beam supports a load P at $x = L/3$. Determine the deflection as a function of x as well as the point of maximum deflection.

13.5 Determine the deflection curve for a simply supported beam subjected to the distributed loading $w(x) = A \sin (\pi x/L)$.

13.6 A uniform beam of length L is simply supported at its ends and midpoint. What is the deflection curve for the following cases in which the total downward load is W:

    a. A concentrated force acts at $x = L/4$.
    b. A uniformly distributed force acts in the region $3L/4 \leq x \leq L$.

13.7 A uniform cantilever beam of length $2L$ is simply supported at its midpoint. The beam weight is W and a concentrated load P acts at the free end. Determine the deflection curve and the ratio P/W for which the deflection at $L/4$ is a minimum.

13.8 Remove the load P in Exercise 13.7 and permit the beam to experience small transverse vibrations. Apply the Rayleigh method to determine the lowest natural frequency. What are the normal modes of vibration given an initial deflection $y(x, 0) = Cx^2(L^2 - x^2)$?

13.9 A uniform cable of length L and mass m is extended between the points $A(-H, -W)$ and $B(H, W)$. What is the shape of the cable?

13.10 Repeat Exercise 13.9 for a cable loaded at a constant rate w per unit horizontal distance.

13.11 Consider a string made up of two segments L_1 and L_2, of respective mass densities ρ_1 and ρ_2. If the ends of this compound string are attached to rigid walls and the tension in the first segment is τ, determine:

    a. the normal modes of vibration.
    b. the reflection properties at the interface between the string segments.

13.12 One end of a semiinfinite string of mass per unit length ρ_L and tension F is vibrated according to $y(0, t) = A \sin \omega t$. If the string is initially at rest and undeformed, determine its subsequent motion.

13.13 A string of length L is fixed at $x = L$. The end at $x = 0$ is vibrated according to $y(0, t) = A \sin \omega t$. If the string is initially undeformed and possesses no initial velocity, determine the motion.

13.14 A string, fixed at both ends, is initially distorted into the shape of a "vee" and released from rest. What are its normal modes of vibration?

13.15 Derive the equation of motion of a stretched string for circumstances under which its weight cannot be neglected.

13.16 A stretched membrane covers a circular drum. Derive the two-dimensional wave equation for the case of small vibration.

13.17 A cylindrical rod of length L and constant cross-sectional area is clamped at one end. The free end is twisted through an initial angle, small to assure elastic deformation, and released. Derive the differential equation of motion for small oscillation and the natural frequencies of oscillation.

13.18 The only force acting on a uniform cantilever beam is its weight. Apply the Rayleigh method to ascertain the lowest natural frequency. Solve the appropriate differential equation to determine all the natural frequencies.

13.19 Apply the Rayleigh method to a cable to determine the fundamental mode of oscillation.

13.20 A steel rod of rectangular cross section (3 by 5 in.), 40 ft in length, is subject to compressive loading. Calculate the buckling load as well as the maximum stress at failure.

13.21 A transmission line is often analyzed by assuming the current and voltage to vary as $e^{i\omega t}$, where ω represents the frequency of voltage impressed at $x = 0$. If at the end of the line, $x = L$, there is a resistor of R ohms, how do the voltage and current vary with x?

13.22 Determine the torque about point O when the water is filled to a height b as shown in Fig. P13.22. The width of the gate is W.

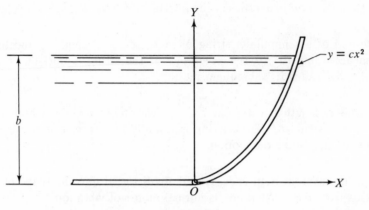

Fig. P13.22

13.23 A wood cylinder of length L and diameter D is pivoted about an end as shown in Fig. P13.23. If the specific gravity of the wood is 0.8 and the fluid in which it is immersed is water, what is the value at equilibrium of the angle α? Determine the period of small oscillation.

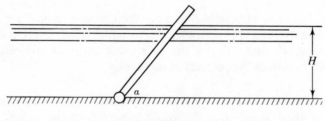

Fig. P13.23

13.24 Write Euler's equation and the equation of continuity in integral form and interpret each term. Apply Gauss' divergence theorem where appropriate.

13.25 A right circular cone of height H, half-angle α, and mass density ρ_c floats in a liquid of density ρ_w. Demonstrate that the cone is stable only if the vertex is pointed vertically upward. Determine the submerged volume of the cone and the frequency of the system for small oscillation.

13.26 Repeat Exercise 13.25 for an ellipsoid of revolution.

13.27 A sealed cubical container, filled half with water, half with air, has a small hole drilled in its base. Determine the velocity of efflux of the water as a function of water level. Assume the water incompressible, possessing zero vapor pressure. Assume also that the entire process is isothermal. Compare the results with those obtained for an open container.

13.28 An ideal incompressible fluid fills a container of constant cross-sectional area A_1 and height H. How much time is required for half the fluid to drain through a small hole of area A_2 located at the bottom.

13.29 Demonstrate that if an ideal incompressible fluid experiences irrotational flow, the velocity can be derived from a scalar potential satisfying Laplace's equation,

$$\nabla^2 \phi = 0$$

where

$$\mathbf{v} = -\nabla \phi$$

and

$$\nabla^2 = \nabla \cdot \nabla = \frac{\partial^2}{\partial x^2} + \frac{\partial^2}{\partial y^2} + \frac{\partial^2}{\partial z^2}$$

13.30 Write Laplace's equation in cylindrical polar and spherical polar coordinates.

13.31 Rewrite Euler's equations of motion in cylindrical polar coordinates.

13.32 Rewrite Euler's equations of motion in spherical polar coordinates.

13.33 The force due to fluid viscosity may be written

$$\mathbf{F} = \int_A \eta \nabla \mathbf{v} \cdot d\mathbf{A}$$

Determine the components of the force per unit volume exerted on a differential cube.

13.34 If $v_x = x^3 y z^2 t^3$ and $v_y = 3x^2 y z^2 t^3$, determine v_z for incompressible flow.

13.35 A cylindrical container partially filled with an incompressible fluid is rotated at constant velocity ω, the fluid rotating with the container and assuming the surface shown in Fig. P13.35. Derive the equations of the fluid surface and the pressure distribution within the fluid.

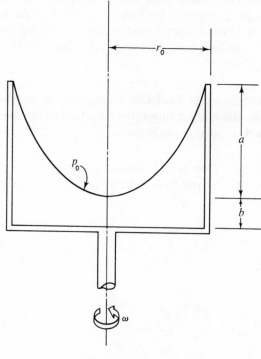

Fig. P13.35

13.36 The velocity distribution for an incompressible fluid in turbulent flow through a pipe of circular cross section is given by

$$v = v_0 \left(1 - \frac{r}{r_0}\right)^{1/7}$$

where v_0 represents the velocity at the center line and r_0 is the radius. Determine the volumetric flow rate.

13.37 The Cartesian components of fluid *vorticity* are given by

$$\zeta_x = \frac{\partial v_z}{\partial y} - \frac{\partial v_y}{\partial z}$$

$$\zeta_y = \frac{\partial v_x}{\partial z} - \frac{\partial v_z}{\partial x}$$

$$\zeta_z = \frac{\partial v_y}{\partial x} - \frac{\partial v_x}{\partial y}$$

a. Demonstrate that only if $\zeta = 0$ is the flow irrotational.

b. Determine the components of ζ in cylindrical polar and spherical polar coordinates.

c. Determine ζ for a velocity field given by

$$\mathbf{v} = ax^2\hat{\mathbf{i}} - ay^2\hat{\mathbf{j}} + z\hat{\mathbf{k}}$$

13.38 Verify Eq. (13.241).

13.39 Air is blown across the top of a pipe of circular cross section (radius r_0, length L) open at both ends. Write the differential equation for the sound waves induced. Use cylindrical polar coordinates. Determine the normal modes along the axis of the pipe employing appropriate conditions. How are the results altered if one end is closed?

CHAPTER 14

VARIATIONAL METHODS

14.1 INTRODUCTION

In this chapter, Newton's laws of motion are extended through the use of variational methods. We are thus led to new formulations such as the equations of Lagrange and Hamilton. The unique insights gained from such approaches offer reason enough for their study, but there are more compelling justifications. For one, the theme central to these methods is the energy concept, which, because of its scalar nature, obviates the need for more complex vector equations, particularly cumbersome in complicated, multicomponent systems. In addition, the methods introduced provide a more encompassing approach to the treatment of constraints. Many simple cases involving constraints have been dealt with without much formality, and we are, at this point, without a general procedure for analyzing constrained dynamical systems. Finally, variational methods may be extended, for conservative systems, through a development not pursued in this text, to arrive at momenta, all of which are constant. The integrations are thus quite routine.

14.2 CONSTRAINTS

A mechanics problem can, in principle, be solved if all the forces acting are known as functions of time and position, and if there are sufficient conditions to specify the initial position and velocity of each particle. Often, however, some of the forces are given only indirectly as constraints of the motion, such as in the case of a mass sliding on a plane or a particle confined within a volume. A constraint thus represents any restriction upon the motion of a particle or system of particles. The forces associated with the constraint are just sufficient to fulfill the conditions of constraint; that is, reaction cannot exceed action. The complete solution of a mechanics problem includes, then, the determination of both the constraint forces and the motion.

There are a wide variety of ways in which constraints may be imposed upon a system, and the following discussion serves to conveniently classify them. When a constraint can be represented by a functional relationship between the coordinates and time, it is termed *holonomic*. The constraint equation need not, however, contain every coordinate or time. A particle moving on a horizontal plane is subject to the holonomic, time-independent constraint $z = $ constant. If such a functional relationship does not exist, the constraint is *nonholonomic*. A relationship involving derivatives of coordinates in addition to the coordinates themselves is nonholonomic. A particle trapped inside a closed rectangular parallelepiped of sides a, b, c, for example, is subject to the nonholonomic constraint: $0 \le x \le a$, $0 \le y \le b$, $0 \le z \le c$.

Constraints that are explicit functions of time are *rheonomic* as, for instance, a bead constrained to move along a straight wire, rotating about an end, as shown in Fig. 14.1. This holonomic, rheonomic constraint is expressed by

$$y = x \tan \omega t \qquad\qquad (14.1)$$

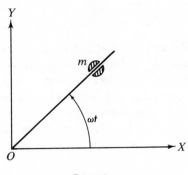

Fig. 14.1

If the volume of the parallelepiped changes with time, we have rheonomic, but nonholonomic constraints upon an internal particle, as described by $0 \le x \le a(t)$, $0 \le y \le b(t)$, $0 \le z \le c(t)$. Finally, *scleronomic constraints* are those which are independent of time, for example, the mass moving on a plane.

14.3 INDEPENDENT COORDINATES

How do constraints influence the number of independent coordinates required to describe the motion of a system? As discussed in Chapter 12, there are a maximum of $3n$ independent coordinates or degrees of freedom associated with an n-particle system. Each holonomic constraint reduces the number of independent coordinates or degrees of freedom by 1, provided, of course, that the constraints are independent. Thus, if there are h constraint equations, there are $3n - h$ independent coordinates. Certain nonholonomic constraints, the subject of a later discussion, also serve to reduce the number of independent coordinates.

14.4 VIRTUAL WORK

Heretofore zero net force has been the criterion for the equilibrium of a particle. The principle of virtual work, now explored, offers an alternative by which the equilibrium of a particle may be expressed. This approach is not only extended to the statics of a system of particles, but, upon introduction of D'Alembert's principle, to the dynamics of multiparticle systems as well.

An unconstrained particle is in equilibrium when

$$\sum_{i=1}^{n} \mathbf{F}_i = \mathbf{F} = 0 \qquad (14.2)$$

where $\mathbf{F}$ represents the resultant of all forces. Now imagine the particle to undergo an arbitrary small finite displacement $\delta\mathbf{r}$. Since this is a fictiticious displacement, it is usually described as *virtual*, as distinct from $\mathbf{dr}$, which is *infinitesimal* but actual. The latter may approach zero, whereas a virtual displacement must always be finite.

Associated with the virtual displacement is the virtual work described by

$$\delta W = \mathbf{F} \cdot \delta\mathbf{r} \qquad (14.3)$$

The net force is zero and it follows that the virtual work is also zero. Note that because the virtual displacement is different from zero and not, in general, perpendicular to $\mathbf{F}$, it cannot, of itself, result in zero virtual work. Therefore, zero virtual work represents both a necessary and sufficient condition for equilibrium of an unconstrained particle. In this instance $\delta W = 0$ implies $\mathbf{F} = \mathbf{0}$.

Consider now a particle subject to constraints. The forces acting may be divided into an applied resultant $\mathbf{F}$ and a resultant associated with the conditions of constraint, $\mathbf{R}$ (Fig. 14.2).

For a holonomic, scleronomic constraint, it is clear that a virtual displacement can no longer be arbitrary lest the constraint be violated. We must therefore redefine the virtual displacement as one that is arbitrary but consistent with the constraints. Associated with each constraint is a constraint force, which we take to be normal to the surface of contact. On this basis, only frictionless constraints are permitted, or alternatively frictional forces must be regarded as applied or external, rather than as reactions.

For equilibrium, including applied and constraint forces,

$$\mathbf{F} + \mathbf{R} = 0 \qquad (14.4)$$

The work associated with a virtual displacement is, in this instance,

$$\delta W = (\mathbf{F} + \mathbf{R}) \cdot \delta\mathbf{r} \qquad (14.5)$$

Since $\mathbf{F} + \mathbf{R} = 0$, $\delta W = 0$ is a necessary condition for equilibrium. Because the constraint force and virtual displacement are perpendicular, $\mathbf{R} \cdot \delta\mathbf{r} = 0$, and therefore the virtual work associated with only the applied forces must likewise be zero:

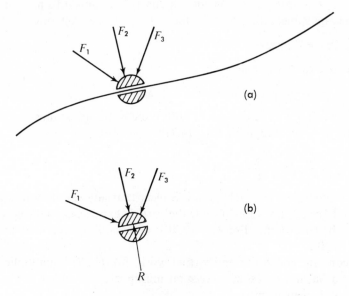

Fig. 14.2

$$\delta W = (\mathbf{F} + \mathbf{R}) \cdot \delta \mathbf{r} = \mathbf{F} \cdot \delta \mathbf{r} = 0 \qquad (14.6)$$

Again recalling that δr is finite, zero virtual work, associated with applied forces only, is both necessary and sufficient to assure equilibrium.

An approach is now developed enabling the principle of virtual work to be related to the equations of constraint, thus embodying the constraints in the equations of motion. For simplicity, assume a single constraint equation of the form $\phi(x, y, z, t) = 0$. As the particle experiences a virtual displacement at *fixed* time t, its coordinates change from x, y, z to $x + \delta x, y + \delta y, z + \delta z$. The constraint equation is now $\phi(x + \delta x, y + \delta y, z + \delta z, t) = 0$, and the corresponding change of ϕ is

$$\delta\phi = \phi(x + \delta x, y + \delta y, z + \delta z, t) - \phi(x, y, z, t)$$

To first order in virtual displacement, $\delta\phi$ may be written

$$\delta\phi = \left[\phi(x, y, z, t) + \frac{\partial\phi}{\partial x}\,\delta x + \frac{\partial\phi}{\partial y}\,\delta y + \frac{\partial\phi}{\partial z}\,\delta z \right] - \phi(x, y, z, t)$$

$$= \frac{\partial\phi}{\partial x}\,\delta x + \frac{\partial\phi}{\partial y}\,\delta y + \frac{\partial\phi}{\partial z}\,\delta z \qquad (14.7)$$

where ϕ has been expanded about the original position in a three-dimensional Taylor series. Equation (14.7) must equal zero on the grounds that the virtual displacement cannot alter the constraint. In other words, subsequent to the virtual displacement, the constraint equation still applies.

To illustrate this point, consider the particle confined to motion on a horizontal plane and subject, therefore, to the holonomic, scleronomic constraint

$$\phi(x, y, z) = 0 = z - C$$

Subsequent to a virtual displacement,

$$\phi(x + \delta x, y + \delta y, z + \delta z) = 0 = z + \delta z - C$$

If the displacement is to be consistent with the constraint, δz must equal zero and consequently $\delta\phi = 0$. Applying Eq. (14.7) for $\delta\phi = 0$,

$$\frac{\partial\phi}{\partial x}\,\delta x + \frac{\partial\phi}{\partial y}\,\delta y + \frac{\partial\phi}{\partial z}\,\delta z = 0$$

Since $\partial\phi/\partial x$ and $\partial\phi/\partial y$ are each zero, it is clear that any variation in x and y may occur (δx and δy need not vanish) while the particle experiences a displacement consistent with the constraint. For $\delta\phi = 0$, it is necessary however that $\delta z = 0$, since $\partial\phi/\partial z = 1\,(\neq 0)$.

To combine constraint and virtual work, Eq. (14.7) is multiplied by λ, an undetermined but not necessarily constant multiplier, and added to Eq. (14.6) expressed in component form:

$$F_x\,\delta x + F_y\,\delta y + F_z\,\delta z + \lambda\frac{\partial\phi}{\partial x}\,\delta x + \lambda\frac{\partial\phi}{\partial y}\,\delta y + \lambda\frac{\partial\phi}{\partial z}\,\delta z = 0 \qquad (14.8a)$$

or

$$\left(F_x + \lambda\frac{\partial\phi}{\partial x}\right)\delta x + \left(F_y + \lambda\frac{\partial\phi}{\partial y}\right)\delta y + \left(F_z + \lambda\frac{\partial\phi}{\partial z}\right)\delta z = 0 \qquad (14.8b)$$

In the absence of constraint, each component of $\boldsymbol{\delta r}$ is independent of any others. Each constraint reduces the number of independent components of $\boldsymbol{\delta r}$ by 1. For the case of a single constraint, only δy and δz, for example, can be treated as independent. The undetermined coefficient λ (there would be one for each constraint equation) is nevertheless arbitrary. Therefore, by appropriate choice of λ, δy, and δz, each term in Eq. (14.8b) may be equated to zero, independently of the others.

To begin with, let us select λ so that

$$\lambda = -\frac{F_x}{\partial\phi/\partial x} \qquad (14.9a)$$

This has the effect of making the first term of Eq. (14.8b) zero regardless of δx. Consider a virtual displacement $\delta y \neq 0$, $\delta z \neq 0$, executed in two phases, as follows: $\delta y \neq 0$, $\delta z = 0$ and $\delta y = 0$, $\delta z \neq 0$. When Eq. (14.9a) is substituted into Eq. (14.8b), together with $\delta y \neq 0$, $\delta z = 0$, we have

$$F_y + \lambda\frac{\partial\phi}{\partial y} = 0 \qquad (14.9b)$$

Similarly, Eq. (14.8b) yields (for $\delta y = 0$, $\delta z \neq 0$)

$$F_z + \lambda \frac{\partial \phi}{\partial z} = 0 \qquad\qquad\qquad \text{(14.9c)}$$

Since the virtual work is zero, any point for which Eqs. (14.9) hold must be a point of equilibrium. It is thus clear that the terms $\lambda(\partial \phi / \partial x)$, and so on, are components of the reaction force, which may be obtained directly from $\lambda(\partial \phi / \partial x)$, and so on, upon determination of λ.

Illustrative Example 14.1

A particle of mass m is constrained to move on the surface of a sphere of radius R_0 by means of its attachment to a rigid massless rod as shown in Fig. 14.3. Determine the position of the particle when in equilibrium.

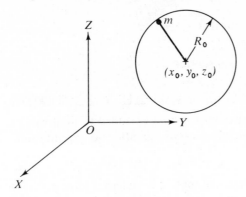

Fig. 14.3

SOLUTION
The constraint equation is

$$\phi(x, y, z, t) = (x - x_0)^2 + (y - y_0)^2 + (z - z_0)^2 - R_0^2 = 0$$

The external force, which is due only to gravity, is simply

$$\mathbf{F} = 0\hat{\mathbf{i}} + 0\hat{\mathbf{j}} - mg\hat{\mathbf{k}}$$

Applying Eq. (14.9a),

$$F_x = -\lambda \frac{\partial \phi}{\partial x} = -\lambda 2(x - x_0)$$

Since λ is arbitrary and not necessarily zero, $x - x_0 = 0$. Now applying Eq. (14.9b),

$$F_y + \lambda \frac{\partial \phi}{\partial y} = 0 + \lambda 2(y - y_0) = 0$$

and therefore $y - y_0 = 0$. Finally, Eq. (14.9c) yields

$$F_z + \lambda \frac{\partial \phi}{\partial z} = -mg + \lambda 2(z - z_0) = 0$$

The equation of constraint together with $x = x_0$ and $y = y_0$, prescribes the value of z: $z = z_0 \pm R_0$. Solving for λ,

$$\lambda = \pm \frac{mg}{2R_0}$$

Thus there are *two* points of equilibrium, one at the top, the other at the bottom of the sphere: $x = x_0, y = y_0, z = z_0 + R_0$; $x = x_0, y = y_0, z = z_0 - R_0$.

The reaction forces are now readily determined:

$$R_x = \lambda \frac{\partial \phi}{\partial x} \bigg|_{x=x_0, y=y_0, z=z_0 \pm R_0} = 0$$

$$R_y = \lambda \frac{\partial \phi}{\partial y} \bigg|_{x=x_0, y=y_0, z=z_0 \pm R_0} = 0$$

$$R_z = \lambda \frac{\partial \phi}{\partial z} \bigg|_{x=x_0, y=y_0, z=z_0 \pm R_0} = mg$$

14.5 VIRTUAL WORK OF n PARTICLES IN EQUILIBRIUM

The principle of virtual work is now extended to multiparticle systems. For a system of particles, there are n vector equations of equilibrium,

$$\mathbf{F}_i + \mathbf{R}_i = \mathbf{0} \qquad (i = 1, 2, \cdots, n) \tag{14.10}$$

where $\mathbf{F}_i$ represents the net applied or external force acting on the ith particle and $\mathbf{R}_i$ is the reaction force associated with system constraints. We now depart from our practice of denoting the force components in terms of coordinate direction x, y, z and select instead $x_1, x_2, \cdots, x_{3n}$, where $3n$ denotes the total number of coordinates required, the degrees of freedom. The coordinates thus selected impart no a priori distinction to one coordinate relative to another. In component form, we have therefore a total of $3n$ scalar equations of equilibrium:

$$F_j + R_j = 0 \qquad (j = 1, 2, \cdots, 3n) \tag{14.11}$$

The virtual work associated with a virtual displacement is

$$\delta W = \sum_{j=1}^{3n} F_j \, \delta x_j + \sum_{j=1}^{3n} R_j \, \delta x_j = 0 \tag{14.12}$$

Since the virtual displacement may not violate the constraints,

$$\sum_{j=1}^{3n} R_j \, \delta x_j = 0$$

and therefore

$$\sum_{j=1}^{3n} F_j \, \delta x_j = 0 \tag{14.13}$$

Once again equilibrium is assured if the work associated with a virtual displacement is zero. As in treating the single particle, this represents both a necessary and sufficient condition.

Assume that h holonomic constraint equations relate the various coordinates of the system:

$$\phi_k(x_1, x_2, \cdots, x_{3n}, t) = 0 \qquad (k = 1, 2, \cdots, h) \qquad \text{(14.14)}$$

Because no constraint may be violated, the variation of the kth constraint equation resulting from a virtual displacement is zero,

$$\delta\phi_k = \sum_{j=1}^{3n} \frac{\partial\phi_k}{\partial x_j} \delta x_j = 0 \qquad \text{(14.15)}$$

Multiplying each of the h constraint equations of the form of Eq. (14.15) by λ_k, an undetermined, nonzero coefficient, and adding to Eq. (14.13) we have

$$\sum_{j=1}^{3n} \left(F_j + \sum_{k=1}^{h} \lambda_k \frac{\partial\phi_k}{\partial x_j} \right) \delta x_j = 0 \qquad \text{(14.16)}$$

By appropriate choice of the terms λ_k and δx_j, each of the $3n$ terms in parentheses in Eq. (14.16) may be equated to zero independently of the remaining terms. The statics of multiparticle systems thus proceeds as with a single particle. The term $\lambda_k(\partial\phi_k/\partial x_j)$ represents the reaction force exerted by the kth constraint in the jth direction.

Illustrative Example 14.2

Consider blocks m_1 and m_2 shown in Fig. 14.4 subject to the following idealizations: The pulley is frictionless, the surfaces of the blocks and the inclines are perfectly smooth and frictionless, and the connecting string is massless and inextensible of length L_0.

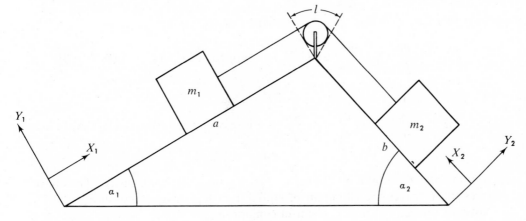

Fig. 14.4

Apply the principle of virtual work to develop a relationship describing the conditions of equilibrium for this two-particle system.

SOLUTION

In terms of the coordinate system shown, the constraint equations are

$$\phi_1 = y_1 = 0$$

$$\phi_2 = y_2 = 0$$

$$\phi_3 = (a - x_1) + (b - x_2) - (L_0 - l) = 0$$

where l represents the remaining length of string. In addition to the constraints described above, there are others implied by geometry, $x_1 < a$ and $x_2 < b$. These restrictions give assurance that the constraints remain holonomic for all possible system configurations.

The applied forces are

$$F_{x1} = -m_1g \sin \alpha_1$$

$$F_{x2} = -m_2g \sin \alpha_2$$

$$F_{y1} = -m_1g \cos \alpha_1$$

$$F_{y2} = -m_2g \cos \alpha_2$$

Applying Eq. (14.16) for $j = 4$ and $h = 3$:

$$F_{x1} + \lambda_1 \frac{\partial \phi_1}{\partial x_1} + \lambda_2 \frac{\partial \phi_2}{\partial x_1} + \lambda_3 \frac{\partial \phi_3}{\partial x_1} = -m_1g \sin \alpha_1 + \lambda_1(0) + \lambda_2(0) + \lambda_3(-1)$$

$$= 0$$

$$F_{x2} + \lambda_1 \frac{\partial \phi_1}{\partial x_2} + \lambda_2 \frac{\partial \phi_2}{\partial x_2} + \lambda_3 \frac{\partial \phi_3}{\partial x_2} = -m_2g \sin \alpha_2 + \lambda_1(0) + \lambda_2(0) + \lambda_3(-1)$$

$$= 0$$

$$F_{y1} + \lambda_1 \frac{\partial \phi_1}{\partial y_1} + \lambda_2 \frac{\partial \phi_2}{\partial y_1} + \lambda_3 \frac{\partial \phi_3}{\partial y_1} = -m_1g \cos \alpha_1 + \lambda_1(1) + \lambda_2(0) + \lambda_3(0)$$

$$= 0$$

$$F_{y2} + \lambda_1 \frac{\partial \phi_1}{\partial y_2} + \lambda_2 \frac{\partial \phi_2}{\partial y_2} + \lambda_3 \frac{\partial \phi_3}{\partial y_2} = -m_2g \cos \alpha_2 + \lambda_1(0) + \lambda_2(1) + \lambda_3(0)$$

$$= 0$$

From the first two equations above, $\lambda_3 = -m_1g \sin \alpha_1 = -m_2g \sin \alpha_2$. The above expression provides the condition of equilibrium $m_1g \sin \alpha_1 = m_2g \sin \alpha_2$, and all consistent values of x_1 and x_2 result in equilibrium.

Recall that $\lambda_k(\partial \phi_k/\partial x_j)$ represents a reaction force, and therefore $\lambda_3(\partial \phi_3/\partial x_1)$

$= \lambda_3(-1) = \lambda_3(\partial\phi_3/\partial x_2)$ is the constraint force acting parallel to the inclines through the string.

The Y-reaction forces given by

$$\lambda_1 \frac{\partial\phi_1}{\partial y_1} \quad \text{and} \quad \lambda_2 \frac{\partial\phi_2}{\partial y_2}$$

are λ_1 and λ_2 (corresponding to m_1 and m_2, respectively). From the third and fourth equations, $\lambda_1 = m_1 g \cos\alpha_1 = Y_1$-reaction and $\lambda_2 = m_2 g \cos\alpha_2 = Y_2$-reaction, as expected.

Although the technique described in the foregoing development is more elaborate and lengthy than the methods of Chapter 11, it has the virtue of prescribing the points at which the system is in equilibrium, and, as is later demonstrated, may be generalized to encompass dynamical situations. In addition, the constraint forces are included in a tractable manner.

14.6 VIRTUAL WORK AND NONHOLONOMIC CONSTRAINTS

Certain types of nonholonomic constraints are compatible with the principle of virtual work as applied to the statics of multiparticle systems. Consider, in this regard, constraints given in the form

$$d'\phi_{k'} = a_{1k'}\, dx_1 + a_{2k'}\, dx_2 + \cdots + a_{3nk'}\, dx_{3n} + b_{k'}\, dt$$

$$= \sum_{j=1}^{3n} a_{jk'}\, dx_j + b_{k'}\, dt = 0 \qquad (k' = 1, 2, \cdots, h') \tag{14.17}$$

where $a_{1k'}$, $b_{k'}$ and so on, are functions of the coordinates and time.

The constraints $d'\phi_{k'}$ are so written because they are presumably not total differentials and therefore not integrable. That is, they do not indicate integrability according to the test

$$\frac{\partial a_{ik'}}{\partial x_j} = \frac{\partial a_{jk'}}{\partial x_i} \qquad \frac{\partial a_{ik'}}{\partial t} = \frac{\partial b_{k'}}{\partial x_i}$$

for all i and j. Hence the apparent differential $d'\phi_{k'}$ is not a total differential, although it relates differentials.

Consider the change in constraint associated with a virtual displacement

$$\delta\phi_{k'} = a_{1k'}\, \delta x_1 + a_{2k'}\, \delta x_2 + \cdots + a_{3nk'}\, \delta x_{3n} \tag{14.18}$$

Since the displacement occurs at constant time, a term involving δt is absent. Furthermore, $\delta\phi_{k'} = 0$ because the virtual displacement must not violate the constraints. Again multiplying each of the h' constraint equations by an undetermined coefficient $\lambda_{k'}$ and adding to Eq. (14.13), we obtain

$$\sum_{j=1}^{3n} \left(F_j + \sum_{k'=1}^{h'} \lambda_{k'} a_{jk'} \right) \delta x_j = 0 \tag{14.19}$$

The method of solution now proceeds as with holonomic constraints. Of course, if both holonomic and nonholonomic constraints of the form of Eq. (14.17) are present, Eqs. (14.16) and (14.19) may be combined:

$$\sum_{j=1}^{3n} \left(F_j + \sum_{k=1}^{h} \lambda_k \frac{\partial \phi_k}{\partial x_j} + \sum_{k'=1}^{h'} \lambda_{k'} a_{jk'} \right) \delta x_j = 0 \tag{14.20}$$

Illustrative Example 14.3

Referring to Fig. 14.5, determine the constraint equations governing the pure rolling of a cylinder on a flat surface.

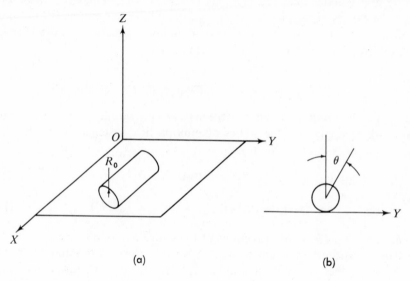

(a)

(b)

Fig. 14.5

SOLUTION

Since no slip occurs, the cylinder is limited to straight-line motion normal to its longitudinal axis. Selecting this axis as parallel to the X-axis, as shown in Fig. 14.5(a), it is clear that the X- and Z-coordinates remain constant for all positions of the cylinder. The cylinder is therefore subject to the constraint equations $x =$ constant and $z =$ constant.

The angular displacement θ of the cylinder is measured relative to a radial vertical reference line drawn through the cylinder. Because the cylinder rolls without slipping, dy and $d\theta$ are related by

$$dy = R_0 \, d\theta$$

or

$$dy - R_0 \, d\theta = 0$$

The above equation is clearly nonholonomic as written, although, for the *particular* case under study, it may be integrated to yield

$$y - R_0\theta = \text{constant}$$

which, since it provides a functional relationship between coordinates of the system, represents a holonomic constraint.

Illustrative Example 14.4

Now consider a disk of radius R_0 and infinitesimal thickness which rolls without slipping on a horizontal surface as the plane of the disk remains vertical, Fig. 14.6. Determine the equations of constraint.

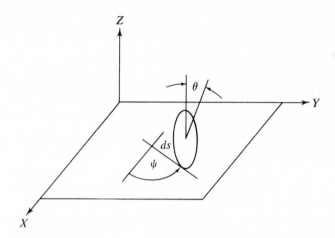

Fig. 14.6

SOLUTION

Because the disk is thin, it is free to move anywhere in the XY-plane. We thus retain one constraint of Illustrative Example 14.3, $z = \text{constant}$, discarding $x = \text{constant}$. When the disk travels a distance ds as it rolls without slipping, the angle θ changes so that

$$ds = R_0\, d\theta$$

where

$$ds = dx \cos\psi + dy \sin\psi$$

and ψ is the angle which the plane of the disk makes with the X-axis, as shown. The disk is therefore subject to the nonholonomic constraint

$$dx \cos\psi + dy \sin\psi = R_0\, d\theta$$

Since no motion occurs normal to the plane of the disk, we have the additional constraint

$$-dx \sin \psi + dy \cos \psi = 0$$

These constraint equations do not prescribe the position and orientation of the disk as was the case in Illustrative Example 14.3. The disk can be made to move in the XY-plane in a circle of any desired radius. When it returns to its original position the angle θ will have changed, whereas x and y will not. In going from one point in the plane to another, the change of θ is determined by the length of the path joining the points. Since an infinite number of paths join two points, the change of θ is not specified by the constraint equations. The constraints are thus nonholonomic in that they relate differentials but are not integrable without additional information.

Although the constraint equations thus derived are nonholonomic, they nevertheless lend themselves to the techniques discussed, because the four variables x, y, θ, and ψ are related by equations of the form of Eq. (14.17).

14.7 D'ALEMBERT'S PRINCIPLE [1]

Newton's second law for an unconstrained particle, $\mathbf{F} = m\ddot{\mathbf{r}}$, written

$$\mathbf{F} - m\ddot{\mathbf{r}} = 0 \tag{14.21}$$

offers an interesting and often useful approach to the solution of dynamics problems. If the term $-m\ddot{\mathbf{r}}$ is regarded as an applied force, equal and opposite to $\mathbf{F}$, the actual net applied force, Eq. (14.21) becomes, in effect, an equation of statics. The concept of thus transforming from dynamical to pseudostatical situations is known as *D'Alembert's principle*. Although the force $-m\ddot{\mathbf{r}}$ may not be known explicitly, we may nevertheless proceed with a further development of the equations of virtual work as though it were.

The virtual work associated with a virtual displacement $\delta\mathbf{r}$, is, for the system of Eq. (14.21),

$$\delta W = (\mathbf{F} - m\ddot{\mathbf{r}}) \cdot \delta\mathbf{r} = 0 \tag{14.22a}$$

or

$$(F_x - m\ddot{x})\,\delta x + (F_y - m\ddot{y})\,\delta y + (F_z - m\ddot{z})\,\delta z = 0 \tag{14.22b}$$

Since the virtual displacement is arbitrary in magnitude and direction, the terms in parentheses in Eq. (14.22b) must vanish independently of one another.

For a constrained particle, the D'Alembert equation is

$$\mathbf{F} + \mathbf{R} - m\ddot{\mathbf{r}} = 0 \tag{14.23}$$

Once again, the virtual displacement cannot violate the constraints. In order for the work done by the reaction forces in dynamics to equal zero, the constraints must be time-independent. Subject to this condition,

$$\mathbf{R} \cdot \delta\mathbf{r} = 0$$

[1] D'Alembert, J., *Traité de dynamique*, Paris, 1843.

The virtual work is therefore

$$\delta W = (\mathbf{F} - m\ddot{\mathbf{r}}) \cdot \delta \mathbf{r} = 0 \qquad (14.24a)$$

or

$$(F_x - m\ddot{x}) \, \delta x + (F_y - m\ddot{y}) \, \delta y + (F_z - m\ddot{z}) \, \delta z = 0 \qquad (14.24b)$$

Given a single time-independent equation of constraint,

$$\phi(x, y, z) = 0 \qquad (14.25)$$

the variation associated with a virtual displacement is

$$\delta \phi = \frac{\partial \phi}{\partial x} \delta x + \frac{\partial \phi}{\partial y} \delta y + \frac{\partial \phi}{\partial z} \delta z \qquad (14.26)$$

Adding $\lambda \, \delta \phi$ to Eq. (14.24b) and associating terms, we have

$$\left(F_x - m\ddot{x} + \lambda \frac{\partial \phi}{\partial x} \right) \delta x + \left(F_y - m\ddot{y} + \lambda \frac{\partial \phi}{\partial y} \right) \delta y$$

$$+ \left(F_z - m\ddot{z} + \lambda \frac{\partial \phi}{\partial z} \right) \delta z = 0 \qquad (14.27)$$

where λ is an undetermined coefficient. As previously noted, appropriate choice of λ, and two of δx, δy, δz, consistent with the constraint, permits the setting to zero of each term in parentheses independently. For example, let

$$\lambda = - \frac{F_x - mx}{\partial \phi / \partial x} \qquad (14.28)$$

and perform the virtual displacement $\delta y = 0$ and $\delta z \neq 0$. This leads to

$$F_z - m\ddot{z} + \lambda \frac{\partial \phi}{\partial z} = 0 \qquad (14.29a)$$

Now perform another virtual displacement, $\delta z = 0$, $\delta y \neq 0$ to yield

$$F_y - m\ddot{y} + \lambda \frac{\partial \phi}{\partial y} = 0 \qquad (14.29b)$$

The constraint equation, together with Eqs. (14.28) and (14.29), may be solved for $\ddot{x}$, $\ddot{y}$, $\ddot{z}$, which are then integrated as functions of time.

Illustrative Example 14.5

A particle moves on the outer surface of a hoop of radius R_0 as shown in Fig. 14.7. Determine its position as a function of time. At what point does the particle leave the hoop?

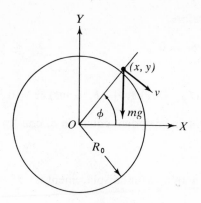

Fig. 14.7

SOLUTION

The constraint equations are

$$\phi(x, y, z) = x^2 + y^2 - R_0^2 = 0$$

and

$$z = 0$$

Then

$$d\phi = 2x\,dx + 2y\,dy = 0$$

and

$$dz = 0$$

The only applied force is that of gravity, $-mg\hat{\mathbf{j}}$. Applying Eq. (14.28),

$$\lambda = -\frac{0 - mx}{2x}$$

From (Eq. 14.29b),

$$-mg - m\ddot{y} + \lambda 2y = 0$$

To solve this pair of equations, multiply the first by $\dot{x}$ and the second by $\dot{y}$, and add to obtain

$$-m\dot{x}\ddot{x} - m\dot{y}\ddot{y} - mg\dot{y} + 2\lambda\dot{x}x + 2\lambda\dot{y}y = 0$$

The first two terms are simply

$$-\frac{d}{dt}\left[\tfrac{1}{2}m(\dot{x}^2 + \dot{y}^2)\right]$$

The last two terms are zero, as verified by taking the time derivative of the constraint equation

$$\frac{d\phi}{dt} = 2x\dot{x} + 2yy = 0$$

What remains is

$$-\frac{d}{dt}\left[\tfrac{1}{2}m(\dot{x}^2 + \dot{y}^2)\right] - mg\dot{y} = 0$$

Since $v^2 = \dot{x}^2 + \dot{y}^2$, the integral of the above expression is

$$-\tfrac{1}{2}mv^2 - mgy = C$$

which is, of course, an energy equation. If the particle starts from rest at the top of the hoop, $\dot{x}(0) = \dot{y}(0) = 0$ and $x(0) = 0$, $y(0) = R_0$. Substitution yields $C = -mgR_0$ and the above equation is written

$$\tfrac{1}{2}mv^2 = mg(R_0 - y)$$

The particle leaves the hoop when the constraint force vanishes. The direction of this force is, for all positions of the mass, perpendicular to the velocity (the direction of any virtual displacement), and therefore colinear with the radius vector running from the origin to the particle. Referring to Fig. 14.7 and introducing polar coordinates,

$$x = R_0 \cos\phi \qquad y = R_0 \sin\phi$$

$$v^2 = \dot{x}^2 + \dot{y}^2 = R_0^2\dot{\phi}^2 \qquad \tfrac{1}{2}mv^2 = \tfrac{1}{2}mR_0^2\dot{\phi}^2 = mg(R_0 - y) = mgR_0(1 - \sin\phi)$$

The constraint force (R-directed) is

$$\left(\lambda\frac{\partial\phi}{\partial x}\right)\cos\phi + \left(\lambda\frac{\partial\phi}{\partial y}\right)\sin\phi = N = m\ddot{x}\cos\phi + m\ddot{y}\sin\phi + mg\sin\phi$$

or

$$N = -\frac{mv^2}{R_0} + mg\sin\phi$$

Since we are concerned with the point at which contact is lost, N is equated to zero, yielding $v^2 = gR_0 \sin\phi$. When this is substituted into the energy equation, we obtain

$$\tfrac{1}{2}mgR_0 \sin\phi = mgR_0(1 - \sin\phi) \qquad \text{or} \qquad \sin\phi = \tfrac{2}{3}$$

This value of ϕ corresponds to

$$y = \tfrac{2}{3}R_0 \qquad \text{and} \qquad x = \sqrt{\tfrac{5}{9}}\,R_0$$

Illustrative Example 14.6
A particle subject to zero external force is constrained to move on the curve $\phi(x, y, z) = 0$. Determine the motion.

SOLUTION

Applying the method of undetermined coefficients, the equations of motion are

$$-m\ddot{x} + \lambda\frac{\partial\phi}{\partial x} = 0$$

$$-m\ddot{y} + \lambda\frac{\partial\phi}{\partial y} = 0$$

$$-m\ddot{z} + \lambda\frac{\partial\phi}{\partial z} = 0$$

Multiplying the first equation by $\dot{x}$, the second by $\dot{y}$, and the third by $\dot{z}$, and adding, we have

$$-\frac{d}{dt}\left[\tfrac{1}{2}m(\dot{x}^2 + \dot{y}^2 + \dot{z}^2)\right] + \lambda\left(\frac{\partial\phi}{\partial x}\dot{x} + \frac{\partial\phi}{\partial y}\dot{y} + \frac{\partial\phi}{\partial z}\dot{z}\right) = 0$$

From the equation of constraint,

$$\frac{d\phi}{dt} = \frac{\partial\phi}{\partial x}\dot{x} + \frac{\partial\phi}{\partial y}\dot{y} + \frac{\partial\phi}{\partial z}\dot{z} = 0$$

and the equation of motion is therefore

$$\frac{d}{dt}\left[\tfrac{1}{2}m(\dot{x}^2 + \dot{y}^2 + \dot{z}^2)\right] = \frac{dT}{dt} = 0$$

where T represents the kinetic energy of the particle. On the basis of this expression, the mass moves along the curve specified by the constraint equation, at the constant speed $v = (\dot{x}^2 + \dot{y}^2 + \dot{z}^2)^{1/2}$.

The acceleration in the X-direction is

$$\ddot{x} = \frac{d}{dt}\dot{x} = \frac{d}{dt}\frac{dx}{ds}\frac{ds}{dt} = \frac{d}{dt}\left(v\frac{dx}{ds}\right) = \frac{dx}{ds}\frac{dv}{dt} + v\frac{d}{dt}\frac{dx}{ds}$$

$$= \frac{dv}{dt}\frac{dx}{ds} + v^2\frac{d}{ds}\frac{dx}{ds} = \frac{dv}{dt}\frac{dx}{ds} + v^2\frac{d^2x}{ds^2}$$

Since v is constant,

$$\ddot{x} = v^2\frac{d^2x}{ds^2}$$

Similarly,

$$\ddot{y} = v^2\frac{d^2y}{ds^2}$$

$$\ddot{z} = v^2\frac{d^2z}{ds^2}$$

Expressing the acceleration vector in terms of components normal and tangent to the path,

$$\mathbf{a} = \dot{v}\hat{\mathbf{e}}_t + \frac{v^2}{\rho}\hat{\mathbf{e}}_n$$

where $\hat{\mathbf{e}}_t$ is a unit vector tangent to the surface of constraint and $\hat{\mathbf{e}}_n$ is a unit vector perpendicular to the surface. Since there is no tangential acceleration ($\dot{v} = 0$),

$$\mathbf{a} = \frac{v^2}{\rho}\hat{\mathbf{e}}_n$$

Thus v^2/ρ must equal the magnitude of the acceleration,

$$v^2\left[\left(\frac{d^2x}{ds^2}\right)^2 + \left(\frac{d^2y}{ds^2}\right)^2 + \left(\frac{d^2z}{ds^2}\right)^2\right]^{\frac{1}{2}}$$

Therefore, the reciprocal of the radius of curvature is given by

$$\frac{1}{\rho} = \left[\left(\frac{d^2x}{ds^2}\right)^2 + \left(\frac{d^2y}{ds^2}\right)^2 + \left(\frac{d^2z}{ds^2}\right)^2\right]^{\frac{1}{2}}$$

14.8 LAGRANGE'S EQUATIONS OF THE FIRST KIND [2]

In a system of n particles subject to h scleronomic constraints, the work associated with virtual displacements consistent with these constraints is zero. Following a procedure similar to that pertaining to the statics of a holonomic multiparticle system (or, equivalently, extending the foregoing derivation applicable to a single particle), we obtain

$$\sum_{j=1}^{3n}\left(F_j - m_j\ddot{x}_j + \sum_{k=1}^{h}\lambda_k\frac{\partial\phi_k}{\partial x_j}\right)\delta x_j = 0 \qquad (14.30a)$$

For a nonholonomic system,

$$\sum_{j=1}^{3n}\left(F_j - m_j\ddot{x}_j + \sum_{k'=1}^{h'}\lambda_{k'}a_{jk'}\right)\delta x_j = 0 \qquad (14.30b)$$

By appropriate choice of the δx_j and the undetermined coefficients, each term of the summation over j in Eqs. (14.30) is zero independently of the others. Thus, corresponding to Eqs. (14.30a) and (14.30b), we have

$$F_j - m_j x_j + \sum_{k=1}^{h}\lambda_k\frac{\partial\phi_k}{\partial x_j} = 0 \qquad (j = 1, 2, \cdots, 3n) \qquad (14.31a)$$

$$F_j - m_j x_j + \sum_{k'=1}^{h'}\lambda_{k'}a_{jk'} = 0 \qquad (j = 1, 2, \cdots, 3n) \qquad (14.31b)$$

---

[2] Lagrange, J. L., *Misc. Taurin* **II** (1760).

Equations (14.31), written in Cartesian coordinates, are known as *Lagrange's equations of the first kind*, and are applied in the ensuing illustrative example.

Illustrative Example 14.7
Analyze the motion of the two-mass double-incline system shown in Fig. 14.4.

SOLUTION
Referring to Illustrative Example 14.2, the equations of constraint are

$$\phi_1 = y_1 = 0$$

$$\phi_2 = y_2 = 0$$

$$\phi_3 = (a - x_1) + (b - x_2) - (L_0 - l) = 0$$

The external forces are

$$F_{x1} = -m_1 g \sin \alpha_1 \qquad F_{y1} = -m_1 g \cos \alpha_1$$

$$F_{x2} = -m_2 g \sin \alpha_2 \qquad F_{y2} = -m_2 g \cos \alpha_2$$

Applying Eq. (14.31a), we obtain the following set of equations:

$$F_{x1} - m_1 \ddot{x}_1 + \lambda_1 \frac{\partial \phi_1}{\partial x_1} + \lambda_2 \frac{\partial \phi_2}{\partial x_1} + \lambda_3 \frac{\partial \phi_3}{\partial x_1}$$

$$= -m_1 g \sin \alpha_1 - m_1 \ddot{x}_1 + \lambda_1(0) + \lambda_2(0) + \lambda_3(-1) = 0$$

$$F_{y1} - m_1 \ddot{y}_1 + \lambda_1 \frac{\partial \phi_1}{\partial y_1} + \lambda_2 \frac{\partial \phi_2}{\partial y_1} + \lambda_3 \frac{\partial \phi_3}{\partial y_1}$$

$$= -m_1 g \cos \alpha_1 - m_1 \ddot{y}_1 + \lambda_1(1) + \lambda_2(0) + \lambda_3(0) = 0$$

$$F_{x2} - m_2 \ddot{x}_2 + \lambda_1 \frac{\partial \phi_1}{\partial x_2} + \lambda_2 \frac{\partial \phi_2}{\partial x_2} + \lambda_3 \frac{\partial \phi_3}{\partial x_2}$$

$$= -m_2 g \sin \alpha_2 - m_2 x_2 + \lambda_1(0) + \lambda_2(0) + \lambda_3(-1) = 0$$

$$F_{y2} - m_2 \ddot{y}_2 + \lambda_1 \frac{\partial \phi_1}{\partial y_2} + \lambda_2 \frac{\partial \phi_2}{\partial y_2} + \lambda_3 \frac{\partial \phi_3}{\partial y_2}$$

$$= -m_2 g \cos \alpha_2 - m_2 \ddot{y}_2 + \lambda_1(0) + \lambda_2(1) + \lambda_3(0) = 0$$

The above expressions reduce to

$$-m_1 g \sin \alpha_1 - m_1 \ddot{x}_1 - \lambda_3 = 0$$

$$-m_1 g \cos \alpha_1 - m_1 \ddot{y}_1 + \lambda_1 = 0$$

$$-m_2 g \sin \alpha_2 - m_2 \ddot{x}_2 - \lambda_3 = 0$$

$$-m_2 g \cos \alpha_2 - m_2 \ddot{y}_2 + \lambda_2 = 0$$

In addition to the above set, the third constraint equation when differentiated twice provides the information $\ddot{x}_1 = -\ddot{x}_2$, which, upon substitution into the above set of equations, yields

$$x_1 = g\frac{m_2 \sin \alpha_2 - m_1 \sin \alpha_1}{m_1 + m_2} = -\ddot{x}_2 = \text{constant} = A$$

The constraint (tension) associated with the rope is

$$\lambda_3\frac{\partial \phi_3}{\partial x_1} = \lambda_3 \qquad \frac{\partial \phi_3}{\partial x_2} = -\lambda_3$$

Substituting for $\ddot{x}_1$ in the first equation of motion, the rope tension is found to be

$$-\lambda_3 = m_1 g \sin \alpha_1 + m_1 \ddot{x}_1 = m_1 g \sin \alpha_1 + m_1 g\frac{m_2 \sin \alpha_2 - m_1 \sin \alpha_1}{m_1 + m_2}$$

$$= \frac{m_1 m_2 g(\sin \alpha_1 + \sin \alpha_2)}{m_1 + m_2}$$

The Y-constraint or reaction forces are given by $\lambda_1(\partial \phi_1 / \partial y_1)$ and $\lambda_2(\partial \phi_2 / \partial y_2)$ which equal, respectively, λ_1 and λ_2:

$$\lambda_1 = m_1 g \cos \alpha_1 + m_1 \ddot{y}_1 = m_1 g \cos \alpha_1 \qquad (\text{since } y_1 = 0)$$

$$\lambda_2 = m_2 g \cos \alpha_2 + m_2 \ddot{y}_2 = m_2 g \cos \alpha_2 \qquad (\text{since } y_2 = 0)$$

It is of interest to point out that if the value $\pi/2$ is assigned α_1 and α_2, the problem reduces to an Atwood's machine with the acceleration and string tension given by

$$a = \frac{g(m_2 - m_1)}{m_1 + m_2}$$

$$\tau = \frac{2m_1 m_2}{m_1 + m_2} g$$

verifying the derivation in Chapter 11.

14.9 RHEONOMIC CONSTRAINTS

The foregoing development of Lagrange's equations of the first kind has been restricted to scleronomic constraints. A procedure is now introduced whereby rheonomic constraints imposed upon a single particle or multiparticle system may be incorporated into Lagrange's equations.

Consider a single particle subject to the time-dependent constraint $\phi(x, y, x, t) = 0$. The equation of motion is

$$\mathbf{F} + \mathbf{R} - m\ddot{\mathbf{r}} = 0 \qquad\qquad (14.32)$$

where **F** represents the sum of the applied forces and **R** the reactions associated with ϕ. The virtual work performed by the reactions need no longer be zero, even if the virtual displacement is consistent with the constraint. This is because, with the passage of time, a change in direction of a constraint force may occur, resulting in a possible displacement parallel to this force.

Even if this contribution to the virtual work were zero, the variation in ϕ,

$$\delta\phi = \frac{\partial\phi}{\partial x}\,\delta x + \frac{\partial\phi}{\partial y}\,\delta y + \frac{\partial\phi}{\partial z}\,\delta z + \frac{\partial\phi}{\partial t}\,\delta t$$

(where δt is the change in time accompanying the virtual displacement) cannot be included, as before, into the equation of virtual work because $(\partial\phi/\partial t)\,\delta t \neq 0$. Only by requiring the virtual displacement to occur at fixed time can rheonomic constraints be incorporated. For this case, since $\delta t = 0$,

$$\delta\phi = \frac{\partial\phi}{\partial x}\,\delta x + \frac{\partial\phi}{\partial y}\,\delta y + \frac{\partial\phi}{\partial z}\,\delta z = \boldsymbol{\nabla}\phi\cdot\boldsymbol{\delta r} \tag{14.33}$$

and therefore the term $\lambda\boldsymbol{\nabla}\phi$ can be made to represent the reaction force. We may thus construct the equation of virtual work involving a rheonomic constraint in a manner similar to the case of a scleronomic constraint:

$$\delta W = (\mathbf{F} + \lambda\boldsymbol{\nabla}\phi - m\ddot{\mathbf{r}})\cdot\boldsymbol{\delta r}$$

$$= \left(F_x + \lambda\frac{\partial\phi}{\partial x} - m\ddot{x}\right)\delta x + \left(F_y + \lambda\frac{\partial\phi}{\partial y} - m\ddot{y}\right)\delta y$$

$$+ \left(F_z + \lambda\frac{\partial\phi}{\partial z} - m\ddot{z}\right)\delta z = 0 \tag{14.34}$$

The virtual work done by the constraint force is now nonzero. Nevertheless, each term of Eq. (14.34) can be made to vanish independently. The formalism is different but the equations are the same as those applicable to the scleronomic case.

For a system of n particles and h constraints, the virtual work is

$$\sum_{j=1}^{3n}\left(F_j + \sum_{k=1}^{h}\lambda_k\frac{\partial\phi_k}{\partial x_j} - m\ddot{x}_j\right)\delta x_j = 0 \tag{14.35}$$

Each term in parentheses in the above expression may be made to vanish independently of the others. The terms $\lambda_k(\partial\phi_k/\partial x_j)$ represent the force exerted by the kth constraint, in the direction of x_j. It is no longer required that this force be workless.

An extension of the above method to constraints of the form of Eq. (14.17) is straightforward and is left to the reader.

Illustrative Example 14.8

Determine the motion of a bead subject to zero applied force, constrained to move on a straight wire rotating at constant angular velocity about an axis at one end, the axis being perpendicular to the wire.

SOLUTION

The constraint equation is

$$\phi(x, y, t) = y - x \tan \omega t = 0$$

where ω is the angular velocity of the wire, Fig. 14.1. Applying Eq. (13.34) we have

$$0 - mx + \lambda \frac{\partial \phi}{\partial x} = -m\ddot{x} - \lambda \tan \omega t = 0$$

$$0 - m\ddot{y} + \lambda \frac{\partial \phi}{\partial y} = -m\ddot{y} + \lambda = 0$$

Eliminating λ,

$$m\ddot{x} + m\ddot{y} \tan \omega t = 0$$

or

$$m\ddot{x} \cos \omega t + m\ddot{y} \sin \omega t = 0$$

Introducing polar coordinates, $x = R \cos \omega t$ and $y = R \sin \omega t$:

$$\dot{x} = \dot{R} \cos \omega t - R\omega \sin \omega t$$

$$\dot{y} = \dot{R} \sin \omega t + R\omega \cos \omega t$$

$$\ddot{x} = \ddot{R} \cos \omega t - R\omega^2 \cos \omega t - 2\dot{R}\omega \sin \omega t$$

$$\ddot{y} = \ddot{R} \sin \omega t - R\omega^2 \sin \omega t + 2\dot{R}\omega \cos \omega t$$

Substitution for the acceleration components in the equation of motion yields

$$m(\ddot{R} \cos \omega t - R\omega^2 \cos \omega t - 2\dot{R}\omega \sin \omega t) \cos \omega t$$

$$+ m(\ddot{R} \sin \omega t - R\omega^2 \sin \omega t + 2\dot{R}\omega \cos \omega t) \sin \omega t = 0$$

or

$$\ddot{R} - \omega^2 R = 0$$

The general solution of the above equation is

$$R = A \cosh \omega t + B \sinh \omega t$$

If at $t = 0$, $R = R_0$ and $\dot{R} = 0$, $A = R_0$ and $B = 0$. Therefore,

$$R = R_0 \cosh \omega t$$

and

$$x = R_0 \cosh \omega t \cos \omega t$$

$$y = R_0 \cosh \omega t \sin \omega t$$

$$\lambda = m\ddot{y} = 2mR_0\omega^2 \sinh \omega t \cos \omega t$$

The X- and Y-components of the constraint force N may now be determined:

$$N_x = \lambda \frac{\partial \phi}{\partial x} = 2mR_0\omega^2 \sinh \omega t \cos \omega t (-\tan \omega t)$$

$$= -2mR_0\omega^2 \sinh \omega t \sin \omega t$$

$$N_y = \lambda \frac{\partial \phi}{\partial y} = 2mR_0\omega^2 \sinh \omega t \cos \omega t (1)$$

and, consequently,

$$N = (N_x^2 + N_y^2)^{\frac{1}{2}} = 2mR_0\omega^2 \sinh \omega t$$

The reaction is in the direction of increasing ωt, normal to R.

Because the constraint is time-dependent, it is no longer workless, the bead energy increasing as a result of the work done by the reaction or constraint force.

14.10 GENERALIZED COORDINATES

Until now, the concepts of virtual work have been formulated in Cartesian coordinates exclusively, and any use of other coordinate systems has been accomplished through subsequent transformation, as in the previous illustrative example. Although the Cartesian system is most familiar, other coordinates, such as polar, cylindrical, and spherical, are often more convenient. Even so, there is nothing sacrosanct about any coordinate system. We are therefore led to seek a more universal approach to the matter of coordinates, one relying less on intuition or recollections of problems past.

Let us define our task as the selection of *3n generalized coordinates* for a system of n masses. By generalized we mean any *dimensionally consistent* set of mixed dynamical variables such as length, angle, sum or difference of two Cartesian coordinates, linear or angular momentum, and energy.

Equations permitting the transfer from generalized to Cartesian coordinates are of the form

$$x_j = x_j(q_1, q_2, \cdots, q_{3n}, t) \tag{14.36}$$

Here $q_1, q_2, \cdots, q_{3n}$ are the $3n$ generalized coordinates and t represents time.

The inverse equations are

$$q_l = q_l(x_1, x_2, \cdots, x_{3n}, t) \tag{14.37}$$

To assure the existence of an inverse set, the *Jacobian* of the transformation must be different from zero everywhere or nearly everywhere. The Jacobian is defined by the determinant

$$\begin{vmatrix} \dfrac{\partial q_1}{\partial x_1} & \cdots & \dfrac{\partial q_{3n}}{\partial x_1} \\[2ex] \vdots & & \vdots \\[2ex] \dfrac{\partial q_1}{\partial x_{3n}} & \cdots & \dfrac{\partial q_{3n}}{\partial x_{3n}} \end{vmatrix}$$

When the Jacobian of the transformation is zero at a point, it usually means that the number of coordinates overspecifies the configuration at that point. For example, consider the transformation involving Cartesian and polar coordinates:

$$x = R \cos \phi \qquad\qquad y = R \sin \phi$$

The inverse equations are

$$R = (x^2 + y^2)^{1/2} \qquad\qquad \phi = \tan^{-1}\left(\frac{y}{x}\right)$$

A particle at the origin has, of course, Cartesian coordinates, $x = 0$, $y = 0$. In polar coordinates, its position is described by $R = 0$ while the angle ϕ is arbitrary. Only the value of R is thus required at the origin. As the reader may verify, the Jacobian for this transformation is zero only at the origin. The inverse transformation therefore exists everywhere except at the origin.

14.11 VIRTUAL WORK IN GENERALIZED COORDINATES

Consider a virtual displacement involving the Cartesian variable x_j. In terms of generalized coordinates, we have, from Eq. (14.36),

$$\delta x_j = \sum_{l=1}^{3n} \frac{\partial x_j}{\partial q_l} \delta q_l \tag{14.38}$$

where δq_l represents the virtual displacement of the lth generalized coordinate. The virtual work associated with virtual displacement δx_j is

$$\delta W = \sum_{j=1}^{3n} F_j\, \delta x_j = \sum_{j=1}^{3n} F_j \sum_{l=1}^{3n} \frac{\partial x_j}{\partial q_l} \delta q_l \tag{14.39}$$

It will prove convenient to define the *generalized force* as follows:

$$Q_l = \sum_{j=1}^{3n} F_j \frac{\partial x_j}{\partial q_l} \tag{14.40}$$

In this manner the virtual work is represented as the product of the generalized force and the generalized virtual displacement:

$$\delta W = \sum_{l=1}^{3n} Q_l\, \delta q_l \tag{14.41}$$

Although the dimensions of generalized force are not necessarily those of conventional force, the product given by Eq. (14.41) does display dimensions of work.

For conservative systems, the potential energy may be expressed in terms of the generalized coordinates

$$V = V(q_1, q_2, \cdots, q_{3n}) \tag{14.42}$$

and therefore the generalized force derivable from this potential *should* be

$$Q_l = -\frac{\partial V}{\partial q_l} \tag{14.43}$$

To demonstrate that the foregoing is indeed valid, consider the change of potential resulting from a virtual displacement δq_l:

$$\delta V = \sum_{l=1}^{3n} \frac{\partial V}{\partial q_l} \delta q_l \tag{14.44}$$

For a conservative system $\delta W = -\delta V$, and therefore

$$\sum_{l=1}^{3n} Q_l \, \delta q_l = -\sum_{l=1}^{3n} \frac{\partial V}{\partial q_l} \delta q_l \tag{14.45}$$

Equation (14.43) is verified by equating like terms of Eq. (14.45).

When the virtual work associated with the virtual displacement of a conservative system is zero, δV is also zero. The potential is then an extremum, and the system is in a state of equilibrium.

14.12 LAGRANGE'S EQUATIONS OF THE SECOND KIND

Let us again consider the virtual work of a system of n unconstrained particles acted upon by external forces. In Cartesian coordinates,

$$\delta W = \sum_{j=1}^{3n} (F_j - m_j \ddot{x}_j) \, \delta x_j = 0 \tag{14.46}$$

It is now our task to represent the virtual work in terms of the generalized coordinates, anticipating an expression of the form

$$\delta W = \sum_{l=1}^{3n} (Q_l - m_l \ddot{q}_l) \, \delta q_l = 0$$

where m_l is a generalized mass associated with the m_j's. We must determine the virtual work carefully, however, applying the transformation equations [Eqs. (14.36)], not trusting to intuition.

Applying Eqs. (14.36) and (14.38), the virtual work becomes

$$\delta W = \sum_{j=1}^{3n} F_j \sum_{l=1}^{3n} \frac{\partial x_j}{\partial q_l} \delta q_l - \sum_{j=1}^{3n} m_j \ddot{x}_j \sum_{l=1}^{3n} \frac{\partial x_j}{\partial q_l} \delta q_l = 0 \tag{14.47}$$

From Eq. (14.40),

$$\delta W = \sum_{l=1}^{3n} Q_l \, \delta q_l - \sum_{j=1}^{3n} m_j \ddot{x}_j \sum_{l=1}^{3n} \frac{\partial x_j}{\partial q_l} \delta q_l = 0 \tag{14.48}$$

To express $\ddot{x}_j$ in terms of the generalized coordinates, we begin by writing the time derivative of one of the Cartesian coordinates:

$$\dot{x}_j = \frac{dx_j}{dt} = \sum_{l=1}^{3n} \frac{\partial x_j}{\partial q_l} \dot{q}_l + \frac{\partial x_j}{\partial t} \tag{14.49}$$

Therefore,

$$\frac{\partial \dot{x}_j}{\partial \dot{q}_l} = \frac{\partial x_j}{\partial q_l} \tag{14.50}$$

since x_j is a function of q_l, not $\dot{q}_l$. The foregoing expression is an example of the so-called "cancellation of the dots." Substituting this result in Eq. (14.48),

$$\delta W = \sum_{l=1}^{3n} Q_l \, \delta q_l - \sum_{l=1}^{3n} \sum_{j=1}^{3n} m_j \ddot{x}_j \frac{\partial \dot{x}_j}{\partial \dot{q}_l} \delta q_l \tag{14.51}$$

The summations have been interchanged in accordance with the commutative property of addition. Now consider the identity

$$\ddot{x}_j \frac{\partial \dot{x}_j}{\partial \dot{q}_l} = \frac{d}{dt}\left(\dot{x}_j \frac{\partial \dot{x}_j}{\partial \dot{q}_l} \right) - \dot{x}_j \frac{d}{dt} \frac{\partial \dot{x}_j}{\partial \dot{q}_l}$$

or, in view of Eq. (14.50),

$$\ddot{x}_j \frac{\partial \dot{x}_j}{\partial \dot{q}_l} = \frac{d}{dt}\left(x_j \frac{\partial \dot{x}_j}{\partial \dot{q}_l} \right) - \dot{x}_j \frac{d}{dt} \frac{\partial x_j}{\partial q_l} \tag{14.52a}$$

$$\ddot{x}_j \frac{\partial \dot{x}_j}{\partial \dot{q}_l} = \frac{d}{dt}\left(\dot{x}_j \frac{\partial \dot{x}_j}{\partial \dot{q}_l} \right) - \dot{x}_j \frac{\partial \dot{x}_j}{\partial \dot{q}_l} \tag{14.52b}$$

The order of the partial and total derivative have been interchanged in the last step; its validity is demonstrated below. Proceeding from Eq. (14.49),

$$\dot{x}_j = \sum_{k=1}^{3n} \frac{\partial x_j}{\partial q_k} q_k + \frac{\partial x_j}{\partial t} \qquad .$$

and

$$\frac{\partial \dot{x}_j}{\partial q_l} = \sum_{k=1}^{3n} \frac{\partial^2 x_j}{\partial q_l \, \partial q_k} \dot{q}_k + \frac{\partial^2 x_j}{\partial q_l \, \partial t} \tag{14.53}$$

But

$$\frac{d}{dt}\left(\frac{\partial x_j}{\partial q_l} \right) = \sum_{k=1}^{3n} \frac{\partial^2 x_j}{\partial q_k \, \partial q_l} \dot{q}_k + \frac{\partial^2 x_j}{\partial t \, \partial q_l} \tag{14.54}$$

Thus, if the coordinates are continuous, as they must be in a physical situation, the partial derivatives may be interchanged. Equations (14.53) and (14.54) are equal, and the interchange of the partial and total derivatives is therefore permitted.

Returning to the problem at hand, Eq. (14.52b) may also be written

$$x_j \frac{\partial x_j}{\partial q_l} = \frac{d}{dt}\left[\frac{\partial}{\partial \dot{q}_l}\left(\tfrac{1}{2}\dot{x}_j^2\right)\right] - \frac{\partial}{\partial q_l}\left(\tfrac{1}{2}\dot{x}_j^2\right) \tag{14.55}$$

as the reader may verify. Substituting this result into Eq. (14.48) and rearranging terms,

$$\delta W = \sum_{l=1}^{3n}\left\{Q_l - \left[\frac{d}{dt}\frac{\partial}{\partial \dot{q}_l}\sum_{j=1}^{3n}\left(\tfrac{1}{2}m_j \dot{x}_j^2\right) - \frac{\partial}{\partial q_l}\sum_{j=1}^{3n}\tfrac{1}{2}m_j \dot{x}_j^2\right]\right\}\delta q_l = 0 \tag{14.56}$$

Since $\sum_{j=1}^{3n}\tfrac{1}{2}m_j \dot{x}_j^2$ represents the kinetic energy T of the system of particles, the virtual work may be written

$$\delta W = \sum_{l=1}^{3n}\left(Q_l - \frac{d}{dt}\frac{\partial T}{\partial \dot{q}_l} + \frac{\partial T}{\partial q_l}\right)\delta q_l = 0 \tag{14.57}$$

where the kinetic energy has been expressed in terms of generalized coordinates. Since the virtual displacements are arbitrary and independent, it follows that for the virtual work to equal zero for *any* set of virtual displacements, each term in parentheses in Eq. (14.57) must be identically equal to zero. Consequently, there are $3n$ equations of the form

$$Q_l - \frac{d}{dt}\frac{\partial T}{\partial \dot{q}_l} + \frac{\partial T}{\partial q_l} = 0 \qquad (l = 1, 2, \cdots, 3n) \tag{14.58}$$

The above expression is known as Lagrange's equation of the second kind for a system of unconstrained particles. Application of Eq. (14.58) requires that the kinetic energy be expressed in generalized coordinates q_l and generalized velocities $\dot{q}_l$, and that the generalized forces be ascertained in accordance with Eq. (14.40). What results are the equations of motion in generalized coordinates. Although partial derivatives are involved, the expressions derived from Eq. (14.58) are a set of second-order total differential equations.

Illustrative Example 14.9
A particle of mass m is acted upon by a force $\mathbf{F}$. Apply Lagrange's equations of the second kind to determine the equations of motion in Cartesian coordinates.

SOLUTION
In Cartesian coordinates, the kinetic energy

$$T = \tfrac{1}{2}m(\dot{x}^2 + \dot{y}^2 + \dot{z}^2)$$

The components of force are F_x, F_y, F_z. From Eq. (14.58),

$$F_x - \frac{d}{dt}\frac{\partial}{\partial \dot{x}}\tfrac{1}{2}m(\dot{x}^2 + \dot{y}^2 + \dot{z}^2) + \frac{\partial}{\partial x}\tfrac{1}{2}m(\dot{x}^2 + \dot{y}^2 + \dot{z}^2) = 0$$

Carrying out the indicated operations,

$$F_x = m\ddot{x}$$

Similarly, for the Y- and Z-directions, $F_y = m\ddot{y}$ and $F_z = m\ddot{z}$.

Illustrative Example 14.10
Apply Lagrange's equations of the second kind to the problem of the one-dimensional, undamped, unforced linear oscillator.

SOLUTION
The kinetic energy $T = \frac{1}{2}m\dot{x}^2$ and the spring force is $-kx$. Employing Eq. (14.58),

$$-kx - \frac{d}{dt}\frac{\partial}{\partial \dot{x}}\tfrac{1}{2}m\dot{x}^2 + \frac{\partial}{\partial x}\tfrac{1}{2}m\dot{x}^2 = 0 \quad \text{or} \quad -kx - m\ddot{x} = 0$$

Illustrative Example 14.11
A particle is acted upon by a central force. Determine the equations of motion in polar coordinates.

SOLUTION
For the present, we shall assume the motion to be planar. In a later section, the application of Lagrange's equations to three-dimensional motion is explored.

In Cartesian coordinates, the "generalized" forces and kinetic energy are

$$Q_x = F_x \qquad Q_y = F_y \qquad T = \tfrac{1}{2}m(\dot{x}^2 + \dot{y}^2)$$

To transform to polar coordinates, the following equations are applied: $x = R \cos \phi$, $y = R \sin \phi$. Employing Eq. (14.40) to determine the generalized forces:

$$Q_R = F_x\frac{\partial x}{\partial R} + F_y\frac{\partial y}{\partial R} = F_x \cos \phi + F_y \sin \phi$$

$$Q_\phi = F_x\frac{\partial x}{\partial \phi} + F_y\frac{\partial y}{\partial \phi} = -F_x R \sin \phi + F_y R \cos \phi$$

$Q_\phi = 0$, however, owing to the specification of a central force. The kinetic energy expressed in generalized coordinates is

$$T = \tfrac{1}{2}m(\dot{R}^2 + R^2\dot{\phi}^2)$$

Lagrange's equations are

$$Q_R - \frac{d}{dt}\frac{\partial}{\partial \dot{R}}\tfrac{1}{2}m(\dot{R}^2 + R^2\dot{\phi}^2) + \frac{\partial}{\partial R}\tfrac{1}{2}m(\dot{R}^2 + R^2\dot{\phi}^2) = 0$$

$$Q_\phi - \frac{d}{dt}\frac{\partial}{\partial \dot{\phi}}\tfrac{1}{2}m(\dot{R}^2 + R^2\dot{\phi}^2) + \frac{\partial}{\partial \phi}\tfrac{1}{2}m(\dot{R}^2 + R^2\dot{\phi}^2) = 0$$

leading to

$$Q_R = m\ddot{R} - mR\dot{\phi}^2$$

$$0 = \frac{d}{dt} mR^2\dot{\phi}$$

which are, of course, two very familiar expressions.

The generalized forces Q_R and Q_ϕ may be readily interpreted by referring to Fig. 14.8. As shown, the component of **F** perpendicular to the radial direction, in the direction of increasing ϕ, is $-F_x \sin \phi + F_y \cos \phi$. In Fig. 14.8(a) **F** makes an

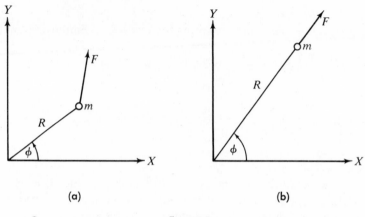

(a)   (b)

Fig. 14.8

arbitrary angle relative to the radial direction, and consequently a nonzero ϕ-component exists. In Fig. 14.8(b), **F** is radial, and consequently $F_\phi = 0$. The generalized force in the R-direction is $F_x \cos \phi + F_y \sin \phi$. In the ϕ-direction, the generalized force possesses dimensions of moment of force and is indeed the moment of force about the origin, as the reader may verify. For a central force, this moment is zero.

Illustrative Example 14.12

Determine the equations of motion of a rigid body experiencing rotation about a fixed point and subject to zero torque.

SOLUTION

It has been demonstrated in Chapter 12 that the kinetic energy of rotation of a rigid body is given by

$$T = \tfrac{1}{2}(I_{xx}\omega_x^2 + I_{yy}\omega_y^2 + I_{zz}\omega_z^2)$$

where I_{xx}, I_{yy}, I_{zz} are the moments of inertia about principal axes.

The Euler angles are most often selected as the generalized coordinates because of the convenience they offer, and the ease with which they relate to a space-fixed coordinate system. In terms of the Euler angles ϕ, θ, ψ, Lagrange's equations are written

$$-\frac{d}{dt}\frac{\partial T}{\partial \dot\phi} + \frac{\partial T}{\partial \phi} = 0$$

$$-\frac{d}{dt}\frac{\partial T}{\partial \dot\theta} + \frac{\partial T}{\partial \theta} = 0$$

$$-\frac{d}{dt}\frac{\partial T}{\partial \dot\psi} + \frac{\partial T}{\partial \psi} = 0$$

The relationships between the angular velocities and the Euler angles are (Chapter 12)

$$\omega_x = \dot\theta \cos\psi + \dot\phi \sin\theta \sin\psi$$

$$\omega_y = -\dot\theta \sin\psi + \dot\phi \sin\theta \cos\psi$$

$$\omega_z = \dot\psi + \dot\phi \cos\theta$$

The resulting expression relating the kinetic energy to the Euler angles is quite complicated. Since $\dot\psi$ appears only in the equation for ω_z, we shall derive the equation of motion for the ψ-coordinate, leaving as an exercise the equations for the θ- and ϕ-motions.

Applying the chain rule for partial differentiation,

$$\frac{\partial}{\partial \dot\psi}(T\omega_x,\ \omega_y,\ \omega_z) = \frac{\partial T}{\partial \omega_x}\frac{\partial \omega_x}{\partial \dot\psi} + \frac{\partial T}{\partial \omega_y}\frac{\partial \omega_y}{\partial \dot\psi} + \frac{\partial T}{\partial \omega_z}\frac{\partial \omega_z}{\partial \dot\psi}$$

$$= I_{xx}\omega_x(0) + I_{yy}\omega_y(0) + I_{zz}\omega_z(1)$$

Therefore,

$$\frac{d}{dt}\frac{\partial T}{\partial \dot\psi} = I_{zz}\dot\omega_z$$

Again applying the chain rule,

$$\frac{\partial T}{\partial \psi} = \frac{\partial T}{\partial \omega_x}\frac{\partial \omega_x}{\partial \psi} + \frac{\partial T}{\partial \omega_y}\frac{\partial \omega_y}{\partial \psi} + \frac{\partial T}{\partial \omega_z}\frac{\partial \omega_z}{\partial \psi}$$

$$= I_{xx}\omega_x(-\dot\theta \sin\psi + \dot\phi \sin\phi \cos\psi)$$

$$+ I_{yy}\omega_y(-\dot\theta \cos\psi - \dot\phi \sin\theta \sin\psi) + I_{zz}\omega_z(0)$$

$$= I_{xx}\omega_x\omega_y + I_{yy}\omega_y(-\omega_x)$$

Combining the results obtained,

$$-I_{zz}\dot\omega_z + \omega_x\omega_y(I_{xx} - I_{yy}) = 0$$

The above expression is one of the familiar Euler equations for torque-free motion. Since the X, Y, Z coordinates are arbitrary, the remaining equations may be determined by permutation of the variables.

14.13 LAGRANGE'S EQUATIONS AND THE LAGRANGIAN

Considerable simplification may be realized when Lagrange's equations of the second kind are applied to conservative systems. Since the generalized force is in this case derivable from a potential, it is possible to define a new function, L, called the *Lagrangian*, in the following manner:

$$L = L(q_1, q_2, \cdots, q_{3n}, \dot{q}_1, \dot{q}_2, \cdots, \dot{q}_{3n}, t)$$

$$= T(q_1, q_2, \cdots, q_{3n}, \dot{q}_1, \dot{q}_2, \cdots, \dot{q}_{3n}, t) - V(q_1, q_2, \cdots, q_{3n}) \qquad (14.59)$$

The Lagrangian is thus the difference between the kinetic and potential energies. Expressed in generalized coordinates, the Lagrangian describing a conservative system may be an explicit function of time when, for example, the system is analyzed in terms of a rotating or accelerating coordinate system.

Referring to Eq. (14.43), Lagrange's equation becomes

$$Q_l - \frac{d}{dt}\frac{\partial T}{\partial \dot{q}_l} + \frac{\partial T}{\partial q_l} = -\frac{\partial V}{\partial q_l} - \frac{d}{dt}\frac{\partial T}{\partial \dot{q}_l} + \frac{\partial T}{\partial q_l} = 0$$

$$= -\frac{d}{dt}\frac{\partial T}{\partial \dot{q}_l} + \frac{\partial(T - V)}{\partial q_l} = 0 \qquad (14.60)$$

Since V is not a function of the generalized velocities, we are free to add the term $(d/dt)(\partial V/\partial \dot{q}_l) = 0$ to the above equation:

$$-\frac{d}{dt}\frac{\partial(T - V)}{\partial \dot{q}_l} + \frac{\partial(T - V)}{\partial q_l} = 0 \qquad (l = 1, 2, \cdots, 3n) \qquad (14.61a)$$

or

$$-\frac{d}{dt}\frac{\partial L}{\partial \dot{q}_l} + \frac{\partial L}{\partial q_l} = 0 \qquad (l = 1, 2, \cdots, 3n) \qquad (41.61b)$$

The above is Lagrange's equation for a conservative system. It is apparent that only energy terms are involved, in addition, of course, to the expression defining the Lagrangian. If nonconservative forces are present as well, Eq. (14.61) may be modified as follows:

$$Q_l' - \frac{d}{dt}\frac{\partial L}{\partial \dot{q}_l} + \frac{\partial L}{\partial q_l} = 0 \qquad (l = 1, 2, \cdots, 3n) \qquad (14.62)$$

where Q_l' represents the nonconservative forces, and L is again the difference between the kinetic and potential energies.

Illustrative Example 14.13

Derive the equation governing the motion of a one-dimensional undamped, unforced linear oscillator.

SOLUTION
From a knowledge of the kinetic and potential energies, the Lagrangian may be immediately written

$$L = T - V = \tfrac{1}{2}m\dot{x}^2 - \tfrac{1}{2}kx^2.$$

Apply Eq. (14.61b),

$$-\frac{d}{dt}\frac{\partial}{\partial \dot{x}}(\tfrac{1}{2}m\dot{x}^2 - \tfrac{1}{2}kx^2) + \frac{\partial}{\partial x}(\tfrac{1}{2}m\dot{x}^2 - \tfrac{1}{2}kx^2) = 0$$

yielding $-m\ddot{x} - kx = 0$.

14.14 APPLICATION TO AN ELECTRIC CIRCUIT

Lagrange's equations may be applied to multiloop inductive-capacitive circuits in which the charge in each loop plays the role of a generalized coordinate. For a single-loop circuit, the kinetic energy $T = \tfrac{1}{2}\pounds\dot{Q}^2$ (the energy associated with the motion of the charge through the inductance) and the potential energy $V = \tfrac{1}{2}Q^2/C$ (the energy stored in the capacitor). The Lagrangian is thus

$$L = T - V = \frac{1}{2}\pounds\dot{Q}^2 - \frac{1}{2}\frac{Q^2}{C}$$

and Lagrange's equation becomes

$$-\frac{d}{dt}\frac{\partial}{\partial \dot{Q}}\left(\frac{1}{2}\pounds\dot{Q}^2 - \frac{1}{2}\frac{Q^2}{C}\right) + \frac{\partial}{\partial Q}\left(\frac{1}{2}\pounds\dot{Q}^2 - \frac{1}{2}\frac{Q^2}{C}\right) = 0$$

$$-\pounds\ddot{Q} - \frac{Q}{C} = 0$$

where we have used $\pounds$ to denote the inductance so as not to confuse it with the Lagrangian. This method can be extended to multiloop circuits which include both resistive elements and impressed voltages (driving forces).

Illustrative Example 14.14
Derive the equations of motion of the one-dimensional oscillator subject to a linear restoring force, velocity damping, and sinusoidal forcing.

SOLUTION
The Lagrangian as before is $L = \tfrac{1}{2}m\dot{x}^2 - \tfrac{1}{2}kx^2$. The forces acting in addition to the spring force are the damping, $-b\dot{x}$, and $F(t) = F_0 \sin \omega t$. Therefore, the generalized force, not derivable from a potential, is

$$Q_i' = F_0 \sin \omega t - b\dot{x}$$

Applying Eq. (14.62) we have

$$F_0 \sin \omega t - b\dot{x} - \frac{d}{dt}\frac{\partial}{\partial \dot{x}}(\tfrac{1}{2}m\dot{x}^2 - \tfrac{1}{2}kx^2) + \frac{\partial}{\partial x}(\tfrac{1}{2}m\dot{x}^2 - \tfrac{1}{2}kx^2) = 0$$

or

$$F_0 \sin \omega t - b\dot{x} - m\ddot{x} - kx = 0$$

14.15 GENERALIZED POTENTIALS

Consider the possibility of deriving nonconservative forces from a potential U, according to the equation

$$Q_l'' = \frac{d}{dt}\frac{\partial U}{\partial \dot{q}_l} - \frac{\partial U}{\partial q_l} \tag{14.63}$$

The potential U is a function of the generalized coordinates and generalized velocities (and possibly time), and cannot represent the potential energy. Substituting Eq. (14.63) into Eq. (14.58), the result is

$$-\frac{d}{dt}\frac{\partial}{\partial \dot{q}_l}(T - U) + \frac{\partial}{\partial q_l}(T - U) = 0 \tag{14.64}$$

Defining a new Lagrangian, $L' = T - U$, Eq. (14.64) is now

$$-\frac{d}{dt}\frac{\partial L'}{\partial \dot{q}_l} + \frac{\partial L'}{\partial q_l} = 0 \qquad (l = 1, 2, \cdots, 3n) \tag{14.65}$$

If, in addition to those forces derivable from U, other conservative and nonconservative forces are present, Lagrange's equations of the second kind are now written

$$Q_l''' - \frac{d}{dt}\frac{\partial L''}{\partial \dot{q}_l} + \frac{\partial L''}{\partial q_l} = 0 \qquad (l = 1, 2, \cdots, 3n) \tag{14.66}$$

where Q_l''' includes all the nonconservative forces not derivable from a potential and $L'' = T - V - U$. Although the foregoing appears somewhat contrived, it does serve to extend the applicability of Lagrange's equations, particularly insofar as electromagnetic phenomena are concerned.

Illustrative Example 14.15

A charge e is moving in a region of free space in which both electric and magnetic fields are present. Apply Maxwell's equations to demonstrate that the force acting on the charge may be derived from an expression of the form of Eq. (14.63), and determine U.

SOLUTION

Maxwell's equations in differential form, written for free space, are

$$\nabla \times \mathbf{E} = -\frac{\partial \mathbf{B}}{\partial t}$$

$$\nabla \cdot \mathbf{B} = 0$$

$$\nabla \cdot \mathbf{E} = 0$$

$$\nabla \times \mathbf{B} = \mu_0 \epsilon_0 \frac{\partial \mathbf{E}}{\partial t}$$

where $\mathbf{E}$ and $\mathbf{B}$ represent respectively, the electric and magnetic fields, and ϵ_0 and μ_0 are the permittivity and permeability of free space. The electric and magnetic potentials ϕ and $\mathbf{A}$ are defined by the expressions

$$\mathbf{E} = -\nabla \phi - \frac{\partial \mathbf{A}}{\partial t}$$

$$\mathbf{B} = \nabla \times \mathbf{A}$$

where $\phi = \phi(x, y, z, t)$ and $\mathbf{A} = \mathbf{A}(x, y, z, t)$. We shall assume ϕ and $\mathbf{A}$ known, and predicate the equations of motion upon these functions.

The Lorentz force on a charge e traveling with velocity $\mathbf{v}$ is therefore

$$\mathbf{F} = e(\mathbf{v} \times \mathbf{B}) + e\mathbf{E} = e\left[\mathbf{v} \times (\nabla \times \mathbf{A}) - \nabla \phi - \frac{\partial \mathbf{A}}{\partial t} \right]$$

Expansion of the magnetic force term as a vector triple product yields

$$\mathbf{v} \times (\nabla \times \mathbf{A}) = \nabla_A(\mathbf{v} \cdot \mathbf{A}) - (\mathbf{v} \cdot \nabla_A)\mathbf{A}$$

where the nomenclature ∇_A indicates that ∇ only operates on vector $\mathbf{A}$. When an inertial frame of reference is selected, the velocity $\mathbf{v}$ is a function of time only, and the triple product may be written

$$\mathbf{v} \times (\nabla \times \mathbf{A}) = \nabla (\mathbf{v} \cdot \mathbf{A}) - (\mathbf{v} \cdot \nabla)\mathbf{A}$$

The electromagnetic force thus becomes

$$\mathbf{F} = e\left[\nabla(\mathbf{v} \cdot \mathbf{A}) - (\mathbf{v} \cdot \nabla)\mathbf{A} - \nabla \phi - \frac{\partial \mathbf{A}}{\partial t} \right]$$

The time derivative of $\mathbf{A}$ is

$$\frac{d\mathbf{A}}{dt} = \frac{\partial \mathbf{A}}{\partial x}\frac{dx}{dt} + \frac{\partial \mathbf{A}}{\partial y}\frac{dy}{dt} + \frac{\partial \mathbf{A}}{\partial z}\frac{dz}{dt} + \frac{\partial \mathbf{A}}{\partial t}$$

$$\frac{d\mathbf{A}}{dt} = x\frac{\partial \mathbf{A}}{\partial x} + \dot{y}\frac{\partial \mathbf{A}}{\partial y} + \dot{z}\frac{\partial \mathbf{A}}{\partial z} + \frac{\partial \mathbf{A}}{\partial t}$$

or $\qquad = (\mathbf{v} \cdot \nabla)\mathbf{A} + \dfrac{\partial \mathbf{A}}{\partial t}$

where we have substituted $\mathbf{v} = \dot{x}\hat{\mathbf{i}} + \dot{y}\hat{\mathbf{j}} + \dot{z}\hat{\mathbf{k}}$ and

$$\nabla A = \frac{\partial A}{\partial x}\hat{\mathbf{i}} + \frac{\partial A}{\partial y}\hat{\mathbf{j}} + \frac{\partial A}{\partial z}\hat{\mathbf{k}}$$

The force is now written

$$\mathbf{F} = e\left[\nabla(\mathbf{v}\cdot\mathbf{A}) - \nabla\phi - \frac{d\mathbf{A}}{dt}\right]$$

In Cartesian coordinates,

$$F_x = e\left[\frac{\partial}{\partial x}(\dot{x}A_x + \dot{y}A_y + \dot{z}A_z) - \frac{\partial\phi}{\partial x} - \frac{dA_x}{dt}\right]$$

$$F_y = e\left[\frac{\partial}{\partial y}(\dot{x}A_x + \dot{y}A_y + \dot{z}A_z) - \frac{\partial\phi}{\partial y} - \frac{dA_y}{dt}\right]$$

$$F_z = e\left[\frac{\partial}{\partial z}(\dot{x}A_x + \dot{y}A_y + \dot{z}A_z) - \frac{\partial\phi}{\partial z} - \frac{dA_z}{dt}\right]$$

Upon careful examination of these components of force, it is possible to ascertain that if a potential U is defined

$$U = e(\phi - \mathbf{v}\cdot\mathbf{A})$$

the forces are derivable from U in accordance with Eq. (14.63). This may be shown by beginning with

$$\frac{d}{dt}\frac{\partial U}{\partial \dot{x}} - \frac{\partial U}{\partial x} = \frac{d}{dt}\frac{\partial}{\partial \dot{x}}[e(\phi - \mathbf{v}\cdot\mathbf{A})] - \frac{\partial}{\partial x}[e(\phi - \mathbf{v}\cdot\mathbf{A})]$$

$$= -e\frac{dA_x}{dt} - e\frac{\partial\phi}{\partial x} + e\mathbf{v}\cdot\frac{\partial\mathbf{A}}{\partial x}$$

$$= -e\frac{dA_x}{dt} - e\frac{\partial\phi}{\partial x} + e\dot{x}\frac{\partial A_x}{\partial x} + e\dot{y}\frac{\partial A_y}{\partial x} + e\dot{z}\frac{\partial A_z}{\partial x}$$

The last expression is identical with F_x above. Similar results obtain for F_y and F_z.

If the form of the potential U did not immediately occur to the reader, he need not be dismayed. Considerable effort lies behind the establishment of the appropriate forms of ϕ, $\mathbf{A}$, and U in order to permit direct incorporation of the electromagnetic forces into Lagrange's equations. Much nineteenth-century scientific thought was devoted to this not so obvious development.

14.16 LAGRANGE'S EQUATIONS OF THE SECOND KIND WITH CONSTRAINTS

Lagrange's equations in generalized coordinates may be suitably modified in the event constraints are present. Recall that each independent holonomic constraint may be employed to reduce the number of coordinates by 1. The remaining coordinates must exactly total the number of degrees of freedom.

Alternatively, the method of undetermined coefficients proves useful, especially when it is required to determine the constraint forces. Once again, only holonomic, or nonholonomic constraints of the form of Eq. (14.17) will be considered. The constraint force exerted by the kth constraint in the lth generalized direction is given by $\lambda_k(\partial\phi_k/\partial q_l)$ or $\lambda_{k'}a_{lk'}$, depending upon the type of constraint. The total constraint force corresponding to the l-direction is then

$$\sum_{k=1}^{h} \lambda_k \frac{\partial\phi_k}{\partial q_l} \quad \text{or} \quad \sum_{k'=1}^{h'} \lambda_{k'}a_{lk'}$$

(or the appropriate sum of the two terms when both types of constraints are present). By treating the constraint force as an external force, Lagrange's equations become

$$Q_l' + \sum_{k=1}^{h} \lambda_k \frac{\partial\phi_k}{\partial q_l} - \frac{d}{dt}\frac{\partial L}{\partial \dot{q}_l} + \frac{\partial L}{\partial q_l} = 0 \quad (l = 1, 2, \cdots, 3n) \tag{14.67}$$

where now Q_l' includes all the nonconstraint forces which are not derivable from a potential. There are now $3n + h$ unknowns, the $3n$ generalized coordinates and the h undetermined multipliers. Fortunately, $(3n + h)$ equations are available in the form of $3n$ Lagrange's equations and h equations of constraint.

Illustrative Example 14.16

Determine the motion of the bead of Illustrative Example 14.8. Apply Lagrange's equations of the second kind in polar coordinates.

SOLUTION

The kinetic energy in polar coordinates is expressed

$$T = \tfrac{1}{2}m(\dot{R}^2 + R^2\dot{\phi}^2) \tag{a}$$

No external forces are present, and the potential energy is zero. The constraint equation is simply

$$\Phi = \phi - \omega t = 0 \tag{b}$$

where now Φ applies to the constraint and ϕ represents the polar angle. Applying Eq. (14.67),

$$0 + \lambda\frac{\partial\Phi}{\partial R} - \frac{d}{dt}\frac{\partial T}{\partial\dot{R}} + \frac{\partial T}{\partial R} = 0 + 0 - m\ddot{R} + mR\dot{\phi}^2 = 0 \tag{c}$$

$$0 + \lambda\frac{\partial\Phi}{\partial\phi} - \frac{d}{dt}\frac{\partial T}{\partial\dot{\phi}} + \frac{\partial T}{\partial\phi} = 0 + \lambda(1) - \frac{d}{dt}(mR^2\dot{\phi}) + 0 = 0 \tag{d}$$

Since $\phi = \omega t$, $\dot{\phi} = \omega$, which when substituted into Eq. (c) yields

$$m\ddot{R} - mR\omega^2 = 0 \quad \text{or} \quad \ddot{R} - \omega^2 R = 0$$

for which the solution is

$$R = A\cosh\omega t + B\sinh\omega t$$

The initial conditions $R(0) = R_0$ and $\dot{R}(0) = 0$ lead to the solution

$$R = R_0 \cosh \omega t$$

Substituting this result into the expression for the ϕ-motion, we have

$$\lambda = \frac{d}{dt}(mR^2\dot{\phi}) = \frac{d}{dt}(m\omega R_0^2 \cosh^2 \omega t)$$

$$= 2m\omega^2 R_0^2 \cosh \omega t \sinh \omega t$$

It is particularly interesting to note that λ represents the generalized force, in this case a torque, exerted by the wire on the bead. To ascertain the force exerted by the wire, we simply employ the fact that the moment arm is R:

$$F_\phi = \frac{\lambda}{R} = 2m\omega^2 R_0 \sinh \omega t$$

Unquestionably, polar coordinates and Lagrange's equations of the second kind have led to a simpler and more rapid solution than previously.

14.17 SMALL OSCILLATIONS

The methods of Lagrange's equations are well adapted to the determination of the frequencies and amplitudes of small oscillations about positions of stable equilibrium in conservative systems. Consideration has already been given small oscillations of a single particle and several specific multiparticle systems. The general case of n particles with $3n$ degrees of freedom is now treated. If holonomic constraints are present, they should be employed to reduce the number of coordinates required to describe the configuration of the system.

For a conservative system of $3n$ degrees of freedom, the potential energy is expressed $V = V(q_1, q_2, \cdots, q_{3n})$. Denoting the equilibrium coordinates by $q_{1,0}$, $\cdots$, $q_{3n,0}$, for small motions about equilibrium, we may write

$$q_l = q_{l,0} + \bar{q}_l \qquad (l = 1, 2, \cdots, 3n) \tag{14.68}$$

where $\bar{q}_l$ represents a *small* displacement from equilibrium. Expanding the potential energy about the equilibrium point in a multidimensional Taylor series, we have

$$V(q_1, q_2, \cdots, q_{3n}) = V(q_{1,0}, q_{2,0}, \cdots, q_{3n,0}) + \sum_{l=1}^{3n} \left(\frac{\partial V}{\partial q_l}\right)_{q_{l,0}} q_l$$

$$+ \frac{1}{2} \sum_{l=1}^{3n} \sum_{m=1}^{3n} \left(\frac{\partial^2 V}{\partial q_l \, \partial q_m}\right)_{q_{l,0}, q_{m,0}} \bar{q}_l \bar{q}_m + \cdots = 0 \tag{14.69}$$

Since the zero of potential energy is arbitrary, the first term in the foregoing expansion may be equated to zero without loss of generality. For equilibrium, $(\partial V/\partial q_l)_{q_{l,0}} = 0$, and the potential energy (to second order) becomes

$$V = V(q_1, q_2, \cdots, q_{3n}) = \frac{1}{2} \sum_{l=1}^{3n} \sum_{m=1}^{3n} \left(\frac{\partial^2 V}{\partial q_l \, \partial q_m}\right)_{q_{l,0}, q_{m,0}} \bar{q}_l \bar{q}_m \qquad (14.70)$$

or

$$V = \frac{1}{2} \sum_{l=1}^{3n} \sum_{m=1}^{3n} V_{lm} \bar{q}_l \bar{q}_m \qquad (14.71)$$

where

$$V_{lm} = \left(\frac{\partial^2 V}{\partial q_l \, \partial q_m}\right)_{q_{l,0}, q_{m,0}} = V_{ml} = \text{constant}$$

Inasmuch as motions about *stable* equilibrium are under consideration, the potential energy must be a minimum. For one-dimensional motion, this means that $\partial V/\partial q = 0$ and $\partial^2 V/\partial q^2 > 0$, where the derivatives are evaluated at the equilibrium point. For a multidimensional system,[3]

$$\frac{\partial V}{\partial q_l} = 0 \qquad (l = 1, 2, \cdots, 3n) \qquad (14.72)$$

$$\frac{\partial^2 V}{\partial q_l^2} > 0 \qquad (l = 1, 2, \cdots, 3n)$$

$$\begin{vmatrix} \dfrac{\partial^2 V}{\partial q_l^2} & \dfrac{\partial^2 V}{\partial q_l \, \partial q_m} \\[3mm] \dfrac{\partial^2 V}{\partial q_l \, \partial q_m} & \dfrac{\partial^2 V}{\partial q_m^2} \end{vmatrix} > 0 \qquad \begin{array}{l} (l = 1, 2, \cdots, 3n, \\ m = 1, 2, \cdots, 3n, \\ l \neq m) \end{array}$$

$$\vdots$$

$$\begin{vmatrix} \dfrac{\partial^2 V}{\partial q_1^2} & \dfrac{\partial^2 V}{\partial q_1 \, \partial q_2} & \cdots & \dfrac{\partial^2 V}{\partial q_1 \, \partial q_{3n}} \\[3mm] \dfrac{\partial^2 V}{\partial q_1 \, \partial q_2} & \dfrac{\partial^2 V}{\partial q_2^2} & \cdots & \dfrac{\partial^2 V}{\partial q_2 \, \partial q_{3n}} \\[3mm] \vdots & \vdots & & \vdots \\[3mm] \dfrac{\partial^2 V}{\partial q_1 \, \partial q_{3n}} & \dfrac{\partial^2 V}{\partial q_2 \, \partial q_{3n}} & \cdots & \dfrac{\partial^2 V}{\partial q_{3n}^2} \end{vmatrix} > 0 \qquad (14.73)$$

---

[3] Osgood, W. F., *Advanced Calculus*, The Macmillan Company, New York, 1925.

Stable equilibrium is also possible if $\partial^2 V/\partial q_l\, \partial q_m = 0$ for all l and m, provided the first nonzero derivatives of the potential are of even order. The methods of this chapter do not, however, apply to such potentials.

The kinetic energy, in Cartesian coordinates, is given by the summation

$$T = \frac{1}{2} \sum_{j=1}^{3n} m_j \dot{x}_j^2 \tag{14.74}$$

Employing the equations governing transformation from Cartesian to generalized coordinates,

$$x_j = x_j(q_1, q_2, \cdots, q_{3n}, t)$$

$$\dot{x}_j = \sum_{l=1}^{3n} \frac{\partial x_j}{\partial q_l} \dot{q}_l + \frac{\partial x_j}{\partial t}$$

The kinetic energy is therefore

$$T = \frac{1}{2} \sum_{j=1}^{3n} m_j \left(\sum_{l=1}^{3n} \frac{\partial x_j}{\partial q_l} \dot{q}_l + \frac{\partial x_j}{\partial t} \right) \left(\sum_{m=1}^{3n} \frac{\partial x_j}{\partial q_m} \dot{q}_m + \frac{\partial x_j}{\partial t} \right) \tag{14.75}$$

By expansion of the velocity terms in Eq. (14.75), it is clear that the kinetic energy is separable into three parts: one quadratic in the generalized velocities, one linear in the generalized velocities, and the remaining term independent of these velocities. By restricting our area of interest to transformation equations which do not contain time explicitly, only terms quadratic in the generalized velocities remain. (Transformation equations for conservative systems involving time explicitly occur when, for example, rotating or translating coordinate systems are employed.) Under these conditions, the kinetic energy may be written

$$T = \frac{1}{2} \sum_{l=1}^{3n} \sum_{m=1}^{3n} \left(\sum_{j=1}^{3n} m_j \frac{\partial x_j}{\partial q_l} \frac{\partial x_j}{\partial q_m} \right) \dot{q}_l \dot{q}_m \tag{14.76}$$

For small oscillations about equilibrium, the term in parentheses becomes

$$\sum_{j=1}^{3n} m_j \left(\frac{\partial x_j}{\partial q_l}\right)_{q_{l,0}} \left(\frac{\partial x_j}{\partial q_m}\right)_{q_{m,0}} + \sum_{j=1}^{3n} m_j \sum_{p=1}^{3n} \frac{\partial}{\partial q_p} \left(\frac{\partial x_j}{\partial q_l} \frac{\partial x_j}{\partial q_m}\right)_{q_{l,0},q_{m,0}} \bar{q}_p + \cdots$$

Since our concern is with small motion and $\dot{q}_l$ and $\dot{q}_m$ ($= \mathring{q}_l$ and $\mathring{q}_m$, respectively) are themselves small, we need retain terms of $\mathring{q}$ in T, only to the same order as $\bar{q}$ in V [Eq. (14.70)]. Consequently,

$$T \cong \frac{1}{2} \sum_{l=1}^{3n} \sum_{m=1}^{3n} T_{lm} \mathring{q}_l \mathring{q}_m \tag{14.77}$$

where

$$T_{lm} = \frac{1}{2} \sum_{j=1}^{3n} m_j \left(\frac{\partial x_j}{\partial q_l}\right)_{q_{l,0}} \left(\frac{\partial x_j}{\partial q_m}\right)_{q_{m,0}} = T_{ml} \tag{14.78}$$

We are now in a position to write the Lagrangian,

$$L = T - V = \frac{1}{2} \sum_{l=1}^{3n} \sum_{m=1}^{3n} (T_{lm}\dot{\bar{q}}_l\dot{\bar{q}}_m - V_{lm}\bar{q}_l\bar{q}_m) \tag{14.79}$$

and Lagrange's equations for the system under study are

$$\sum_{m=1}^{3n} (T_{lm}\ddot{\bar{q}}_m + V_{lm}\bar{q}_m) = 0 \qquad (l = 1, 2, \cdots, 3n) \tag{14.80a}$$

or

$$T_{l1}\ddot{\bar{q}}_1 + T_{l2}\ddot{\bar{q}}_2 + \cdots + T_{l,3n}\ddot{\bar{q}}_{3n}$$
$$+ V_{l1}\bar{q}_1 + V_{l2}\bar{q}_2 + \cdots + V_{l,3n}\bar{q}_{3n} = 0 \qquad (l = 1, 2, \cdots, 3n) \tag{14.80b}$$

Before proceeding with the general solution of these $3n$ linear, coupled second-order differential equations, it is well to consider a special case, in order to deduce the form of solutions to Eqs. (14.80). Suppose only one generalized coordinate exists. Then, from Eqs. (14.80),

$$T_{11}\ddot{\bar{q}}_1 + V_{11}\bar{q}_1 = 0 \tag{14.81}$$

The general solution of this equation is

$$\bar{q}_1 = A \cos\left[\left(\frac{V_{11}}{T_{11}}\right)^{\frac{1}{2}} t + \alpha\right] \tag{14.82}$$

where A is the amplitude of the oscillation and α, a phase angle. Of course, if V_{11} is negative, the motion is not oscillatory, since $\cos ia = \cosh a$. In this eventuality,

$$\bar{q}_1 = A' \cosh\left(\frac{V_{11}}{T_{11}}\right)^{\frac{1}{2}} t + B' \sinh\left(\frac{V_{11}}{T_{11}}\right)^{\frac{1}{2}} t \tag{14.83}$$

Under the condition described above, the motion grows without bound and violates the small oscillation restriction. To ensure "small" oscillation,

$$V_{11} = \left(\frac{\partial^2 V}{\partial q_1^2}\right)_{q_{1,0}} > 0$$

which, for a single-degree-of-freedom system, simply represents the condition for stable equilibrium. (Negative values of T_{11} are, of course, prohibited.)

On the basis of what we have learned from the solution of the one-dimensional case, let us judiciously "guess" that each coordinate $\bar{q}_m$ in Eqs. (14.80) varies as

$$\bar{q}_m = A_m \cos(\omega t + \alpha_m) \tag{14.84}$$

where A_m and α_m are the amplitude and phase angle, determined from the initial conditions, and ω, the natural circular frequency, is ascertained on the basis of the system constants. Substituting Eq. (14.84) into Eqs. (14.80), we obtain

$$\sum_{m=1}^{3n} [-T_{lm}\omega^2 A_m \cos(\omega t + \alpha_m) + V_{lm}A_m \cos(\omega t + \alpha_m)] = 0$$

$$(l = 1, 2, \cdots, 3n) \quad \textbf{(14.85)}$$

In order that our "guess" be, in fact, a solution, all the α_m must be identical for a given ω. In this way $\cos(\omega t + \alpha_m)$ may be factored from each term in Eq. (14.85):

$$\cos(\omega t + \alpha) \sum_{m=1}^{3n} (-\omega^2 T_{lm}A_m + V_{lm}A_m) = 0 \quad (l = 1, 2, \cdots, 3n) \quad \textbf{(14.86)}$$

Since $\cos(\omega t + \alpha)$ is not, in general, equal to zero, the summation of Eq. (14.86) must vanish to ensure the existence of Eq. (14.84) as a solution to Eq. (14.80). A total of $3n$ linear homogeneous algebraic equations in A_m and ω are thus available:

$$(-\omega^2 T_{11} + V_{11})A_1 + (-\omega^2 T_{12} + V_{12})A_2 + \cdots + (-\omega^2 T_{1,3n} + V_{1,3n})A_{3n} = 0$$
$$\vdots$$
$$(-\omega^2 T_{3n,1} + V_{3n,1})A_1 + (-\omega^2 T_{3n,2} + V_{3n,2})A_2 + \cdots \quad \textbf{(14.87)}$$
$$+ (-\omega^2 T_{3n,3n} + V_{3n,3n})A_{3n} = 0$$

Once again, a nontrivial solution demands that the determinant of the coefficients of A_m vanishes:

$$\begin{vmatrix} (-\omega^2 T_{11} + V_{11}) & (-\omega^2 T_{12} + V_{12}) & \cdots & (-\omega^2 T_{1,3n} + V_{1,3n}) \\ \vdots & \vdots & & \vdots \\ (-\omega^2 T_{3n,1} + V_{3n,1}) & (-\omega^2 T_{3n,2} + V_{3n,2}) & \cdots & (-\omega^2 T_{3n,3n} + V_{3n,3n}) \end{vmatrix} = 0$$

$$\textbf{(14.88)}$$

The above determinant results in a $3n$-degree polynomial in ω^2. Each of the $3n$ roots of the equation represents, as we have noted in Chapter 11, a different circular frequency of oscillation. The general solution for small amplitudes of the multi-dimensional oscillator is thus

$$\bar{q}_l = \sum_{k=1}^{3n} A_{kl} \cos(\omega_k t + \alpha_k) \quad \textbf{(14.89)}$$

where A_{kl} and α_k are found from the initial conditions and ω_k are the roots of the above determinant.

If ω^2 is negative, ω is complex and small oscillations do not occur. In the event $\omega^2 = 0$, q_k is apparently constant. This corresponds to no oscillation (translation or rotation of the entire system), similar to rigid-body motion under zero net force or torque. It remains as an exercise for the reader to verify that the condition of positive ω^2 is equivalent to a minimum in the potential energy $V(q_1, q_2, \cdots, q_{3n})$ at $q_l = q_{l,0}$.

To determine A_{kl}, related to one another by the algebraic equations [Eq.

(14.87)], it is necessary to substitute each value of ω_k separately. For homogeneous linear equations, all but one of the A_{kl} can, in this manner, be found. It is usual to determine the ratios

$$\frac{A_{k2}}{A_{k1}}, \quad \frac{A_{k3}}{A_{k1}}, \quad \cdots, \quad \frac{A_{k,3n}}{A_{k1}}$$

There are thus $6n$ constants (A_{k1} and α_k) to be evaluated from $6n$ required initial conditions.

By appropriate algebraic manipulation of Eq. (14.89), new generalized coordinates may be determined, each of which displays a cosinusoidal variation at a single frequency. That is, we must solve for a set of coordinates, related to $\bar{q}_l$, which satisfy the equations

$$\ddot{\eta}_k + \omega_k^2 \eta_k = 0 \qquad (k = 1, 2, \cdots, 3n) \tag{14.90}$$

Recall that the η_k are termed *normal coordinates*, each involving only one *normal frequency* of oscillation, ω_k. Because of the nature of Eqs. (14.89) it is possible to ascertain the normal coordinates through algebraic operations.

The possibility of *degeneracy* has not yet been broached. By this is meant the occurrence of two or more identical frequencies. In general, a repeated root calls for a term $t \cos \omega t$. For *small* oscillations, the coefficient of such a term must be zero, since the amplitude may not grow with time.

Illustrative Example 14.17
Determine the amplitude and frequency for small oscillations of the double pendulum of Fig. 14.9. Assume all motion is planar.

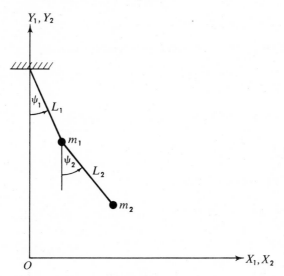

Fig. 14.9

SOLUTION

The kinetic and potential energies of the system, expressed in Cartesian coordinates, are

$$T = \tfrac{1}{2}m_1(\dot{x}_1^2 + \dot{y}_1^2) + \tfrac{1}{2}m_2(\dot{x}_2^2 + \dot{y}_2^2)$$

$$V = m_1gy_1 + m_2gy_2$$

Because the lengths L_1 and L_2 are constant, considerable simplification is realized when T and V are expressed in terms of polar variables. The transformation equations are

$$x_1 = L_1 \sin \psi_1$$

$$x_2 = L_1 \sin \psi_1 + L_2 \sin \psi_2$$

$$y_1 = L_1(1 - \cos \psi_1) + L_2$$

$$y_2 = L_1(1 - \cos \psi_1) + L_2(1 - \cos \psi_2)$$

The Cartesian components of particle velocity are

$$\dot{x}_1 = L_1\dot{\psi}_1 \cos \psi_1$$

$$\dot{x}_2 = L_1\dot{\psi}_1 \cos \psi_1 + L_2\dot{\psi}_2 \cos \psi_2$$

$$\dot{y}_1 = L_1\dot{\psi}_1 \sin \psi_1$$

$$\dot{y}_2 = L_1\dot{\psi}_1 \sin \psi_1 + L_2\dot{\psi}_2 \sin \psi_2$$

Substituting the above into T and V we obtain

$$T = \tfrac{1}{2}m_1(L_1^2\dot{\psi}_1^2) + \tfrac{1}{2}m_2[L_1^2\dot{\psi}_1^2 + L_2^2\dot{\psi}_2^2 + 2L_1L_2\dot{\psi}_1\dot{\psi}_2$$

$$\times (\cos \psi_1 \cos \psi_2 + \sin \psi_1 \sin \psi_2)]$$

$$V = m_1gL_1(1 - \cos \psi_1) + m_2gL_1(1 - \cos \psi_1) + m_2gL_2(1 - \cos \psi_2)$$

The minimum potential energy occurs when $\psi_1 = \psi_2 = 0$. Expanding T and V about this point of stable equilibrium, to second order in ψ_1, ψ_2, $\dot{\psi}_1$, and $\dot{\psi}_2$:

$$T = \tfrac{1}{2}m_1L_1^2\dot{\psi}_1^2 + \tfrac{1}{2}m_2(L_1^2\dot{\psi}_1^2 + L_2^2\dot{\psi}_2^2 + 2L_1L_2\dot{\psi}_1\dot{\psi}_2)$$

$$V = m_1g\left(L_2 + \frac{L_1\psi_1^2}{2}\right) + m_2g\left(\frac{L_1\psi_1^2}{2} + \frac{L_2\psi_2^2}{2}\right)$$

The equations of motion are therefore

$$\frac{d}{dt}\frac{\partial}{\partial \dot{\psi}_1}(T - V) - \frac{\partial}{\partial \psi_1}(T - V) = 0 = m_1L_1^2\ddot{\psi}_1 + m_2L_1^2\ddot{\psi}_1 + m_2L_1L_2\ddot{\psi}_2$$

$$+ m_1gL_1\psi_1 + m_2gL_1\psi_1$$

$$\frac{d}{dt}\frac{\partial}{\partial \dot{\psi}_2}(T - V) - \frac{\partial}{\partial \psi_2}(T - V) = 0 = m_2L_2^2\ddot{\psi}_2 + m_2L_1L_2\ddot{\psi}_1 + m_2gL_2\psi_2$$

These expressions are coupled because of the term $m_2 L_1 L_2 \ddot{\psi}_2$ in the first equation and $m_2 L_1 L_2 \ddot{\psi}_1$ in the second. Select the trial solutions

$$\psi_1 = A_1 \cos(\omega t + \alpha)$$

$$\psi_2 = A_2 \cos(\omega t + \alpha)$$

Substituting into the equations of motion, and dividing by $L_1 \cos(\omega t + \alpha)$ and $m_2 L_2 \cos(\omega t + \alpha)$, respectively, we obtain

$$[(m_1 + m_2)(g - L_1\omega^2)]A_1 - (m_2 L_2 \omega^2)A_2 = 0$$

$$- L_1 \omega^2 A_1 + (g - L_2 \omega^2) A_2 = 0$$

For the assumed solution to be valid, the determinant of the coefficients of A_1 and A_2 must vanish:

$$\begin{vmatrix} (m_1 + m_2)(g - L_1\omega^2) & -m_2 L_2 \omega^2 \\ -L_1 \omega^2 & g - L_2 \omega^2 \end{vmatrix} = 0$$

Expanding and arranging terms in powers of ω^2,

$$[(m_1 + m_2)L_1 L_2 - m_2 L_1 L_2](\omega^2)^2 - (m_1 + m_2)(L_2 g + L_1 g)\omega^2$$
$$+ (m_1 + m_2)g^2 = 0$$

or

$$m_1 L_1 L_2 \omega^4 - (m_1 + m_2)(L_1 + L_2)g\omega^2 + (m_1 + m_2)g^2 = 0$$

This algebraic equation presents no difficulties and has already been treated in Chapter 11 in connection with a two-mass system in rectilinear motion. For the special case $L_1 = L_2 = L$ and $m_1 = m_2 = m$, the equation in ω^2 becomes

$$mL^2 \omega^4 - 4Lmg\omega^2 + 2mg^2 = 0$$

or

$$(\omega^2)^2 - \frac{4g}{L}\omega^2 + 2\left(\frac{g}{L}\right)^2 = 0$$

The roots are $\omega_1^2 = (2 + \sqrt{2})(g/L)$ and $\omega_2^2 = (2 - \sqrt{2})(g/L)$, leading to the following equations of motion:

$$\psi_1 = A_1 \cos\left\{\left[(2 + \sqrt{2})\frac{g}{L}\right]^{1/2} t + \alpha_1\right\} + A_1' \cos\left\{\left[(2 - \sqrt{2})\frac{g}{L}\right]^{1/2} t + \alpha_1'\right\}$$

$$\psi_2 = A_2 \cos\left\{\left[(2 + \sqrt{2})\frac{g}{L}\right]^{1/2} t + \alpha_1\right\} + A_2' \cos\left\{\left[(2 - \sqrt{2})\frac{g}{L}\right]^{1/2} t + \alpha_1'\right\}$$

From either equation relating A_1, A_2, and ω, substitution for ω results in expressions for the ratios A_1/A_2 and A_1'/A_2'. From $\omega^2 = (2 + \sqrt{2})(g/L)$ we obtain $A_1/A_2 =$

$-\sqrt{2}/2$ and from $\omega^2 = (2 - \sqrt{2})(g/L)$, $A_1'/A_2' = \sqrt{2}/2$. The motion is now described as

$$\psi_1 = A_1 \cos(\omega_1 t + \alpha_1) + A_1' \cos(\omega_2 t + \alpha_1')$$

$$\psi_2 = \frac{\sqrt{2}}{2}[-A_1 \cos(\omega_1 t + \alpha_1) + A_1' \cos(\omega_2 t + \alpha_1')]$$

There remain four constants, A_1, A_1', α_1, α_1', to be evaluated on the basis of the initial conditions.

Consider now the following new coordinates, which are expressed below as functions of ψ_1 and ψ_2:

$$\Psi_1 = \frac{\sqrt{2}}{2}\psi_1 + \psi_2 = \sqrt{2}\, A_1 \cos(\omega_1 t + \alpha_1)$$

$$\Psi_2 = \frac{\sqrt{2}}{2}\psi_1 - \psi_2 = \sqrt{2}\, A_1' \cos(\omega_2 t + \alpha_1')$$

where Ψ_1 and Ψ_2 are the normal coordinates. Substitution of these coordinates into the original equations of motion yields

$$\ddot{\Psi}_1 + \omega_1^2 \Psi_1 = 0$$

$$\ddot{\Psi}_2 + \omega_2^2 \Psi_2 = 0$$

representing the uncoupled motions.

14.18 TENSOR METHODS FOR SMALL OSCILLATIONS

The problem of small oscillations can be described and solved more elegantly by employing the techniques of tensor analysis. Consider Eqs. (14.87) written in the form

$$\sum_{m=1}^{3n} V_{lm}A_m = \omega^2 \sum_{m=1}^{3n} T_{lm}A_m \qquad (l = 1, 2, \cdots, 3n) \tag{14.91}$$

which is identical to the $3n$ linear equations:

$$V_{11}A_1 + V_{12}A_2 + \cdots + V_{1,3n}A_{3n} = \omega^2 T_{11}A_1 + \omega^2 T_{12}A_2 + \cdots$$
$$+ \omega^2 T_{1,3n}A_{3n}$$
$$\vdots \tag{14.92}$$
$$V_{3n,1}A_1 + V_{3n,2}A_2 + \cdots + V_{3n,3n}A_{3n} = \omega^2 T_{3n,1}A_1 + \omega^2 T_{3n,2}A_2 + \cdots$$
$$+ \omega^2 T_{3n,3n}A_{3n}$$

The foregoing expressions [Eqs. (14.91) and (14.92)] may be written as the tensor equation

$$(V - \omega^2 T)A = 0 \qquad (14.93)$$

where V and T are described by the square arrays

$$V = \begin{bmatrix} V_{11} & V_{12} & \cdots & V_{1,3n} \\ \vdots & & & \vdots \\ V_{3n,1} & V_{3n,2} & \cdots & V_{3n,3n} \end{bmatrix}$$

$$(14.94)$$

$$T = \begin{bmatrix} T_{11} & T_{12} & \cdots & T_{1,3n} \\ \vdots & & & \vdots \\ T_{3n,1} & T_{3n,2} & \cdots & T_{3n,3n} \end{bmatrix}$$

A is the column vector:

$$\begin{bmatrix} A_1 \\ A_2 \\ \vdots \\ A_{3n} \end{bmatrix}$$

There is a vector A_k corresponding to each normal frequency ω_k. Determination of the normal coordinates and frequencies is a task equivalent to ascertaining that transformation of the equation

$$(V - \omega^2 T)A = 0$$

which results in the form

$$(V' - \omega^2 T')A = 0$$

where

$$V' - \omega^2 T' = \begin{bmatrix} V'_{11} - \omega^2 T'_{11} & 0 & 0 & \cdots & 0 \\ 0 & V'_{22} - \omega^2 T'_{22} & 0 & \cdots & 0 \\ 0 & 0 & V'_{33} - \omega^2 T'_{33} & \cdots & 0 \\ \vdots & \vdots & \vdots & & \vdots \\ 0 & 0 & 0 & & V'_{3n,3n} - \omega^2 T'_{3n,3n} \end{bmatrix}$$

V' and T' are thus diagonalized matrices. The existence of such a transformation is dependent upon the nature of T and V. The condition for simultaneous diagonalization which *must* hold is that T and V be symmetrical (assuming all elements are real) and either V or T be positive definite. The problem of diagonalization of a single tensor has already been encountered in rigid and deformable continua in connection with the inertia and stress tensors and corresponds to determination of certain principal axes.

14.19 SMALL OSCILLATIONS, APPLIED FORCES, AND NONCONSERVATIVE SYSTEMS

Having so far discussed only conservative systems in connection with small oscillation theory, we now introduce techniques applicable to particular nonconservative systems.

If frictional forces are present, they may sometimes be represented $-b\dot{x}_j$ in Cartesian coordinates. To obtain the correct Lagrange equations, a quantity $\mathfrak{F}$, termed the *Rayleigh dissipation function*,[4] is introduced. This quantity is defined

$$\mathfrak{F} = \frac{1}{2} \sum_{j=1}^{3n} b_j \dot{x}_j \tag{14.95}$$

and the modified equations become

$$\frac{d}{dt}\frac{\partial L}{\partial \dot{x}_j} - \frac{\partial L}{\partial x_j} + \frac{\partial \mathfrak{F}}{\partial \dot{x}_j} = 0 \qquad (j = 1, 2, \cdots, 3n) \tag{14.96}$$

In generalized coordinates $\mathfrak{F}$ is written

$$\mathfrak{F} = \frac{1}{2} \sum_{l=1}^{3n} \sum_{m=1}^{3n} b_{lm}\dot{q}_l\dot{q}_m \tag{14.97}$$

and Lagrange's equations are

$$\frac{d}{dt}\frac{\partial L}{\partial \dot{q}_l} - \frac{\partial L}{\partial \bar{q}_l} + \frac{\partial \mathfrak{F}}{\partial \dot{q}_l} = 0 \qquad (l = 1, 2, \cdots, 3n) \tag{14.98}$$

The equations of motion are consequently,

$$\sum_{m=1}^{3n} (T_{lm}\ddot{q}_m + V_{lm}\bar{q}_m + b_{lm}\dot{q}_m) = 0 \qquad (l = 1, 2, \cdots, 3n) \tag{14.99}$$

For the one-dimensional case we have

$$T_{11}\ddot{q}_1 + V_{11}\bar{q}_1 + b_{11}\dot{q}_1 = 0 \tag{14.100}$$

which is the equation describing the damped motion of a particle (see Chapter 6). The nature of its solution depends upon whether $(b_{11}/2T_{11})^2$ is greater than, less than, or equal to V_{11}/T_{11}. In any event, $\bar{q}_1$ takes the form $A_1 e^{-\gamma t}$, where γ may be complex.

For the three-dimensional case, it is clear that one should assume solutions

$$\bar{q}_m = A_m e^{-\gamma t} \qquad \text{real } (\gamma) > 0 \qquad (m = 1, 2, \cdots, 3n) \tag{14.101}$$

---

[4] Lord Rayleigh, *The Theory of Sound*, Dover Publications, Inc., New York, 1945.

In the event motion is overdamped, there are $6n$ values of γ, and $6n$ values of A corresponding to $6n$ initial conditions. The case of critical damping is rare and has been treated in Chapter 6. When underdamping occurs, γ is complex, and $\bar{q}_m$ may then be represented

$$\bar{q}_m = \sum_{k=1}^{3n} A_{mk} e^{-\gamma_k t} \cos(\omega_k t + \alpha_k) \tag{14.102}$$

There are $3n$ values of γ_k and ω_k, and $6n$ constants A_{km}, α_k to be evaluated from the initial conditions. Only under the most fortuitous of circumstances do normal coordinates exist when damping is present, and we shall not, therefore, dwell upon this matter.

A further modification occurs when an external force is applied to a mechanical system, initially in equilibrium. This is analogous to a group of electrical charges, acted upon by an electromagnetic wave,

$$\mathbf{E} = \mathbf{E}_0 \cos \omega_d t$$
$$\mathbf{B} = \mathbf{B}_0 \sin \omega_d t \tag{14.103}$$

where ω_d is a given driving frequency.

In general, the force is represented as $F_j = F_j(t)$, or in generalized coordinates $Q_m = Q_m(t)$. The force may be decomposed into a Fourier series, and the principle of superposition of linear systems applied. The resulting equations of motion in generalized coordinates are

$$\frac{d}{dt}\frac{\partial L}{\partial \dot{q}_l} - \frac{\partial L}{\partial \bar{q}_l} + \frac{\partial \mathfrak{F}}{\partial q_l} = \sum_k G_{lk} \cos(\omega_k t + \beta_k) \tag{14.104}$$

where L, as before, includes those forces derivable from a potential and G_{lk} is the amplitude of kth component of the generalized applied force. The corresponding equations of motion are

$$\sum_{m=1}^{3n} (T_{lm}\ddot{\bar{q}}_m + b_{lm}\dot{\bar{q}}_m + V_{lm}\bar{q}_m) = \sum_k G_{lk} \cos(\omega_k t + \beta_k)$$
$$(l = 1, 2, \cdots, 3n) \tag{14.105}$$

For a single-degree-of-freedom system,

$$T_{11}\ddot{\bar{q}}_1 + b_{11}\dot{\bar{q}}_1 + V_{11}\bar{q}_1 = \sum_k G_k \cos(\omega_k t + \beta_k) \tag{14.106}$$

The transient solution ($G_k = 0$), for damped motion, approaches zero for moderately short times and is taken to be zero. Again, considering a single applied frequency, the steady-state solution is

$$\bar{q}_1 = A_k \cos(\omega_k t + \beta) \tag{14.107}$$

This problem should have a familiar "ring," being the case of forced oscillations treated in Chapter 6. The values of A_k and β are determined by substituting Eq. (14.107) into the equation of motion.

For a $3n$-degree-of-freedom system, each coordinate oscillates at the driving frequency ω_k:

$$\bar{q}_m = A_m \cos(\omega_k t + \beta_m) \tag{14.108}$$

Substituting this trial solution into the equations of motion, we have

$$\sum_{m=1}^{3n} [-T_{lm}A_m\omega_k^2 \cos(\omega_k t + \beta_m) - b_{lm}A_m\omega_k \sin(\omega_k t + \beta_m)$$

$$+ V_{lm}A_m \cos(\omega_k t + \beta_m)]$$

$$= G_k \cos(\omega_k t + \beta_k) \quad (l = 1, 2, \cdots, 3n) \tag{14.109}$$

The cosine and sine terms of Eq. (14.109) are expanded and arranged according to whether they are multiplied by $\cos \omega_k t$ or $\sin \omega_k t$. The coefficient of each of these terms must vanish independently if Eq. (14.108) is to represent a solution valid for all values of time. There are thus $6n$ equations for the $6n$ unknowns, the A_m and the β_m. As an alternative approach, the complex notation of Chapter 6 may be employed:

$$e^{i\omega t} = \text{real } \{\cos \omega t\}$$

or

$$q_m = B_m e^{i\omega_k t}$$

These are now $3n$ unknowns, B_m, each of which is a complex number. The amplitude of the oscillation is the real part of B_m.

These methods are particularly applicable to electric circuit analysis and to mechanical system analysis by means of electrical analogues. The impressed voltages play the role of applied forces and the resistances, that of b_{lm}.

Illustrative Example 14.18
Determine the current in the electrical circuit shown in Fig. 14.10.

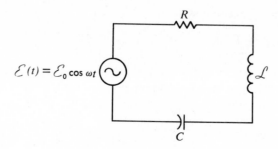

$$\mathcal{E}(t) = \mathcal{E}_0 \cos \omega t$$

Fig. 14.10

SOLUTION

Let us select the charge Q as the generalized coordinate. The kinetic energy is then $T = \frac{1}{2}\mathcal{L}\dot{Q}^2$ and the potential energy $V = \frac{1}{2}(Q^2/C)$. The Rayleigh dissipation function is $\mathfrak{F} = \frac{1}{2}R\dot{Q}^2$ and the generalized force $E_0 \cos \omega_d t$. Assuming small oscillations, so that $\mathcal{L}$, C, and R are constant, Eq. (14.104) becomes

$$\frac{d}{dt}\frac{\partial}{\partial \dot{Q}}\left(\frac{1}{2}\mathcal{L}\dot{Q}^2 - \frac{1}{2}\frac{Q^2}{C}\right) - \frac{\partial}{\partial Q}\left(\frac{1}{2}\mathcal{L}\dot{Q}^2 - \frac{1}{2}\frac{Q^2}{C}\right) + \frac{\partial}{\partial \dot{Q}}\frac{1}{2}(R\dot{Q}^2) = E_0 \cos \omega_d t$$

or

$$\mathcal{L}\ddot{Q} + R\dot{Q} + \frac{Q}{C} = E_0 \cos \omega_d t$$

For charge, we select the form

$$Q = A \cos(\omega_d t + \beta)$$

and substitute into the above differential equation:

$$-\mathcal{L}A\omega_d^2 \cos(\omega_d t + \beta) - RA\omega_d \sin(\omega_d t + \beta) + \frac{A}{C}\cos(\omega_d t + \beta) = E_0 \cos \omega_d t$$

Expanding $\cos(\omega_d t + \beta)$ and $\sin(\omega_d t + \beta)$ and collecting coefficients of $\cos \omega_d t$ and $\sin \omega_d t$,

$$\left(-\mathcal{L}A\omega_d^2 \cos \beta - RA\omega_d \sin \beta + \frac{A}{C}\cos \beta - E_0\right)\cos \omega_d t$$

$$+ \left(\mathcal{L}A\omega_d^2 \sin \beta - RA\omega_d \cos \beta - \frac{A}{C}\sin \beta\right)\sin \omega_d t = 0$$

The quantities in parentheses must be zero independently in order that the assumed expression for Q be a solution of the equation of motion. Thus

$$-\mathcal{L}A\omega_d^2 \cos \beta - RA\omega_d \sin \beta + \frac{A}{C}\cos \beta - E_0 = 0$$

$$\mathcal{L}A\omega_d^2 \sin \beta - RA\omega_d \cos \beta - \frac{A}{C}\sin \beta = 0$$

From the second expression,

$$\tan \beta = \frac{R}{\omega_d \mathcal{L} - \dfrac{1}{\omega_d C}} \qquad \sin \beta = \frac{R}{\left[R^2 + \left(\omega_d \mathcal{L} - \dfrac{1}{\omega_d C}\right)^2\right]^{\frac{1}{2}}}$$

$$\cos \beta = \frac{\omega_d \mathcal{L} - \dfrac{1}{\omega_d C}}{\left[R^2 + \left(\omega_d \mathcal{L} - \dfrac{1}{\omega_d C}\right)^2\right]^{\frac{1}{2}}}$$

From the first equation,

$$A = \frac{E_0}{-\mathcal{L}\omega_d^2 \cos\beta - R\omega_d \sin\beta + \frac{1}{C}\cos\beta} = \frac{E_0/\omega_d \cdot \left(\omega_d\mathcal{L} - \frac{1}{\omega_d C}\right)^2 - R^2}{\left[\left(\omega_d\mathcal{L} - \frac{1}{\omega_d C}\right)^2 + R^2\right]^{\frac{1}{2}}}$$

Thus

$$Q = \frac{-E_0/\omega_d}{\left[\left(\omega_d\mathcal{L} - \frac{1}{\omega_d C}\right)^2 + R^2\right]^{\frac{1}{2}}} \cos(\omega_d t + \beta)$$

and the current

$$I = \dot{Q} = \frac{E_0}{\left[\left(\omega_d\mathcal{L} - \frac{1}{\omega_d C}\right)^2 + R^2\right]^{\frac{1}{2}}} \sin(\omega_d t + \beta)$$

For a nondissipative system $R = 0$ and $\beta = 0$; the charge becomes

$$Q = \frac{-(E_0/\omega_d)\cos\omega_d t}{\omega_d\mathcal{L} - \frac{1}{\omega_d C}} = \frac{-(E_0/\mathcal{L})\cos\omega_d t}{\omega_d^2 - \omega^2}$$

where $\omega = (1/\mathcal{L}C)^{\frac{1}{2}}$ is the natural circular frequency of oscillation. At $\omega = \omega_d$, the resonant frequency, the maximum amplitude of Q is attained. The value of Q, in general, depends upon E_0, the driving voltage, and the closeness of the applied frequency to the resonant frequency.

14.20 IGNORABLE COORDINATES AND CONSTANTS OF THE MOTION

There are instances in which the Lagrangian is independent of one or more of the generalized coordinates, q_α. For such coordinates, one may write

$$\frac{dL}{dq_\alpha} = 0 \quad \text{and therefore} \quad \frac{d}{dt}\frac{\partial L}{\partial \dot{q}_\alpha} = 0$$

The quantity $\partial L/\partial \dot{q}_\alpha$, a first integral of an equation of motion, is thus a constant of the motion.

We now define a generalized momentum, conjugate to coordinate q_l, as follows:

$$p_l = \frac{\partial L}{\partial \dot{q}_l} \tag{14.110}$$

For the case of q_α, the coordinate is called *ignorable* or *cyclic*. For cyclic coordinates, the following general conservation principle applies: If a coordinate q_α does not appear in the Lagrangian, the conjugate momentum, p_α, is conserved.

Applying the foregoing concepts to a free particle, we begin by writing the kinetic and potential energies: $T = \frac{1}{2}m(\dot{x}^2 + \dot{y}^2 + \dot{z}^2)$ and $V = 0$. Consequently,

$$\frac{d}{dt}\frac{\partial L}{\partial \dot{x}} = \frac{d}{dt}(m\dot{x}) = 0$$

$$\frac{d}{dt}\frac{\partial L}{\partial \dot{y}} = \frac{d}{dt}(m\dot{y}) = 0$$

$$\frac{d}{dt}\frac{\partial L}{\partial \dot{z}} = \frac{d}{dt}(m\dot{z}) = 0$$

and we have the constant momenta:

$$p_x = m\dot{x} \qquad p_y = m\dot{y} \qquad p_z = m\dot{z}$$

The principle of conservation of linear momentum of a particle in the absence of net applied force is thus verified from yet another viewpoint.

When generalized coordinates are employed, the conjugate momenta need not, of course, possess dimensions of linear momentum. Consider a particle subject to a conservative central force $F(r) = -\partial V/\partial r$. We shall employ spherical polar coordinates, and therefore the following equations are required to transform the kinetic energy, $T = \frac{1}{2}m(\dot{x}^2 + \dot{y}^2 + \dot{z}^2)$:

$$x = r \sin \theta \cos \phi \qquad \dot{x} = \dot{r} \sin \theta \cos \phi + r\dot{\theta} \cos \theta \cos \phi - r\dot{\phi} \sin \theta \sin \phi$$

$$y = r \sin \theta \sin \phi \qquad \dot{y} = \dot{r} \sin \theta \sin \phi + r\dot{\theta} \cos \theta \sin \phi + r\dot{\phi} \sin \theta \cos \phi$$

$$z = r \cos \theta \qquad \dot{z} = \dot{r} \cos \theta - r\dot{\theta} \sin \theta$$

Substituting for the Cartesian components of velocity,

$$T = \frac{1}{2}m(\dot{r}^2 + r^2\dot{\theta}^2 + r^2\dot{\phi}^2 \sin^2 \theta)$$

Lagrange's equations are, for this case,

$$\frac{d}{dt}\frac{\partial L}{\partial \dot{r}} - \frac{\partial L}{\partial r} = \frac{d}{dt}m\dot{r} - mr\dot{\theta}^2 - mr\dot{\phi}^2 \sin^2 \theta + \frac{\partial V}{\partial r} = 0$$

$$\frac{d}{dt}\frac{\partial L}{\partial \dot{\theta}} - \frac{\partial L}{\partial \theta} = \frac{d}{dt}mr^2\dot{\theta} - mr^2\dot{\phi}^2 \sin \theta \cos \theta = 0$$

$$\frac{d}{dt}\frac{\partial L}{\partial \dot{\phi}} - \frac{\partial L}{\partial \phi} = \frac{d}{dt}mr^2\dot{\phi} \sin^2 \theta = 0$$

On the basis of the last expression, we may immediately write a first integral of the motion:

$$\frac{\partial L}{\partial \dot{\phi}} = mr^2\dot{\phi} \sin^2 \theta = p_\phi = \text{constant}$$

Note that the Lagrangian is not a function of ϕ. When this result is substituted into the θ-equation of motion, the following obtains:

$$\frac{d}{dt} mr^2\dot{\theta} - \frac{\partial}{\partial\theta} \frac{1}{2} p_\phi \dot{\phi} = \frac{d}{dt} mr^2\dot{\theta} = 0$$

and another constant momentum is determined:

$$\frac{\partial L}{\partial\dot{\theta}} = mr^2\dot{\theta} = p_\theta = \text{constant}$$

There are, therefore, two constants of the motion, p_θ and p_ϕ. When these are appropriately inserted into the r-equation, the result is

$$\frac{d}{dt} m\dot{r} - \frac{p_\theta^2}{mr^3} - \frac{p_\phi^2}{mr^3 \sin^2\theta} + \frac{\partial V}{\partial r} = 0$$

It should be recalled that in the treatment of central force motion (Chapter 8) the XY-plane was chosen as the plane of motion. Consequently, $p_\theta = 0$ ($\dot{\theta} = 0$) and $p_\phi = mr^2\dot{\phi} = H$. This agrees with the conclusions then drawn with respect to the conservation of angular momentum in central force motion.

The term "angular momentum" may be explained as the generalized momentum, conjugate to the angular coordinate, and we thus have a rationale for the use of this expression in place of the somewhat more descriptive "moment of linear momentum."

We now verify that energy is a constant of the motion in conservative systems. Subject to the restrictions of time-independent constraints and motion referred to an inertial frame of reference, the Lagrangian is not an explicit function of time and is expressed

$$L = L(q_1, q_2, \cdots, q_{3n}, \dot{q}_1, \dot{q}_2, \cdots, \dot{q}_{3n}) \tag{14.111}$$

The time derivative of Eq. (14.111) is

$$\frac{dL}{dt} = \sum_{l=1}^{3n} \left(\frac{\partial L}{\partial q_l} \dot{q}_l + \frac{\partial L}{\partial\dot{q}} \ddot{q}_l \right) \tag{14.112}$$

Consider now Lagrange's equation expressed as follows:

$$\frac{d}{dt} \frac{\partial L}{\partial\dot{q}_l} = \frac{\partial L}{\partial q_l} = \frac{d}{dt} p_l = \dot{p}_l \tag{14.113}$$

where $\partial L/\partial\dot{q}_l = p_l$. Substitution of Eq. (14.113) into Eq. (14.112) yields

$$\frac{dL}{dt} = \sum_{l=1}^{3n} (\dot{p}_l\dot{q}_l + p_l\ddot{q}_l) = \frac{d}{dt} \sum_{l=1}^{3n} p_l\dot{q}_l \tag{14.114}$$

Thus

$$\frac{d}{dt} \left(\sum_{l=1}^{3n} p_l\dot{q}_l - L \right) = 0 \tag{14.115}$$

$$\sum_{l=1}^{3n} p_l \dot{q}_l - L = H = \text{constant} \tag{14.116}$$

where H is the *Hamiltonian*.

To demonstrate that H is the total energy for a *conservative system*, consider a Cartesian representation of the generalized coordinates. The momentum $p_l = m_l \dot{x}_l$ and $\dot{q}_l = \dot{x}_l$. The Hamiltonian is now written

$$H = \sum_{l=1}^{3n} (m_l \dot{x}_l^2 - \tfrac{1}{2} m_l \dot{x}_l^2 + V) = \sum_{l=1}^{3n} \tfrac{1}{2} m_l \dot{x}_l^2 + V = E$$

Since H is a constant, any transformation to generalized coordinates does not alter its meaning or magnitude.

When a time-dependent Lagrangian is encountered, we retain, for generality, the definition of H introduced above, although H is no longer constant. Let Eq. (14.116) serve as a transformation relating the Lagrangian (as a function of generalized coordinates, generalized velocities, and time) to the Hamiltonian (as a function of the generalized coordinates, conjugate momenta, and time):

$$L = L(q_1, q_2, \cdots, q_{3n}, \dot{q}_1, \dot{q}_2, \cdots, \dot{q}_{3n}, t) \tag{14.117}$$

$$H = H(q_1, q_2, \cdots, q_{3n}, p_1, p_2, \cdots, p_{3n}, t) \tag{14.118}$$

Consider the total differential of H, determined from Eq. (14.116):

$$dH = \sum_{l=1}^{3n} p_l \, d\dot{q}_l + \sum_{l=1}^{3n} \dot{q}_l \, dp_l - \sum_{l=1}^{3n} \frac{\partial L}{\partial q_l} dq_l - \sum_{l=1}^{3n} \frac{\partial L}{\partial \dot{q}_l} d\dot{q}_l - \frac{\partial L}{\partial t} dt \tag{14.119}$$

On the basis of the definition, $p_l = \partial L / \partial \dot{q}_l$, the first and fourth terms cancel, leaving

$$dH = \sum_{l=1}^{3n} \dot{q}_l \, dp_l - \sum_{l=1}^{3n} \frac{\partial L}{\partial q_l} dq_l - \frac{\partial L}{\partial t} dt \tag{14.120}$$

From Eq. (14.118),

$$dH = \sum_{l=1}^{3n} \frac{\partial H}{\partial q_l} dq_l + \sum_{l=1}^{3n} \frac{\partial H}{\partial p_l} dp_l + \frac{\partial H}{\partial t} dt \tag{14.121}$$

Terms in Eqs. (14.120) and (14.121), multiplied by like differentials, may be equated inasmuch as each coordinate and momentum varies independently of all the others,

$$\dot{q}_l = \frac{\partial H}{\partial p_l} \tag{14.122a}$$

$$-\frac{\partial L}{\partial q_l} = \frac{\partial H}{\partial q_l} \tag{14.122b}$$

$$-\frac{\partial L}{\partial t} = \frac{\partial H}{\partial t} \tag{14.122c}$$

Limiting our discussion to systems devoid of generalized forces, except for those derivable from a potential function V or U,

$$\frac{d}{dt}\frac{\partial L}{\partial \dot{q}_l} = \dot{p}_l = \frac{\partial L}{\partial q_l} \qquad (14.123)$$

Equation (14.122b) is therefore

$$-\dot{p}_l = \frac{\partial H}{\partial q_l} \qquad (14.124)$$

The expressions given by Eqs. (14.122a), (14.122c), and (14.124) are known as *Hamilton's equations of motion*.[5] These are $6n$ first-order differential equations in the generalized coordinates and momenta rather than the $3n$ second-order equations associated with Lagrange's equations.

To further interpret the Hamiltonian, consider a particle in one-dimensional motion, subject to a force derivable from a time-independent potential, $F = -dV/dx$. The Lagrangian, momentum, and Hamiltonian are

$$L = \tfrac{1}{2}m\dot{x}^2 - V(x)$$

$$p = \frac{\partial L}{\partial \dot{x}} = m\dot{x}$$

$$H = p\dot{x} - L = p\dot{x} - \tfrac{1}{2}m\dot{x}^2 + V(x)$$

Since H is a function of x and p and not of $\dot{x}$ [Eq. (14.118)], we substitute $\dot{x} = p/m$ and the Hamiltonian is now expressed

$$H = \frac{p^2}{m} - \frac{p^2}{2m} + V(x) = \frac{p^2}{2m} + V(x) = E$$

For the case at hand, Hamilton's equations yield

$$\frac{\partial H}{\partial q} = \frac{\partial H}{\partial x} = \frac{\partial V}{\partial x} = -F = -\dot{p}$$

which is, of course, Newton's second law, and

$$\frac{\partial H}{\partial p} = \frac{p}{m} = \dot{x}$$

which provides no new information but is merely a restatement of the definition of momentum.

In summary, the application of Hamilton's equations to a system of particles is expedited by following the procedure outlined below:

a. Begin with the most comfortable coordinate system (usually Cartesian), and establish the Lagrangian. By applying appropriate transformation equations, the Lagrangian is then written in generalized coordinates.

b. Apply Eq. (14.110) to define the generalized momenta.

c. Construct the Hamiltonian by eliminating the generalized velocities.

---

[5] Hamilton, W. R., *Brit. Assoc. Rept.*, p. 513 (1834).

d. Employ Eqs. (14.122) and (14.124) to determine Hamilton's equations of motion.

The strength and utility of Hamilton's equations lie not with the fact that they yield yet another approach to Newtonian mechanics, but rather in the transformation theory of mechanics to which they lead and in the relationship they offer between classical and modern physics. As for the former, we shall state only that for a given problem it may be possible to determine a transformation to new generalized coordinates and momenta, all of which are constants. It is developed in more advanced works that a generating function exists, related to the Hamiltonian, which is an additional constant of the motion. Once determined, the transformation may, of course, be applied again in problems of a similar nature.

14.21 QUANTUM MECHANICS

As alluded to above, a compelling reason for developing Hamilton's equations rests with the smooth transition to quantum mechanics flowing directly from the classical Hamiltonian. In the final pages of this text, it is fitting that the way be pointed (although no claim is made to presenting a definitive treatment) to quantum mechanics so that something be shown of the sweep of mechanics.

As we have observed, the Hamiltonian represents the total energy of a conservative system, so that in quantum mechanics, as in classical mechanics, $H = E$. The distinction is now made, however, that in quantum mechanics H and E serve as differential operators, manifesting no physical meaning in and of themselves. The significance of these terms arises from their operation on a mathematical function Ψ, called the wave function, leading to the following expression, known as the *Schrödinger wave equation*[6]:

$$H\Psi = E\Psi \tag{14.125}$$

The Hamiltonian of quantum mechanics is formulated from that of classical physics by substituting for p_j, the Cartesian momentum, the term $-i\hbar(\partial/\partial x_j)$ where $i = \sqrt{-1}$ and $\hbar$ is Planck's constant divided by 2π. The energy operator is $E = i\hbar(\partial/\partial t)$. Consequently, the Hamiltonian of a single particle, transformed from classical Cartesian to quantum representation, is written

$$H = \frac{p_x^2}{2m} + \frac{p_y^2}{2m} + \frac{p_z^2}{2m} + V(x, y, z)$$

$$= -\frac{\hbar^2}{2m}\frac{\partial^2}{\partial x^2} - \frac{\hbar^2}{2m}\frac{\partial^2}{\partial y^2} - \frac{\hbar^2}{2m}\frac{\partial^2}{\partial z^2} + V(x, y, z)$$

and the wave equation is thus a complex second-order partial differential equation in four variables:

$$\left[-\frac{\hbar^2}{2m}\frac{\partial^2}{\partial x^2} - \frac{\hbar^2}{2m}\frac{\partial^2}{\partial y^2} - \frac{\hbar^2}{2m}\frac{\partial^2}{\partial z^2} + V(x, y, z) \right]\Psi = i\hbar\frac{\partial\Psi}{\partial t} \tag{14.126}$$

---

[6] Schrödinger, E., *Ann. Physik* **79**, 361 (1926); **80**, 437 (1926); **81**, 109 (1926).

The solution of this equation does not end the mechanics problem because the wave function, Ψ, is an unobservable quantity without, in itself, physical meaning. The first interpretation of Ψ was offered by M. Born[7]: The product of the wave function and its complex conjugate Ψ^*, integrated over a region in space, is equal to the probability of finding the particle, associated with Ψ, in that region. Thus $\Psi^*\Psi\, dV$ represents the probability of finding the particle within dV. It follows that this probability is unity, if the integration is carried out over all space. Thus

$$\int_{space} \Psi^*\Psi\, dV = 1 \tag{14.127}$$

Illustrative Example 14.19

A particle confined within a three-dimensional box is subject to no forces except those occurring as a result of interactions with the walls. Determine the wave function.

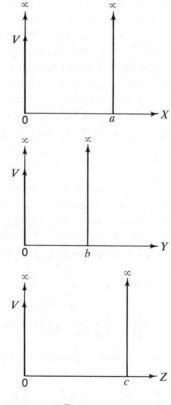

Fig. 14.11

---

[7] Born, M., Z. Physik **37**, 863 (1926); **38**, 803 (1926).

SOLUTION

We shall, of course, require some information regarding the nature of the potential energy. This quantity is zero everywhere within the box ($0 < x < a$, $0 < y < b$, $0 < z < c$), but must be such as to prevent the particle from penetrating the walls regardless of particle energy. This may be expressed by stating that the potential at the walls is infinite; the wall exerts a force upon the particle only when contact occurs (see Fig. 14.11).

Since the particle must be located within the box, the boundary conditions applied to Ψ are

$$\Psi(0, y, z, t) = \Psi(a, y, z, t) = 0$$

$$\Psi(x, 0, z, t) = \Psi(x, b, z, t) = 0$$

$$\Psi(x, y, 0, t) = \Psi(x, y, c, t) = 0$$

The wave equation for the region within the box (where $V = 0$) is

$$-\frac{\hbar^2}{2m}\left(\frac{\partial^2\Psi}{\partial x^2} + \frac{\partial^2\Psi}{\partial y^2} + \frac{\partial^2\Psi}{\partial z^2}\right) = i\hbar\frac{\partial\Psi}{\partial t}$$

We attempt a solution by separation of variables, beginning with the product solution

$$\Psi(x, y, z, t) = X(x)Y(y)Z(z)T(t)$$

Making appropriate substitutions for the various partial derivatives, the wave equation, after dividing by $\Psi = XYZT$, becomes

$$-\frac{\hbar^2}{2m}\left(\frac{1}{X}\frac{d^2X}{dx^2} + \frac{1}{Y}\frac{d^2Y}{dy^2} + \frac{1}{Z}\frac{d^2Z}{dz^2}\right) = \frac{i\hbar}{T}\frac{dT}{dt}$$

As in the case of the vibrating string, each side of the above equation must equal a constant, ϵ. The following equations result:

$$-\frac{\hbar^2}{2m}\left(\frac{1}{X}\frac{d^2X}{dx^2} + \frac{1}{Y}\frac{d^2Y}{dy^2} + \frac{1}{Z}\frac{d^2Z}{dz^2}\right) = \epsilon$$

and

$$\frac{i\hbar}{T}\frac{dT}{dt} = \epsilon$$

Multiplying both sides of the equation for T by $-iT/\hbar$, the solution is readily found to be

$$T = Ce^{-i\epsilon t/\hbar}$$

In the remaining equation, the terms

$$\frac{1}{X}\frac{d^2X}{dx^2} \qquad \frac{1}{Y}\frac{d^2Y}{dy^2} \qquad \frac{1}{Z}\frac{d^2Z}{dz^2}$$

are each functions of a single independent variable, and consequently the sum of these quantities can equal a constant only if each is constant. Since m and h are positive, the separation constant must be negative, r representing the energy, a positive number.

Denoting the separation constants as $-\alpha_x^2$, $-\alpha_y^2$, and $-\alpha_z^2$, we are led to the following equations:

$$\frac{1}{X} \frac{d^2 X}{dx^2} = -\alpha_x^2$$

$$\frac{1}{Y} \frac{d^2 Y}{dy^2} = -\alpha_y^2$$

$$\frac{1}{Z} \frac{d^2 Z}{dz^2} = -\alpha_z^2$$

for which the solutions are

$$X(x) = A_x \cos \alpha_x x + B_x \sin \alpha_x x$$

$$Y(y) = A_y \cos \alpha_y y + B_y \sin \alpha_y y$$

$$Z(z) = A_z \cos \alpha_z z + B_z \sin \alpha_z z$$

Applying the boundary conditions at $x = y = z = 0$, it is concluded that $A_x = A_y = A_z = 0$. From the conditions at $x = a$, $y = b$, $z = c$,

$$B_x \sin \alpha_x a = 0$$

$$B_y \sin \alpha_y b = 0$$

$$B_z \sin \alpha_z c = 0$$

If B_x, B_y, or B_z is zero, $\Psi = 0$ inside the box, and a trivial solution results. We conclude therefore that the constants are different from zero, and consequently

$$\sin \alpha_x a = 0 \qquad \alpha_x = \frac{n_x \pi}{a} \qquad n_x = 1, 2, \cdots$$

$$\sin \alpha_y b = 0 \qquad \alpha_y = \frac{n_y \pi}{b} \qquad n_y = 1, 2, \cdots$$

$$\sin \alpha_z c = 0 \qquad \alpha_z = \frac{n_z \pi}{c} \qquad n_z = 1, 2, \cdots$$

The quantities n_x, n_y, and n_z are related to ϵ by the expression

$$\frac{\hbar^2}{2m} \left(\frac{n_x^2 \pi^2}{a^2} + \frac{n_y^2 \pi^2}{b^2} + \frac{n_z^2 \pi^2}{c^2} \right) = \epsilon$$

Thus not all ϵ are permitted, but only those prescribed by integral values of n_x, n_y, n_z.

Assembling the components of Ψ, we have

$$\Psi = B_x B_y B_z C \sin \frac{n_x \pi x}{a} \sin \frac{n_y \pi y}{b} \sin \frac{n_z \pi z}{c} e^{-i\epsilon t/\hbar}$$

$$\Psi = B \sin \frac{n_x \pi x}{a} \sin \frac{n_y \pi y}{b} \sin \frac{n_z \pi z}{c} e^{-i\epsilon t/\hbar}$$

where constants have been lumped.

Since the particle must exist within the box,

$$\int_0^a \int_0^b \int_0^c \Psi^* \Psi \, dx \, dy \, dz = \int_0^a \int_0^b \int_0^c |\Psi|^2 \, dx \, dy \, dz = 1$$

and

$$\int_0^a \int_0^b \int_0^c |B|^2 \sin^2 \frac{n_x \pi x}{a} \sin^2 \frac{n_y \pi y}{b} \sin^2 \frac{n_z \pi z}{c} \, dx \, dy \, dz = 1$$

The result of this integration is

$$|B|^2 \frac{abc}{8} = 1$$

or

$$|B| = \left(\frac{8}{abc} \right)^{\frac{1}{2}}$$

Note that the constant B cannot be uniquely ascertained, since it may be a complex number. In addition, since the wave equation is linear, the complete solution is the sum of all possible solutions, a triply infinite set described by

$$\Psi = \left(\frac{8}{abc} \right)^{\frac{1}{2}} \sum_{n_x} \sum_{n_y} \sum_{n_z} \sin \left(\frac{n_x \pi x}{a} \right) \sin \left(\frac{n_y \pi y}{b} \right) \sin \left(\frac{n_z \pi z}{c} \right) e^{-i[(\epsilon t/\hbar) - \gamma]}$$

where γ, a constant phase factor associated with B, has been introduced and n_x, n_y, n_z are positive integers.

Returning to the energy ϵ, expressed in terms of n_x, n_y, n_z, recall that only discrete values of ϵ are permitted. Here we have a fundamental distinction between classical and quantum physics. Although we have observed quantized effects in classical mechanics, such as the permitted wavelengths of a vibrating string, the quantized energy states of a particle represent a new and seemingly strange concept.

A reconciliation between classical and quantum physics may be found in a careful examination of the energy. Although until now we have used literal representations of energy, let us now employ some numbers to emphasize several points of importance.

Suppose an electron is placed within a box 1 mm on a side. According to the equation for ϵ, its minimum energy, corresponding to $n_x = n_y = n_z = 1$, is

$$\epsilon = \frac{\pi^2 \hbar^2}{2m} \frac{3}{a^2} \cong 1.6 \times 10^{-29} \text{ J}$$

or approximately 10^{-10} eV. This is an apparently immeasurable quantity. Now suppose $n_x = n_y = n_z = 10^6$. The electronic energy is now 100 eV and the energy of an adjacent electronic state ($n_x = n_y = 10^6$, $n_z = 10^6 + 1$) is indistinguishable from that corresponding to $n_x = n_y = n_z = 10^6$. It is clear, therefore, that for high quantum numbers the energy levels appear continuous. Furthermore, if one examines the wave functions associated with a particle in a one-dimensional box, it is apparent that the probability of finding the particle within a length dx, $|\psi|^2 \, dx$, becomes independent of where dx is located (between 0 and a) provided the n_x is large. For low quantum numbers, this is not the case.

The reader may be inclined to argue that 1 mm is too large a dimension to select (since ϵ varies as $1/a^2$). If a series of measurements are to be made along the direction of a, this distance is indeed realistic in terms of an experiment involving an electron. If, instead, a had been selected as, say, 10^{-15} m, which, not coincidentally, is approximately a nuclear radius, the minimum electron energy would have been found to be so large that even the nucleus could not contain it, and the assumption regarding the impenetrability of the container walls would lose all physical validity.

The theory and conclusions of quantum mechanics may be difficult to accept (especially in the context of the superficial treatment here presented), but this may apply to the special theory of relativity as well. The outstanding success of quantum mechanics in predicting the results of experiments in atomic physics, and the fact that results of classical mechanics may be demonstrated as approximating those of quantum mechanics, are strong supporting arguments. Quantum mechanics is not very successful, however, in predicting most nuclear phenomena. Nevertheless, from its foundations one hopes will develop the theories pertinent to the behavior of nuclear forces and interactions.

References

Corben, H. C., and Stehle, P., *Classical Mechanics*, Second Edition, John Wiley & Sons, Inc., New York, 1960.

Eisberg, R. M., *Fundamentals of Modern Physics*, John Wiley & Sons, Inc., New York, 1961.

Fong, P., *Elementary Quantum Mechanics*, Addison-Wesley Publishing Company, Inc., Reading, Mass., 1962.

Goldstein, H., *Classical Mechanics*, Addison-Wesley Publishing Company, Inc., Reading, Mass., 1950.

Kilmister, C. W., *Hamiltonian Dynamics*, John Wiley & Sons, Inc., New York, 1964.

Lanczos, C., *The Variational Principles of Mechanics*, University of Toronto Press, Toronto, 1949.

Osgood, W. F., *Advanced Calculus*, The Macmillan Company, New York, 1925.

Osgood, W. F., *Mechanics*, The Macmillan Company, New York, 1946.

Symon, K. R., *Mechanics*, Second Edition, Addison-Wesley Publishing Company, Inc., Reading, Mass., 1960.

Whittaker, E. T., *Analytical Dynamics*, Fifth Edition, Dover Publications, Inc., New York, 1944.

Zajac, A., *Basic Principles and Laws of Mechanics*, D. C. Heath and Company, Boston, 1966.

EXERCISES

14.1 A hoop of radius R_0 and mass m_1 is constrained to move on a horizontal plane. Apply the principle of virtual work to a uniform rod of length R_0 and mass m_2, itself constrained to move within the hoop, maintaining contact with the inner surface at all times. Determine the equilibrium angle of the rod.

14.2 An ellipse is described by the expression

$$\frac{x^2}{a^2} + \frac{y^2}{b^2} = 1$$

Determine the angle which a rod of length L and variable mass per unit length ρ makes with the horizontal if constrained to remain in contact with the inner surface of the ellipse.

14.3 A particle of mass m is acted upon by a force

$$\mathbf{F} = F_x\hat{\mathbf{i}} + F_y\hat{\mathbf{j}} + F_z\hat{\mathbf{k}}$$

Determine the equations of motion in spherical polar coordinates and interpret the generalized forces Q_r, Q_θ, Q_ϕ.

14.4 A small sphere of radius R_1 rolls without slipping over a second sphere of radius R_2. If the smaller sphere starts from rest at the highest point of the larger sphere, at what point do they separate?

14.5 A simple pendulum is suspended inside a hoop as shown in Fig. P14.5. Derive the equations of motion for the system as the hoop rolls without slipping down the incline.

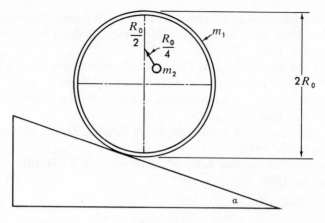

Fig. P14.5

14.6 A particle of mass m is constrained to move on the inner surface of a cone of angle $\alpha/2$ as shown in Fig. P14.6. Determine the motion and demonstrate that the Z-coordinate varies between the limits z_1 and z_2. Assume frictionless constraint.

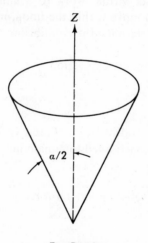

Fig. P14.6

14.7 Apply Lagrange's equations to determine the motion of an Atwood's machine. Do not neglect pulley inertia.

14.8 Apply Lagrange's equations to the motion of a double Atwood's machine.

14.9 Determine the motion of each mass of an Atwood's machine. Include the effects of pulley inertia and the mass of the connecting rope (total length L, mass m).

14.10 Employ the Lagrangian to formulate the equations of motion of the spherical pendulum.

14.11 Establish the Lagrangian for the double spherical pendulum.

14.12 Derive the equations of motion of point masses m_1 and m_2 subject only to their mutual gravitational influences. For generalized coordinates, select the coordinates of the center of mass and the reduced mass.

14.13 Employ the Lagrangian to derive the equations of motion of the system shown in Fig. P14.13.

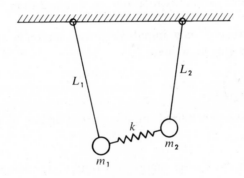

Fig. P14.13

14.14 A bead of mass m is constrained to move on a branch of a hyperbola. Determine the equations of motion as well as the forces of constraint.

14.15 A particle under the influence of gravity is constrained to move on the surface of an ellipsoid of revolution. Apply the method of undetermined coefficients in determining the points of equilibrium and reaction forces.

14.16 A particle in a gravitational field is constrained to move in a plane along a cycloidal curve. Apply the method of undetermined coefficients to describe the motion, the points of equilibrium, and the reactions.

14.17 A particle of mass m moves along the inner surface of a paraboloid of revolution. What angular velocity must a radius vector running from the center line to the particle have in order for the plane of this vector to remain horizontal?

14.18 A ladder of mass m and length L rests against a frictionless wall and is inclined at an angle ϕ relative to the vertical. A man of mass $3m$ climbs the ladder at a uniform velocity. Apply Lagrange's equations to determine the equations of motion and the point at which the ladder begins to slip. At what angle will the ladder leave the wall if the man stops climbing at the instant slip begins? The coefficients of static and kinetic friction are μ_s and μ_k, respectively.

14.19 Repeat Exercise 14.1 for the case in which a constant horizontal force F is applied to the hoop.

14.20 Apply Lagrange's equations to a symmetrical top. Determine the equations of motion and the generalized momenta.

14.21 A wire is bent into the shape of a circle and rotated about a point on its circumference with constant angular velocity. What are the equations of motion and forces of constraint pertaining to a bead constrained to move on the wire?

14.22 Demonstrate that if only impulsive forces act on a system of particles, Lagrange's equations may be written

$$\left(\frac{\partial L}{\partial \dot{q}_l}\right)_{\text{after}} - \left(\frac{\partial L}{\partial \dot{q}_l}\right)_{\text{before}} = I_l$$

where I_l represents the generalized impulse corresponding to generalized coordinate q_l.

14.23 Beginning with the kinetic energy of a rigid body rotating about the center of mass written in Cartesian coordinates, derive the equations for the angular rates $\dot{\psi}, \dot{\theta}, \dot{\phi}$ (the time derivatives of the Euler angles) for the case of torque-free motion. Use Lagrange's equations.

14.24 Demonstrate that for small oscillations the restriction stating that the transformation equations cannot be explicit functions of time is too severe, since transformation equations involving two inertial frames of reference yield the same differential equations of motion.

14.25 Demonstrate that $\omega^2 > 0$, in the general problem of small oscillations, is consistent with a minimum in potential energy.

14.26 Determine the normal coordinates and normal frequencies for small oscil-

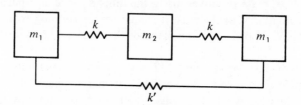

Fig. P14.26

lations of the linear system of three masses shown in Fig. P14.26. This model may be regarded as representing a triatomic linear molecule, such as CO_2.

14.27 Repeat Exercise 14.26 for the system depicted in Fig. P14.27. Assumed all motion to be planar. There are six degrees of freedom but only three nonzero normal frequencies.

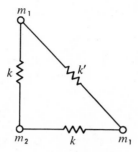

Fig. P14.27

14.28 A mass m, constrained to move along a horizontal line, has connected to it a simple pendulum of mass m_p and length L, as shown in Fig. P14.28. If m is itself connected to a rigid wall by means of a linear spring, determine the Lagrangian of the system as well as the normal modes of oscillation.

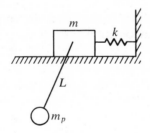

Fig. P14.28

14.29 Derive the equations of motion of the two masses of an Atwood's machine if one of the masses experiences small oscillations as a simple pendulum with slowly varying length.

14.30 Two pendulous masses m are connected by a spring of negligible mass. What is the Lagrangian of the system? Determine the normal modes for small oscillation.

14.31 Assume a uniform spring to be composed of a finite number of equal masses, uniformly spaced, and connected to one another by identical springs. If the masses are restricted to small transverse motion, demonstrate that as the number of masses grows without limit (and the individual masses each approach zero), the equations of motion reduce to the wave equation of a continuous string.

14.32 Demonstrate that for a relativistic particle, the Lagrangian, defined as

$$L = m_0 c^2 \left[\frac{1}{(1 - v^2/c^2)^{1/2}} - 1 \right] - V$$

where v is the velocity and V the potential energy, does not generate the correct equations of motion, but that when it is defined

$$L = -m_0 c^2 \left[1 - \frac{v^2}{c^2} \right]^{1/2} - V$$

the correct results obtain.

14.33 Apply the Lagrangian of Exercise 14.32 to develop expressions for the relativistic momentum and the Hamiltonian.

14.34 Determine the Hamiltonian for a charged particle in a time-dependent electromagnetic field.

14.35 Apply Hamilton's equations to the motion of a spherical pendulum.

14.36 Apply Lagrange's equations to the Foucault pendulum. Demonstrate, using the Hamiltonian, that one component of angular momentum is constant.

BIBLIOGRAPHY

ELEMENTARY MECHANICS

Feynman, R. P., Leighton, R. B., and Sands, M., *The Feynman Lectures on Physics*, Vol. 1, Addison-Wesley Publishing Company, Inc., Reading, Mass., 1963.

Halliday, D., and Resnick, R., *Physics*, Part 1, John Wiley & Sons, Inc., New York, 1966.

Kittel, C., Knight, W. D., and Ruderman, M. A., *Mechanics*, McGraw-Hill, Inc., New York, 1965.

Sears, F. W., and Zemansky, M. W., *University Physics*, Third Edition, Addison-Wesley Publishing Company, Inc., Reading, Mass., 1963.

Taylor, E. F., *Introductory Mechanics*, John Wiley & Sons, Inc., New York, 1964.

INTERMEDIATE MECHANICS

Becker, R. A., *Introduction to Theoretical Mechanics*, McGraw-Hill, Inc., New York, 1954.

Blass, G. A., *Theoretical Physics*, Appleton-Century-Crofts, New York, 1962.

Christie, D. E., *Vector Mechanics*, Second Edition, McGraw-Hill, Inc., New York, 1964.

Constant, F. W., *Theoretical Physics*, Addison-Wesley Publishing Company, Inc., Reading, Mass., 1954.

Fowles, G. R., *Analytical Mechanics*, Holt, Rinehart and Winston, Inc., New York, 1962.

Fox, E. A., *Mechanics*, Harper & Row, Publishers, New York, 1967.

Goodman, L. E., and Warner, W. H., *Dynamics*, Wadsworth Publishing Co., Inc., Belmont, Calif., 1963.

Greenwood, D. T., *Principles of Dynamics*, Prentice-Hall, Inc., Englewood Cliffs, N.J., 1965.

Housner, G. W., and Hudson, D. E., *Applied Mechanics, Statics*, D. Van Nostrand Company, Inc., Princeton, N.J., 1954.

Housner, G. W., and Hudson, D. E., *Applied Mechanics, Dynamics*, Second Edition, D. Van Nostrand Company, Inc., Princeton, N.J., 1959.

Lindsay, R. B., *Physical Mechanics*, Third Edition, D. Van Nostrand Company, Inc., Princeton, N.J., 1961.

Lindsay, R. B., and Margenau, H., *Foundations of Physics*, John Wiley & Sons, Inc., New York, 1936.

Long, R. R., *Engineering Science Mechanics*, Prentice-Hall, Inc., Englewood Cliffs, N.J., 1963.

Marion, D. B., *Classical Dynamics of Particles and Systems*, Academic Press, Inc., New York, 1965.

Meriam, J. L., *Dynamics*, John Wiley & Sons, Inc., New York, 1966.

Nara, H. R., *Vector Mechanics for Engineers*, Parts 1 and 2, John Wiley & Sons, Inc., New York, 1962.

Osgood, W. F., *Mechanics*, The Macmillan Company, New York, 1946.

Pletta, D. H., and Frederick, D., *Engineering Mechanics*, The Ronald Press Company, New York, 1964.

Shames, I. H., *Engineering Mechanics*, Second Edition, Prentice-Hall, Inc., Englewood Cliffs, N.J., 1967.

Stephenson, R. J., *Mechanics and Properties of Matter*, Second Edition, John Wiley & Sons, Inc., New York, 1960.

Symon, K. R., *Mechanics*, Second Edition, Addison-Wesley Publishing Company, Inc., Reading, Mass., 1960.

Synge, J. L., and Griffith, B. A., *Principles of Mechanics*, Third Edition, McGraw-Hill, Inc., New York, 1959.

Timoshenko, S. P., and Young, D. H., *Advanced Dynamics*, McGraw-Hill, Inc., New York, 1948.

Wangsness, R. K., *Introduction to Theoretical Physics*, John Wiley & Sons, Inc., New York, 1963.

ADVANCED MECHANICS

Corben, H. C., and Stehle, P., *Classical Mechanics*, Second Edition, John Wiley & Sons, Inc., New York, 1960.

Goldstein, H., *Classical Mechanics*, Addison-Wesley Publishing Company, Inc., Reading, Mass., 1950.

Joos, G., *Theoretical Physics*, Third Edition, Hafner Publishing Company, New York, 1956.

Kilmister, C. W., *Hamiltonian Dynamics*, John Wiley & Sons, Inc., New York, 1964.

Lanczos, C., *The Variational Principles of Mechanics*, University of Toronto Press, Toronto, 1949.

Landau, L. D., and Lifshitz, E. M., *Mechanics*, Addison-Wesley Publishing Company, Inc., Reading, Mass., 1960.

Sommerfeld, A., *Mechanics*, Translated by M. O. Stern, Academic Press, Inc., New York, 1957.

Whittaker, E. T., *Analytical Dynamics*, Fifth Edition, Dover Publications, Inc., New York, 1944.

Zajac, A., *Basic Principles and Laws of Mechanics*, D. C. Heath and Company, Boston, 1966.

ELASTICITY

Cottrell, A. H., *The Mechanical Properties of Matter*, John Wiley & Sons, Inc., New York, 1964.

Den Hartog, J. P., *Advanced Strength of Materials*, McGraw-Hill, Inc., New York, 1952.

Nadeau, G., *Introduction to Elasticity*, Holt, Rinehart and Winston, Inc., New York, 1964.

Rogers, G. L., *Mechanics of Solids*, John Wiley & Sons, Inc., New York, 1964.

Sokolnikoff, I. S., *Mathematical Theory of Elasticity*, McGraw-Hill, Inc., New York, 1956.

Timoshenko, S. P., *Theory of Elastic Stability*, McGraw-Hill, Inc., New York, 1936.

Timoshenko, S. P., *Elements of Strength of Materials*, D. Van Nostrand Company, Inc., Princeton, N.J., 1956.

Timoshenko, S. P., and Young, D. H., *Theory of Structures*, McGraw-Hill, Inc., New York, 1945.

Wang, C., *Applied Elasticity*, McGraw-Hill, Inc., New York, 1953.

ELECTRICITY AND MAGNETISM

Bitter, F., *Currents, Fields and Particles*, John Wiley & Sons, Inc., New York, 1960.

Cheston, W. B., *Elementary Theory of Electric and Magnetic Fields*, John Wiley & Sons, Inc., New York, 1964.

Halliday, D., and Resnick, R., *Physics*, Part 2, John Wiley & Sons, Inc., New York, 1962.

Marion, J. B., *Classical Electromagnetic Radiation*, Academic Press, Inc., New York, 1965.

Matveyev, A. N., *Principles of Electrodynamics*, Translated by L. F. Landovitz, Reinhold Publishing Corporation, New York, 1966.

Winch, R. P., *Electricity and Magnetism*, Second Edition, Prentice-Hall, Inc., Englewood Cliffs, N.J., 1963.

FLUID MECHANICS

Lamb, H., *Hydrodynamics*, Dover Publications, Inc., New York, 1945.

Li, W. H., and Lam, S. H., *Principles of Fluid Mechanics*, Addison-Wesley Publishing Company, Inc., Reading, Mass., 1964.

Longwell, P. A., *Mechanics of Fluid Flow*, McGraw-Hill, Inc., New York, 1966.

Pao, R. H. F., *Fluid Mechanics*, John Wiley & Sons, Inc., New York, 1961.

Streeter, V. L., *Fluid Mechanics*, Fourth Edition, McGraw-Hill, Inc., New York, 1966.

MATHEMATICS

Arfken, G., *Mathematical Methods for Physicists*, Academic Press, Inc., New York, 1966 (vector analysis, matrices, differential equations).

Bak, T. A., and Lichtenberg, J., *Mathematics for Scientists*, W. A. Benjamin, Inc., New York, 1966 (vectors, matrices).

Churchill, R. V., *Fourier Series and Boundary Value Problems*, McGraw-Hill, Inc., New York, 1952 (Fourier series, partial differential equations).

Kaplan, W., *Advanced Calculus*, Addison-Wesley Publishing Company, Inc., Reading, Mass., 1952 (vector analysis, line integrals, functions of more than one variable).

Kreyszig, E., *Advanced Engineering Mathematics*, John Wiley & Sons, Inc., New York, 1962 (differential equations, partial differential equations, matrices, Fourier series).

Margenau, H., and Murphy, G. M., *The Mathematics of Physics and Chemistry*, Second Edition, D. Van Nostrand Company, Inc., Princeton, N.J., 1956 (vector analysis, differential equations, partial differential equations, matrices, Fourier series).

Mathews, J., and Walker, R. L., *Mathematical Methods of Physics*, W. A. Benjamin, Inc., New York, 1961 (elliptic integrals, Fourier series, differential equations, partial differential equations).

Janke, E., and Emde, F., *Tables of Functions*, Dover Publications, Inc., New York, 1943 (elliptic integrals).

Johnson, D. E., and Johnson, J. R., *Mathematical Methods in Engineering and Physics*, The Ronald Press Company, New York, 1965 (Fourier series, partial differential equations).

Von Kármán, T., and Biot, M. A., *Mathematical Methods in Engineering*, McGraw-Hill, Inc., New York, 1940 (Fourier series, differential equations, partial differential equations).

Osgood, W. F., *Advanced Calculus*, The Macmillan Company, New York, 1925 (maxima and minima, infinite series, partial differential equations).

Sagan, H., *Boundary and Eigenvalue Problems in Mathematical Physics*, John Wiley & Sons, Inc., New York, 1961 (partial differential equations).

Schwartz, L., *Mathematics for the Physical Sciences*, Addison-Wesley Publishing Company, Inc., Reading, Mass., 1966 (partial differential equations).

Sokolnikoff, I. S., and Redheffer, R. M., *Mathematics of Physics and Modern Engineering*, McGraw-Hill, Inc., New York, 1958 (differential equations, partial differential equations, vector analysis).

Woods, F., *Advanced Calculus*, Ginn & Company, Boston, 1934 (vector analysis, partial differential equations).

OSCILLATION

Haag, J., *Oscillatory Motions*, Wadsworth Publishing Co., Inc., Belmont, Calif., 1962.

Lindsay, R. B., *Mechanical Radiation*, McGraw-Hill, Inc., New York, 1960.

Tse, F. S., Morse, I. E., and Hinkle, R. T., *Mechanical Vibrations*, Allyn and Bacon, Inc., Boston, 1963.

QUANTUM MECHANICS

Bohm, D., *Quantum Theory*, Prentice-Hall, Inc., Englewood Cliffs, N.J., 1951.

Born, M., *Atomic Physics*, Sixth Edition, Hafner Publishing Company, New York, 1956.

Eisberg, R. M., *Fundamentals of Modern Physics*, John Wiley & Sons, Inc., New York, 1961.

Fong, P., *Elementary Quantum Mechanics*, Addison-Wesley Publishing Company, Inc., Reading, Mass., 1962.

RELATIVITY

Born, M., *Einstein's Theory of Relativity*, Revised Edition, Dover Publications, Inc., New York, 1962.

Eddington, A. S., *The Mathematical Theory of Relativity*, University Press, Cambridge, Mass., 1924.

Einstein, A., Lorentz, H. A., Minkowski, H., and Weyl, H., *The Principle of Relativity*, Dover Publications, Inc., New York, 1952.

MISCELLANEOUS

Bowden, F. P., and Tabor, D., *The Friction and Lubrication of Solids*, Oxford University Press, New York, 1950.

Bowden, F. P., and Tabor, D., *The Friction and Lubrication of Solids*, Part 2, Oxford University Press, New York, 1964.

Bridgman, P. W., *Dimensional Analysis*, Yale University Press, New Haven, Conn., 1931.

Bryant, P. J., Lavik, M., and Salomon, G., *Mechanisms of Solid Friction*, Elsevier Publishing Company, Amsterdam, 1964.

Buckingham, E., "On Physically Similar Systems, Illustrations of the Use of Dimensional Equations," *Phys. Rev.* 4, 345 (1914).

Glasstone, S., and Edlund, M. C., *The Elements of Nuclear Reactor Theory*, D. Van Nostrand Company, Inc., Princeton, N.J., 1952.

Goldsmith, W., *Impact*, Edward Arnold, Ltd., London, 1960.

INDEX

INDEX